In Memoriam

Eugene M. Shoemaker

April 28, 1928—July 18, 1997

Charles B. Hunt

August 9, 1906—September 3, 1997

CONTENTS

LACCOLITH COMPLEXES OF SOUTHEASTERN UTAH: TIME OF EMPLACEMENT AND TECTONIC SETTING— WORKSHOP PROCEEDINGS

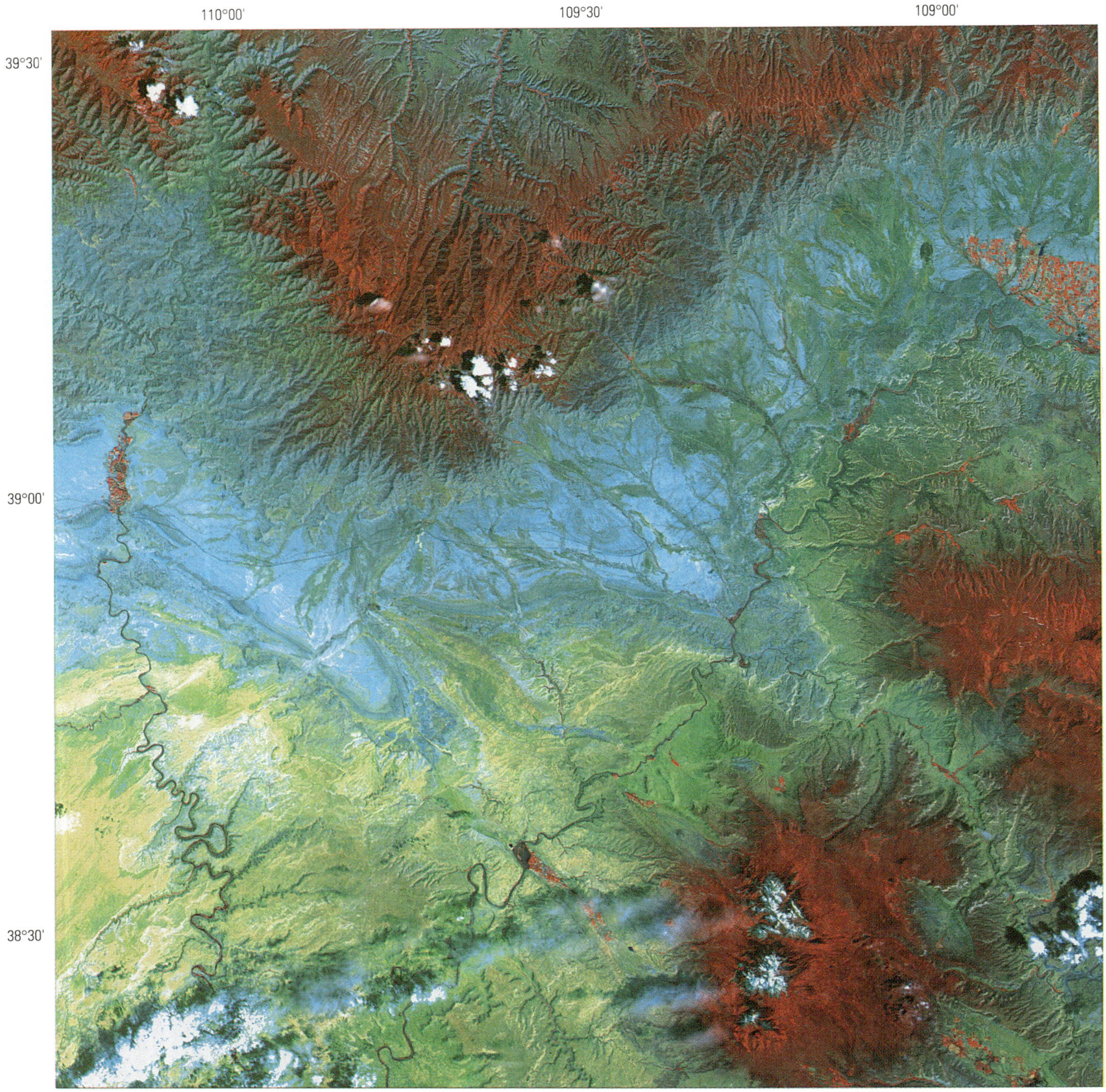

Frontispiece. Landsat 5 Thematic Mapper color-infrared composite image of the northern part of the Paradox Basin and the La Sal Mountain laccolith complex. See page 4 for generalized geologic map of the area.

Laccolith Complexes of Southeastern Utah: Time of Emplacement and Tectonic Setting—Workshop Proceedings

By Jules D. Friedman *and* A. Curtis Huffman, Jr., *Coordinators*

U.S. GEOLOGICAL SURVEY BULLETIN 2158

Application of new geochemical dating techniques, geophysics, and geologic mapping results to the classic problem of the nature and geologic setting of these well-known complexes in order to clarify time of Tertiary volcanism and relationship to regional tectonics

UNITED STATES GOVERNMENT PRINTING OFFICE, WASHINGTON : 1998

U.S. DEPARTMENT OF THE INTERIOR
BRUCE BABBITT, Secretary

U.S. GEOLOGICAL SURVEY
Mark Schaefer, Acting Director

For sale by U.S. Geological Survey, Information Services
Box 25286, Federal Center
Denver, CO 80225

Library of Congress Cataloging-in-Publication Data

Laccolith complexes of southeastern Utah : time of emplacement and tectonic setting : workshop proceedings / [edited] by Jules D. Friedman and A. Curtis Huffman, Jr., coordinators.
p. cm. – (U.S. Geological Survey bulletin ; 2158)
Report of a workshop held in Denver, Colo., July 6–7, 1992.
Includes bibliographical references.
Supt. of Docs. no.: I 19.3:2158
1. Laccoliths—Utah—Congresses. 2. Geology, Stratigraphic—Tertiary—Congresses. 3. Geology—Utah—Congresses.
I. Friedman, Jules D. II. Huffman, A.C. III. Series.
QE75.B9 no. 2158
[QE611.5.U5]
557.3 s—dc20
[551.8′8} 96-26993
CIP

Introduction and Overview

By Jules D. Friedman *and* A. Curtis Huffman, Jr.

Three well-known laccolith complexes of southeastern Utah, the Henry, La Sal, and Abajo Mountains, have similar petrogenesis, tectonic control, and lithophile hydrothermal mineralization. These complexes were the principal subjects of a U.S. Geological Survey sponsored workshop on July 6 and 7, 1992, in Denver, Colo., convened by Jules D. Friedman and A. Curtis Huffman, Jr. The purpose of the workshop was to present and clarify new field and laboratory data in order to (1) provide an analog for Cordilleran laccolith fields on intracratonic blocks; (2) narrow the age range of Colorado Plateau laccoliths by consideration of errors in earlier dating attempts in relation to recent high-precision $^{40}Ar/^{39}Ar$ and fission-track dates; (3) relate distribution of southeastern Utah intrusive centers to tectonics and structures; (4) associate emplacement of laccolith complexes on the Colorado Plateau with specific tectonomagmatic events and possible thermal events of the mantle; and (5) define possible economic ore-deposit implications. This report summarizes much of the material presented at the workshop and provides an extensive list of selected references pertaining to geochronology and tectonics of the Colorado Plateau to facilitate further research.

Eugene Shoemaker (USGS) set the stage for a discussion of tectonic control of the laccolith complexes by presenting evidence for Precambrian movement on northeast- and northwest-trending discontinuities in the Grand Canyon region that extend into southeastern Utah. Movement on the northeast-trending set, including the Bright Angel and Sinyala fault zones, was earlier than that on the northwest-trending systems. Geophysical and subsurface data presented by Richard Blank (USGS) and William Butler (USGS) support the importance and regional nature of the northeastern and northwestern trends. Northwesterly structural trends of the Paradox and Uncompahgre fault zone and basement uplift are strongly expressed in both magnetic and isostatic gravity data, whereas the northeast-striking fault systems that enter the region from northern Arizona are somewhat less so. The gravity data indicate that most known or geophysically inferred laccolithic centers are on the margin or noses of basement uplifts.

Recent field mapping in the southern Henry Mountains by Marie Jackson (USGS) reaffirms that the major intrusive bodies of southeastern Utah are true laccoliths, floored at shallow depths. Her structural analysis indicates that the Oligocene stress field in the vicinity of the Henry Mountains laccoliths was isotropic. Field mapping and structural analysis of the La Sal Mountains by Michael Ross (UGS) and the Paradox fold and fault belt by Hellmut Doelling (UGS) indicate that the northwest-striking faults that control the Paradox Valley–Castle Valley–Salt Valley salt diapirs are possibly connected by a northeast-striking fault zone or ramp-monocline structure forming a kink in the tectonic boundary between two parts of the late Paleozoic Paradox Basin. The La Sal intrusions were emplaced at the intersection of these fault systems.

Whether the other laccolith complexes were emplaced at similar nodes in an orthogonal northwest-northeast fault grid is not nearly as certain although it seems likely, and this hypothesis appears to be supported by correlation of northeasterly and northwesterly lineament trends evident in remote sensing, geologic, and geophysical data discussed by Jules Friedman (USGS). This argument is further strengthened by demonstrated episodic movement on both northeast- and northwest-striking basement faults that are part of an orthogonal grid mapped in the San Juan and Paradox Basins by Curtis Huffman (USGS) and David Taylor (USGS) using reflection seismic data. Movement on some of these faults influenced deposition of sediments throughout the Phanerozoic. Earl Verbeek (USGS) and Marilyn Grout (USGS) documented local control of joints by basement faults but went on to point out that widely distributed post-Laramide regional joint sets commonly developed independent of basement control.

Petrologic studies by Stephen Nelson (UCLA) and Jon Davidson (UCLA) indicate that all three laccolith complexes are petrogenically bimodal but proportionately variable; these quartz monzonite porphyries probably evolved from mantle-derived magma ponded in the deep crust. New high precision $^{40}Ar/^{39}Ar$ and fission-track ages reported by Nelson and Davidson and by Kim Sullivan (Brigham Young University) place the southeastern Utah intrusive activity during the interval from 31.2 to 23.3 Ma, hence late Oligocene to early Miocene. The laccoliths are thus partly correlative with much of the late Oligocene to early Miocene volcanic belt that extends 1,100 km sublatitudinally from western Nevada through southern Utah to southwestern Colorado, and from there south to west Texas. Charles Chapin and William McIntosh (both New Mexico Bureau of Mines and Mineral Resources) and coworkers pointed out that the laccolith clusters of southeastern Utah were emplaced during the second of two major episodes of ash-flow volcanism in the Mogollon-Datil volcanic belt within this vast region (this paper is not included in this proceedings volume).

Peter Rowley (USGS), Charles Cunningham (USGS), Tom Steven (USGS) and coauthors demonstrated that the calc-alkaline intrusive rocks of the Pioche-Marysvale and Delamar–Iron Springs igneous belts (in southwestern Utah and southeastern Nevada) are generally similar in age and composition to the laccolith complexes of southeastern Utah and to most other parts of the sublatitudinal igneous province that spans the Great Basin, Colorado Plateau, and adjacent areas. Nelson, Davidson, Rowley and coauthors inferred that the calc-alkaline magmas of mostly intermediate composition probably evolved from mantle-derived basaltic magmas emplaced into the deep crust and modified by fractional crystallization and subordinate assimilation of crustal components; crustal melting (or, alternatively, large-scale assimilation by mantle-derived magmas, according to other workers) may have formed younger silicic rocks in the province. They concluded that the Oligocene and Miocene calc-alkaline part of the igneous province probably originated by subduction of oceanic lithosphere beneath western North America and that east-northeast extensional deformation accompanied the magmatism.

The workshop ended on a thought-provoking, enigmatic note with the suggestion by Felix Mutschler (E. Washington Univ.) and coauthors Edwin Larson (Univ. of Colorado) and David Gaskill (USGS, Ret.) that the Colorado Plateau is an isolated block of Proterozoic craton that was reduced in size by the lateral encroachment of a series of Late Cretaceous to Holocene *passive* hotspots. They also see a modest potential for alkaline-rock related gold mineralization in all of the laccolith complexes, a point that was also made by Michael Ross (UGS) in relation to the La Sal Mountains.

No attempt was made by the coordinators to achieve a consensus on controversial points presented by different authors.

Description of Landsat Thematic Mapper Image of Northern Part of the Paradox Basin and the La Sal Mountains Laccolith Complex

By K. Eric Livo, Jules D. Friedman, *and* Shirley L. Simpson

CONTENTS

FIGURES

ABSTRACT

A Landsat 5 Thematic Mapper color-infrared composite image highlights the geographic and geologic features of the northern part of the Paradox Basin including the La Sal Mountains laccolith complex. This image depicts the outcrop pattern of Paleozoic, Mesozoic, and Cenozoic stratigraphic units, using color distinctions based on spectral characteristics of individual lithologies. Structures and lithologies, presented in plan view, can be correlated with geologic maps. Anticlinal and synclinal forms, surface faults, domes and collapse structures, major river courses, and intrusive laccolith centers align with major lineaments. The image also yields evidence for the geologic processes of superposition, antecedence, and subsurface salt solution with concomitant valley floor subsidence, all of which have shaped the region.

DESCRIPTION OF IMAGE

Many geographical and geologic features of the northern part of the Paradox Basin are clearly visible on the accompanying satellite image (see frontispiece to this volume). Feature locations are shown in the accompanying sketch map (fig. 1).

The color-infrared composite (CIR) image was acquired by the Landsat 5 Thematic Mapper (TM) on June 14, 1985 (Path 36, Row 33, Scene ID 45047017262x0). The volume frontispiece is a subset of the TM scene. It has 4,834 scan lines and 4,673 pixels per scan line, and it covers an area of 20,000 square kilometers at a resolution of 30 meters per pixel. It has been saturation enhanced, using bands 4, 3, and 2 colored red, green, and blue, respectively, to increase its color density. Saturation enhancement increases color purity from pastel (colors with white added) to vibrant. The TM

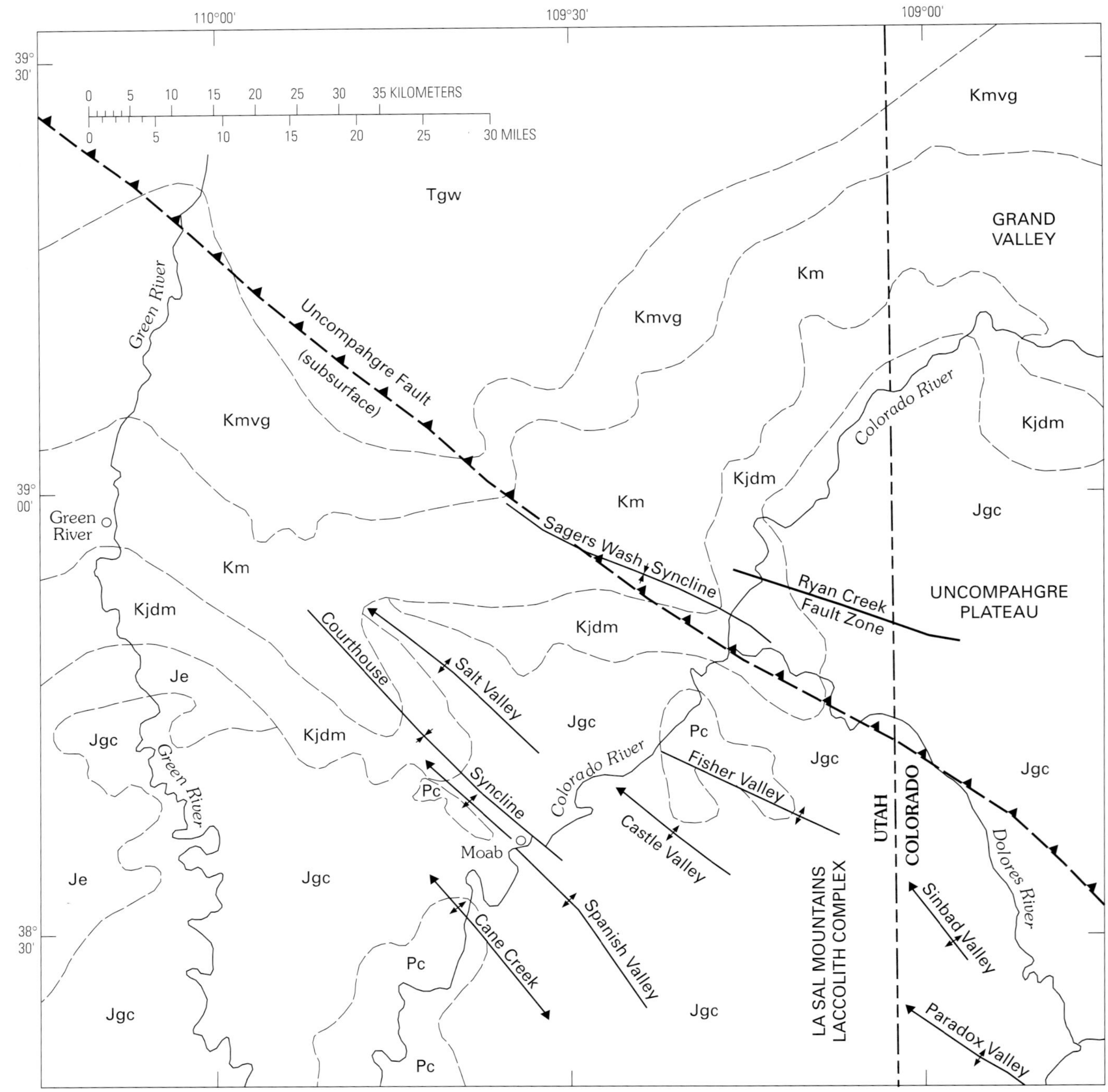

EXPLANATION

Tgw	Green River and Wasatch Formations (Eocene)
Kmvg	Mesaverde Group (Upper Cretaceous)
Km	Mancos Shale (Upper Cretaceous)
KJdm	Dakota Sandstone (Upper Cretaceous) and Morrison Formation (Upper Jurassic)
Je	Entrada Sandstone (Middle Jurassic)
Jgc	Glen Canyon Group (Lower Jurassic)
Pc	Cutler Formation (Lower Permian)

Contact, approximately located

Axis of anticline with plunge direction

Axis of syncline with plunge direction

Thrust fault—Sawteeth on upthrown block

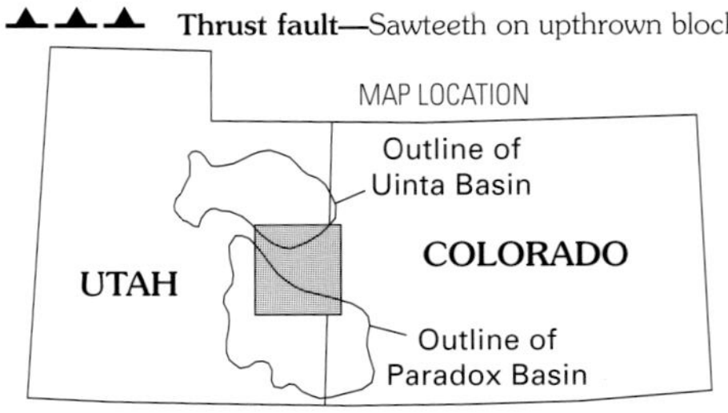

Figure 1 (facing page). Generalized geology of the northern part of the Paradox Basin and the La Sal Mountains laccolith complex. See frontispiece to this volume for color-infrared composite image of the area.

image has been rectified and registered to a Universal Transverse Mercator (UTM) grid, compatible with that used for the USGS 1:250,000 geologic map series. However, because of space restrictions, the image is reproduced here at a scale of only 1:750,000. Crossticks on the image show latitude and longitude coordinates at 30-minute intervals.

LITHOLOGY

Specific lithologies of the northern part of the Paradox Basin (Williams, 1964; Williams and Hackman, 1971; Cashion, 1973) correlate well with the color mapping on the TM image. Shades of blue, green, and yellow are used to identify various sedimentary units. Areas displayed in red delineate regions with abundant vegetation; slightly vegetated regions tend to be lighter in red. White and black areas are clouds with shadows. Snow on the crests of the La Sal Mountains is also mapped as white.

Sparsely vegetated shale, siltstone, mudstone, and marlstone exposures of the Morrison Formation (Upper Jurassic), Mancos Shale (Upper Cretaceous), Wasatch Formation (Eocene), and Green River Formation (Eocene) are mapped in various shades of blue. Shale, siltstone, and mudstone exposures of the Morrison Formation appear as a dark-blue mapped zone that trends east-west through the mid-part of the image, marks the south side of the Grand Valley, forms embayments in the Sagers Wash syncline and the Courthouse syncline, and farther west, passes south of the town of Green River, Utah. Further exposures of the Morrison are present from the La Sal Mountains east into the Uncompahgre uplift, although these exposures are mostly masked (mapped red) by vegetation. Exposures of the Mancos Shale, mapped as light blue, trend east-west through the north-central part of the image, following the Colorado River and the south edge of the Uinta Basin, just north of the exposures of the Morrison Formation, through Green River, Utah. North of the outcrop area of the Mancos Shale, the Wasatch and Green River Formations cover an extensive region and are mapped as purple to dark blue.

Major sandstone units within this CIR image are distinguishable according to the abundance of ferruginous minerals in the rock, using the mapping colors green to yellow. Ferric oxide-bearing red beds, such as the Cutler Formation (Lower Permian), appear as green; less ferruginous sandstones are mapped as green through greenish yellow (Lower Jurassic Glen Canyon Group, Middle Jurassic Entrada Sandstone, and the Upper Cretaceous Dakota Sandstone and Mesaverde Group). Non-ferruginous sandstones and unconsolidated sands are mapped as light yellow to white. The color-infrared composite image translates the visually red color of the Cutler Formation into the bright-green image colors immediately north, northeast, and west of the La Sal Mountains. The Glen Canyon Group dominates the southern half of the image; it includes the Wingate Sandstone and the Kayenta Formation, which are usually dark to light olive green in the image, and the Navajo Sandstone, which appears light yellowish green. The Entrada Sandstone is exposed in the San Rafael Desert, at the far southwest edge of the TM image. It is partly covered by active sand dunes, and appears light yellow to white. Both the Dakota Sandstone, which forms thin hogbacks south of the Mancos Shale, and the sandstone units in the Mesaverde Group, exposed as dip slopes north of the Mancos Shale, are mapped dull brownish olive-green.

STRUCTURE

Structurally controlled geomorphic features are prominent in the TM image, and they clearly show the dominant northwest and northeast trends found within the Paradox Basin. Geomorphic features include salt-cored anticlines and associated synclines, linear features in the clastic rocks, doming around laccoliths, and collapse structures.

Baars and Stevenson (1981), Lee (1988), Friedman and Simpson (1980), and Friedman and others (1994) have mapped a series of northwest-trending lineaments using Landsat Multispectral Scanner (MSS) data. The features that compose these lineaments are visible on the TM image, especially a prominent system of salt-cored anticlines and associated synclines, which formed from late Paleozoic time to the present. One conspicuous example, near the center of the area, is the breached Salt Valley anticline, which is flanked by the Courthouse syncline to the southwest and the Sagers Wash syncline to the northeast. Other principal northwest-trending landforms north of the Colorado River are the Moab anticline, the linear trend of the Green River, and the eastern part of the Grand Valley. Just south of the Colorado River are the Cane Creek, Spanish Valley, Castle Valley, and Fisher Valley anticlines. Farther southeast, beyond the La Sal Mountains, red cultivated fields mark the northwest-trending Paradox Valley anticline, and to the north of it are two other northwest-trending features: the Sinbad Valley anticline and the Dolores River alignment. Also trending northwest are faults and joints produced during Mesozoic and Cenozoic time by tectonic and halokinetic processes, and by valley-floor subsidence over the axial zones of salt-cored anticlines (Hite and Lohman, 1973).

Many of the northeast-trending lineaments represent joints, faults, and other surface fractures (Lee, 1988; Friedman and others, 1994). These linear features are expressed within the present TM image as small discontinuous trends to the northeast. An exception is the Colorado River lineament, which trends northeast parallel to or coincident with the Colorado River Valley throughout this image.

The La Sal Mountains laccolith complex was emplaced at the intersection of major northeast- and northwest-trending lineaments (Friedman and others, 1994). The La Sal Mountain peaks trend along a line to the northeast, whereas most of the individual laccoliths, as well as the axes of folding in the overlying sedimentary rocks, follow the underlying structural trend to the northwest (Hunt, 1983). The North La Sal Mountain laccolith center, now exposed through erosion, is elongated along the Castle Valley and Paradox Valley anticlines.

PROCESSES

Tectonic, sedimentary, erosional, and igneous processes have shaped the region. Tectonic uplift has created the Uncompahgre Plateau northeast of the tectonic line marked by the Uncompahgre fault, while crustal downwarping has created the Paradox and Uinta Basins. The evolving marine and lacustrine environments have controlled sedimentation of facies ranging from evaporites to sandstone to shale. In this area, the shallow northward regional dip and differential erosion have resulted in a pattern that appears as a younging of the formations from the south edge to the north edge of the TM image area.

Uplift of the Colorado Plateau during the Laramide orogeny has led to downcutting and differential erosion. Antecedent cross-axial drainage is demonstrated by the greater incision of streams and rivers to the southwest (Hunt, 1983), where the Green and Colorado Rivers join. Differential erosion has formed massive sandstone cliffs and mesas to the south and has left the more readily eroded thick shale units exposed in protected valley slopes and bottoms to the north.

The locations of major river channels were controlled by several processes. The Green, Colorado, and Dolores Rivers, as well as the southwest-flowing streams that exhibit a trellis drainage pattern north and east of the Ryan Creek fault zone on the east-central part of the image area, have been locked into position along zones of structural weakness (Maarouf, 1983). These rivers follow deep-seated structures but have local deviations. Locally, the meanders of the Green River have been incised into the Glen Canyon Group (southwestern part of image area); farther north, the Green River's course is horizontally displaced by faulting (south of Green River, Utah). The Colorado River locally flows around the noses of plunging anticlines (Case, 1991). Erosion and downcutting by the Dolores River has breached the Sinbad Valley salt-cored anticline, causing collapse of the clastic rock beds over the anticlinal core and producing vertical walls. Erosion has also exposed the laccolith complex within the La Sal Mountains, stripping back the cover of clastic rock units; this feature is especially prominent on the northeast flank of the domal uplift of the North La Sal Mountain laccoliths.

SUMMARY

The CIR image can be used to visualize, confirm, and supplement features mapped by geologic field methods and photogeology. In addition, many lithologies are distinguishable by spectral color differences, which also sharply outline areas of regional dip, laccolithic doming, and anticlinal collapse.

REFERENCES CITED

Baars, D.L., and Stevenson, G.M., 1981, Tectonic evolution of the Paradox Basin, Utah and Colorado, *in* Wiegand, D.L., ed., Geology of the Paradox Basin: Rocky Mountain Association of Geologists Field Conference, 1981, Guidebook, p. 23–31.

Case, J.E., 1991, Geologic map of the northwestern part of the Uncompahgre Uplift, Grand County, Utah, and Mesa County, Colorado, with emphasis on Proterozoic rocks: U.S. Geological Survey Miscellaneous Investigations Series Map I–2088, 2 sheets and 16-p. pamphlet, scale 1:24,000.

Cashion, W.B., 1973, Geologic and structure map of the Grand Junction Quadrangle, Colorado and Utah: U.S. Geological Survey Miscellaneous Investigations Series Map I–736, scale 1:250,000.

Friedman, J.D., Case, J.E., and Simpson, S.L., 1994, Tectonic trends of the northern part of the Paradox Basin, southeastern Utah and southwestern Colorado, as derived from Landsat Multispectral Scanner imaging and geophysical and geologic mapping, chap. C *of* Huffman, A.C., Jr., project coordinator, Evolution of sedimentary basins—Paradox Basin: U.S. Geological Survey Bulletin 2000–C, p. C1–C30.

Friedman, J.D., and Simpson, S.L., 1980, Lineaments and geologic structure of the northern Paradox Basin, Colorado and Utah: U.S. Geological Survey Miscellaneous Field Studies Map MF–1221, 2 sheets.

Hite, R.J., and Lohman, S.W., 1973, Geologic appraisal of Paradox Basin salt deposits for waste emplacement: U.S. Geological Survey Open-File Report 73–114, 75 p.

Hunt, C.B., 1983, Development of the La Sal and other laccolithic mountains on the Colorado Plateau, *in* Averett, W.R., ed., Northern Paradox Basin–Uncompahgre Uplift: Grand Junction Geological Society Field Trip, 1983, Guidebook, p. 29–32.

Lee, Keenan, 1988, Remote sensing study in support of mineral resource appraisal of wilderness study areas near Moab, Utah—Dolores River Canyon wilderness study area, Montrose and San Miguel Counties, Colorado; Lost Spring Canyon wilderness study area, Grand County, Utah; Behind the Rocks wilderness study area, Grand and San Juan Counties, Utah; Indian Creek wilderness study area, San Juan County, Utah; and Butler Wash wilderness study area, San Juan County, Utah: U.S. Geological Survey Open-File Report 88–219, 12 p.

Maarouf, Abdelrahman, 1983, Relationship between basement faults and Colorado Plateau drainage, *in* Averett, W.R., ed., Northern Paradox Basin–Uncompahgre Uplift: Grand Junction Geological Society Field Trip, 1983, Guidebook, p. 59–62.

Williams, P.L., 1964, Geology, structure, and uranium deposits of the Moab Quadrangle, Colorado and Utah: U.S. Geological Survey Miscellaneous Investigations Series Map I–360, 2 sheets, scale 1:250,000.

Williams, P.L., and Hackman, R.J., 1971, Geology, structure, and uranium deposits of the Salina Quadrangle, Utah: U.S. Geological Survey Miscellaneous Investigations Series Map I–591, 2 sheets, scale 1:250,000.

Neogene Uplift and Radial Collapse of the Colorado Plateau—Regional Implications of Gravity and Aeromagnetic Data

By H. Richard Blank, William C. Butler, *and* Richard W. Saltus

CONTENTS

FIGURES

TABLE

ABSTRACT

Residual isostatic gravity and aeromagnetic total-intensity maps of a 5°×10° region centered on the northern Colorado Plateau reveal a first-order correlation of basement anomalies with structural (topographic) relief on the surface of Precambrian crystalline basement over the plateau's interior. Anomaly trends reflect a mosaic of generally northeast and northwest striking basement features. A pattern of arcuate structures peripheral to the plateau is interpreted as a manifestation of radial outward tectonic extension superposed on Laramide structures and the mainly westerly to southwesterly extension of neighboring provinces. The radial component may be attributable to gravitational collapse concomitant with mainly Neogene regional uplift. An earlier mid-Tertiary plateau uplift of lesser amount, accompanied by voluminous peripheral magmatism, was probably related to Laramide subduction, but Neogene activity is believed to be a consequence of the rise and lateral spreading of anomalously hot, low-density upper-mantle asthenosphere that is the source of long-wavelength topographic, geoidal, and Bouguer gravity anomalies of the U.S. western interior. This relatively stationary mantle feature, perhaps indirectly a product of several hundred million years of east-dipping subduction, is also the cause of the Yellowstone hotspot and the Alvarado Ridge of Colorado and New Mexico.

INTRODUCTION

Geophysical investigations of the Colorado Plateau and its margins have utilized data from seismic reflection and refraction profiles, earthquake traveltime studies, heat-flow measurements, paleomagnetic studies, in situ stress measurements, and regional stress analysis based on structural data, magnetotelluric soundings, and reconnaissance potential-field surveys. Results of most work prior to about 1978 have been discussed by Thompson and Zoback (1979), and comprehensive overviews of more recent work can be found in a series of papers on the geophysical framework of the conterminous United States (Pakiser and Mooney, 1989). In this report we consider mainly the residual isostatic gravity and aeromagnetic total-intensity anomaly fields plotted from data of the North American Data Set (available from the National Geophysical Data Center (NGDC), Boulder, Colo.) for the northern Colorado Plateau and parts of adjacent provinces in Colorado, Utah, Nevada, Arizona, and New Mexico (roughly an area bounded by long 105°–115° W. and lat 36°–41° N.). First we describe the residual fields and comment on some of the most prominent anomalies. We compare plateau basement anomalies with known structural (topographic) relief on the surface of the Precambrian (Butler, 1991), searching for indications of deep-seated influences on the disposition of Tertiary laccolithic centers. Next we examine anomaly trends both on the plateau and in surrounding regions. Finally, we remark on the possible significance of these trends with respect to the tectonic evolution of the region and mid- to late-Cenozoic magmatism.

For geologic interpretation of the potential-field data we used State geologic maps published at scales of 1:500,000 (Wilson and others, 1969; Stewart and Carlson, 1978; Tweto, 1979a; Hintze, 1980) and 1:1,000,000 (Hintze, 1975; Stewart and Carlson, 1977; New Mexico Geological Society, 1982; Tweto, 1987; Reynolds, 1988; Western Geographics/Colorado Geological Survey, 1991; Western Geographics/Geological Survey of Wyoming, 1991), as well as regional geophysical maps at these scales (Sauck and Sumner, 1970a, b; Zietz and Kirby, 1972a, b; Aiken, 1975; Zietz and others, 1976, 1977; Cordell and others, 1982; Cordell, 1984; Keller and Cordell, 1984; Saltus, 1988; Hildenbrand and Kucks, 1988; Cook and others, 1989; Abrams, 1993). These included maps of the Bouguer and isostatic gravity anomalies, aeromagnetic total-intensity anomaly, aeromagnetic anomaly reduced to pole, and selected derivatives of these parameters. Smaller-scale maps that were used include (1) wavelength filtered gravity maps of Hildenbrand and others (1982) and Kane and Godson (1989); (2) a wavelength terrain map of Kane and Godson (1989); and (3) basement and tectonic compilations of Bayley and Muehlberger (1968), Bayer (1983), Muehlberger (1992), and Reed (1993). Reed's (1993) basement compilation is also overprinted on isostatic residual gravity and magnetic anomaly maps of the conterminous United States (Jachens and others, 1993, and Committee for the Magnetic Anomaly Map of North America, 1993, respectively).

ACKNOWLEDGMENTS

We are indebted to Viki Bankey, Tien Grauch, and Gerda Abrams for providing preliminary copies of their current regional compilations, and to Viki Bankey, Curtis Huffman, Peter Rowley, and Dwight Schmidt for provocative reviews that encouraged us to look beyond immediate sources of anomalies and investigate more fully their tectonic and magmatic implications. An editorial review by Peter L. Martin considerably improved the manuscript.

GRAVITY FIELD

Residual isostatic gravity anomalies for the study area are shown on the color-contour map of figure 1*A*. This map is based on an Airy-Heiskanen isostatic model that uses a crustal thickness of 35 km, a topographic load density of 2.67 g/cm^3, and a density contrast across the isostatic root of 0.35 g/cm^3 (Simpson and others, 1986), computed at a grid interval of 4 km. This figure for crustal thickness is probably low, considering the average thickness of 40–45 km determined from long-range seismic refraction profiles (Prodehl and Lipman, 1989) and the 50- to 52-km average derived from COCORP deep reflection data (Hauser and Lundy, 1989). However, the residual anomalies over the plateau are only weakly model-dependent, so as a matter of expediency we did not recompute using the larger figure. On the other hand, the data supplied at 4-km grid spacing were regridded at 2 km to improve the visual effect of the contoured product. The map projection used for this and for figures 1*B*, 2*A*, and 2*B* is Albers Conical Equal Area, with standard parallels of 29.5° N. and 45.5° N.

The anomaly field as rendered in figure 1*A*, with a color interval of 5 mGal, emphasizes only the most strongly positive (yellow and red) and most strongly negative (dark blue and purple) areas; anomalies with amplitudes in the range −25 to about +10 mGal are much less conspicuous, although perhaps no less significant.

Figure 1*B* shows loci of maximum horizontal anomaly gradients computed from data of figure 1*A*, using software developed by Blakely and Simpson (1986). The gradient maxima approximately coincide with surface projections of steep upper-crustal density discontinuities. Notably strong gradients are distinguished in the figure by hachures on the side of lower anomaly values. This figure also provides selected geologic and geographic information, as explained in table 1. The thick dashed line on the figure is the outline of the northern plateau according to Bayer (1983).

As expected, broad anomaly lows mark the principal sedimentary basins of the northern Colorado Plateau—the Uinta (UB), Piceance Creek (PB), Paradox (PXB), San Juan (SJB), and Kaiparowits (KB) Basins (fig. 1*B*). The lows are caused by thick, relatively low density sedimentary sections above structural depressions in the basement. Major basins peripheral to the plateau, including the North Park (NP), Salt Wash (SWB), Washakie (WB), and Bear River (BRB) Basins, and basins of the Virgin River depression (VRD), also have associated gravity lows, presumably of similar origin. However, two of the largest negative anomalies, both in breadth and amplitude, are not over depositional basins but over mid-Tertiary volcanic fields—the San Juan field (SJV) and the West Elk and Thirtynine Mile fields (WEV and 39MV). These lows are at least in part due to penecontemporaneous granitoid batholiths in the substrate—the roots of the volcanic systems. To an unknown extent, the lows probably also reflect the presence of intrusions of Laramide and Proterozoic age (Steven, 1975; also see Case, 1965; Plouff and Pakiser, 1972; Tweto and Case, 1972). The lows correspond closely to Phanerozoic "perforations" of the Precambrian basement as inferred by Tweto (1987). Prominent lows in the upper Sevier River basin region (USB), and in the Caliente (CCC), Indian Peak (IPCC), and Marysvale (MV) volcanic centers of the Pioche-Marysvale belt in southeastern Nevada and southwestern Utah (see Rowley and others, this volume) are probably also caused in part by concealed intrusives of mid-Tertiary age.

Areas of exposed Precambrian rock generally coincide with gravity highs. However, Precambrian rocks that crop out along the southwest front of the Uncompahgre uplift (UU) typically coincide with relative gravity lows, because of the low-angle contact between overthrust crystalline rock and low-density sedimentary rock of the Paradox Basin (PXB). (For seismic evidence of this relationship see Frahme and Vaughn, 1983.) Gravity is also relatively low in the Needle Mountains (NM) of southwestern Colorado and in much of the Front Range (FR) and Rampart Range (RR). Anomalously low gravity values in Precambrian crystalline terrane may reflect a predominance of silicic intrusive or metaclastic rock in the section. Where the exposed Precambrian section consists of sedimentary rocks yet is associated with an isostatic anomaly high, as in the Uinta Mountains (UM) and in the Canyon Mountains (CYM) of the eastern Great Basin, the residual isostatic high can be attributed to undercompensation of the range (regional rather than local compensation) or to a mean density of topography in excess of the reduction density (2.67 g/cm^3).

AEROMAGNETIC FIELD

A map of the anomalous aeromagnetic total-intensity field of the region is shown in figure 2*A*. Data are from the North American compilation (Committee for the Magnetic Anomaly Map of North America, 1987) and were supplied as Definitive Geomagnetic Reference Field (DGRF) residuals at a 2-km grid interval and subsequently reduced to the north geomagnetic pole with the assumption of magnetization parallel to the inducing field. It is conceivable that some anomalies that appear on the map are strictly artifacts, the result of errors introduced during digitization of analog records (most data were obtained from surveys performed in the 50's and 60's), for example, or during the manual merging of data from blocks flown at different times with different specifications, which can produce spurious anomalies of any spatial wavelength. Also, anomalies with short

Figure 1 (following pages). Isostatic residual gravity features of the northern Colorado Plateau and vicinity. *A*, Anomaly map. Color interval 5 milligals. Reductions based on Airy-Heiskanen model using the following parameters: crustal density ($\Delta\rho$) (topography) = 2.67 g/cm^3, density contrast ($\Delta\rho$) (crust/upper mantle) = 0.35 g/cm^3; crustal thickness (T) = 35 km (Simpson and others, 1986). Data from Geophysical Data Center, National Oceanic and Atmospheric Administration, Boulder, CO (North American Data Set, 4-km grid, regridded at 2 km). *B*, Loci of maximum horizontal gradients computed from isostatic residual gravity anomaly data of *A*. Teeth point in direction of lower anomaly values where gradients are strong. Symbols explained in table 1.

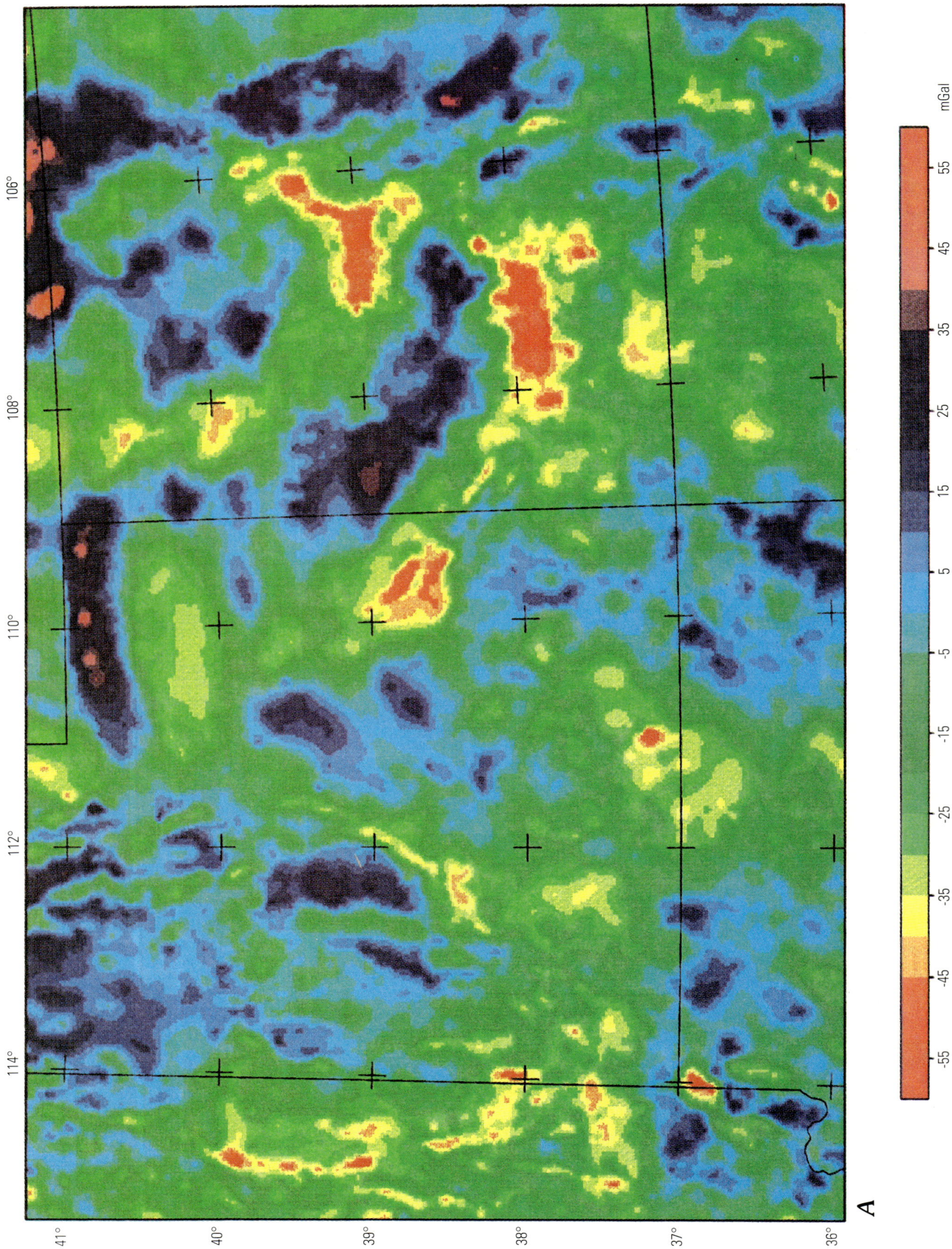
106°
108°
110°
112°
114°
41°
40°
39°
38°
37°
36°
A
mGal
55
45
35
25
15
5
-5
-15
-25
-35
-45
-55

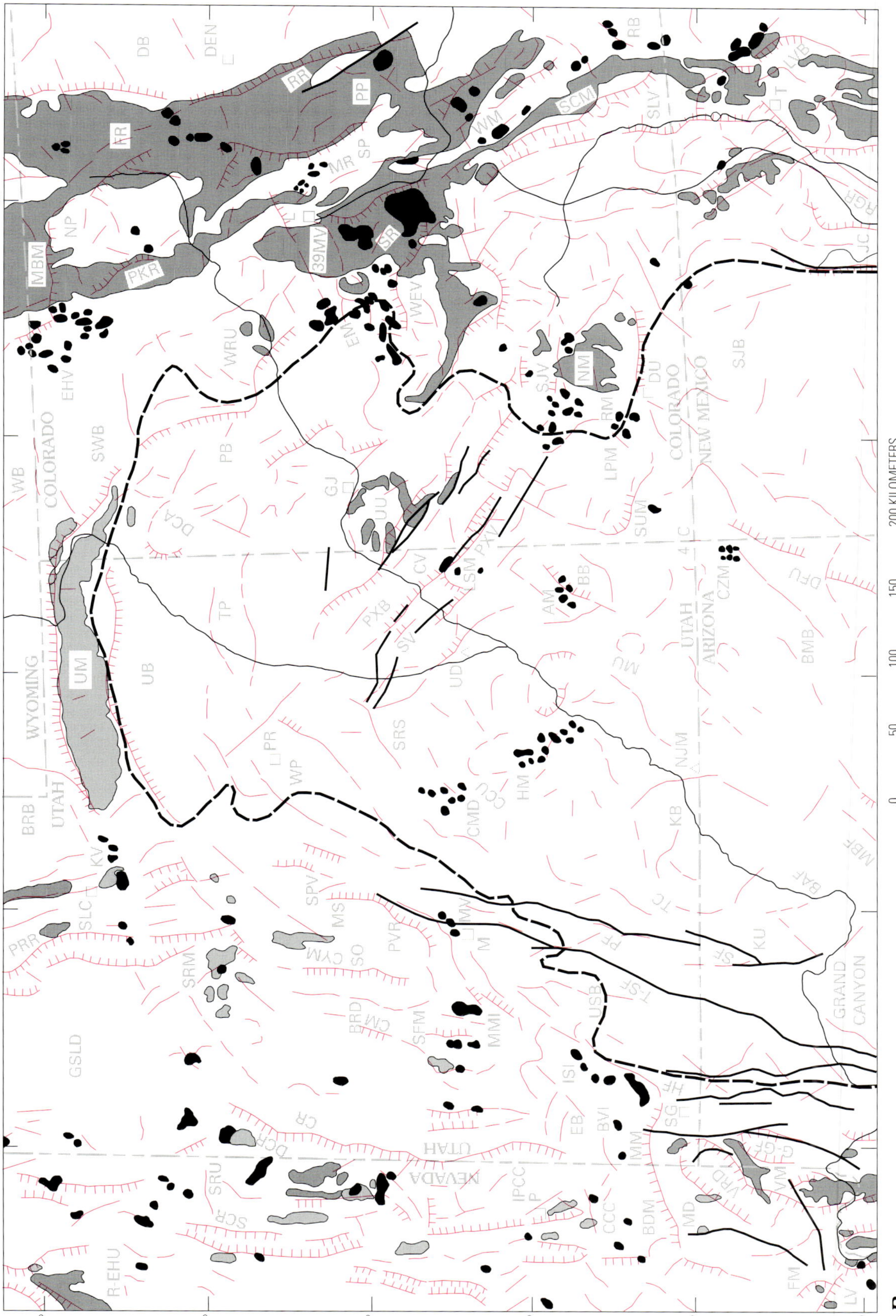
WYOMING
COLORADO
UTAH
ARIZONA
NEW MEXICO
NEVADA
GRAND CANYON
106°
108°
110°
112°
114°
41°
40°
39°
38°
37°
36°
0
50
100
150
200 KILOMETERS

B

TABLE 1. Explanation of patterns and symbols used on figures 1*B* and 2*B*.

[Names of cities and towns in italics]

——	Fault
– – –	Outline of Colorado Plateau
■	Cenozoic and Mesozoic intrusive rock
▒	Precambrian sedimentary rock
░	Precambrian crystalline rock
□	City
AM	Abajo Mountains
BAF	Bright Angel Fault
BB	Blanding Basin
BDM	Beaver Dam Mountains
BMB	Black Mesa Basin
BRB	Bear River Basin
BRD	Black Rock Desert
BVI	Bull Valley intrusion
CCC	Caliente caldera complex
CCU	Circle Cliffs Uplift
CM	Cricket Mountains
CMD	Cedar Mountains dikes
CR	Confusion Range
CV	Castle Valley
CYM	Canyon Mountains
CZM	Carrizo Mountains
DB	Denver Basin
DCA	Douglas Creek Arch
DCR	Deep Creek Range
DEN	*Denver*
DFU	Defiance Uplift
DU	*Durango*
EB	Escalante Basin
EHV	Elkhead volcanic field
EM	Elk Mountains
FM	Frenchman Mountain
FR	Front Range
G-GF	Grand Wash–Gunlock Fault
GJ	*Grand Junction*
GSLD	Great Salt Lake Depression
HF	Hurricane Fault
HM	Henry Mountains
IPCC	Indian Peak caldera complex
ISI	Iron Springs intrusions
JC	Jemez caldera complex
KB	Kaiparowits Basin
KU	Kaibab Uplift
KV	Keetley volcanic field
L	*Leadville*
LPM	La Plata Mountains
LSM	La Sal Mountains
LV	*Las Vegas*
M	*Marysvale*
MBF	Mesa Butte Fault
MBM	Medicine Bow Mountains
MD	Mormon dome
MM	Mineral Mountain
MMI	Mineral Mountains intrusion
MR	Mosquito Range
MS	Middle Sevier Valley
MU	Monument Uplift
MV	Marysvale volcanic field
NJM	Navajo Mountain
NM	Needle Mountains
NP	North Park
P	*Pioche*
PB	Piceance Creek Basin
PF	Paunsaugunt Fault
PKR	Park Range
PP	Pikes Peak batholith
PR	*Price*
PRR	Promontory Range
PVR	Pavant Range
PXB	Paradox Basin
PXV	Paradox Valley
RB	Raton Basin
R-EHU	Ruby–East Humboldt uplift
RGR	Rio Grande Rift
RM	Rico Mountains
RR	Rampart Range
SCM	Sangre de Cristo Mountains
SCR	Schell Creek Range
SF	Sinyala Fault
SFM	San Francisco Mountains
SG	*St. George*
SJB	San Juan Basin
SJV	San Juan volcanic field
SLC	*Salt Lake City*
SLV	San Luis Valley
SO	Sevier oval[1]
SP	South Park
SPV	San Pitch Valley
SR	Sawatch Range
SRM	Sheep Rock Mountains
SRS	San Rafael Swell
SRU	Snake Range Uplift
SUM	Sleeping Ute Mountain
SV	Salt Valley
SWB	Salt Wash Basin
T	*Taos*
TC	The Coxscomb
TP	Tavaputs Plateau
T-SF	Toroweap-Sevier Fault
UB	Uinta Basin
UD	Upheaval Dome
UM	Uinta Mountains
USB	Upper Sevier River Basin
UU	Uncompahgre Uplift
VM	Virgin Mountains
VRD	Virgin River depression
WB	Washakie Basin
WEV	West Elk volcanic field
WM	Wet Mountains
WP	Wasatch Plateau
WRU	White River Uplift
39MV	Thirty-nine Mile volcanic field
4C	Four Corners

[1] Informal designation by T.A. Steven, *in* Mabey and Budding (1987).

wavelengths relative to the flight-line spacing (typically 2–5 km over the Colorado Plateau) probably do not accurately reflect the areal extent of their source rocks; indeed, disturbances due to some intensely magnetized but very localized and depth-limited sources may have been missed entirely, depending upon source location with respect to the flight path. A further filtering effect is imposed by the 2-km gridding.

With these caveats, the map of figure 2*A* can serve to illuminate many important structural and lithologic relations not evident on the gravity map (fig. 1*A*) nor on geologic compilations. Color is a useful adjunct; the interval employed here is 50 nanoteslas (nT). We caution, however, that choice of any particular color scheme can result in visual emphasis or de-emphasis of certain features.

Figure 2*B* shows trends of horizontal gradient maxima of the magnetic field, as estimated by inspection of figure 2*A*. The trends are depicted on the same geologic and geographic base as used for figure 1*B*. We elected not to compute a new gradient map, in view of the constraints of the initial data sets and the compositing procedure, with their implied uncertainties in gradient magnitudes. Figure 2*B* can be considered to indicate approximate margins of major magnetic source bodies, although, analogous to the gravity case, the maximum gradient will directly overlie the upper edge of the boundary only where the magnetization boundary is vertical (and magnetization is parallel to the Earth's field). The depicted trends are commonly discontinuous, with many short segments that in aggregate create a somewhat "noisy" pattern overall, corresponding to the abundance of short-wavelength anomalies on the total-intensity map. (See below.) Aspects of this pattern are addressed in the section on regional trends.

Although the northern Colorado Plateau is predominantly a region of essentially nonmagnetic rocks in surface exposures (upper Paleozoic to Cenozoic sedimentary strata), it displays a remarkably rich and varied fabric of aeromagnetic anomalies (fig. 2*A*). Moreover, its mean residual-intensity level is higher than that of neighboring provinces to the west and east. (See fig. 2*A*.) Overall higher intensity here than in the Basin and Range province has been attributed to a higher effective crustal susceptibility or, alternatively, to a deeper regional Curie isotherm in the Colorado Plateaus province (Shuey and others, 1973). A regional magnetic-intensity boundary occurs well within the physiographic province boundary of the west side of the plateau, and approximately coincides with major lateral changes in other geophysical parameters (seismic refraction, geomagnetic variation) and geochemistry, as pointed out by Shuey and others (1973). The magnetic boundary also closely follows the Basin and Range–Colorado Plateau structural boundary; that is, it is at the eastern limit of basin-range-style extension. On the map of figure 2*A* the magnetic boundary appears as a set of *en echelon* color breaks that coincide, in part, with the Hurricane (HF)–Toroweap-Sevier (T-SF)–Paunsaugunt (PF) family of high-angle normal faults that produce successive down-to-the-west offsets of the Precambrian crystalline basement. "Basement" in much of the eastern Great Basin probably consists of Mesozoic and Cenozoic granitoid batholithic rocks emplaced above the relatively more magnetic Precambrian crystalline basement (Eaton and others, 1978).

East of the plateau, in the Southern Rockies, a lower mean intensity level is indicated by the prevalence of blues and deep greens in figure 2*A*. The extensive areas of very low field intensity in this province include areas inferred to be underlain by granitic batholiths and the deep basins of the northern Rio Grande Rift system (RGR). For example, concealed mid-Tertiary batholiths are probably responsible for low aeromagnetic intensity (as well as for low gravity anomaly values; see above) over much of the San Juan and West Elk volcanic fields (SJV, WEV) in western Colorado, and the partly concealed Proterozoic Pikes Peak batholith (PP) produces a substantial aeromagnetic anomaly depression. Sources of less intense long-wavelength lows in the Southern Rockies include thick blankets of sedimentary and volcanic strata, and weakly magnetic metamorphic belts of the Precambrian crystalline basement.

Comparison of the aeromagnetic and gravity anomaly fields (figs. 1*A* and 2*A*) over exposed Precambrian terrain in the Southern Rocky Mountains province shows that highs and lows of the two sets most commonly do not coincide. Several factors probably combine to account for this lack of agreement: the magnetic data are reduced to the pole but are not pseudogravity anomalies, and therefore depth-limited sources should produce flanking negative anomalies; the orientation of the magnetization vector is not necessarily parallel to the present Earth's field, due to remanence; anomalous magnetic fields are more sensitive to source depths than are anomalous gravity fields; and above all, magnetic heterogeneity obviously need not always correspond directly to density variations. The same general characteristics should also apply to fields produced by the concealed basement complex beneath the Colorado Plateau, and thus it is not surprising that discrete aeromagnetic and gravity anomalies on the plateau also typically do not coincide. However, broad zones of strong aeromagnetic anomaly relief on the plateau roughly correspond to zones of high gravity, and both data

Figure 2 (following pages). Aeromagnetic features of the northern Colorado Plateau and vicinity. *A*, Aeromagnetic residual total-intensity map (data reduced to the pole). Color interval 50 nanoteslas. Data from Geophysical Data Center, National Oceanic and Atmospheric Administration, Boulder, Colo. (North American Data Set, 2-km grid). *B*, Loci of maximum horizontal gradients visually estimated from residual aeromagnetic total-intensity data of *A*. Teeth point in direction of lower anomaly values where gradients are strong. Symbols explained in table 1.

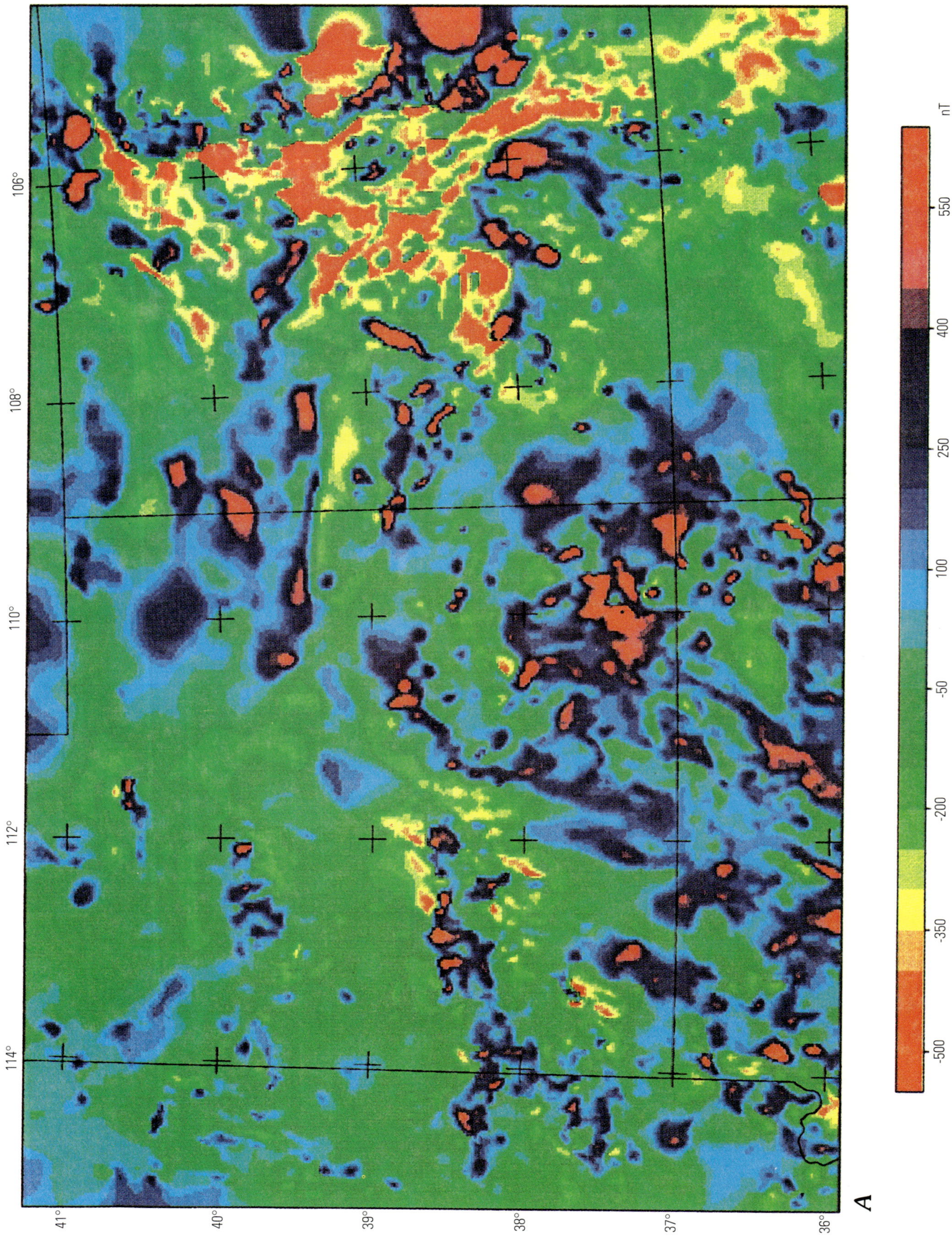
106°
108°
110°
112°
114°
41°
40°
39°
38°
37°
36°
A
nT
550
400
250
100
-50
-200
-350
-500

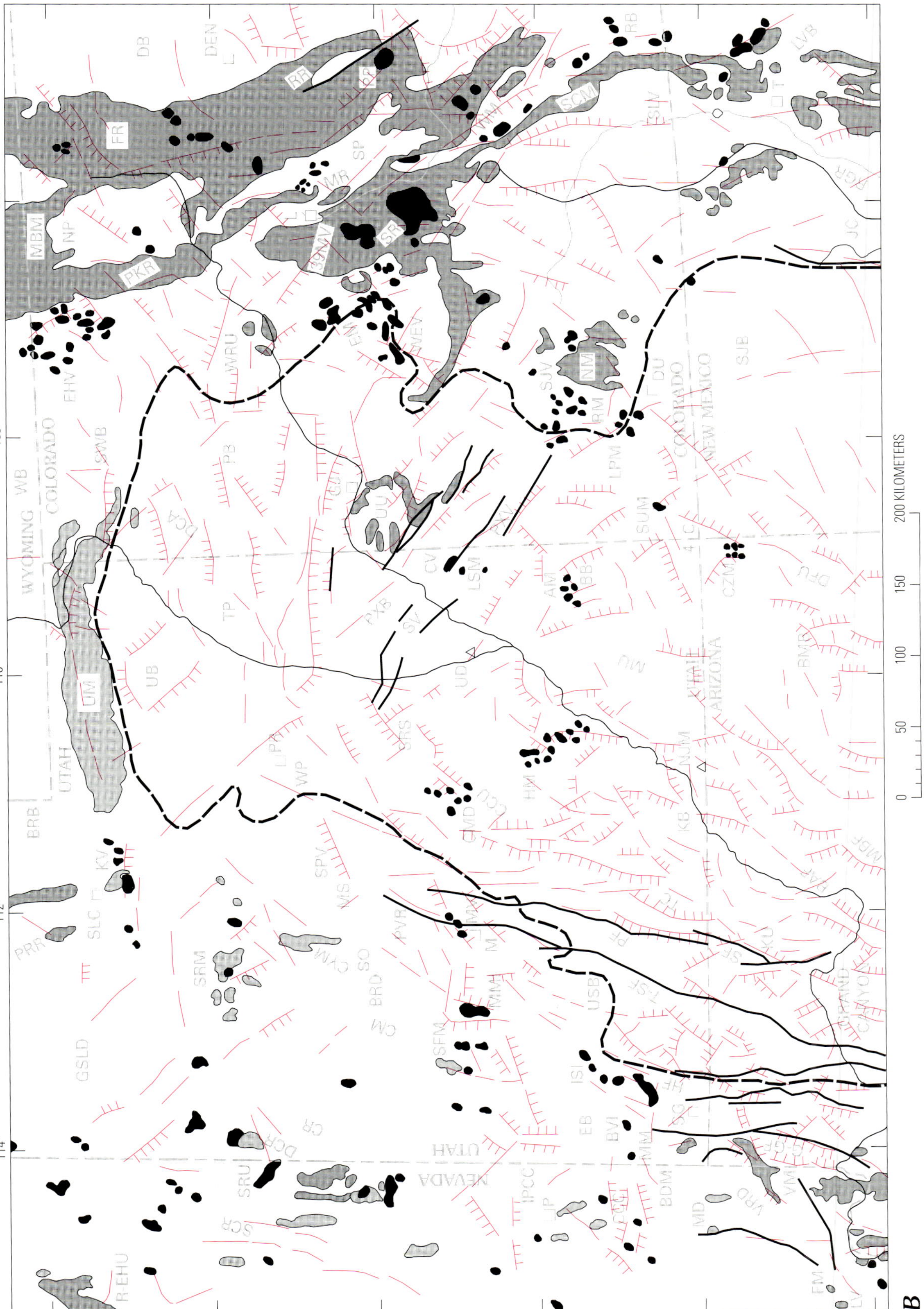
106°
108°
110°
112°
114°
41°
40°
39°
38°
37°
36°
WYOMING
COLORADO
UTAH
NEVADA
ARIZONA
NEW MEXICO
GRAND CANYON
0
50
100
150
200 KILOMETERS

B

sets tend to reflect large-scale structural relief on the surface of the crystalline basement, as will be documented in the following section.

In general, high-frequency (short spatial wavelength) subcircular anomalies are more conspicuous on the aeromagnetic map than on the corresponding gravity map, because of the widespread occurrence of surface or near-surface rocks that have much stronger local magnetization contrasts than density contrasts. The aeromagnetic map itself can be subdivided on the basis of relative abundance of short-wavelength anomalies. The field of the Colorado Plateau south of about lat 39° N., for example, seems to contain more fine structure than does the field of terrain farther north. Short-wavelength anomalies with steep gradients in the southerly domain suggest a relative abundance there of strongly magnetic sources above the basement. Such anomalies commonly coincide with hypabyssal intrusions, including mafic volcanic rocks and plugs as well as granitoid stocks and laccolithic bodies. Local anomalies in the Henry (HM) and La Sal (LSM) Mountains can be unequivocally attributed to exposed laccoliths, as can the anomaly over Sleeping Ute Mountain (SUM) in southwestern Colorado. Navajo Mountain (NJM), on the Utah-Arizona border, is represented by only a weak local anomaly, even though it also appears to consist largely of intrusive rock; the structure of Navajo Mountain is domoform, and syenite is exposed near the mountain crest (Condie, 1964). Strong residual positive anomalies in northeasternmost Arizona (the subcircular, short-wavelength features south and west of the Carrizo Mountains, CZM) are due to intrusive basaltic plugs near Canyon de Chelly and Monument Valley (Sumner, 1985). Many other examples can be cited to illustrate the relation of local anomalies to exposed igneous bodies on the plateau, yet many such sources are not exposed and remain speculative. We consider that most supra-basement sources that have no surface expression are likely to be highly magnetic Tertiary hypabyssal intrusions that failed to breach the level of the present erosion surface.

On the Colorado Plateau another, perhaps unexpected source may contribute to the mapped field—namely clinker, the remains of subterranean coal fires. Pyrometamorphism of overlying sedimentary rocks (Cosca and others, 1989) has been known to produce extremely strong local magnetic anomalies (for example, see Hasbrouck and Hadsell, 1978; Bartsch-Winkler and others, 1988, p. 11), which could register on an airborne detector if the clinker is overflown.

Twin west-northwest-aligned, intense, subcircular anomaly highs near the confluence of the Green and Colorado Rivers in Utah (fig. 2*A*, *B*), are of special interest among speculative sources because the more western of the two almost exactly coincides with Upheaval Dome (UD), which has been interpreted as an impact structure (Shoemaker and Henenhoff, 1984; also see Huntoon and Shoemaker, unpublished manuscript, 1994 GSA Rocky Mountain Section Field Trip no. 5). However, some workers attribute the structure to salt diapirism or salt extrusion (see Mattox, 1975, and Schultz-Ela and others, 1994). In contrast to Upheaval Dome, terrane around the eastern (Grays Pasture) anomaly is not structurally disrupted. Early (Joesting and Plouff, 1958) and recent (V.J.S. Grauch and G.A. Swayze, U.S. Geological Survey, oral commun., 1994) investigations have indicated that the anomaly sources crest within the sedimentary section, well above the general level of the Precambrian crystalline basement. The alignment of the dumbbell-shaped combined twin anomalies is parallel to the regional structural grain. Thus they may be due to fault-controlled, possibly laccolithic, intrusions and may have only a fortuitous relation to Upheaval Dome if indeed that structure is an astrobleme. Alternatively the structural disturbance may prove to have been produced by salt mobilization triggered by igneous intrusion.

BASEMENT OF THE COLORADO PLATEAU

In figure 3 we reproduce part of a newly completed map of the central and southern Colorado Plateau and vicinity showing elevation on the surface of the Precambrian basement (Butler and Kirkpatrick, in press). This map is an updated version of an earlier release (Butler, 1991). In the current version, basement elevations have been determined at a total of 3,763 control points, using outcrops of Precambrian rock on the plateau periphery (about a quarter of all data points), borehole data (340 out of 533 well logs examined yielded basement intercepts, and basement depths for the remaining wells were estimated by extrapolation of the stratigraphic data), seismic picks (from about 50 record sections), and depth estimated from a variety of published maps and stratigraphic sections. The irregularly spaced data were then gridded and computer contoured. As these maps were prepared primarily for use in petroleum resource assessment, elevations are shown in thousands of feet rather than meters. The map scale for figure 3 is the same as for the potential-field maps of figures 1 and 2, but coverage does not extend north of the Paradox Basin and about 40 percent more area to the south of the current study area is included.

A gentle overall northeasterly tilt of the plateau basement can be discerned from the contours; this tilt conforms to the structure of the plateau as a whole, which resembles a tilted saucer. (See Hunt, 1969, fig. 62. Hunt's structure contour map was drawn on the top of the Lower Permian Kaibab Limestone.) Where figure 3 overlaps with the geophysical data, we find a convincing correlation of basement structural highs with gravity anomaly highs, particularly in the Defiance (DFU) and Monument (MU) Uplifts near the Four Corners, the San Rafael Swell (SRS) in the north, and the upthrown sides of major north-northeast-trending normal faults on the west, such as the Hurricane and

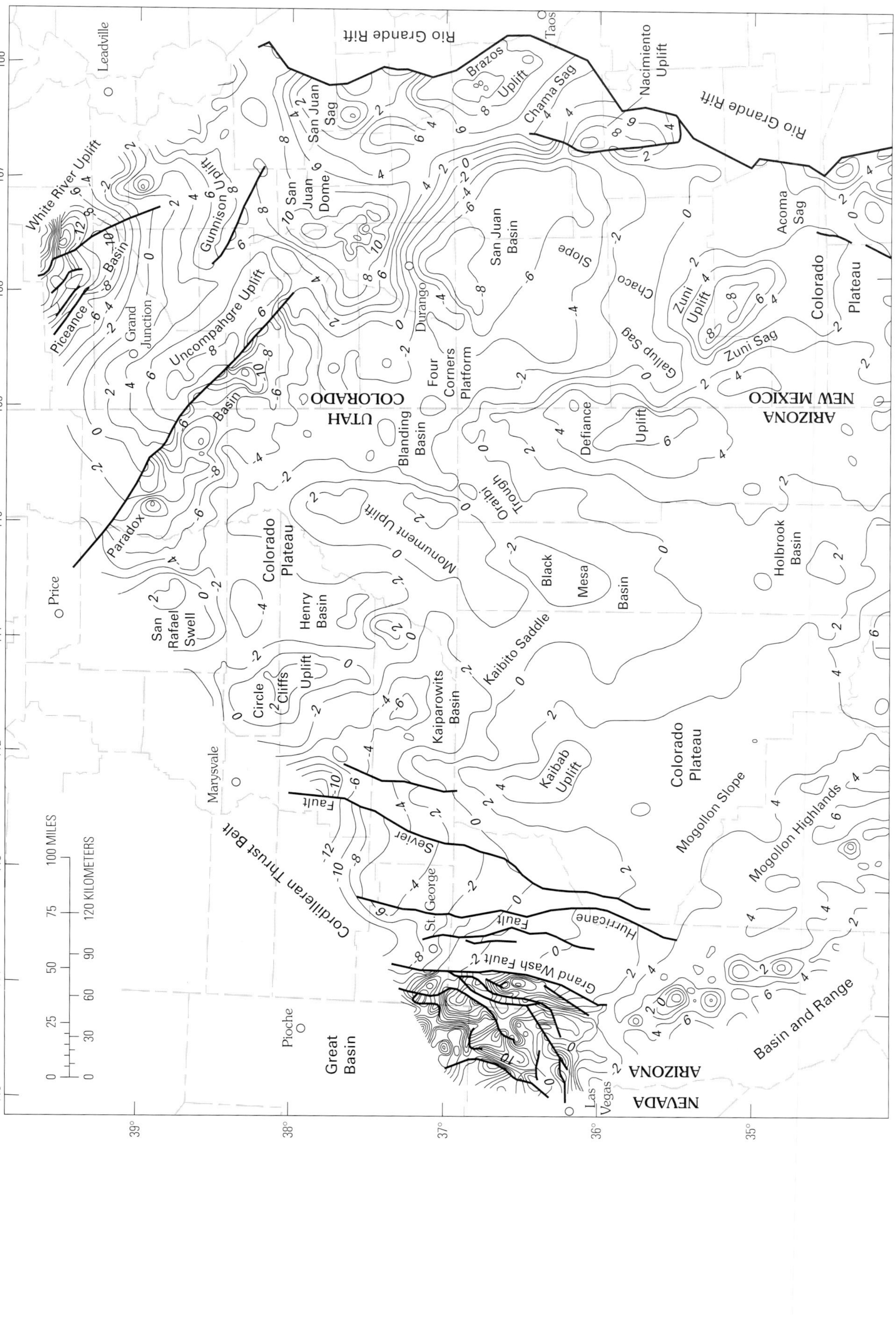

Figure 3. Structure contours showing elevation of surface of Precambrian crystalline basement for part of Colorado Plateau. Numbers give elevation in thousands of feet above sea level; contour interval 2,000 feet (610 m). Modified from Butler (1991).

Toroweap-Sevier zones. Correlation of basement highs with magnetic anomaly highs is less distinct, probably as a result of significant lateral heterogeneity of magnetization in the Precambrian section. However, the mapped southwestern margin of the Uncompahgre crystalline block has a well-defined aeromagnetic signature, even though this margin typically has a negligible gravity expression. (See discussion of gravity anomalies, above.) The signature results from strongly magnetized upper-plate crystalline rocks at or near the surface. Most basement uplifts on the plateau originated as Laramide compressional strucures (Coney, 1976), strongly influenced however by the preexisting tectonic fabric. On the western margins, where basin-range-style extension has encroached on the plateau, basement uplifts have developed as a result of tilting of large blocks on normal faults. Many plateau uplifts were later beveled by erosion and then rejuvenated.

The influence of deep-seated basement structures on the disposition of Tertiary laccolithic igneous centers of the northern Colorado Plateau is at most only weakly expressed at the map scale of figures 1–3. From these maps, we surmise that most igneous centers are located on the flanks of, or between, major basement uplifts rather than on their crests, which suggests structural control by deep-seated faults on the flanks of the uplifts rather than by flexure axial planes. However, when more detailed aeromagnetic maps are consulted, for instance maps at scales of 1:500,000 and larger, the influence of older, regional structures on laccolith emplacement is sometimes quite evident. For example, intrusions in the La Sal Mountains are elongated northwesterly, following the axial trend of the Castle and Paradox Valleys (CV, PXV) and other trends in the anomaly field that reflect the structural grain of the Precambrian basement (Joesting and Case, 1960; Case and Joesting, 1972). Also, centers in the Henry Mountains and dikes in the Cedar Mountains complex (CMD) are aligned north-northwesterly, in agreement with regional basement trends deduced from the aeromagnetics.

REGIONAL ANOMALY TRENDS

Loci of horizontal anomaly gradient maxima displayed in figures 1*B* and 2*B* delineate relatively steep density and magnetization contrasts, and therefore tend to highlight lithologic boundaries such as those produced by high-angle faults or steep-sided igneous intrusions. Where the causal structures dip at significantly less than vertical angles, the loci are nevertheless indicative of structural trends. In the following paragraphs we discuss groups of anomaly trends with more or less common orientation.

Northwesterly trends are conspicuous on both gravity and aeromagnetic maps (figs. 1*B* and 2*B*; refer also to figs. 1*A* and 2*A*) and occur throughout the study area. They are probably most prominent in the central part of the gravity map (fig. 1*B*), particularly in the Southern Rockies. On the Colorado Plateau, they are parallel to (or in some cases coincide with) mapped faults of Paradox Valley and the Uncompahgre Front, including structures on which the earliest movement dates at least as far back as late Paleozoic time (Stokes, 1982) and perhaps earlier (Larson and others, 1985).

Northeasterly trends are also found on both data sets (figs. 1*B*, 2*B*) but are most strongly expressed in the aeromagnetics (fig. 2*B*). In the southern and southeastern parts of the study area, northeasterly aeromagnetic trends align with the Sinyala, Mesa Butte, and Bright Angel fault zones and their inferred extensions across southeastern Utah and into Colorado (Case and Joesting, 1972; Shoemaker and others, 1978; Sumner, 1985). These fault zones and their northeastern extensions, along with parallel but less prominent structures, compose a broad northeast-trending belt (fig. 4) referred to as the Colorado lineament (Warner, 1978, 1980). Approximately at the southeast margin of the lineament belt is the Laramide and post-Laramide Colorado mineral belt (Lovering and Goddard, 1950; Tweto, 1968). Northeasterly anomaly trends also track the Jemez (Springerville-Jemez-Raton) lineament of northern New Mexico (Chapin and others, 1978; Aldrich, 1986; Lipman, 1980; see also Cordell and Keller, 1984). The Jemez lineament (J, fig. 4; also see fig. 6), about 200 km southeast of and parallel to the Colorado lineament (Colorado mineral belt), consists of a northeast alignment of several Neogene volcanic fields, including the Jemez caldera complex (JC). The significance of the Jemez and other northeast-trending Neogene volcanic lineaments of the Cordillera will be addressed in the next section. Sources of northeasterly anomaly trends on the Colorado Plateau include rejuvenated Precambrian and Paleozoic structures. (See, for example, Tweto and Sims, 1963.) Probably most structures delineated by trends of either northeast or northwest orientation involve offset of plateau strata (See, for example, Davis, 1978; Stevenson and Baars, 1986), but many do not closely correspond to any mapped features and may involve only rocks of the crystalline basement.

Northerly trends are present mainly in the eastern and western marginal zones of the northern plateau and in bordering domains, reflecting structures associated with the northern Rio Grande Rift on the east (the source of both gravity and aeromagnetic trends) and with Basin and Range faulting on the west (especially well expressed by gravity trends). On both sides of the plateau these trends tend to be arcuate and concave inward, toward the medial region of the plateau.

Trends oriented approximately east-west are well expressed by potential-field data over the east and west flanks of the Colorado Plateau in belts that include the axis of the San Juan volcanic field (Steven, 1975; Steven and others, 1984) and the Pioche-Marysvale igneous belt (P to M on fig. 2*B*) (Shawe and Stewart, 1976; Stewart and others, 1977; Rowley and others, 1978). Broad, irregular but east-west

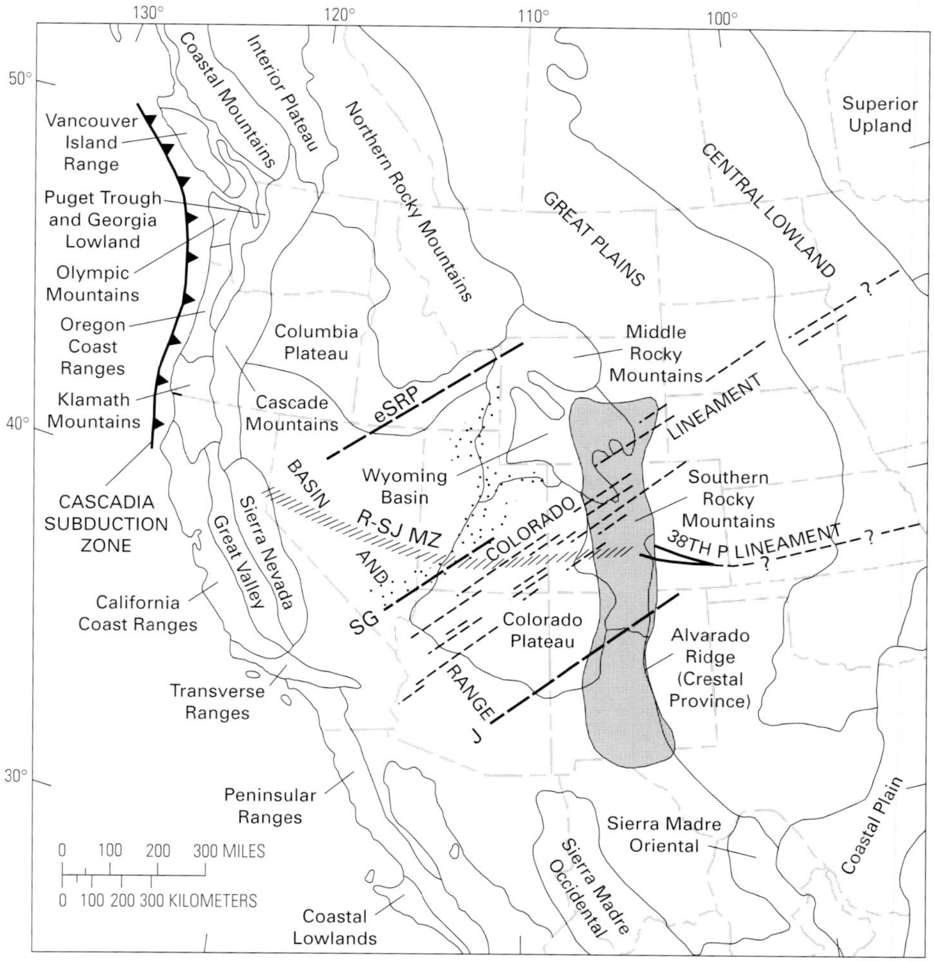

Figure 4. Index map of a part of western North America, showing location of some large-scale tectonic features mentioned in the text. The Colorado lineament is sketched from Warner (1980, fig. 3); eSRP (eastern Snake River Plain–Yellowstone), SG (St. George), and J (Springerville-Jemez-Raton) lineaments (heavy dashed lines) are upper-mantle seismic low-velocity zones that coincide with Neogene volcanic trends, as indicated by Humphreys and Dueker (1994a, fig. 2); the Reno–San Juans magmatic zone (hachured band, R-SJ MZ) is after Nelson and others (1992, fig. 1); the 38th Parallel lineament (queried where highly uncertain) is after Heyl (1972, fig. 1); and the Cascadia subduction zone, a surviving remnant of the Pacific subduction zone after 30 Ma, is from Atwater (1989, fig. 9). The crestal province of the Alvarado Ridge (shaded area bounded by solid line) is from Eaton (1986, fig. 1). Stippled areas denote the central and southern sectors of the Intermountain seismic belt, after Smith and others (1989, fig. 21). Base map with physiographic provinces is adapted from Stewart (1978, fig. 1–2).

elongated gravity lows are associated with root structures of the San Juan volcanic field and the Pioche-Marysvale igneous belt. The approximate alignment of these features with igneous centers between Pioche and Reno, Nevada, and the penecontemporaneity of their magmatic activity with that of laccolithic centers in southern Utah have suggested to some workers the existence of a more or less continuous, Oligocene to early Miocene magmatic zone extending east-west through Nevada, Utah, and western Colorado, termed the "Reno–San Juans magmatic zone" (Sullivan and others, 1991; Nelson and others, 1992; see also Rowley and others, this volume). If the full width of this zone (fig. 4) can be considered to approach two degrees of latitude (about 36°45' N. to 38°45' N.), then it encompasses all known laccolithic centers of the northern Colorado Plateau. Further, it is aligned with the 38th Parallel lineament (fig. 4), a zone of wrench faulting and other structural disturbances, locally accompanied by igneous activity and mineralization, which extends westward through the Eastern and Central United States and has been intermittently active from Cambrian until at least early Tertiary time (Heyl, 1972, 1983; Lidiak and Zietz, 1976). No clearly defined geophysical trends indicate structural continuity between either Marysvale and the San Juans, or the San Juans and the 38th Parallel lineament, however, and the implications of their alignment are unclear. We note that the 1,000-km-long Reno–San Juans magmatic zone is oriented roughly normal to the mid-Tertiary to present Cascadia subduction zone (fig. 4) and by inference, to the corresponding segment of the Pacific subduction zone that preceded it, as well as to the orientation of the associated mid-Tertiary magmatic arc. Thus it may be an intracontinental analog to a leaky transform zone of seafloor-spreading models.

Other prominent trends of the east-west set occur well north of Colorado Plateau laccolith centers, in the north-central part of the study area (Uinta Mountains, Uinta Basin, and Tavaputs Plateau). Long, arcuate, concave-to-the-south trends on the gravity map (fig 1*B*) delineate master faults of the Uinta block, which have produced substantial offsets of the Precambrian basement. Faults of the Tavaputs Plateau and associated blocks (including the southern margin of the Uinta Basin) are reflected in strong magnetic trends that are less continuous than those bounding the Uinta block, and mostly straight, with little surface expression.

In sum, northeasterly (Colorado mineral belt) and northwesterly (Paradox Basin) trends are prevalent over large areas, both inside and outside the plateau margins; northerly trends are best developed outside the plateau in the extreme east (Front Range; Rio Grande Rift) and extreme west (Basin-Range); and easterly trends are also most strongly expressed outside the plateau (San Juan Mountains and Pioche-Marysvale areas). However, in many areas no trend dominates but, rather, trends are mostly curvilinear. Arcuate anomaly trends form a swath that envelops the northern plateau. The swath includes elements internal as well as external to the plateau, but their arcuate character is much less conspicuous in the interior. The center of curvature of arcuate elements seems to be in a region west or southwest of the Four Corners, near the Carrizo Mountains (CZM), where gravity trends are weak and most magnetic trends are northeasterly to northwesterly, reflecting the structural fabric of the basement. This region is also near the geographical center of the plateau.

The symmetry of the pattern of arcuate trends with respect to the northern plateau, particularly those derived from gravity data (fig. 1*A*, *B*), suggests a genetic relationship between the causal structures and the tectonic evolution of the plateau. In many places, the sources are clearly identified as mapped faults or fold axes. (For example, compare Hamilton, 1988, fig. 1.) Steep density discontinuities indicated by prominent horizontal gradients of gravity commonly are located at or near mountain fronts (they are produced by the contrast between basin fill and bedrock of any lithology), and thus can be directly related to structural margins of uplifts and basins, or horsts and grabens, that make up the present topography. Sharply delineated grabens in places merge along strike with less well delineated basinal depressions, or with gaps within the Precambrian crystalline basement that are inferred to be largely occupied by granitoid mid-Tertiary plutons concealed beneath volcanic cover. In many such cases, the geophysical data suggest a structural continuity not otherwise evident.

The north- to north-northwest-trending Rio Grande Rift is illustrative. As a continuous system of grabens, it terminates near Leadville, Colo. (L), although a narrower belt of related block faulting continues north almost to the Wyoming border (Tweto, 1979b, 1980). West-northwest to westerly Neogene trends in northern and northwestern Colorado and northeastern Utah have previously been associated with a major east-west tectonic zone, rather than with the predominantly north trending Rio Grande Rift system. (See, for instance, Tweto, 1979b.) But elements of these two tectonic zones are not necessarily unrelated. Near Leadville, combined gravity and aeromagnetic trends suggest structural linkage of the Rio Grande Rift with a broad depression beneath the Sawatch Range (SR), the Elk Mountains (EM), and the West Elk Mountains (coincides with WEV). This depression, largely occupied by middle to late Tertiary volcanic rocks and their subjacent plutons, flares to the northwest; its gravity signature is contiguous with that of the Piceance Creek Basin (PB). As noted previously, plutons in this region include granitoid rocks of Proterozoic and Laramide as well as Tertiary age. Together these rocks produce a composite regional gravity and aeromagnetic anomaly low. The gravity lows of the volcanic-plutonic zone and the Piceance Creek Basin represent generally northwest trending structural depressions and probably imply significant southwest-directed Neogene extension.

A nearly continuous system of gravity lows, interpreted as structural depressions, links the following features in a grand arc (clockwise from southwestern Utah; see fig. 1*A*, *B*): upper Sevier Basin (USB), Marysvale volcanic field (MV), San Pitch–middle Sevier Valleys (between Pavant Range (PVR) and Wasatch Plateau (WP)), Uinta Basin (UB), Piceance Creek Basin (PB), West Elk volcanic field (WEV), San Luis Valley (SLV), and northern Rio Grande Rift (RGR). This system of depressions is interrupted by the Douglas Creek Arch (DCA), a Laramide uplift whose subsequent structural development was related to differential subsidence between the Uinta and Piceance Creek Basins (Johnson and Finn, 1986). Other prominent arcuate structures that bound depressions, as inferred from the gravity trends, include steep marginal faults of the Schell Creek Range (SCR) in Nevada; the Confusion Range (CR), Uinta Mountains (UM), Tavaputs Plateau (TP), and San Rafael Swell (SRS) in Utah; and North Park (NP) and the Medicine Bow Mountains (MBM), Front Range (FR), and Rampart Range (RR) in Colorado.

Fault-bounded depressions on the east and west flanks of the northern Colorado Plateau (grabens of the Rio Grande

Rift and the Basin and Range province), which trend north to north-northwesterly or north to north-northeasterly, respectively, are well-known manifestations of late Tertiary (Neogene) extension. Basement depressions or perforations now occupied by plutons in the substrate of the Elkhead, West Elk, and San Juan volcanic fields (EHV, WEV, and SJV) of western Colorado also reflect Neogene extension.

The Uinta and Piceance Creek Basins and other elements of the arcuate system in northwest Colorado and northeast Utah are older structures with a complex history of recurrent deformation. Surface mapping, well logs, and seismic reflection data all support the interpretation that major northwest-trending faults in this area are Laramide compressional structures (high-angle reverse faults), many of which have Neogene extensional overprints of lesser magnitude (A.C. Huffman, written commun., 1994). Northwesterly to northerly trending reverse faults comprise a northward-widening splay pattern (fig. 1*A*, *B*) from New Mexico into Colorado; the pattern has been attributed to a northerly increase in total crustal shortening resulting from the clockwise rotation of the Colorado Plateau of 4° or more about an Euler pole in central New Mexico during the Laramide (Steiner, 1986; Hamilton, 1981, 1989; Van der Voo, 1989).

However, a tensional stress field has probably prevailed in Neogene time. (For example, see Verbeek and Grout, 1993.) Normal faults with late Cenozoic movement tend to be localized along zones of Laramide faulting that had movement in the opposite sense (Izett, 1975). Axes of Miocene uplifts and downwarps, as recorded in the distribution of clastic sedimentary formations such as the Browns Park (upper Oligocene and Miocene) and North Park (upper Miocene) Formations (Izett, 1975), tend to be roughly parallel to Laramide axes or, as in the eastern Uintas, to an Archean-Proterozoic terrane boundary (Hansen, 1986; Bryant and Nichols, 1988). In northwest Colorado and northeast Utah, Neogene and older structures alike are roughly parallel to the present northern margin of the Colorado Plateau, which makes this segment of the overall arcuate pattern difficult to resolve in terms of Neogene extension. Nevertheless, on the basis of gravity data and known structures, it appears that the northern plateau is bordered by an envelope of extensional deformation, which is variable in magnitude and possibly discontinuous.

Trends of aeromagnetic anomaly gradient maxima (fig. 2*B*) tend to be less continuous than the gravity trends (fig. 1*B*), and less easily related to mapped structures, structural trends, and topography. This lack of obvious correlation with surface features is at least partly due to strong magnetization contrasts within the concealed Precambrian crystalline basement of the plateau, as noted earlier. Therefore the sources of most magnetic anomalies over the plateau are unidentified. For example, one of the most prominent and continuous gradient trends—that representing the east-west-striking anomaly step north of Grand Junction (figs. 2*A*, 2*B*)—corresponds to no surface feature, nor to any major gravity trend, although it does coincide with very weak gravity gradients (not shown on the trend map of fig. 1*B*). Parallelism of this step with gravity and magnetic anomaly steps over the Tavaputs Plateau and Uinta Basin to the north suggests that all of these east-west anomaly trends are related: they may reflect intrabasement structures, but they could also be produced by high-angle basement faults (block faults) that did not propagate through the suprajacent plateau strata.

North to northeasterly aeromagnetic trends predominate on the west side of the northern plateau and in the transitions to the eastern Basin and Range province, where, on structures such as the Paunsaugunt fault, the anomaly offsets are clearly related to offsets of crystalline basement rocks that result from Neogene normal faulting. On the east side of the plateau and in the Southern Rockies, north to northwesterly trends predominate. Except in a few localities, though, their interpretation as signatures of Neogene extensional structures can be debated. However, the aeromagnetic trend pattern for the region as a whole is very similar to that of the gravity trends: an envelope of arcuate trends encompasses the northern Colorado Plateau, and in most places represents a system of Neogene high-angle normal faults that have produced extensional displacements of the crystalline basement.

DISCUSSION

The Colorado Plateau has been called an "enigmatic crustal block," or microplate, as it has survived nearly intact though much of Phanerozoic time despite being translated, rotated, elevated, and contiguous to provinces of voluminous magmatism. Gravity and aeromagnetic anomalies of the northern Colorado Plateau and vicinity delineate the Precambrian basement fabric and loci of Cenozoic magmatism. In addition, they reveal the regional continuity of broadly arcuate features that can be interpreted as horst-graben structures circumferential to the plateau interior, a pattern that suggests a middle to late Cenozoic stress regime of radial outward extension. To the east and west of the present plateau margins, these structures include the Rio Grande Rift and normal faults of the easternmost Basin and Range province, respectively; in parts of northern Utah and northwestern Colorado, they include arcuate Laramide or earlier structures reactivated in the Neogene with normal (extensional) displacements.

Figure 5*A* (from fig. 3 of Zoback and Zoback, 1989; see also fig. 1 of the same paper) depicts the generalized results of a compilation of probable least-principal-stress directions from various Quaternary indicators, including earthquake focal mechanisms, elliptical well-bore enlargements ("breakouts"), *in situ* stress measurements, and young vent alignments and fault offsets. Tensile stresses are oriented north-northeast near the northeastern and southwestern

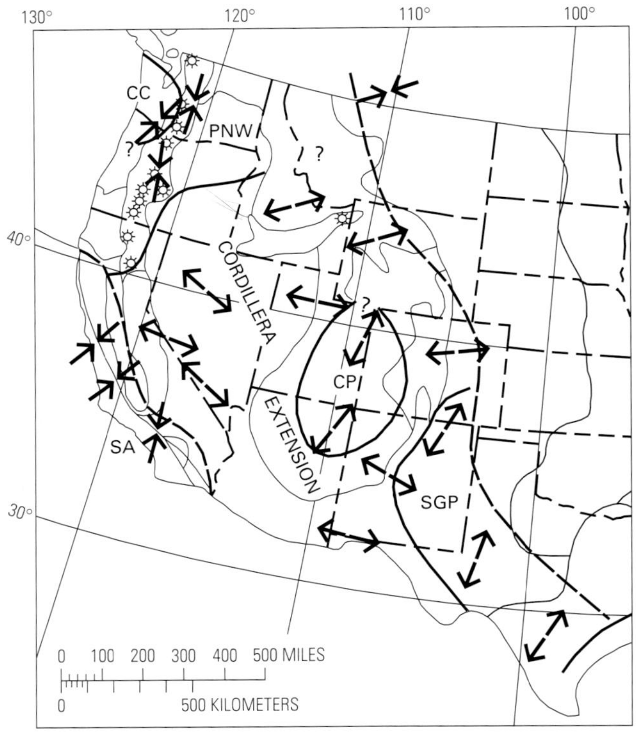

Figure 5. Generalized stress map for Western U.S. Outward-pointing arrows indicate areas of extensional deformation; inward-pointing arrows show areas dominated by compressional tectonics. Heavy lines are stress-province boundaries. CP, Colorado Plateau interior; CC, Cascade convergent province; PNW, Pacific Northwest; SA, San Andreas province; SGP, Southern Great Plains. From Zoback and Zoback (1989, fig. 3).

margins of the plateau and nearly east-west on the western and eastern margins. Thus, at least on the basis of the sparsely distributed data presently available, the late Cenozoic stress field for the Colorado Plateau and its margins seems to accord with radial extension, perhaps superimposed on the predominantly east-west or southwest-northeast least principal stress expressed in neighboring provinces.

Radial outward extension of the plateau, if it happened, may simply indicate an ongoing process of gravitational collapse and lateral spreading more or less concurrent with uplift. The plateau is situated on the axial crest and west flank of the regional elevation anomaly of the Western United States (the Cordilleran "geanticline" of Hunt, 1956). This long-wavelength topographic feature corresponds very closely to the 1,000-km low-pass Bouguer gravity anomaly depicted on the map of figure 6 (modified from Hildenbrand and others, 1982, and Kane and Godson, 1989; a color plate accompanying the latter paper also displays a map of the elevation anomaly). This map shows the deep low that remains after removal from the Bouguer data set of all spectral components with wavelengths less than 1,000 km. Also shown on the map are an outline of the Colorado Plateau (bold line) and three Neogene volcanic trends that coincide with upper-mantle seismic low-velocity zones (after Humphreys and Dueker, 1994a, fig. 2). These trends will be discussed later.

The regional Bouguer anomaly (fig. 6) is seen as a bulbous trough extending from Canada into Mexico; on continent-scale maps it appears as a local widening of a linear negative Bouguer anomaly that coincides with highly elevated terrain along the full length of the Cordillera in western North America inboard of the convergent plate margin. The anomaly minimum in this wide zone in the U.S. Cordillera is located in westernmost Colorado, over the northeastern part of the Colorado Plateau, where it corresponds to a regional elevation maximum of about 2.5 km. The axis of the regional elevation high/Bouguer gravity low approximately coincides with the axis of maximum flexural subsidence and Late Cretaceous sedimentation in the Sevier foreland. (See, for instance, Pang and Nummedal, 1995; Bird, 1984.)

The plateau bears a similar relation to the regional geoidal high of the Western United States, which also has a local maximum in western Colorado (Milbert, 1991).

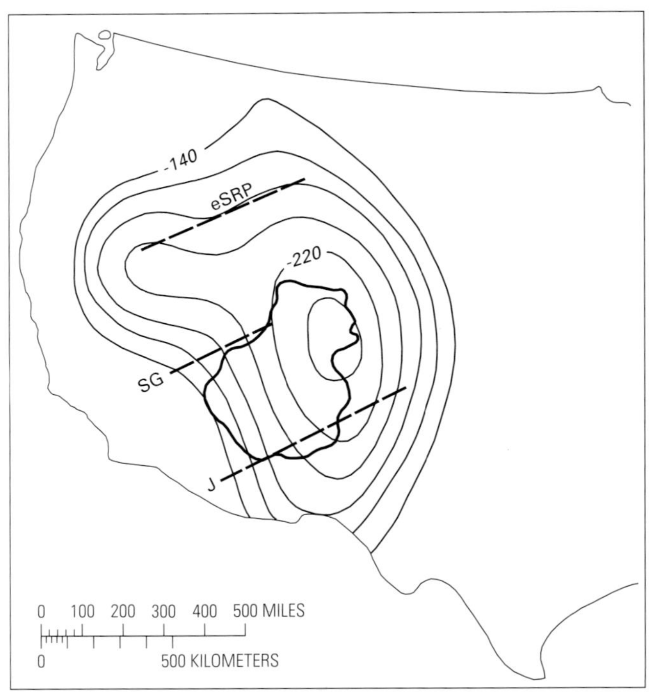

Figure 6. Long-wavelength (λ >1,000 km) regional Bouguer gravity anomaly map of the Western United States, showing disposition of Colorado Plateau (bold outline) with respect to gravity minimum (after Kane and Godson, 1989, fig. 4). Contour interval is 20 milligals, but values higher than –140 milligals are not shown. Dashed lines are Neogene volcanic trends that coincide with upper-mantle seismic low-velocity zones (from Humphreys and Dueker, 1994a, fig. 2). eSRP, eastern Snake River Plains trend; SG, St. George trend; J, Jemez trend.

However, the regional geoidal anomaly is more nearly equidimensional than the other two anomalies and crests exactly over the Yellowstone hotspot (Pierce and Morgan, 1992), a point to which we shall return later in this section.

Various mechanisms of plateau uplift are reviewed in McGetchin and Merrill (1979). It has been emphasized elsewhere that in the case of the Colorado Plateau, uplift must be considered in the context of tectonic evolution of the entire western interior Cordilleran region. That is, uplift of the plateau must be closely related to development of the huge anomalous "root" zone responsible for the long-wavelength gravity, geoid, and elevation anomalies (as in Bird, 1984; Beghoul and Barazangi, 1989; Parsons and others, 1994). Models that specifically seek to explain the 2-km average elevation of the Colorado Plateau include (1) those that require the presence of a mantle plume (such as Wilson, 1973, and Parsons and others, 1994; see also Parsons and McCarthy, 1995) or group of plumes (Sbar and Sykes, 1973) beneath the plateau and its environs, resulting in thermal thinning of the lithosphere and, possibly, crustal thickening by magmatic underplating (as described by Morgan and Swanberg, 1985), and (2) those that ascribe the uplift to processes related directly to subhorizontal subduction during the Laramide orogeny. Models of this latter class generally require delamination of the overriding North American plate (Bird, 1979) or of the subducted Farallon plate (Bird, 1988; Beghoul and Barazangi, 1989), leading to the replacement of lithosphere by hot asthenosphere, and probable crustal thickening. Bird (1988) proposed large-scale shear decoupling and translation of lithosphere eastward to the anomalous Cordilleran "root" zone. However, simple buoyancy of a flattish Farallon slab was invoked by Lowell (1974) and Gries (1983) to explain the plateau uplift.

In support of the concept of low-angle subductions, modern analogs with many characteristics similar to those of the U.S. western interior have been documented for the Andean margin of South America, where terrane above a seismically delineated low-angle slab far inboard of the convergent plate margin has been uplifted, deformed, and subjected to magmatism (Isacks and Barazangi, 1977). The low subduction angle has been attributed to a high convergence rate (Engebretson and others, 1984), to youthfulness (and therefore relatively high temperature) of the slab, or to a slab containing thick crustal elements such as aseismic ridges or oceanic plateaus. (See Livaccari and others, 1981, for instance.) Possibly all of these factors contributed to the flattening.

Eaton (1986, 1987) investigated the highly elevated region between Wyoming and El Paso, Texas (in his terminology, the Southern Rockies *sensu lato*), and noted that it straddles a much broader, north-south elongated topographic swell, for which he proposed the name "Alvarado Ridge" (fig. 4; only the crestal province of the ridge is shown in the figure). This swell appears to be virtually identical to the long-wavelength Cordilleran elevation/Bouguer gravity anomaly, although Eaton pointed out that axes of the two features are somewhat displaced and have slightly different orientations. The High Plains (part of the Great Plains physiographic province, fig. 4) are on the east flank of the ridge, the Colorado Plateau is on its west flank, and the Rio Grande Rift coincides with its axial zone. Eaton showed that transverse profiles of the Alvarado Ridge bear a striking resemblance to those of mid-ocean spreading ridges, which led him to suggest that the ridge is, in fact, a "continental rise." From the sedimentary record, including age and distribution of the Ogallala Formation (Miocene), he inferred that major uplift of the Southern Rockies *sensu lato* occurred in the interval from about 17 Ma to between 7 and 4 Ma. That is, the rise of the Alvarado Ridge is mainly a late Cenozoic phenomenon, although some uplift of the ridge, and initiation of the Rio Grande Rift, probably occurred as early as mid-Oligocene time.

Possible origins of the long-wavelength Cordilleran anomaly, in consideration of gravity and seismic data, have been discussed by Kane and Godson (1989). In their view the Bouguer gravity low cannot be accounted for by crustal variations alone and must involve a buoyant upper mantle (75 percent of the buoyancy residing in the upper 160 km) in order to maintain a condition of regional isostatic equilibrium as implied by near-zero average (long-wavelength) free-air anomalies (as shown by Simpson and others, 1987, for instance). Because the regional gravity low corresponds to generally low velocities of both compressional (P_n) and shear (S) waves, Kane and Godson postulated the existence of anomalously high mean upper mantle temperatures to account for the required low density.

In an analysis of all available teleseismic p-wave travel-time residuals, Dueker and Humphreys (1990) and Humphreys and Dueker (1994a, b) demonstrated that both high- and low-velocity domains are present in low-density upper mantle of the U.S. western interior. Humphreys and Dueker (1944a, b) concluded that elevated mantle temperatures locally produce partial melts that depress the P-wave velocities and lead to observed lateral variations in velocity structure. Three principal low-velocity zones were delineated (see figs. 4 and 6), each trending northeasterly and each corresponding to a belt of generally northeast younging Neogene (mainly) bimodal volcanism: the eastern Snake River Plain–Yellowstone zone (eSRP, figs. 4 and 6), the St. George (Utah) zone (SG, figs. 4 and 6), and the Jemez (Springerville-Jemez-Raton) zone (J, figs. 4 and 6) across northwestern New Mexico (Humphreys and Dueker, 1994a). The latter two zones are on or near the northwest and southeast borders, respectively, of the Colorado Plateau. All three low-velocity zones are on the west flank of the long-wavelength Cordilleran anomaly, and all three coincide with belts of relatively steep crustal isopach gradients that separate regions of thick crust (40 to 50 km in central Idaho and western Montana and in the Colorado Plateau) from regions of thinner crust, according to data from seismic refraction profiles. (See Allenby and Schnetzler, 1983, fig. 2.) Also, all three are parallel to or coincident with known Proterozoic terrane boundaries (Bowring and Karlstrom, 1990), and lie on great circle arcs through the Eulerian pole of relative motion between the North American and Pacific tectonic plates (Eaton, 1979, fig. 1*B*). The significance of these relations lies in the following: (1) ancient deep-crustal flaws are implicated in the localization of late Cenozoic magmatism over a large region of the Cordillera; and (2) the parallelism of the zones and the general tendency for Neogene volcanism to young northeastward within each zone (see below) suggest common underlying factors in their tectono-magmatic evolution. The common factors may be (1) the presence of a mantle plume beneath the wide (transverse axis >>1,000 km) central part of the long-wavelength Cordilleran anomaly, and (2) dominantly southwestward movement of the North American plate across the plume head during Neogene time. In this scenario, late Cenozoic uplift of the entire interior Cordilleran region is attributable to the presence of a plume in the broadest sense—that is, a vast convective upwelling of hot asthenosphere, regardless of its relationship to subduction or the maximum depth of mantle involved in the total convective system. The Colorado Plateau remains high because it has been relatively little extended.

Of the three Humphreys-Dueker zones, the eastern Snake River Plain–Yellowstone zone offers the most compelling case for a hotspot track and underlying mantle plume (Morgan, 1972; Pierce and Morgan, 1992; Smith and Braile, 1993; Parsons and others, 1994). The orientation of the volcanic axis and the migration of magmatism at a rate of about 3 cm yr^{-1} (well determined for the interval 10–2 Ma) in a northeasterly direction (about N. 54° E.) to its culmination at Yellowstone (Pierce and Morgan, 1992) are in good agreement with independent global reconstructions that yield the absolute motion of the North American plate. (For example, see Stock and Molnar, 1988; Atwater, 1989.) Furthermore, the pattern of deformation and the amplitude and wavelength of the Yellowstone geoidal anomaly (about +10 m and 1,000 km, respectively) are similar to properties of hotspots worldwide (Crough, 1983; Sleep, 1990; Duncan and Richards, 1991), although most known examples occur over oceanic, rather than continental crust. The St. George zone is only weakly delineated as a magmatic belt but, together with the western Grand Canyon area, shows a distinct northeasterly younging (as documented, for instance, by Nealey and Sheridan, 1989, and by unpublished data of Nealey). This zone follows the northeasterly trend of the Intermountain Seismic Belt, which bifurcates in central Utah, one branch continuing northeasterly (fig. 4); the zone may also mark a mini-plate boundary (Suppe and others, 1975). The Springerville-Jemez-Raton zone, which extends from near the middle of Arizona's eastern border across northern New Mexico into southern Colorado (Chapin and others, 1978; Smith and Luedke, 1984), exhibits no systematic age migration as a whole (Lipman, 1980), but within each discrete field of the zone the age progression is northeasterly (Nealey, oral commun., 1995). Magmatism in these latter two zones appears to be more diffusely distributed than on the well-defined Snake River Plain axis, but in all three cases, the fundamental structural control may consist of first-order preexisting tectonic lineaments oriented approximately in the direction of absolute motion of the lithosphere.

Interpretation of the long-wavelength Cordilleran topographic/gravity anomaly as an upwelling of hot asthenospheric material, or a very broad mantle plume, provides a genetic linkage for the three northeast-trending Humphreys-Dueker low-velocity and magmatic zones, and places the Yellowstone hotspot and its presumed source plume in a wider context. The proposition that several hotspots can be identified in the Cordillera is discussed elsewhere in this volume (see Mutschler and others). Hotspot magmatism is attributed to decompression melting of asthenospheric mantle ascending in a narrow conduit ("chimney") above a fixed plume (Duncan and Richards, 1991; Richards and others, 1989; see also White and McKenzie, 1989). There is no apparent *a priori* reason why a broad plume head should not produce multiple chimneys, provided favorable structural environments are present. It seems possible that extensional faults peripheral to the Colorado Plateau, particularly where they intersect major structural flaws in the basement complex, could facilitate egress of magma. For the Basin and Range province it has been shown that magmatism is most closely related in space and time to crustal extension of 100 percent or more (Leeman and Harry, 1993). Thus the relatively weak magmatic activity on the Colorado Plateau (in contrast to that on its periphery) may be due to a lack of adequate crustal preconditioning. (See, for example, Best and Christiansen, 1991.)

Whereas a very broad mantle upwelling (plume) may be the ultimate source of Cordillera-wide Neogene uplift and bimodal magmatism, the existence of such a feature prior to about 17 Ma is highly speculative, and any connection with earlier post-Laramide deformation and magmatism is tenuous. The timing of early uplift on the plateau is also uncertain (Lucchitta, 1979). Most workers agree that the Colorado Plateau and its surroundings were uplifted only a kilometer or so during post-Laramide Paleogene time—probably in the Oligocene, coinciding with voluminous magmatism in neighboring areas (Hunt, 1969; Bird, 1988). This amount of uplift contrasts with Neogene uplift that has probably exceeded 3 km.

Products of Laramide (Late Cretaceous to early Eocene) to mid-Tertiary magmatism peripheral to the plateau typically have calc-alkaline affinities (Steven, 1975; Lipman and others, 1978; Lipman, 1981; Rowley and others, this volume) and are likely derived from partial melting of the mantle lithosphere together with crustal anatexis, which suggests a close association with Laramide subduction. In contrast, many Neogene bimodal products throughout the

Cordilleran region have asthenospheric signatures, particularly those products emplaced within the last 6 million years. (See Fitton and others, 1991, for example.) This suggests a deeper origin of the later products, and a temporal evolution of crust-mantle plutonism related to the evolution of the postulated mantle plume. The plume need not have originated as deep as the core-mantle boundary (see Anderson, 1994; but also see Grand, 1994, and Zhong and Gurnis, 1995), but seems likely to have involved large-scale overturning in a deep-mantle convection system in conjunction with several hundred million years of east-dipping subduction (as reviewed by Atwater, 1989) beneath what is now western North America.

SUMMARY AND CONCLUSIONS

1. Regional gravity anomalies and, less distinctly, regional aeromagnetic anomalies over the Colorado Plateau primarily reflect relief on the surface of the Precambrian basement as determined from stratigraphic data, borehole data, and seismic reflection picks, rather than basement inhomogeneity. In the plateau interior, this relief is a result mainly of Laramide compression.

2. Gravity and aeromagnetic anomaly fields of the northern Colorado Plateau and vicinity delineate northwest- and northeast-trending structures that correspond to the well-known Paradox Basin and Colorado lineament trends, respectively, and transect the plateau and its environs.

3. The nearly east-west Pioche-Marysvale and San Juan anomaly trends strike into the plateau from either side but do not appear to transect the plateau interior, although they are essentially collinear with one another and with the 38th Parallel lineament. The plateau thus may represent a major hiatus in east-west-trending magmatic and hydrothermal activity brought about through a lack of adequate "tectonic conditioning."

4. Other anomaly trends in the vicinity of the northern plateau delineate arcuate structures that wrap around the plateau and its margins. Many of these structures were initially compressional but have subsequently been reactivated with normal (extensional) movement; others are purely extensional. Taken together they suggest a radial outward extension of the plateau as a whole relative to the plateau interior. A northwest-trending structural depression connecting the northern Rio Grande Rift with the Piceance Creek Basin is a key element of the circumferential pattern.

5. Radial outward extension is in general agreement with least-principal-stress orientations for the late Cenozoic as determined from other indicators.

6. The Colorado Plateau is situated near the crest of the long-wavelength topographic high and Bouguer gravity anomaly low of the U.S. western interior. Radial outward extension of the plateau may be primarily a manifestation of gravitational collapse and lateral spreading penecontemporaneous with some 3 km of Neogene uplift.

7. The preceding (mid-Tertiary) uplift amounted to only about a third as much but may have initiated radial extension; the swath of arcuate normal faults around the northern plateau includes loci of voluminous mid-Tertiary magmatism in crust that perhaps has been "tectonically conditioned" on the plateau periphery.

8. The interior Cordilleran topographic high/gravity low, in turn, is probably caused by upwelling of relatively hot, light asthenosphere—a broad mantle "plume." Rise of the plume produced the Alvarado Ridge of Colorado and New Mexico and the Yellowstone hotspot in the Neogene, and the changing character of the plume promoted the evolution of crust-mantle magmatism from calc-alkaline to bimodal in character during middle to late Cenozoic time, as subduction-related processes came to be dominated by deep-mantle convection.

REFERENCES CITED

Abrams, Gerda, 1993, Complete Bouguer gravity anomaly map of the state of Colorado: U.S. Geological Survey Miscellaneous Field Studies Map MF–2263, scale 1:500,000.

Aiken, C.L.V., 1975, Residual Bouguer gravity anomaly map of Arizona: Arizona University Geophysical Lab., Tucson, Ariz., scale 1:1,000,000.

Aldrich, M.J., Jr,. 1986, Tectonics of the Jemez lineament in the Jemez Mountains and Rio Grande rift: Journal of Geophysical Research, v. 91, p. 1753–1762.

Allenby, R.J., and Schnetzler, C.C., 1983, United States crustal thickness: Tectonophysics, v. 93, p. 13–31.

Anderson, D.L., 1994, Superplumes or supercontinents?: Geology, v. 22, p. 39–42.

Anderson, D.L., Tanimoto, T., and Zhang, Y. S., 1992, Plate tectonics and hotspots: Science, v. 256, p. 1645–1651.

Atwater, Tanya, 1989, Plate tectonic history of the northeast Pacific and western North America, *in* Winterer, E.L., Hussong, D.M., and Decker, R.W., eds., The Geology of North America; DNAG Series, v. N: The Eastern Pacific Ocean and Hawaii: Boulder, Colo., Geological Society of America, p. 21–72.

Bartsch-Winkler, Susan, Barton, H.N., Cady, J.W., Cook, K.L., and Martin, C.M., 1988, Mineral resources of the Fifty Mile Mountain Wilderness Study Area, Kane County, Utah: United States Geological Survey Bulletin 1747–A, 20 p.

Bayer, K.C., 1983, Generalized structural, lithologic, and physiographic provinces in the fold and thrust belts of the United States (exclusive of Alaska and Hawaii): Washington, D.C., U.S. Geological Survey, map scale 1:2,500,000 (3 sheets).

Bayley, R.W., and Muehlberger, W.R., 1968, Basement rock map of the United States exclusive of Alaska and Hawaii: Washington, D.C., U.S. Geological Survey, scale 1:2,500,000 (2 sheets).

Beghoul, N., and Barazangi, M., 1989, Mapping high P_n velocity beneath the Colorado Plateau constrains uplift models: Journal of Geophysical Research 94, v. B6, p. 7083–7104.

Best, M.G., and Christiansen, E.H., 1991, Limited extension during peak Tertiary volcanism, Great Basin of Nevada and Utah: Journal of Geophysical Research, v. 96, no. B8, p. 13509–13528.

Bird, Peter, 1979, Continental delamination and the Colorado Plateau: Journal of Geophysical Research, v. 84, no. B13, p. 7561–7571.

———1984, Laramide crustal thickening event in the Rocky Mountain foreland and Great Plains: Tectonics, v. 3, p. 741–758.

———1988, Formation of the Rocky Mountains, western United States—A continuum computer model: Science, v. 239, p. 1501–1507.

Blakely, R.J., and Simpson, R.W., 1986, Approximating edges of source bodies from magnetic or gravity anomalies: Geophysics, v. 51, no. 7, p. 1494–1498.

Bowring, S.A., and Karlstrom, K.E., 1990, Growth, stabilization and reactivation of Proterozoic lithosphere in the southwestern United States: Geology, v. 18, p. 1203–1206.

Bryant, Bruce, and Nichols, D.J., 1988, Late Mesozoic and Early Tertiary reactivation of an ancient crustal boundary along the Uinta trend and its interaction with the Sevier orogenic belt, *in* Schmidt, C.J., and Perry, W.J., Jr., eds., Interaction of the Rocky Mountains Foreland and the Cordilleran Thrust Belt: Geological Society of America Memoir 171, p. 411–430.

Butler, W.C., 1991, Elevation data for the Precambrian surface in the central and southern Colorado Plateau and vicinity: U.S. Geological Survey Open-File Report 92–194, 75 p., with map, scale 1:1,250,000.

Butler, W.C., and Kirkpatrick, J.C., in press, Topographic contour map on top of Precambrian age rocks, central and southern Colorado Plateau and vicinity: U.S. Geological Survey Miscellaneous Field Studies Map MF–2310.

Case, J.E., 1965, Gravitational evidence for a batholithic mass of low density along a segment of the Colorado Mineral Belt: Geological Society of America Special Paper 82, Abstracts for 1964, p. 26.

Case, J.E., and Joesting, H.R., 1972, Regional geophysical investigations in the central Colorado Plateau: U.S. Geological Survey Professional Paper 736, 31 p.

Chapin, C.E., Chamberlin, R.M., Osburn, G.R., White, D.W., and Sanford, A.R., 1978, Exploration framework of the Socorro geothermal area, New Mexico, *in* Chapin, C.E., and Elston, W.E., eds., Field guide to selected cauldrons and mining districts of the Datil-Mogollon volcanic field, New Mexico: New Mexico Geological Society Special Paper 7, p. 115–130.

Committee for the Magnetic Anomaly Map of North America, 1987, Magnetic anomaly map of North America: Geological Survey of America, 4 sheets, scale 1:5,000,000.

Committee for the Magnetic Anomaly Map of North America, 1993, Magnetic anomaly map of part of the conterminous United States and some adjacent parts of Canada, scale 1:5,000,000; plate 3 *in* Reed, J.C., Jr., Bickford, M.E., Houston, R.S., Link, P.K., Rankin, D.W., Sims, P.K., and Van Schmus, W.R., eds., DNAG Series, The Geology of North America, vol. C–2, Precambrian—Conterminous U.S.: Boulder, Colo., Geological Society of America, 657 p.

Condie, K.C., 1964, Crystallization Po_2 of syenite porphyry from Navajo Mountain, southern Utah: Geological Society of America Bulletin, v. 75, p. 359–362.

Coney, P., 1976, Plate tectonics and the Laramide orogeny: New Mexico Geological Society Spec. Publication 6, p. 5–10.

Cook, K.L., Bankey, Viki, Mabey, D.R., and DePangher, Michael, 1989, Complete Bouguer gravity anomaly map of Utah: Utah Geological and Mineral Survey Map 122, scale 1:1,000,000.

Cordell, Lindrith, 1983, Composite residual total intensity aeromagnetic map of New Mexico: New Mexico State University Energy Institute, Geothermal Resources of New Mexico—Scientific Map Series, overlay 2, scale 1:500,000. (Available from National Geophysical Data Center, Boulder, CO, 80303, as report 84–TGB–05.)

Cordell, Lindrith, and Keller, G.R., 1984, Regional structural trends inferred from gravity and aeromagnetic data in the New Mexico–Colorado border region: New Mexico Geological Society Guidebook 35, p. 21–23.

Cordell, Lindrith, Keller, G.R., and Hildenbrand, T.G., 1982, Bouguer gravity anomaly map of the Rio Grande rift, Colorado, New Mexico, and Texas: U.S. Geological Survey Geophysical Investigations Map GP–949, scale 1:1,000,000.

Cosca, M.A., Essene, E.J., Geissman, J.W., Simmons, W.B., and Coates, D.A., 1989, Pyrometamorphism of rocks associated with natural burns in coal seams, Powder River Basin: American Mineralogist, v. 74, p. 85–100.

Crough, T.S., 1983, Hotspot swells, *in* Wetherill, G.W., Arden, A.L., and Stehli, F.G., eds., Annual Review of Earth and Planetary Sciences, v. 11, 1983: Palo Alto, Calif., Annual Reviews, Inc., p. 165–194.

Davis, G.H., 1978, Monoclinal fold pattern of the Colorado Plateau, *in* Matthews, Vincent III, ed., Laramide folding associated with basement block faulting in the western United States: Geological Society of America Memoir 151, p. 215–234.

Dueker, K.G., and Humphreys, E.D., 1990, Upper-mantle velocity structure of the Great Basin: Geophys. Research Letters, v. 17, p. 1327–1330.

Duncan, R.A., and Richards, M.A., 1991, Hotspots, mantle plumes, flood basalts, and true polar wander: Reviews of Geophysics, v. 29, p. 31–50.

Eaton, G.P., 1979, A plate-tectonic model for late Cenozoic crustal spreading in the western United States, *in* Riecker, R.E., ed., 1979, Rio Grande Rift—Tectonics and magmatism: Washington, D.C., American Geophysical Union, p. 7–32.

———1986, A tectonic redefinition of the Southern Rocky Mountains: Tectonophysics, v. 132, p. 163–194.

———1987, Topography and origin of the southern Rocky Mountains and Alvarado Ridge, *in* Coward, M.P., Dewey, J.F., and Hancock, P.L., eds., Continental extension tectonics: Geological Society of London, Special Publication 28, p. 355–369.

Eaton, G.P., Wahl, R.R., Prostka, H.J., Mabey, D.R., and Kleinkopf, M.D., 1978, Regional gravity and tectonic patterns—Their relation to late Cenozoic epeirogeny and lateral spreading in the western Cordillera, *in* Smith, R.B., and Eaton, G.P., eds., Cenozoic tectonics and regional geophysics of the Western Cordillera: Geological Society of America Memoir 152, p. 51–92.

Engebretson, D.C., Cox, Allan, and Thompson, G.A., 1984, Correlation of plate motions with continental tectonics—Laramide to Basin and Range: Tectonics, v. 3, no. 2, p. 115–119.

Fitton, J.G., James, Dodie, and Leeman, W.P., 1991, Basic magmatism associated with late Cenozoic extension in the western

United States—Compositional variations in space and time: Journal of Geophysical Research, v. 96, no. B8, p. 13693–13711.

Frahme, C.W., and Vaughn, E.B., 1983, Proterozoic geology and seismic stratigraphy of the northern Uncompahgre front, Grant County, Utah, *in* Lowell, J.D., ed., Rocky Mountain foreland basins and uplifts: Denver, Colo., Rocky Mountain Association of Geologists, p. 201–211.

Grand, S.P., 1994, Mantle shear structure beneath the Americas and surrounding oceans: Journal of Geophysical Research, v. 99, no. B6, p. 11591–11621.

Gries, Robbie, 1983, North-south compression of Rocky Mountain structures, *in* Lowell, J.D., ed., Rocky Mountain foreland basins and uplifts: Denver, Colo., Rocky Mountain Association of Geologists, p. 9–32.

Hamilton, W.B., 1981, Plate-tectonic mechanism of Laramide deformation, *in* Boyd, D.W., and Lillegraven, J.A., eds., Rocky Mountain foreland basement tectonics: University of Wyoming Contributions to Geology, v. 19, p. 87–92.

———1988, Laramide crustal shortening, *in* Schmidt, C.J., and Perry, W.J., Jr., eds., Interaction of the Rocky Mountains Foreland and the Cordilleran Thrust Belt: Geological Society of America Memoir 171, p.27–39.

———1989, Crustal geologic processes of the United States, *in* Pakiser, L.C., and Mooney, W.D., eds., Geophysical framework of the continental United States: Geological Society of America Memoir 172, p. 743–782.

Hansen, W.R., 1986, History of faulting in the eastern Uinta Mountains, Colorado and Utah, *in* Stone, D.S., ed., New interpretations of northwestern Colorado geology: Denver, Colo., Rocky Mountain Association of Geologists, p. 5–17.

Hasbrouck, W.P., and Hadsell, F.A., 1978, Geophysical techniques for coal exploration and development, *in* Hodgson, H.E., ed., Proceedings of the Second Symposium on the Geology of Rocky Mountain Coal—1977: Denver, Colo., Colorado Geological Survey, Department of Resources, p. 187–218.

Hauser, E.C., and Lundy, J., 1989, COCORP deep reflections—Moho at 50 km (16s) beneath the Colorado Plateau: Journal of Geophysical Research, v. 94, no. B6, p. 7071–7081.

Heyl, A.V., 1972, The 38th Parallel lineament and its relation to ore deposits: Economic Geology, v. 67, p. 879–894.

———1983, Some major lineaments reflecting deep-seated fracture zones in the central United States, and mineral districts related to the zones: Global Tectonics and Metallogeny, v. 2, nos. 1 & 2, p. 75–89.

Hildenbrand, T.G., and Kucks, R.W., 1988, Filtered magnetic anomaly maps of Nevada: Nevada Bureau of Mines and Geology Map 93B, scales 1:1,000,000 and 1:2,500,000 (5 sheets).

Hildenbrand, T.G., Simpson, R.W., Godson, R.H., and Kane, M.F., 1982, Digital colored residual and regional Bouguer gravity maps of the conterminous United States with cut-off wavelengths of 250 km and 1000 km: U.S. Geological Survey Geophysical Investigations Map GP–953–A, scale 1:7,500,000 (2 sheets).

Hintze, L.F., 1975, Geological highway map of Utah: Provo, Utah, Brigham Young University, scale 1:1,000,000.

Hintze, L.F., compiler, 1980, Geologic map of Utah: Salt Lake City, Utah, Utah Geological and Mineral Survey, scale 1:500,000 (2 sheets).

Humphreys, E.D., and Dueker, K.G., 1994a, Western United States upper mantle structure: Journal of Geophysical Research, v. 99, no. B5, p. 9615–9634.

———1994b, Physical state of the western U.S. upper mantle: Journal of Geophysical Research, v. 99, no. B5, p. 9635–9650.

Hunt, C.B., 1956, Cenozoic geology of the Colorado Plateau: U.S. Geological Survey Professional Paper 279, 99 p.

———1969, Geologic history of the Colorado River: U.S. Geological Survey Professional Paper 669–C, p. C59–C130.

Isacks, B.L., and Barazangi, Muawia, 1977, Geometry of Benioff zones—Lateral segmentation and downward bending of the subducted lithosphere, *in* Talwani, Manik, and Pitman, W.C., III, eds., Island arcs, deep sea trenches and back-arc basins: Washington, D.C., American Geophysical Union, p. 99–111.

Izett, G.A., 1975, Late Cenozoic sedimentation and deformation in northern Colorado and adjoining areas, *in* Curtis, B.F., ed., Cenozoic History of the Southern Rocky Mountains: Geological Society of America Memoir 144, p. 179–209.

Jachens, R.C., Simpson, R.W., Blakely, R.J., and Saltus, R.W., 1993, Isostatic residual gravity map of the conterminous United States and some adjacent parts of Canada, scale 1:5,000,000, plate 4 *in* Reed, J.C., Jr., Bickford, M.E., Houston, R.S., Link, P.K., Rankin, D.W., Sims, P.K., and Van Schmus, W.R., 1993, eds., The Geology of North America, vol. C–2, Precambrian: Conterminous U.S.: Boulder, Colo., Geological Society of America, 657 p.

Joesting, H.R., and Case, J.E., 1960, Salt anticlines and deep-seated structures in the Paradox basin, Colorado and Utah, *in* Short papers in the geological sciences: U.S. Geological Survey Professional Paper 400–B, p. B252–B256.

Joesting, H.R., and Plouff, Donald, 1958, Geophysical studies of the Upheaval Dome area, San Juan County, Utah, *in* Sanborn, A.F., ed., International Association of Petroleum Geologists Ninth Annual Field Conference: Guidebook to the Geology of the Paradox Basin, p. 86–92.

Johnson, R.C., and Finn, T.M., 1986, Cretaceous through Holocene history of the Douglas Creek arch, Colorado and Utah, *in* Stone, D.S., ed., New interpretations of northwestern Colorado Geology: Denver, Colo., Rocky Mountain Association of Geologists, p. 75–77.

Kane, M.F., and Godson, R.H., 1989, A crust/mantle structural framework of the conterminous United States based on gravity and magnetic trends, *in* Pakiser, L.C., and Mooney, W.D., eds., Geophysical framework of the continental United States: Geological Society of America Memoir 172, p. 383–404.

Keller, G.R., and Cordell, Lindrith, 1983, Bouguer gravity anomaly map of New Mexico: New Mexico State University Energy Institute, Geothermal Resources of New Mexico—Scientific Map Series, overlay 1, scale 1:500,000. (Available from National Geophysical Data Center, Boulder, CO 80303, as report 84–TGB–05.)

Larson, E.E., Patterson, P.E., Curtis, G., Drake, R., and Mutschler, F.E., 1985, Petrologic, paleomagnetic, and structural evidence of a Proterozoic rift system in Oklahoma, New Mexico, Colorado and Utah: Geological Society of America Bulletin, v. 96, no. 11, p. 1364–1372.

Leeman, W.P., and Harry, D.L., 1983, A binary source model for extension-related magmatism in the Great Basin, western North America: Science, v. 262, p. 1550–1554.

Lidiak, E.G., and Zietz, Isidore, 1976, Interpretation of aeromagnetic anomalies between latitude 37°N and 38°N in eastern and central United States: Geological Society of America Special Paper 167, 37 p.

Lipman, P.W., 1980, Cenozoic volcanism in the western United States—Implications for continental tectonics, *in* Studies in geophysics—Continental tectonics: Washington, D.C., National Academy of Sciences, p. 161–174.

———1981, Volcano-tectonic setting of Tertiary ore deposits, southern Rocky Mountains, *in* Dickinson, W.R., and Payne, W.D., eds., Relations of tectonics to ore deposits in the southern Cordillera: Arizona Geological Society Digest, v. 14, p. 199–213.

Lipman, P.W., Doe, B.R., Hedge, C.E., and Steven, T.A., 1978, Petrologic evolution of the San Juan volcanic field, southwestern Colorado—Pb and Sr isotopic evidence: Geological Society of America Bulletin, v. 89, p. 59–82.

Livacarri, R.F., Burke, Kevin, and Sengor, A.M.C., 1981, Was the Laramide orogeny related to subduction of an oceanic plateau?: Nature, v. 289, p. 276–278.

Lovering, T.S., and Goddard, E.N., 1950, Geology and ore deposits of the Front Range, Colorado: U.S. Geological Survey Professional Paper 223, 319 p.

Lowell, J.D., 1974, Plate tectonics of foreland basement deformation: Geology, v. 2, p. 275–278.

Lucchitta, Ivo, 1979, Late Cenozoic uplift of the southwestern Colorado Plateau and adjacent lower Colorado River region: Tectonophysics, v. 61, p. 63–95.

Mabey, D.R., and Budding, K.E., 1987, High-temperature geothermal resources of Utah: Utah Geological and Mineral Survey Bulletin 123.

Mattox, R.B., 1975, Upheaval Dome, a possible salt dome in the Paradox Basin, Utah, *in* Fassett, J.E., ed., Canyonlands Country: Four Corners Geological Society Guidebook, Eighth Field Conference, p. 225–234.

McGetchin, T.R., and Merrill, R.B., eds., 1979, Plateau uplift—Mode and mechanism: Tectonophysics (special issue), v. 61, 336 p.

Milbert, D.G., 1991, GEOID90—A high-resolution geoid for the United States: Eos, v. 72, no. 49, p. 545, 554.

Morgan, P., and Swanberg, C.A., 1985, On the Cenozoic uplift and tectonic stability of the Colorado Plateau: Journal of Geodynamics, v. 3, p. 39–63.

Morgan, W.J., 1972, Deep mantle convection plumes and plate motions: American Association of Petroleum Geologists Bulletin, v. 56, p. 203–213.

Muehlberger, W.R., compiler, 1992, Tectonic map of North America: Tulsa, Okla., American Association of Petroleum Geologists, scale 1:5,000,000 (2 sheets).

Nealey, L.D., and Sheridan, M.F., 1989, Post-Laramide volcanic rocks of Arizona and northern Sonora, Mexico, and their inclusions, *in* Jenny, J.P., and Reynolds, S.J., eds., Geological evolution of Arizona: Tucson, Arizona, Arizona Geological Society Digest, v. 17, p. 609–647.

Nelson, S.T., Davidson, J.P., and Sullivan, K.R., 1992, New age determinations of central Colorado Plateau laccoliths, Utah; Reorganizing disturbed K-Ar systematics and re-evaluating tectonomagmatic relationships: Geological Society of America Bulletin, v. 104, no. 12, p. 1547–1560.

New Mexico Geological Society, 1982, New Mexico highway geologic map: Socorro, N. Mex., New Mexico Bureau of Mines and Mineral Resources, scale 1:1,000,000.

Pakiser, L.C., and Mooney, W.D., eds., 1989, Geophysical framework of the Continental United States: Geological Society of America Memoir 172, 826 p.

Pang, Ming, and Nummedal, Dag, 1995, Flexural subsidence and basement tectonics of the Cretaceous Western Interior basin, United States: Geology, v. 23, no. 1, p. 173–176.

Parsons, Tom, and McCarthy, Jill, 1995, The active southwest margin of the Colorado Plateau: uplift of mantle origin: Geological Society of America Bulletin, v. 107, no. 2, p. 139–147.

Parsons, Tom, Thompson, G.A., and Sleep, N.H., 1994, Mantle plume influence on the Neogene uplift and extension of the U.S. western Cordillera?: Geology, v. 22, p. 83–86.

Pierce, K.L., and Morgan, L.A., 1992, The track of the Yellowstone hot spot: Volcanism, faulting, and uplift, *in* Link, P.K., Kuntz, M.A., and Platt, L.B., eds., Regional geology of eastern Idaho and western Wyoming: Geological Society of America Memoir 179, p. 1–53.

Plouff, Donald, and Pakiser, L.C., 1972, Gravity study of the San Juan Mountains, Colorado: U.S. Geological Survey Professional Paper 800–B, p. B183–B190.

Prodehl, Claus, and Lipman, P.W., 1989, Crustal structure of the Rocky Mountain region, *in* Pakiser, L.C., and Mooney, W.D., eds., Geophysical Framework of the continental United States: Geological Society of America Memoir 172, p. 249–284.

Reed, J.C., Jr., 1993, compiler, Precambrian rocks of the conterminous United States, scale 1:5,000,000, plate 1 *in* Reed, J.C., Jr., Bickford, M.E., Houston, R.S., Link, P.K., Rankin, D.W., Sims, P.K., and Van Schmus, W.R., 1993, eds., The Geology of North America, vol. C–2, Precambrian; Conterminous U.S.: Boulder, Colo., Geological Society of America, 657 p.

Reynolds, S.J., 1988, Geologic map of Arizona: Arizona Geological Survey Map 26, scale 1:1,000,000.

Richards, M.A., Duncan, R.A., and Courtillot, V.E., 1989, Flood basalts and hot spot tracks; plume heads and tails: Science, v. 246, p. 103–107.

Rowley, P.D., Lipman, P.W., Mehnert, H.H., Lindsey, D.A., and Anderson, J.J., 1978, Blue Ribbon lineament, an east-trending structural zone within the Pioche mineral belt of southwestern Utah and eastern Nevada: U.S. Geological Survey Journal of Research, v. 6, no. 2, p. 175–192.

Saltus, R.W., 1988, Regional, residual, and derivative gravity maps of Nevada: Nevada Bureau of Mines and Geology Map 94B, scales 1:1,000,000 and 1:2,500,000 (4 sheets).

Sauck, W.A., and Sumner, J.S., 1970a, Residual aeromagnetic map of Arizona: Tucson, Univ. of Arizona, scale 1:1,000,000.

_____1970b, Total intensity aeromagnetic map of New Mexico: Tucson, Univ. of Arizona, scale 1:1,000,000.

Sbar, M.L., and Sykes, L.R., 1973, Contemporary compressive stress and seismicity in eastern North America; An example of intraplate tectonics: Geological Society of America Bulletin 84, p. 1861–1882.

Schultz-Ela, D.D., Jackson, M.P.A., Hudec, M.R., Fletcher, R.C., Porter, M.L., and Watson, I.A., 1994, Constraints on the origin of Upheaval Dome, Paradox Basin, Utah: American Association of Petroleum Geologists 1994 Annual Convention,

Denver, Colorado, June 12–15, 1994, Abstracts with Programs, p. 253.

Shawe, D.R., and Stewart, J.H., 1976, Ore deposits as related to tectonics and magmatism, Nevada and Utah: American Inst. of Mining, Metallurgical, and Petroleum Engineers Transactions, v. 260, p. 225–232.

Shoemaker, E.M., and Henenhoff, K.E., 1984, Upheaval Dome impact structure, Utah [abs.]: Lunar and Planetary Science Conference, v. 15, pt. 2, p. 778–779.

Shoemaker, E.M., Squires, R.L., and Abrams, M.J., 1978, The Bright Angel and Mesa Butte fault systems of northern Arizona, *in* Smith, R.B., and Eaton, G.P., eds., Cenozoic tectonics and regional geophysics of the Western Cordillera: Geological Society of America Memoir 152, p. 341–367.

Shuey, R.T., Schellinger, D.K., Johnson, E.H., and Alley, L.B., 1973, Aeromagnetics and the transition between the Colorado Plateau and Basin Range provinces: Geology, v. 1, no. 11, p. 107–110.

Simpson, R.W., Hildenbrand, T.G., Godson, R.H., and Kane, M.F., 1987, Digital colored Bouguer gravity, free-air gravity, station location, and terrain maps for the conterminous United States: U.S. Geological Survey Geophysical Investigations Map GP–953–B, scale 1:7,500,000 (2 sheets).

Simpson, R.W., Jachens, R.C., Blakely, R.J., and Saltus, R.W., 1986, A new isostatic residual gravity map of the conterminous United States with a discussion on the significance of isostatic residual anomalies: Journal of Geophysical Research, v. 91, p. 8348–8372.

Sleep, N.H., 1990, Hotspots and mantle plumes; Some phenomenology: Journal of Geophysical Research, v. 95, no. B5, p. 6715–6736.

Smith, R.B., and Braile, L.W., 1993, Topographic signature, space-time evolution, and physical properties of the Yellowstone–Snake River Plain volcanic system; The Yellowstone hotspot, *in* Snoke, A.W., Steidtmann, J.R., and Roberts, S.M., eds., Geology of Wyoming: Geological Survey of Wyoming Memoir No. 5, p. 694–754.

Smith, R.B., Nagy, W.C., Julander, K.A. (Smith), Viveiros, J.J., Barker, C.A., and Gants, D.G., 1989, Geophysical and tectonic framework of the eastern Basin and Range–Colorado Plateau–Rocky Mountain transition, *in* Pakiser, L.C., and Mooney, W.D., eds., Geophysical framework of the continental United States: Geological Society of America Memoir 172, p. 205–234.

Smith, R.L., and Luedke, R.G., 1984, Potentially active volcanic lineaments and loci in western conterminous United States, *in* Explosive volcanism—Inception, evolution, and hazards: Washington, D.C., National Academy Press, Studies in Geophysics, p. 47–66.

Steiner, M.B., 1986, Rotation of the Colorado plateau: Tectonics, v. 5, p. 649–660.

Steven, T.A., 1975, Middle Tertiary volcanic field in the Southern Rocky Mountains, *in* Curtis, B.F., ed., Cenozoic history of the Southern Rocky Mountains: Geological Society of America Memoir 144, p. 75–94.

Steven, T.A., Rowley, P.D., and Cunningham, C.G., 1984, Calderas of the Marysvale volcanic field, west-central Utah: Journal of Geophysical Research, v. 89, no. B10, p. 8751–8764.

Stevenson, G.M., and Baars, D.L., 1986, The Paradox—A pull-apart basin of Pennsylvanian age, *in* Peterson, J.A., ed., Paleotectonics and sedimentation: American Association of Petroleum Geologists Memoir 41, p. 513–539.

Stewart, J.H., 1978, Basin-range structure in western North America—A review, *in* Smith, R.B., and Eaton, G.P., eds., Cenozoic tectonics and regional geophysics of the western Cordillera: Geological Society of America Memoir 152, p. 1–32.

Stewart, J.H., and Carlson, J.E., 1977, Geologic map of Nevada, one million scale set: Nevada Bureau of Mines and Geology Map 57, scale 1:1,000,000.

Stewart, J.H., and Carlson, J.E., compilers, 1978, Geologic map of Nevada: U.S. Geological Survey, scale 1:500,000 (2 sheets).

Stewart, J.H., Moore, W.J., and Zietz, Isidore, 1977, East-west patterns of Cenozoic igneous complexes, aeromagnetic anomalies, and mineral deposits: Geological Society of America Bulletin 88, p. 67–77.

Stock, J., and Molnar, P., 1988, Uncertainties and implications of the late Cretaceous and Tertiary position of North America relative to the Farallon, Kula and Pacific plates: Tectonics, v. 7, no. 6, p. 1339–1384.

Stokes, W.L., 1982, Geologic comparisons and contrasts, Paradox and Arapien Basins, *in* Nielson, D.L., ed., Overthrust belt of Utah; 1982 Symposium and Field Conference: Salt Lake City, UT, Utah Geological Association Publication 10, p. 1–12.

Sullivan, K.R., Kowallis, B.J., and Mehnert, H.H., 1991, Isotopic ages of igneous intrusions in southeastern Utah—Evidence for a mid-Cenozoic Reno–San Juan magmatic zone: Brigham Young University Geology Studies, v. 37, p. 139–144.

Sumner, J.S., 1985, Crustal geology of Arizona as interpreted from magnetic, gravity, and geologic data, *in* Hinze, W.J., ed., The utility of regional gravity and magnetic anomaly maps: Tulsa, Okla., Society of Exploration Geophysicists, p. 164–180.

Suppe, J., Powell, J.C., and Berry, R., 1975, Regional topography, seismicity, Quaternary volcanism and the present-day tectonics of the western United States: American Journal of Science, v. 275–A, p. 397–436.

Thompson, G.A., and Zoback, M.L., 1979, Regional geophysics of the Colorado Plateau: Tectonophysics, v. 61, p. 149–181.

Tweto, O.L., 1968, Geologic setting and interrelationships of mineral deposits in the mountain province of Colorado and south-central Wyoming, *in* Ridge, J.D., ed., Ore deposits of the United States, 1933–1967 (Graton-Sales Volume): New York, American Institute of Mining, Metall., and Petroleum Engineers, v. 1, p. 551–558.

———1979a, Geologic map of Colorado: Reston, VA, U.S. Geological Survey, scale 1:500,000.

———1979b, The Rio Grande Rift system in Colorado, *in* Riecker, R.E., ed., 1979, Rio Grande Rift—Tectonics and magmatism: Washington, D.C., American Geophysical Union, p. 33–56.

———1980, Precambrian geology of Colorado, *in* Kent, H.C., and Porter, K.W., eds., Colorado Geology: Denver, Colo., Rocky Mountain Association of Geologists, p. 37–46.

———1987, Rock units of the Precambrian basement in Colorado: U.S. Geological Survey Professional Paper 1321–A, 54 p.; with colored geologic map, scale 1:1,000,000.

Tweto, O.L., and Case, J.E., 1972, Gravity and magnetic features as related to geology in the Leadville 30-minute quadrangle,

Colorado: U.S. Geological Survey Professional Paper 726–C, 31 p.

Tweto, O.L., and Sims, P.K., 1963, Precambrian ancestry of the Colorado Mineral Belt: Geological Society of America Bulletin, v. 74, no. 8, p. 991–1014.

Van der Voo, Rob, 1989, Paleomagnetism of North America; the craton, its margins, and the Appalachian Belt, *in* Pakiser, L.C., and Mooney, W.D., eds., Geophysical Framework of the continental United States: Geological Society of America Memoir 172, p. 449–470.

Verbeek, E.R., and Grout, M.A., 1993, Geometry and structural evolution of gilsonite dikes in the eastern Uinta Basin, Utah: United States Geological Survey Bulletin 1787–HH, 42 p.

Warner, L.A., 1978: The Colorado lineament—A middle Precambrian wrench fault system: Geological Society of America Bulletin 89, p. 161–171.

———1980, The Colorado lineament, *in* Kent, H.C., and Porter, K.W., eds., Colorado geology: Denver, Colo., Rocky Mountain Association of Geologists, p. 11–21.

Western Geographics/Colorado Geological Survey, 1991, Colorado geological highway map: Canon City, Colo., Western Geographics, scale 1:1,000,000.

Western Geographics/Geological Survey of Wyoming, 1991, Wyoming geological highway map: Canon City, Colo., Western Geographics, scale 1:1,000,000.

White, R.S., and McKenzie, D., 1989, Magmatism at rift zones; The generation of volcanic continental margins and flood basalts: Journal of Geophysical Research, v. 94, p. 7685–7729.

Wilson, E.D., Moore, R.T., and Cooper, J.R., 1969, Geologic map of Arizona, scale 1:500,000: Washington, D.C., U.S. Geological Survey.

Wilson, J.T., 1973, Mantle plumes and plate motions: Tectonophysics, v. 19, p. 149–164.

Zhong, Shijie, and Gurnis, Michael, 1995, Mantle convection with plates and mobile, faulted plate margins: Science, v. 267, p. 838–843.

Zietz, Isidore, and Kirby, J.R., Jr., 1972a, Aeromagnetic map of Colorado: U.S. Geological Survey Geophysical Investigations Map GP–836, scale 1:500,000.

———1972b, Aeromagnetic map of Colorado: U.S. Geological Survey Geophysical Investigations Map GP–880, scale 1:1,000,000.

Zietz, Isidore, Shuey, Ralph, and Kirby, J.R., Jr., 1976, Aeromagnetic map of Utah: U.S. Geological Survey Geophysical Investigations Map GP–907, scale 1:1,000,000.

Zietz, Isidore, Stewart, J.H., Gilbert, F.P., and Kirby, J.R., Jr., 1977, Aeromagnetic map of Nevada: U.S. Geological Survey Miscellaneous Field Studies Map MF–902, scale 1:500,000 (2 sheets, with accompanying text).

Zoback, M.L., and Zoback, M.D., 1989, Tectonic stress field of the continental United States, *in* Pakiser, L.C., and Mooney, W.D., eds., Geophysical framework of the continental United States: Geological Society of America Memoir 172, p. 523–540.

Isotopic Ages of Igneous Intrusions in Southeastern Utah

By Kim R. Sullivan[1]

CONTENTS

TABLES

Previously determined ages for the Abajo, Henry, and La Sal Mountains in southeastern Utah (table 1) lack consistency and reflect the problems associated with K-Ar age determination on certain materials (such as hornblende, pyroxene, whole rocks, or altered mafic minerals). To provide a more consistent geochronology of these laccolithic centers, samples were collected from multiple intrusive bodies in each range. Zircon and sphene were separated, prepared, and analyzed according to standard fission-track techniques (Naeser, 1978). The resulting ages (table 2) show that these laccolithic centers were emplaced between about 30 and 20 Ma, making them contemporaneous with the Reno-Marysvale and San Juan volcanic zones to the west and east. Ranges of zircon ages for these intrusions are as follows: 22.6±2.2 to 28.6±3.4 Ma (Abajo Mountains); 20.0±1.9 to 29.2±2.3 Ma (Henry Mountains); 28.7±2.7 Ma (La Sal Mountains, one sample). These data suggest the existence of an essentially continuous, intracontinental magmatic zone extending from Reno to the San Juan Mountains during the Oligocene and early Miocene. The length of this zone (more than 1,000 km) and its orientation perpendicular to the trend of the subduction zone along the western coast of North America are additional constraints on the subduction models generally used to explain mid-Cenozoic igneous activity in the western United States.

[1] Department of Geology, Brigham Young University, Provo, Utah 84602.

Table 1. Previous age determinations of Utah laccolithic centers

[All K–Ar ages corrected for new age constants (Dalrymple, 1979). Do., ditto (same as above)]

Laccolithic center	Lithology	Material dated	Method	Age (Ma)	Reference
Abajo Mountains	Altered hornblende-diorite porphyry.	Altered mafic rock (chlorite).	K–Ar	28.1	Armstrong (1969).
		Whole rock minus heavy fraction.	K–Ar	28.9	Do.
Henry Mountains	Hornblende-monzodiorite porphyry.	Hornblende (17 percent rock).	K–Ar	[1] 44.8	Do.
		Whole rock minus heavy fraction.	K–Ar	49.2	Do.
La Sal Mountains	Hornblende-diorite porphyry.	Pyroxene-hornblende (5 percent rock).	K–Ar	24.1	Do.
	Soda syenite porphyry ---	Aegerine augite --------	K–Ar	26.1±2.6	Stern and others (1965).
		Zircon -------------------	U–Pb[2]	32 ±2	Do.
	Monzonite porphyry -----	Aegerine augite --------	K–Ar	23.1±3.3	Do.
		Zircon -------------------	U–Pb[2]	32 ±2	Do.
	Diorite porphyry ----------	Hornblende -------------	K–Ar	[3] 56.2±1.5	Do.
		Zircon -------------------	U–Pb[2]	494 ±20	Do.

[1] Armstrong notes the possibility of xenolithic contamination in hornblende diorites from other intrusive centers on the Colorado Plateau but does not suggest the same explanation for this older age (Armstrong, 1969, p. 2082).

[2] Pb^{206}/U^{238} ages reported here. Pb^{207}/U^{235} and Pb^{208}/U^{232} ages were also obtained but were not concordant for any sample.

[3] Stern and others (1965, p. 1503) interpret this age as "due to incomplete degassing of Precambrian hornblende and the presence of Precambrian zircon xenocrysts."

REFERENCES CITED

Armstrong, R.L., 1969, K-Ar dating of laccolithic centers of the Colorado Plateau and vicinity: Geological Society of America Bulletin, v. 80, p. 2081–2086.

Dalrymple, G.B., 1979, Critical tables for conversion of K-Ar ages from old to new constants: Geology, v. 7, p. 558–560.

Naeser, C.W., 1978, Fission track dating: U.S. Geological Survey Open-File Report 76–190, 87 p.

Naeser, C.W., Zimmerman, R.A., and Cebula, G.T., 1981, Fission track dating of apatite and zircon—An interlaboratory comparison: Nuclear Tracks, v. 5, p. 65–72.

Stern, T.W., Newell, M.F., Kistler, R.W., and Shawe, D.R., 1965, Zircon uranium-lead ages and mineral potassium-argon ages of La Sal Mountains rocks: Journal of Geophysical Research, v. 70, p. 1503–1507.

Table 2. New fission-track age determinations for Colorado Plateau laccoliths

[All samples from diorite porphyry except as noted. No. = number; tr = tracks; n = neutrons. Fision-track constants (Naeser and others, 1981): $\lambda_d = 1.55\times10^{-10}\,yr^{-1}$; $\lambda_f = 7.03\times10^{-10}\,yr^{-1}$; $U^{235}/U^{238} = 0.00725$]

Sample number	Field number	North latitude	West longitude	Mineral	Age ± 2σ (Ma)	Fossil-track density		Induced-track density		Neutron fluence		No. of grains
						tr/cm^2 $\times10^6$	(No. of tr)	tr/cm^2 $\times10^6$	(No. of tr)	n/cm^2 $\times10^{15}$	(No. of n)	
A1	MONT–1	37°48'11"	109°27'09"	Zircon	28.1±2.1	5.82	(2049)	5.02	(3534)	0.8114	(3204)	6
				Sphene	26.8±3.7	0.438	(290)	1.56	(2068)	3.202	(2484)	8
A2	MONT–2	37°50'42"	109°29'58"	Zircon	28.6±3.4	2.84	(671)	2.41	(1138)	0.8114	(3204)	7
				Sphene	32.3±4.6	0.906	(288)	2.68	(1706)	3.202	(2484)	8
A3	MONT–3	37°51'08"	109°27'09"	Zircon	22.6±2.2	2.59	(892)	2.77	(1914)	0.8114	(3204)	9
A4	MONT–4	37°51'08"	109°28'33"	Zircon	24.7±2.2	5.55	(1155)	5.44	(2264)	0.8114	(3204)	4
A5	LIN–1	37°49'34"	109°30'12"	Zircon	24.2±2.4	5.62	(966)	5.63	(1938)	0.8114	(3204)	7
				Sphene	26.9±3.8	0.623	(273)	2.22	(1942)	3.202	(2484)	6
H1	ELLN–1	38°03'55"	110°47'22"	Zircon	20.0±1.9	3.87	(979)	4.70	(2376)	0.8114	(3204)	9
H2	PENL–1	37°59'43"	110°48'05"	Zircon	23.9±2.1	3.38	(1276)	3.44	(2592)	0.8114	(3204)	6
				Sphene	26.4±2.5	1.07	(673)	3.88	(4880)	3.202	(2484)	9
H3	HIL–1	37°57'11"	110°34'36"	Zircon	29.2±2.3	1.37	(1790)	1.14	(2978)	0.8114	(3204)	22
H4	HIL–2	37°55'06"	110°36'00"	Zircon	21.2±2.4	3.99	(655)	4.56	(1496)	0.8114	(3204)	4
				Sphene	25.6±3.3	0.394	(332)	1.47	(2480)	3.202	(2484)	10
H5	BULL–1	38°08'34"	110°43'31"	Zircon	28.6±2.4	5.32	(1356)	5.23	(1334)	0.9620	(3555)	5
[1] L1	CSTV–2	38°32'36"	109°16'24"	Zircon	28.7±2.7	8.90	(1122)	7.52	(1896)	0.8114	(3204)	7
				Sphene	30.3±3.2	0.831	(567)	2.63	(3582)	3.202	(2484)	10
[2] L2	LAS–1	38°28'57"	109°17'05"	Sphene	28.5±3.8	0.583	(322)	1.96	(2160)	3.202	(2484)	6

[1] Biotite from this sample also yielded a K–Ar date of 28.21±1.16 Ma, based on determinations of 0.51–0.52% Na_2O, 6.30–6.32% K_2O, 2.5826 moles/g ^{40}Ar (49.1% of total Ar), and the following decay constants: $\lambda_\varepsilon = 0.581\times10^{-10}\,yr^{-1}$; $\lambda_\beta = 4.962\times10^{-10}\,yr^{-1}$; $^{40}K/K = 1.167\times10^{-4}$. Age obtained by Harald Mehnert, U.S. Geological Survey, Denver, Colo.

[2] Sample from monzonite porphyry.

Reevaluation of the Central Colorado Plateau Laccoliths in the Light of New Age Determinations

By Stephen T. Nelson[1]

CONTENTS

FIGURE

TABLE

This report summarizes the work of Nelson, Heizler, and Davidson (1992) and of Nelson, Davidson, and Sullivan (1992). It has been included in this volume to provide a complete record of the proceedings of the laccolith conference. The new $^{40}Ar/^{39}Ar$ ages presented here for the laccoliths of the Henry and La Sal Mountains (fig. 1; table 1) give much tighter constraints on the age of magmatism. Also, the new $^{40}Ar/^{39}Ar$ data allow the assessment of processes that disturb apparent K-Ar systematics.

K-Ar (Stern and others, 1965; Armstrong, 1969; Dubiel and others, 1990) and fission-track (Sullivan and others, 1991) ages for laccoliths of the Henry and La Sal Mountains appear to be contradictory (fig. 1). The K-Ar ages are bimodal (41–56 and 24–29 Ma), although the fission-track ages range exclusively from 20 to 30 Ma. It appears that either some fission-track ages are too young, some K-Ar ages are too old, or a sampling problem exists.

The Eocene K-Ar ages imply an episode of magmatism at 40–56 Ma during Laramide deformation and

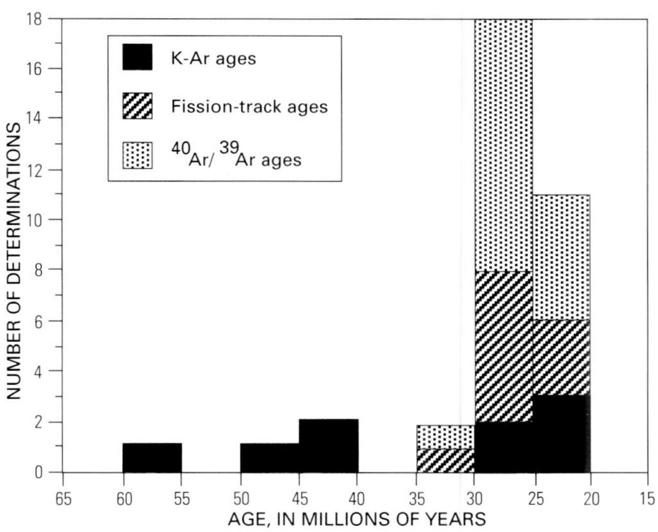

Figure 1. Histogram of age determinations for intrusive rocks of the Henry and La Sal Mountains, Utah. Fission-track data include sphene and zircon age determinations where available from the same sample. The 40- to 60-Ma ages, all determined by the K-Ar method, are interpreted to be the result of excess argon and partially outgassed hornblende xenocrysts in samples that are actually 20–30 Ma. From Nelson, Davidson, and Sullivan (1992, fig. 2).

[1] Department of Geology, 673 WIDB, Brigham Young University, Provo, Utah 84602.

Table 1. $^{40}Ar/^{39}Ar$ age determinations from the Henry and La Sal Mountains intrusions, Utah.

[n.d., not determined. See Nelson, Heizler, and Davidson (1992) for further details of age determinations]

Intrusive center	Sample	North latitude	West longitude	$^{40}Ar/^{36}Ar$	$^{40}Ar/^{39}Ar$	J[1]	MSWD[2]	Age (Ma) ± 1 s.d.[3]	Sample type[4]	Lithology[5]	Comments
						Henry Mountains					
Mount Ellen	BULL–1	38°9'6"	110°44'23"	n.d.	n.d.	0.0050261	n.d.	<31.9±0.4	HB	PHP	Xenocryst contamination: Fission-track age of 28.6±1.2 Ma is considered best estimate of crystallization age.
	ELLN–4	38°4'39"	110°45'21"	n.d.	n.d.	0.0050319	n.d.	<39.0±0.7	HB	PHP	Xenocryst contamination.
	ELLN–10 [6]	38°4'29"	110°45'26"	320±15	3.30863 ± 0.03171	0.0049700	173	29.4±0.3	HB	PHP	
	ELLN–7	38°0'10"	110°47'36"	n.d.	n.d.	0.0047909	n.d.	<28.4±0.5	HB	PHP	Xenocryst contamination.
				302.3±2.2	2.70879 ± 0.02869	0.0047909	0.75	23.3±0.2	HB	PHP	Results from laser fusions.
Mount Pennell	PENL–14	37°57'58"	110°45'33"	690±51	2.61151 ± 0.23623	0.0048134	148	22.5±1.9	HB	PHP	Age of PENL–14 must be greater than about 25 Ma based on crosscutting relationships. This is concordant within analytical error.
	PENL–13 [6]	37°57'0"	110°46'11"	293.1±1.7	2.82955 ± 0.00656	0.0049043	7.5	24.9±0.1	HB	PHP	
	PENL–12 [6]	37°57'1"	110°47'50"	283±12	2.72196 ± 0.05494	0.0050004	9.7	24.4±0.5	HB	SP	
	PENL–9 [6]	37°57'13"	110°47'50"	305.5±3.3	2.79397 ± 0.00445	0.0050738	10.8	25.4±0.1	AF	SP	
	PENL–3 [6]	37°56'54"	110°46'44"	300.4±4.8	2.74228 ± 0.02704	0.0050156	36	24.6±0.2	HB	SP	
Mount Hillers	HILL-10KRS	37°51'26"	110°42'10"	464±11	3.44632 ± 0.03944	0.00480005	0.64	29.6±0.4	HB	PHP	Two sets of data are reported because of two distinct trapped excess argon components in the sample. Interpreted age is 29.35±0.33 Ma.
				316.8±9.9	3.28337 ± 0.03468	0.0048005	7.3	29.1±0.3	HB	PHP	
Mount Ellsworth	ELL–1KRS	37°44'50"	110°37'19"	310.8±2.4	3.44058 ± 0.11118	0.0050770	6.97	31.2±1.0	HB	PHP	Excess argon.
						La Sal Mountains					
North Mountain	MW–13	38°32'34"	109°14'5"	n.d.	n.d.	0.0049390	n.d.	27.9–31.5	HB	PHP	Xenocryst contamination. Age constrained by crosscutting relationships with MW–2 and MW–17.
				n.d.	n.d.	n.d.	n.d.	27.5±1.4	HB	PHP	Results from laser fusions. Mean of three Oligocene ages.
	MW–2 [6]	38°32'19"	109°14'46"	302±10	3.08077 ± 0.02991	0.0050530	2.3	27.9±0.3	AF	SP	
	MW–17 [6]	38°31'19"	109°13'29"	296.7±4.9	2.13906 ± 0.00122	0.007287	0.9	27.9±0.03	AF	SP	
Middle Mountain	TUK–1 [6]	38°27'23"	109°15'57"	285±5.9	2.94181 ± 0.01607	0.0050676	3.7	26.7±0.2	AF	SP	
	TUK–2	38°28'25"	109°16'5"	n.d.	n.d.	0.0048854	n.d.	<62.9±2.7	HB	PHP	Xenocryst contamination.
				n.d.	n.d.	n.d.	n.d.	25.1±4.1	HB	PHP	Results from laser fusions. Mean of nine Oligocene ages.
South Mountain	TUK–6	38°24'51"	109°15'45"	n.d.	n.d.	0.0049223	n.d.	<56.5±0.3	HB	PHP	Xenocryst contamination.
				n.d.	n.d.	n.d.	n.d.	26.0±2.4	HB	PHP	Results from laser fusions. Mean of two Oligocene ages.

[1]Irradiation parameter *J*.
[2]Mean square of weighted deviations.
[3]Ages determined by weighted linear regression (York, 1969) of inverse correlation data.
[4]HB = hornblende; AF = alkali feldspar.
[5]PHP = plagioclase-hornblende porphyry; SP = syenite porphyry.
[6]Samples with flat age spectra and well-correlated inverse correlation plots.

accompanying flat subduction (as suggested by Bird, 1988; Hamilton, 1988). However, Oligocene-Miocene igneous activity in the Henry and La Sal Mountains could be related regionally to contemporaneous voluminous calc-alkaline magmatism of the ignimbrite flareup. Therefore, it is important to determine the true age of magmatism in order to understand the laccoliths in their tectonic context.

The results reported here indicate that igneous activity in the Henry and La Sal Mountains was exclusively middle to late Oligocene (fig. 1; table 1). Magmatism was contemporaneous and appears to have spanned several million years in the Henry (31.2–23.3 Ma) and La Sal (27.9–25.1 Ma) Mountains. About half of the samples make up a well-correlated data set: they define a linear array on inverse correlation diagrams, give flat age spectra, and have a trapped component of argon that is atmospheric in composition. Many of the other samples, however, are disturbed as a result of either excess ^{40}Ar or contamination by xenocrysts. Both processes result in the incorporation of ^{40}Ar that does not result from the in situ decay of ^{40}K in the sample, thereby increasing the sample's apparent age.

Excess ^{40}Ar is easily recognized by the $^{40}Ar/^{39}Ar$ inverse correlation technique (Heizler and Harrison, 1988). Moreover, the $^{40}Ar/^{39}Ar$ method could also be a powerful tool in identifying late-stage xenocrystic contamination of magmas by just a few weight percent, especially if the xenocrysts are much older than their host, and provided they retain a modest proportion of their radiogenic ^{40}Ar. Xenocrysts were identified by single-crystal laser fusion analysis and by electron microprobe analysis of hornblendes from the laccoliths. Mixing calculations based upon the known K content of igneous and xenocrystic hornblende and the age of the basement rocks (≈1,800 Ma) indicate that only a small proportion of xenocrysts (1–4 percent) is required to increase the apparent age of the sample from Oligocene to Eocene, even if the xenocrysts are outgassed by 75 percent. Greater age, greater K concentration, and a smaller degree of degassing of the xenocrysts all tend to maximize the effect of contamination. Calculations for diffusion over short length scales (80 µm) indicate that at magmatic temperatures of 850°–900°C, xenocrysts could retain as much as 10 percent of their radiogenic ^{40}Ar after 25–100 yr. Therefore, it is not unreasonable to expect hornblende xenocrysts to have retained some radiogenic ^{40}Ar, given their inherent retentivity and the fact that diffusion rates would have decreased exponentially as the laccoliths began to cool.

REFERENCES CITED

Armstrong, R.L., 1969, K-Ar dating of laccolithic centers of the Colorado Plateau and vicinity: Geological Society of America Bulletin, v. 80, p. 2081–2086.

Bird, Peter, 1988, Formation of the Rocky Mountains, western United States—A continuum computer model: Science, v. 239, p. 1501–1507.

Dubiel, R.F., Bromfield, C.S., Church, S.E., Kemp, W.M., Larson, M.J., Peterson, Fred, and Pierson, C.T., 1990, Mineral resources of the Mount Pennell Wilderness Study Area, Garfield County, Utah: U.S. Geological Survey Bulletin 1751–D, p. D1–D17.

Hamilton, W.B., 1988, Laramide crustal shortening, *in* Schmidt, C.J., and Perry, W.J., eds., Interaction of the Rocky Mountain foreland and the Cordilleran thrust belt: Geological Society of America Memoir 171, p. 27–39.

Heizler, M.T., and Harrison, T.M., 1988, Multiple trapped argon isotope components revealed by $^{40}Ar/^{39}Ar$ isochron analysis: Geochimica et Cosmochimica Acta, v. 52, p. 1295–1303.

Nelson, S.T., Davidson, J.P., and Sullivan, K.R., 1992, New age determinations of central Colorado Plateau laccoliths, Utah; Recognizing disturbed K-Ar systematics and re-evaluating tectonomagmatic relationships: Geological Society of America Bulletin, v. 104, p. 1547–1560.

Nelson, S.T., Heizler, M.T., and Davidson, J.P., 1992, New age determinations of intrusive rocks from the Henry and La Sal Mountains, Utah: Utah Geological Survey Miscellaneous Publication 92–2, 24 p.

Stern, T.W., Newell, M.F., Kistler, R.W., and Shawe, D.R., 1965, Zircon uranium-lead and thorium-lead ages and mineral potassium-argon ages of La Sal Mountains rocks, Utah: Journal of Geophysical Research, v. 70, p. 1503–1507.

Sullivan, K.R., Kowallis, B.J., and Mehnert, H.H., 1991, Isotopic ages of igneous intrusions in southeastern Utah—Evidence for a mid-Cenozoic Reno–San Juan magmatic zone: Brigham Young University Geology Studies, v. 37, p. 139–144.

York, Derek, 1969, Least squares fitting of a straight line with correlated errors: Earth and Planetary Science Letters, v. 5, p. 320–324.

Relationship of Basement Faulting to Laccolithic Centers of Southeastern Utah and Vicinity

By A. Curtis Huffman, Jr., *and* David J. Taylor

CONTENTS

FIGURE

Recent interpretation of more than 3,500 km of seismic lines in the San Juan and southeastern Paradox Basins has enabled us to map a large number of faults that have measurable offset at the top of the basement and to speculate about possible structural control on intrusive activity. Basement faults throughout the region form an orthogonal pattern: in the San Juan Basin they have average trends of N. 60°–70° W. and N. 30°–40° E. and a typical spacing of 6–16 km, whereas those in the southeastern Paradox Basin are more closely spaced and are rotated 10°–15° clockwise. Apparent vertical displacement in the plane of section is commonly 50–75 m, measured at the top of the basement. Strike-slip movement on many faults is also suggested by the map pattern, but no reliable measurements have yet been made.

The faults probably originated in the Precambrian but had major episodic movement again in the late Paleozoic, in Jurassic to Early Cretaceous time, in Late Cretaceous to early Tertiary time, and perhaps in mid-Tertiary time. The present basement offset represents the sum of all previous movements, but the actual sense of movement on any particular fault was governed by its orientation in the stress field existing at the time. The offset of most faults is not detectable above Permian strata on the available seismic sections; however, the sections do show draping, measurable differences in thickness, and lithologic changes in Mesozoic rocks above many fault zones.

Basement fault patterns and fault movement are reflected by depositional patterns throughout the Phanerozoic section. Many of the Pennsylvanian through Jurassic depositional patterns can be explained as a result of differential movement on large blocks, which typically are 50–80 km on a side and are bounded by dominant faults. Basin subsidence in the Late Cretaceous was apparently controlled to some extent by movement along some of the northwest-trending faults, as sandstone buildups marking many of the transgressive and regressive shorelines overlie and parallel the basement faults. Recent seismicity and present-day drainage patterns suggest that the basement faults still influence tectonics, sedimentation, and erosion in the San Juan Basin. Fault patterns and movement histories in the southeastern Paradox Basin are nearly identical to those of the San Juan Basin, and we believe that this is also true throughout the remainder of the Paradox Basin, although we do not have sufficient data to fully document such an argument.

Baars (1966) first called attention to the large northwest-trending structures (Grenadier and Sneffels Horsts) in the San Juan Mountains north of the San Juan Basin and projected them into some of the large salt structures of the Paradox Basin. There is little direct evidence in the outcrops of the San Juans to define the geometry of the bounding faults of these structures, so many of the published reconstructions are somewhat speculative. Although the extensions of these fault systems to the northwest in the central Paradox Basin are better known because of seismic surveys and drilling for oil and gas, the fault geometries and movement histories are still poorly understood. In both areas, available information can be interpreted in several different ways; the late Paleozoic stress field remains ambiguous.

Farther to the northwest along the basin margin, in the vicinity of Cisco, Utah, Frahme and Vaughn (1983) showed

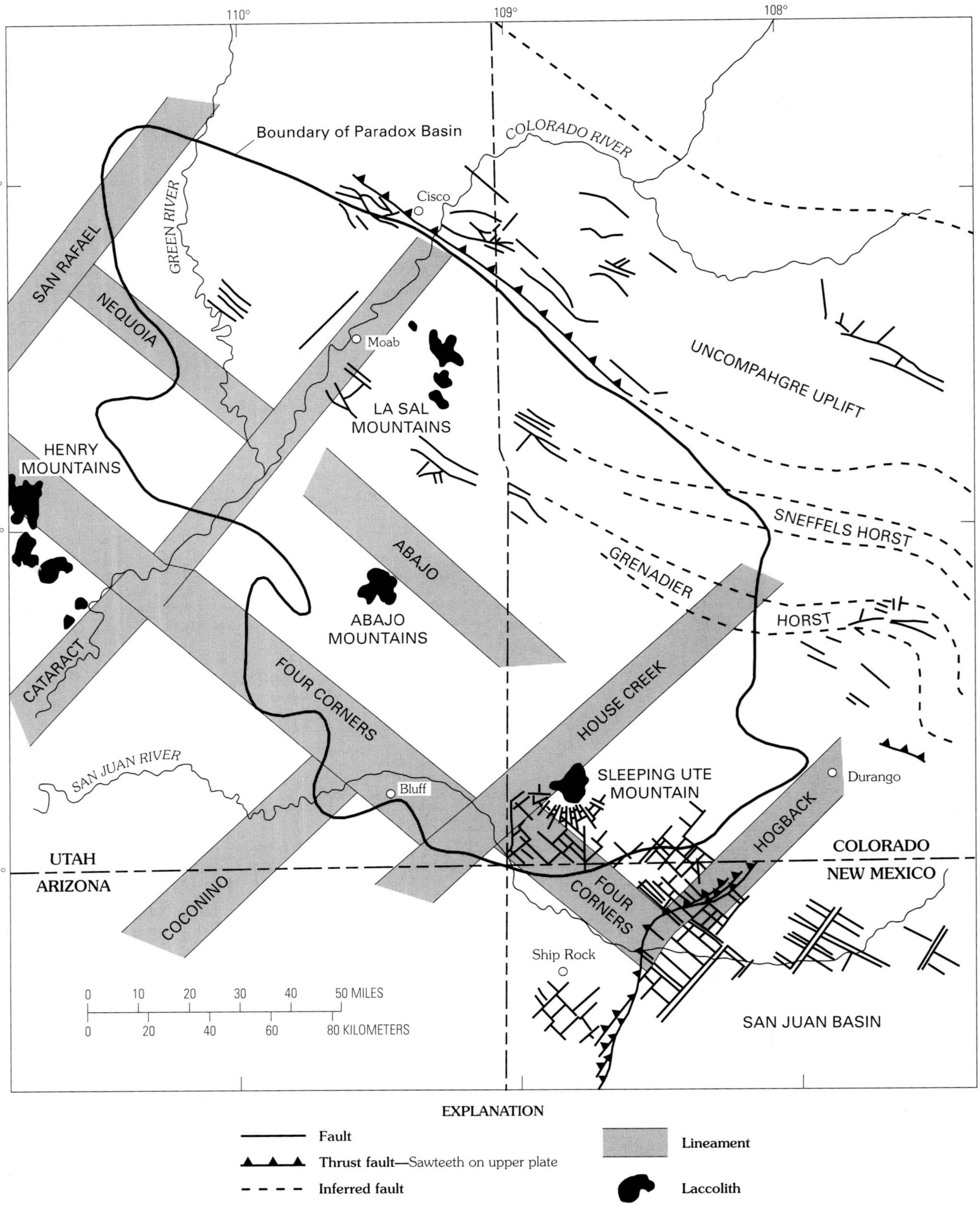

Figure 1. Location of faults in rocks older than the Paradox Formation in the Paradox Basin and vicinity. Lineaments and Grenadier and Sneffels structures from Stevenson and Baars (1986).

that the Uncompahgre uplift had a large southwestward thrust component from late Desmoinesian time through Wolfcampian time. They concluded that this thrusting caused folding and faulting in the Paradox Basin and coincided with major salt movement. Even in this area, however, the actual movement on the faults has not yet been demonstrated adequately.

Movement on these fault systems during the mid-Tertiary period of laccolith emplacement is even more conjectural because of the lack of Oligocene and younger sedimentary deposits in the Paradox and San Juan Basins. Recent analyses of the Oligocene stress field based on dike orientations and other criteria (Delaney and others, 1986; Ren and others, 1987; Tingey and others, 1990) suggested that the least principal horizontal stress was oriented northwest to nearly east-west in the Colorado Plateau and north-south to north-northeast in the eastern Basin and Range province and the Wasatch Plateau. Delaney and others (1986) discussed the difficulties inherent in this type of analysis and noted that the magnitudes of the horizontal regional stresses may have been nearly equal in the vicinity of Ship Rock in northwestern New Mexico. A similar conclusion was reached by Marie Jackson (USGS, oral commun., 1992) for the stress field during emplacement of the Henry Mountains laccoliths. Without more information, it is difficult to determine whether the area of the Paradox and San Juan Basins was under compression or extension in mid-Tertiary time, but the presence of large numbers of dikes, as on the Wasatch Plateau, argues for extension.

Under extensional conditions, particularly north-south or east-west directed, nearly any of the mapped basement faults and especially their intersections would be appropriate sites for intrusive activity. Compression would limit the likely areas of intrusion, as would any significant wrench component. The master faults that bound the large blocks would have been the most likely sites in any stress regime. Figure 1 suggests possible correlations of laccolithic centers with major lineaments and with the extended Sneffels and Grenadier structures proposed by Stevenson and Baars (1986).

Our interpretation of the seismic data has not yet proceeded far enough into the Paradox Basin to evaluate many of the proposed lineaments or fault zones properly. We have found little evidence to suggest major movement on the southeastern end of the Four Corners lineament, but the Hogback lineament has been a major structure, intermittently active, at least since Pennsylvanian time (Taylor and Huffman, 1988). Work is currently underway to evaluate several of the other structures.

REFERENCES CITED

Baars, D.L., 1966, Pre-Pennsylvanian paleotectonics—Key to basin evolution and petroleum occurrences in Paradox Basin, Utah and Colorado: American Association of Petroleum Geologists Bulletin, v. 50, no. 10, p. 2082–2111.

Delaney, P.T., Pollard, D.D., Ziony, F.I., and McKee, E.H., 1986, Field relations between dikes and joints—Emplacement processes and paleostress analysis: Journal of Geophysical Research, v. 91, no. B5, p. 4920–4938.

Frahme, C.W., and Vaughn, E.B., 1983, Paleozoic geology and seismic stratigraphy of the northern Uncompahgre front, Grant County, Utah, *in* Lowell, J.D., ed., Rocky Mountain foreland basins and uplifts: Denver, Colo., Rocky Mountain Association of Geologists, p. 201–211.

Ren, X., Kowallis, B.J., and Best, M.G., 1987, Stress rotation in the eastern Basin and Range, Utah and Nevada from healed microcrack orientation data: Geological Society of America Abstracts with Programs, v. 19, no. 7, p. 816.

Stevenson, G.M., and Baars, D.L., 1986, The Paradox—A pull-apart basin of Pennsylvanian age, *in* Peterson, J.A., ed., Paleotectonics and sedimentation: American Association of Petroleum Geologists Memoir 41, p. 513–539.

Taylor, D.J., and Huffman, A.C., Jr., 1988, Overthrusting in the northwestern San Juan Basin, New Mexico—A new interpretation of the Hogback Monocline, *in* Carter, L.M.H., ed., USGS research on energy resources, 1988: U.S. Geological Survey Circular 1025, p. 60–61.

Tingey, D.G., Christiansen, E.H., and Best, M.G., 1990, Late Oligocene and Miocene Minette and olivine nephelinite dikes, Wasatch Plateau, Utah: Geological Society of America Abstracts with Programs, v. 22, no. 3, p. 88.

Coincidence of N. 50°–58° W. Trends in Geologic Mapping, Magnetic and Gravity Anomalies, and Lineaments in the Northern Paradox Basin, Utah and Colorado

By Jules D. Friedman

CONTENTS

FIGURES

Previous attempts to explain the coincidence and near-coincidence of structural, geophysical and lineament trends in the region of salt-cored anticlines in the Paradox Basin (fig. 1) relied heavily on concepts of propagation of older (Precambrian) structural patterns through a thick sequence of Paleozoic and Mesozoic sedimentary rocks (including the salt units of the Paradox Formation). Propagation of fault patterns by reactivation of Precambrian structures has been widely suggested to explain both northwest- and northeast-trending fault systems (Case and others, 1963; Case and Joesting, 1973; Shoemaker and others, 1978; Friedman and others, 1994).

The presence of salt units within the Paradox Formation has provided the basis for a major objection to this concept. The incompetent salt units, most likely, would not effectively transmit stress from the lower strata into overlying beds.

Here, I attempt to explain the coincidence of some structural trends in the sedimentary strata overlying the Paradox Formation by means of a passive form of structural control exerted by block faulting of the Precambrian basement and resulting differential concentration and solution of salt in the northern part of the Paradox Basin. This model does not depend on the transmission of stress by incompetent salt units to explain the reiteration at the surface of older structural patterns. Rather, it credits this effect to deformation due to the differential thickness of salt controlled by the reactivation of earlier structures.

Several types of linear features in this area show strongly correlated azimuthal trends toward the northwest (fig. 2). Lineaments in both the gravity and magnetic-field data have peak distributions at N. 55° W., and major throughgoing lineaments have a sharp defined peak at N. 50° W. These peaks also coincide with the N. 50°–58° W. peak distribution of geologically mapped faults. Because the magnetic field represents the orientation of Precambrian basement discontinuities, the congruence or coincidence of magnetic trends with

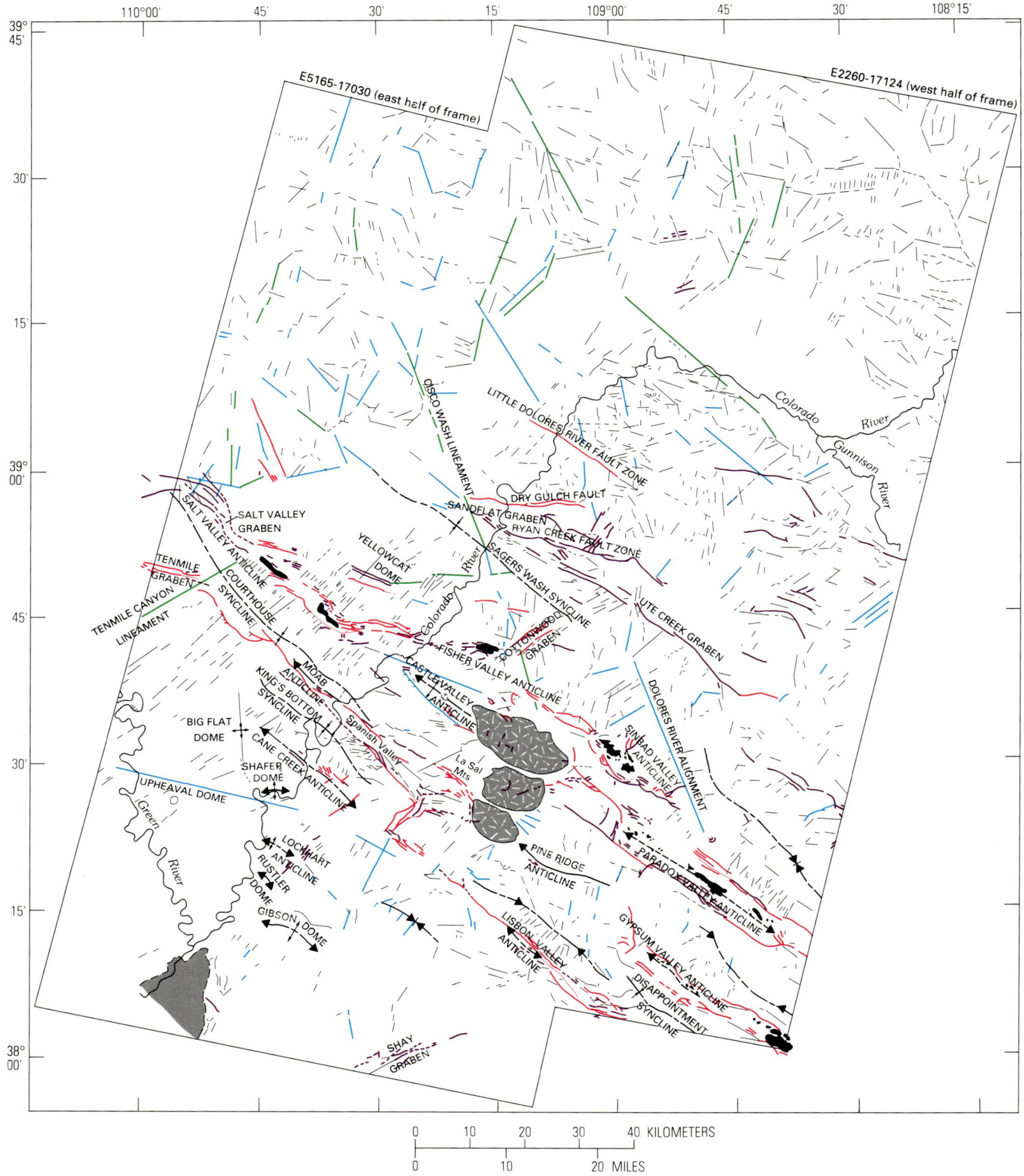

Figure 1 (above and facing page). Tectonic features of the northern part of the Paradox Basin, southeastern Utah and southwestern Colorado. Landsat multispectral scanner images 5165–17030 and 2260–17124 were used to map some lineaments and alignments. From Friedman and others (1994, fig. 2).

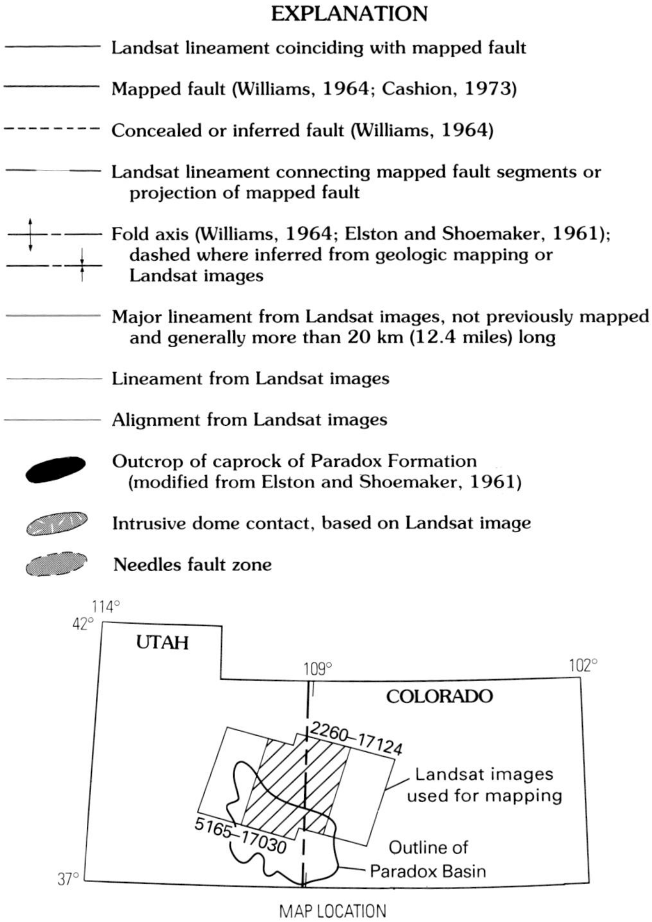

geologically mapped structures is strong evidence for basement control of mapped faults. It further suggests that the Precambrian crystalline basement has been involved in pre-Laramide, Laramide, and post-Laramide tectonic episodes. The northwest alignment of the salt anticlines of the Paradox Basin suggests that the crystalline basement, already block-faulted in the late Paleozoic, may have been further deformed during Laramide compression. It is possible, but less likely, that northwest-trending, parallel antiforms and synforms were then formed in the Precambrian crystalline rocks and that the synforms provided the locus for enhanced thickening of the deep-seated keels of the salt anticlinal cores (Friedman and others, 1994).

I suggest here that a multistage process controlled the azimuth and position of fault blocks in the Precambrian basement. The steep and relatively continuous magnetic and gravity gradients along the subsurface Uncompahgre Fault Zone, which underlies the northeastern boundary of the Paradox Basin, suggest that this zone marks a fundamental boundary within the Precambrian basement. This boundary, here termed the proto-Uncompahgre tectonic line (Cashion and others, 1990), has been the site of recurrent thrusting (Frahme and Vaughn, 1983). Movement along this zone may have been the earliest discernible event in the multistage process culminating in the tectonic development of parallel fault blocks in the Precambrian basement.

Stone (1977) placed the first recognizable tectonic activity along the proto-Uncompahgre line of structural weakness in late Precambrian time. The resulting faults, bounding the proto-Uncompahgre uplift, probably determined the trend of the deep-seated northwest-striking faults bounding blocks in the Precambrian basement rocks. The structural position of the basement blocks in turn controlled the positions and northwest trends of the major salt-cored anticlines of the Paradox Basin (Witkind, 1991; Friedman and others, 1994).

The coincidence of azimuthal trends in magnetic data, gravity data, surface faults, and lineaments (longer than 20 km) is the result of a threefold sequence of structural and tectonic events: (1) basement block faulting (as reflected in the magnetic field), (2) parallel alignment of the thick keels of salt-cored anticlines along the downdropped fault blocks (as reflected in the gravity field), and (3) listric and extensional faulting as a result of differentially greater salt solution and subsidence of clastic-rock units overlying the thicker salt keels of the anticlines, parallel to basement fault-block margins (Doelling, 1985). The resulting faults and folds are mapped at the surface, where they make up part of a northwest-trending lineament system. Many of the northwest-trending lineaments and extensional faults (fig. 3), and some shorter ones trending northeast, terminate approximately at the zero isolith of subsurface salt of the Paradox Formation. This coincidence is hardly fortuitous.

The emplacement of the laccolith complex and domal uplift of the La Sal Mountains and several other laccolith complexes at intersections of some of the major northwest and northeast lineaments of the Paradox Basin is of special tectonic significance (Friedman and others, 1994). The La Sal, Henry, and Abajo Mountains laccolith complexes, among others, may have been intruded at nodes of a northwest-northeast lineament grid. Recent studies indicate a mantle source for the laccolith magmas (Friedman and others, 1994), and this finding suggests faulting down to the depth of the crust and the deep-seated nature of the N. 40°–60° W. and N. 40°–50° E. discontinuities.

SELECTED REFERENCES

Baars, D.L., and Stevenson, G.M., 1981, Tectonic evolution of the Paradox Basin, Utah and Colorado, *in* Wiegand, D.L., ed., Geology of the Paradox Basin: Rocky Mountain Association of Geologists Field Conference, 1981, Guidebook, p. 23–31.

Case, J.E., and Joesting, H.R., 1973, Regional geophysical investigations on the central Colorado Plateau: U.S. Geological Survey Professional Paper 736, 31 p.

Case, J.E., Joesting, H.R., and Byerly, P.C., 1963, Regional geophysical investigations in the La Sal Mountains area, Utah and Colorado: U.S. Geological Survey Professional Paper 316–F, p. 91–116.

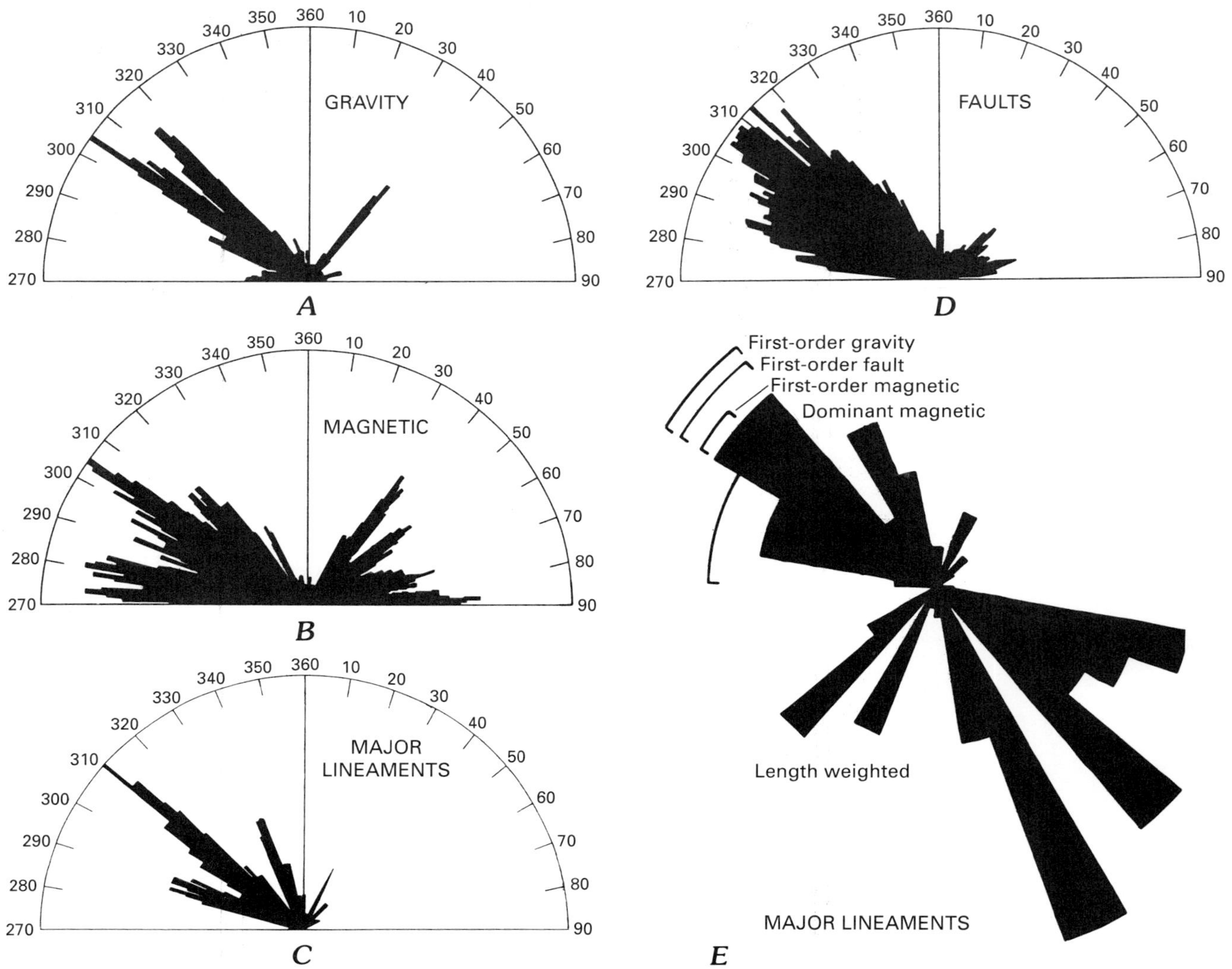

Figure 2. Rose diagrams showing frequency of azimuthal trends of linear features in the northern part of the Paradox Basin, southeastern Utah and southwestern Colorado. *A*, Gravity-field lineaments; 89 lines. *B*, Magnetic-field lineaments; 281 lines. *C*, Major lineaments (longer than 20 km) mapped from Landsat Multispectral Scanner images; 77 lines. *D*, Geologically mapped faults; 916 lines. *E*, Composite diagram showing relation of major lineaments (from diagram *C*) to gravity and magnetic lineaments (from diagrams *A* and *B*). Figures from Friedman and others (1994, figs. 9, 7, and 10).

Cashion, W.B., 1973, Geologic and structure map of the Grand Junction Quadrangle, Colorado and Utah: U.S. Geological Survey Miscellaneous Investigations Series Map I–736, scale 1:250,000.

Cashion, W.B., Kilburn, J.E., Barton, H.N., Kelley, K.D., Kulik, D.M., and McDonnell, J.R., Jr., 1990, Mineral resources of the Desolation Canyon, Turtle Canyon, and Floy Canyon Wilderness Study Areas, Carbon, Emery, and Grand Counties, Utah: U.S. Geological Survey Bulletin 1753, p. B18.

Colman, S.M., 1983, Influence of the Onion Creek diapir on the late Cenozoic history of Fisher Valley, southeastern Utah: Geology, v. 11, p. 240–243.

Doelling, H.H., 1985, Geology of Arches National Park: Utah Geological and Mineral Survey Map 74, 15 p.

Elston, D.P., and Shoemaker, E.M., 1961, Preliminary structure contour map on top of salt in the Paradox Member of the Hermosa Formation in the salt anticline region, Colorado and Utah: U.S. Geological Survey Oil and Gas Investigations Map OM–209, scale 1:250,000.

Frahme, C.W., and Vaughn, E.B., 1983, Paleozoic geology and seismic stratigraphy of the northern Uncompahgre Front, Grand County, Utah, *in* Lowell, J.D., and Gries, Robbie, eds., Rocky Mountain foreland basins and uplifts: Rocky Mountain Association of Geologists, Field Conference, Steamboat Springs, Colo., 1983, Guidebook, p. 201–211.

Friedman, J.D., Case, J.E., and Simpson, S.L., 1994, Tectonic trends of the northern part of the Paradox Basin, southeastern Utah and southwestern Colorado, as derived from Landsat Multispectral Scanner imaging and geophysical and geologic

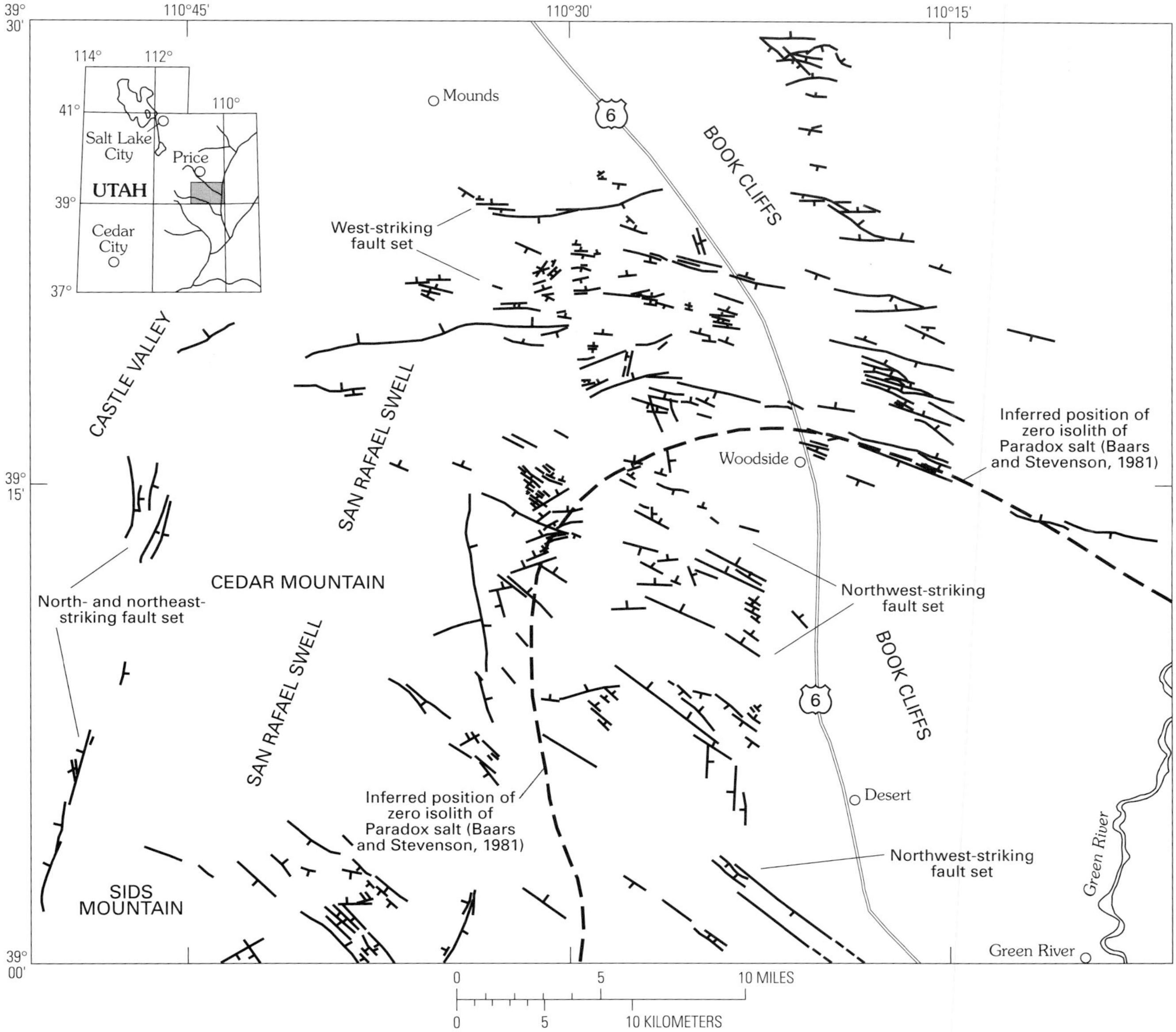

Figure 3. Distribution of known faults and fractures in the northern Paradox Basin, compared to the inferred position of the zero isolith of salt in the Paradox Formation. From Witkind (1991) and earlier sources.

mapping, chap. C *of* Huffman, A.C., Jr., project coordinator, Evolution of sedimentary basins—Paradox Basin: U.S. Geological Survey Bulletin 2000–C, p. C1–C30.

Shoemaker, E.M., Squires, R.L., and Abrams, M.J., 1978, Bright Angel and Mesa Butte fault systems of northern Arizona, *in* Smith, R.B., and Eaton, G.P., eds., 1978, Cenozoic tectonics and regional geophysics of the western Cordillera: Geological Society of America Memoir 152, p. 341–367.

Stone, D.S., 1977, Tectonic history of the Uncompahgre Uplift, *in* Veal, H.K., ed., Exploration frontiers of the central and southern Rockies: Rocky Mountain Association of Geologists Field Conference, 1977, Guidebook, p. 23–30.

Williams, P.L., 1964, Geology, structure, and uranium deposits of the Moab Quadrangle, Colorado and Utah: U.S. Geological Survey Miscellaneous Investigations Series Map I–360, 2 sheets, scale 1:250,000.

Witkind, I.J., 1991, Implications of distinctive fault sets in the San Rafael Swell and adjacent areas, east-central Utah, *in* Chidsey, T.C., Jr., ed., Geology of east-central Utah: Utah Geological Association Publication 19, p. 141–148.

Processes of Laccolithic Emplacement in the Southern Henry Mountains, Southeastern Utah

By Marie Jackson[1]

CONTENTS

FIGURES

A sequence of sedimentary rocks about 4 km thick was bent, stretched, and uplifted during the growth of three igneous domes in the southern Henry Mountains (fig. 1). Mount Holmes, Mount Ellsworth, and Mount Hillers are all about 12 km in diameter, but the amplitudes (total uplifts) of the domes are about 1.2, 1.85, and 3.0 km, respectively (fig. 2). These mountains record successive stages in the growth of near-surface magma chambers. K-Ar dating of Henry Mountains diorite porphyry by Armstrong (1969) gives Eocene ages (40–48 Ma). Sullivan (1987), however, has found younger ages for these rocks (20–29 Ma), using fission-track methods.

[1] 114 North San Francisco Street, Suite 108, Flagstaff, AZ 86001
email: mjackson@futureone.com

New geologic mapping (Jackson and Pollard, 1988) demonstrates that the sedimentary strata over the domes have a doubly hinged shape, consisting of a concave-upward lower hinge and a concave-downward upper hinge (fig. 2). A limb of approximately constant dip connects these two hinges and dips 20° at Mount Holmes, 50°–55° at Mount Ellsworth, and 75°–80° at Mount Hillers. The distal portion of each dome is composed of a gently dipping peripheral limb 3–4 km long, presumably underlain by sills and minor laccoliths. The host rocks deformed along networks of outcrop-scale faults or along deformation bands marked by crushed grains, consolidation of the porous sandstone, and small displacements of the sedimentary rocks (Jackson and Pollard, 1990). Zones of deformation bands oriented

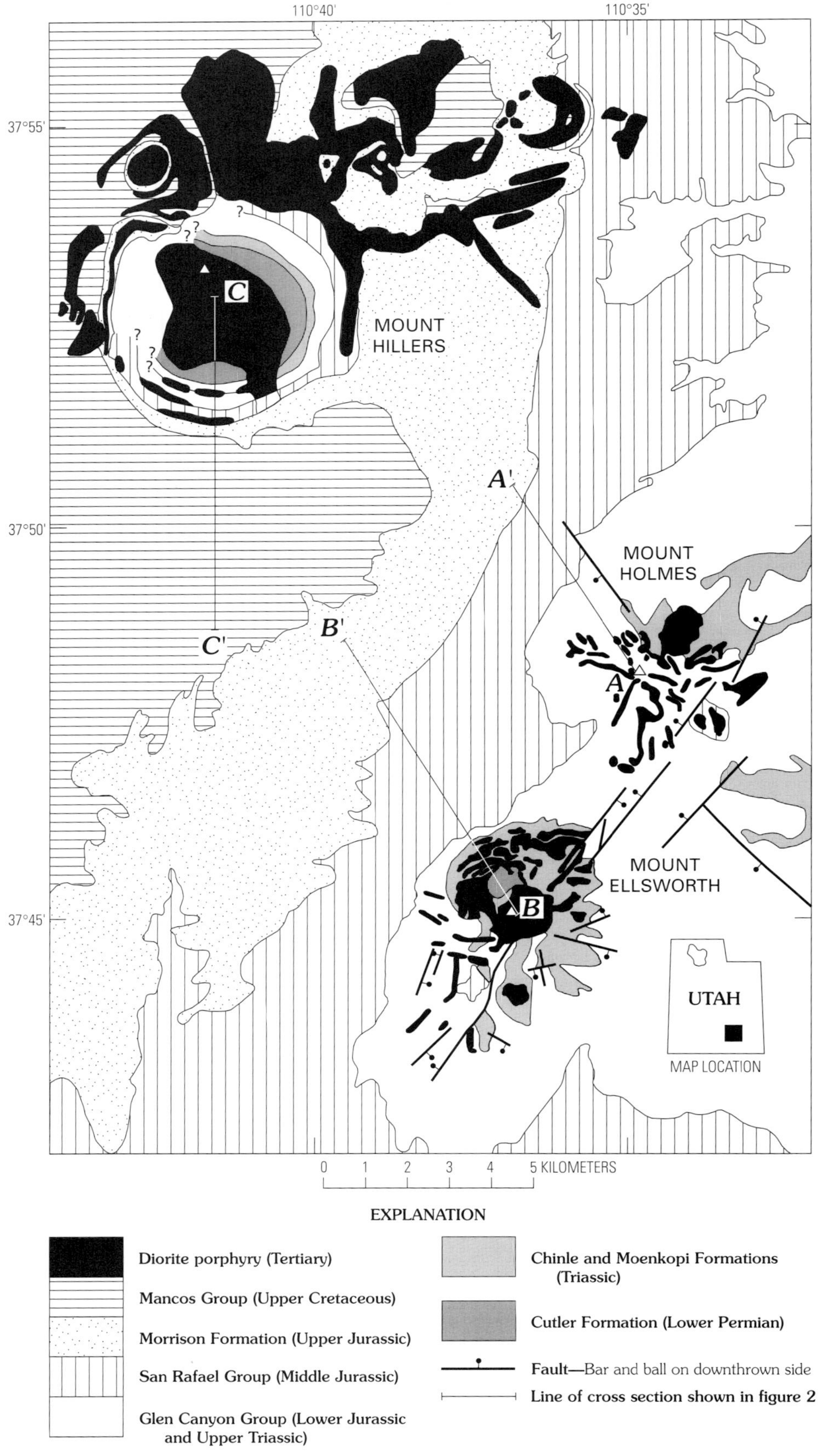

Figure 1. Simplified geologic map of the southern Henry Mountains modified from Hunt and others (1953) and Jackson and Pollard (1988). Younger rocks crop out progressively from east to west. Open triangle, mountain summit. Traverses *A–A′*, *B–B′*, and *C–C′* refer to cross sections in figure 2. Modified from Jackson and Pollard (1990, fig. 1).

parallel to the beds and formation contacts subdivided the overburden into thin layers that slipped over one another during doming.

Measurements of outcrop-scale fault populations at the three mountains reveal a network of faults that strikes at high angles to the sedimentary beds, which themselves strike tangentially about the domes (Jackson and Pollard, 1990). These faults have normal and reverse components of slip that accommodated bending and stretching strains within the strata. An early stage of this deformation is observed at Mount Holmes, where states of stress computed from measurements on three faults correlate with the theoretical distribution of stresses resulting from the bending of thin, circular, elastic plates. Bedding-plane slip and layer flexure were important components of the early deformation, as shown by field observations, by analysis of frictional driving stresses acting on horizontal planes above an opening-mode dislocation, and by paleostress analysis of the faulting. As the amplitude of doming increased, radial and circumferential stretching of the strata and rotation of the older faults in the steepening limbs of the domes increased the complexity of the fault patterns. Steeply dipping map-scale faults with dip-slip displacements indicate a late-stage jostling of major blocks over the central magma chamber. Radial dikes pierced the dome and accommodated some of the circumferential stretching.

At all three domes, porphyritic diorite sills are concordantly interleaved with the outward-dipping, arcuate beds of sedimentary host rock (fig. 2). More than 10 of these sills crop out at Mount Hillers, where they dip nearly vertically. Thermal demagnetization of cored specimens from five of these sills reveals that their partial thermoremanent magnetization has a broad range of blocking temperatures, ranging from 20°C to 590°C (Jackson and Champion, 1987). Figure 3 shows an example of the typical demagnetization data from one of the sills. When rotated to correct for the strike and dip of the beds, the high-temperature component of magnetization (450°C to 590°C) of these sills has about the same orientation as the expected Oligocene declination and inclination (358° and 56°, respectively (Irving and Irving, 1982)). This finding indicates that the sills were emplaced at nearly horizontal orientations and were later tilted. At intermediate blocking temperatures, between 377°C and 194°C, the paleomagnetic vectors recorded by the sample show a range of orientations, indicating that the sill was being rotated as it cooled, during growth of the central intrusion. At lower blocking temperatures, the stability and direction of the magnetization of the samples in their current position suggests the Oligocene magnetization rather than a recent viscous overprint. Apparently, the originally rotated low-temperature magnetization was reset by an *in situ* thermal remagnetization when the main intrusion at the center of the dome reheated the sills. The sills closest to the contact with the main intrusion show this reheating to a greater extent than sills higher in the section.

Whether the central intrusions underlying the domes are laccoliths or stocks has been the subject of controversy. According to G.K. Gilbert (1877), the central intrusions are direct analogs of the much smaller floored intrusions exposed on the flanks of the domes, which grew from sills by lifting and bending of a largely concordant overburden (fig. 4). Gilbert (1877) hypothesized that the intrusion of numerous sills preceded the inflation of an underlying laccolith, as shown by his cross section of Mount Hillers (fig. 4*B*). According to Hunt and others (1953), the central intrusions are cylindrical stocks, sheathed with a zone of shattered sedimentary rocks, and the small flanking sills and laccoliths grew laterally as tongue-shaped masses from the discordant sides of these stocks (fig. 5).

Although geologic cross sections and aeromagnetic data for the three domes are consistent with floored, laccolithic intrusions, these data do not rule out the possibility of a stock at depth (Jackson and Pollard, 1988). The paleomagnetic data from Mount Hillers (as in fig. 3, for example) indicate that the sills intruded during the first stages of doming, as Gilbert suggested. The sills cooled through their high blocking temperatures while horizontal and were then tilted with their host rock on the flanks of the growing dome. This sequence of events is not consistent with the emplacement of a stock and subsequent or contemporaneous lateral growth of sills and laccoliths. Growth in the diameter of a stock from about 300 m at Mount Holmes to nearly 3 km at Mount Hillers, as Hunt and others suggested (fig. 5), should have been accompanied by considerable radial shortening of the sedimentary strata and a style of folding that is not observed.

The following concepts define the difference between stocks and laccoliths (Jackson and Pollard, 1988, p. 117):

1. Laccoliths may be low in height relative to their horizontal dimensions, and they range from circular to tongue-shaped in plan form. Stocks have greater height relative to a roughly constant diameter, and they approximate a tall upright cylinder.

2. Laccoliths have a local feeder, such as a dike or stock, which has a very different size and mechanism of formation from those of the laccolith. Stocks do not have a

Figure 2 (following pages). Interpretive radially oriented cross sections through Mounts Holmes, Ellsworth, and Hillers. See figure 1 for lines of sections. Cross sections are based on a laccolithic model of the central intrusions. Roof contacts are drawn at base of the Triassic section, the deepest rocks exposed on these mountains. True shapes of these contacts are probably much more complex than the concordant shapes we extrapolated from the surficial geology. The tapered peripheries of the central intrusions are zones where satellite diorite dikes, sills, and thin laccoliths are distributed through the stratigraphic section. Sill marked *A* (in cross section *C–C′*) was the source of the sample plotted in figure 3. From Jackson and Pollard (1988, fig. 10).

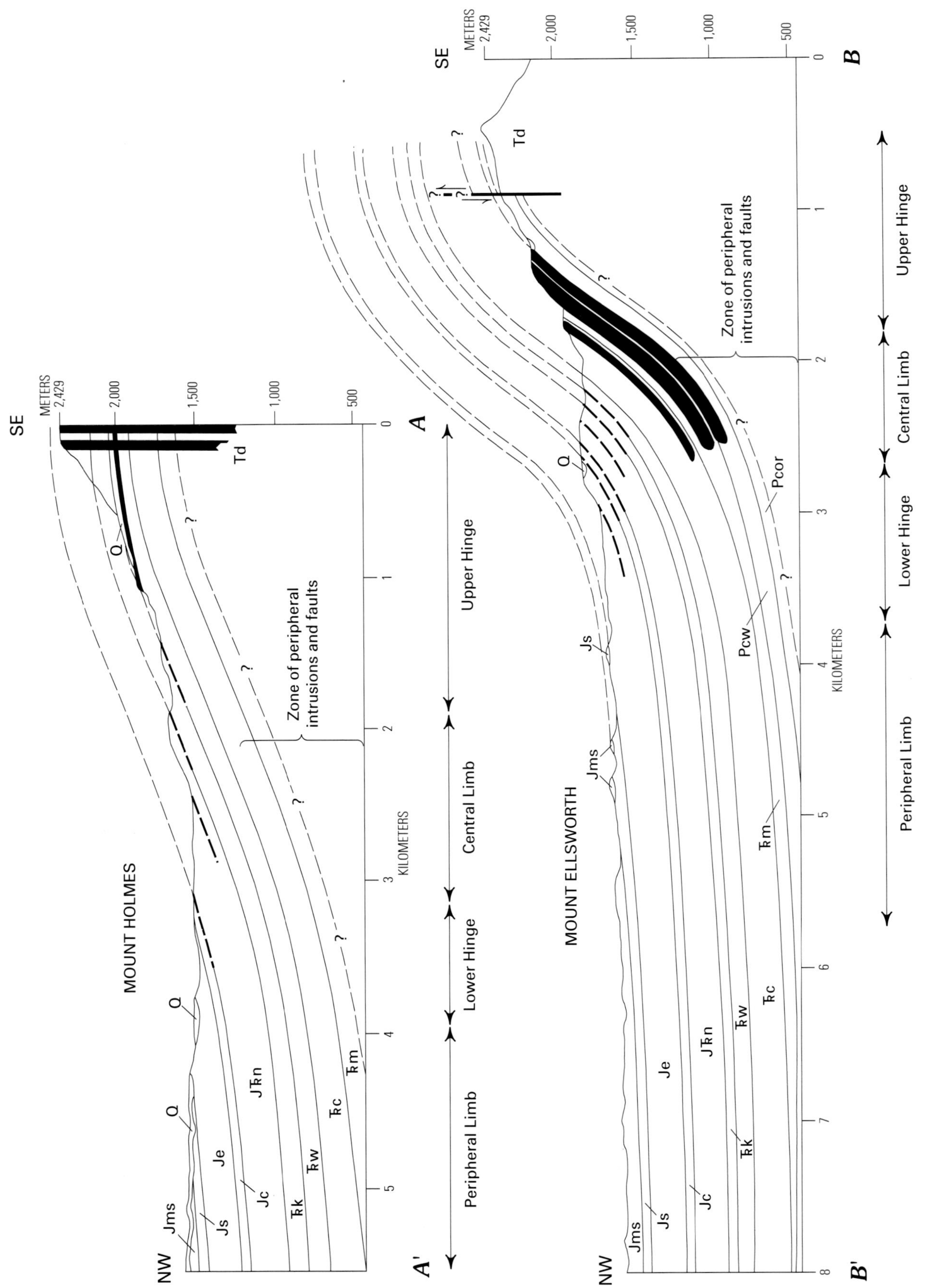
MOUNT HOLMES
A'
A
NW
SE
METERS
2,429
2,000
1,500
1,000
500
0
KILOMETERS
1
2
3
4
5
Zone of peripheral intrusions and faults
Td
Q
Jms
Js
Je
Jc
JŦn
Ŧk
Ŧw
Ŧc
Ŧm
Upper Hinge
Central Limb
Lower Hinge
Peripheral Limb
MOUNT ELLSWORTH
B'
B
NW
SE
METERS
2,429
2,000
1,500
1,000
500
0
KILOMETERS
1
2
3
4
5
6
7
8
Zone of peripheral intrusions and faults
Td
Q
Js
Jms
Je
Jc
JŦn
Ŧk
Ŧw
Ŧc
Ŧm
Pcw
Pcor
Upper Hinge
Central Limb
Lower Hinge
Peripheral Limb

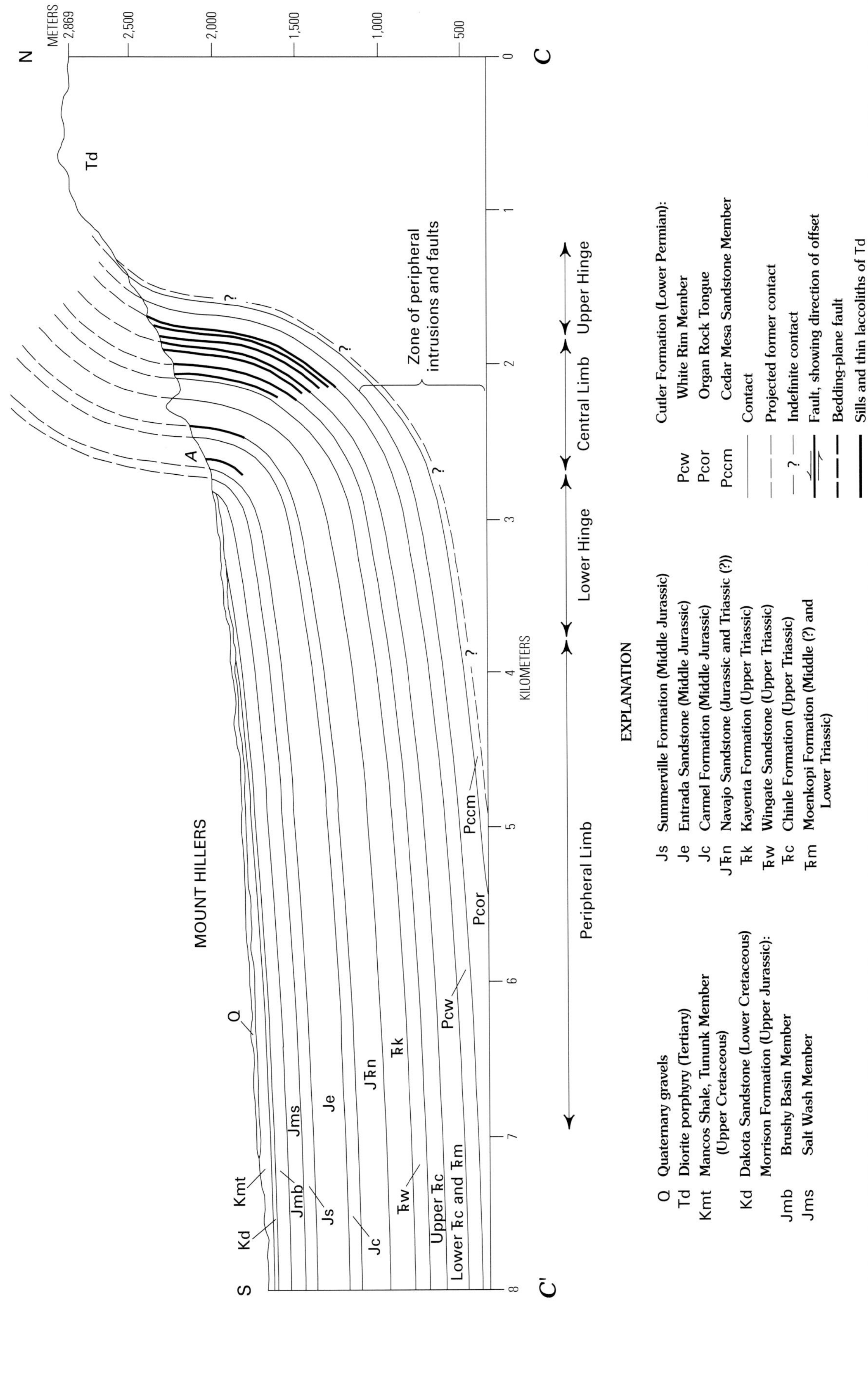

N
S
C
C'
METERS
2,869
2,500
2,000
1,500
1,000
500
0
1
2
3
4
5
6
7
8
KILOMETERS
Td
A
MOUNT HILLERS
Zone of peripheral intrusions and faults
Upper Hinge
Central Limb
Lower Hinge
Peripheral Limb
Q
Kd
Kmt
Jmb
Jms
Js
Je
Jc
JŦn
Ŧk
Ŧw
Upper Ŧc
Lower Ŧc and Ŧm
Pcw
Pcor
Pccm
?
EXPLANATION
Q Quaternary gravels
Td Diorite porphyry (Tertiary)
Kmt Mancos Shale, Tununk Member (Upper Cretaceous)
Kd Dakota Sandstone (Lower Cretaceous)
Morrison Formation (Upper Jurassic):
Jmb Brushy Basin Member
Jms Salt Wash Member
Js Summerville Formation (Middle Jurassic)
Je Entrada Sandstone (Middle Jurassic)
Jc Carmel Formation (Middle Jurassic)
JŦn Navajo Sandstone (Jurassic and Triassic (?))
Ŧk Kayenta Formation (Upper Triassic)
Ŧw Wingate Sandstone (Upper Triassic)
Ŧc Chinle Formation (Upper Triassic)
Ŧm Moenkopi Formation (Middle (?) and Lower Triassic)
Cutler Formation (Lower Permian):
Pcw White Rim Member
Pcor Organ Rock Tongue
Pccm Cedar Mesa Sandstone Member
Contact
Projected former contact
? Indefinite contact
Fault, showing direction of offset
Bedding-plane fault
Sills and thin laccoliths of Td

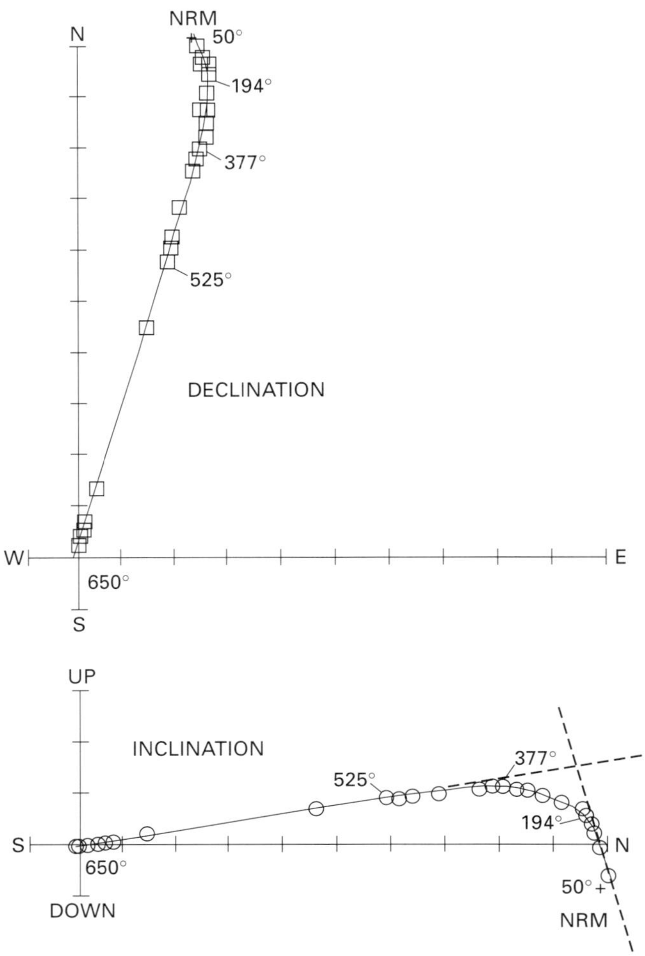

Figure 3. Zijderveldt plots showing the progressive demagnetization of a diorite porphyry sample from sill *A* in cross section C–C′ (fig. 2). Axis=2.36 emu/g. From the NRM (natural remanent magnetization) to the 194°C demagnetization step, declination is 332° and inclination is 70°. From 374°C to 600°C declination is 17° and inclination is –9°. When corrected for strike and dip of adjacent beds, these components become 6° and 77°, respectively.

local feeder; they are continuous to great depth, perhaps extending to a deep magma reservoir.

3. Stocks grow upward, perhaps by stoping, zone melting, and (or) diapiric piercement, so they are not floored and may be largely discordant. Laccoliths grow from a thin sill that thickens into a floored body, and so they are largely concordant. Distinguishing between laccoliths and stocks is made more difficult because laccoliths can attain great height by peripheral faulting. This process produces bismaliths, bodies that have discordant sides but are floored.

Geologic and geophysical data and mechanical models suggest the following sequence of events for the formation of the domes (Jackson and Pollard, 1988). The first stage involved the intrusion of many horizontal diorite sills, some of which grew to be small laccoliths (fig. 6*A*). Many of these sills were elliptical or tongue shaped and formed a radiating pattern around the incipient dome. The feeders for the sills may have been radially oriented dikes that were distributed around the incipient dome much like the conspicuous minette dikes that form a radiating pattern around the Ship Rock volcanic neck in northwestern New Mexico (Delaney and Pollard, 1981). At some point, one such sill grew to sufficiently great radius, probably between 1 and 3 km, to thicken by bending the overburden. The growth of this laccolith was enhanced by its great radius, its circular plan shape, local heating of the host rock by the older sills, and a continuing supply of magma.

As the central intrusion began to inflate (fig. 6*B*), host-rock flexure passed through an early bending stage of deformation, such as that observed at Mount Holmes (fig. 2, *A–A′*). The overburden behaved as a stack of layers that slipped over one another on bedding-plane faults, and the

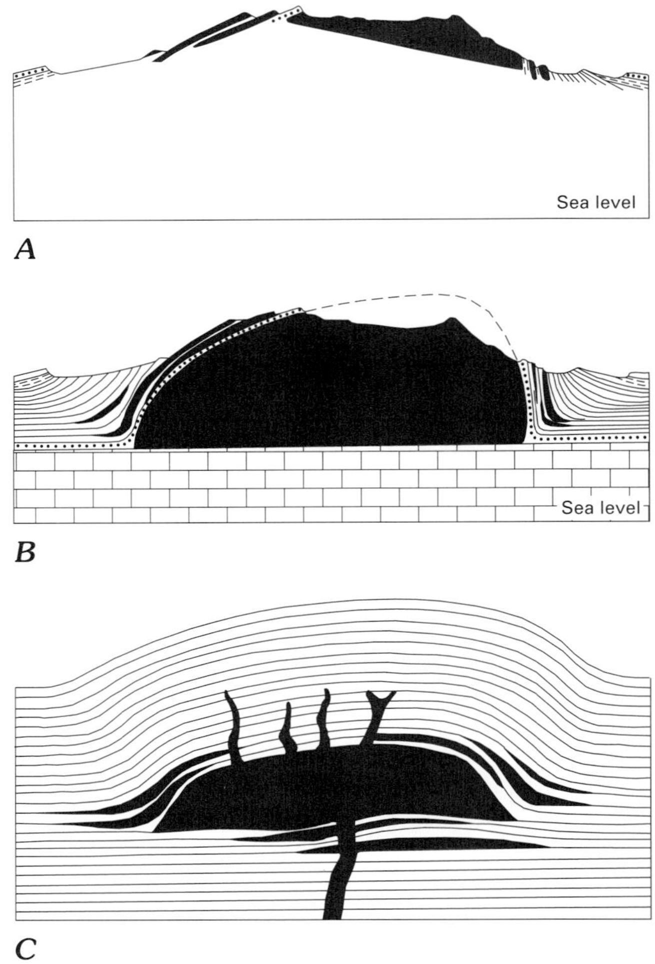

Figure 4. G.K. Gilbert's concept of laccoliths in the Henry Mountains (modified from Gilbert, 1877). *A*, Geologic cross section of Mount Hillers, striking N. 35°W. Diorite in black. *B*, Gilbert's interpretation of subsurface structure of Mount Hillers. *C*, Idealized laccolithic intrusion with a narrow feeder at its base.

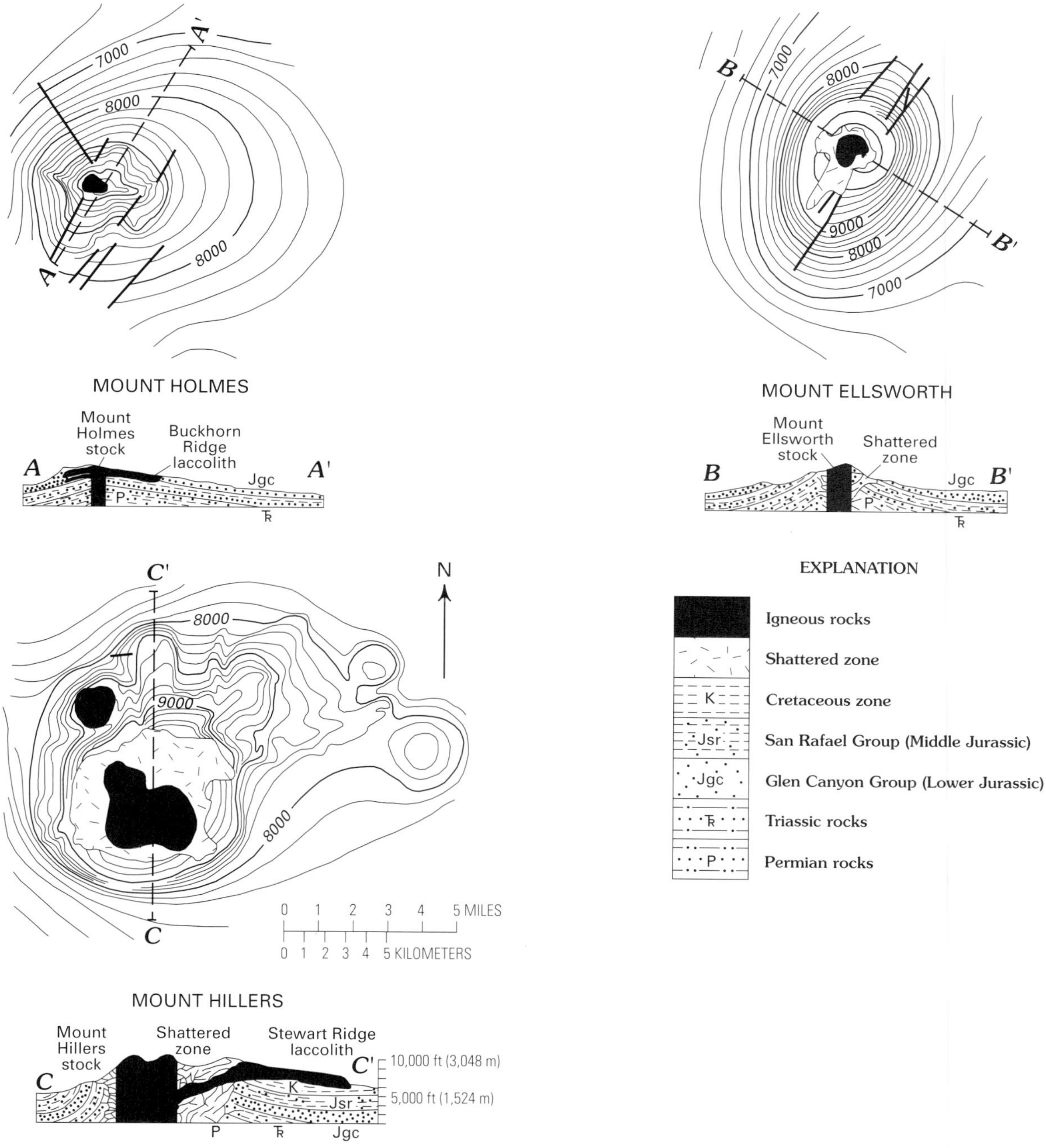

Figure 5. Structure contour maps and cross sections of Mounts Holmes, Ellsworth, and Hillers, illustrating C.B. Hunt's concept of relationships between the stocks and uplift of the beds (modified from Hunt and others, 1953). Contour interval 200 ft (≈60 m).

sills were gently rotated. In addition the layers of host rock were slightly stretched over the dome as the amplitude of deflection increased. Multiple injections of dikes and sills extended laterally beyond the sides of the central intrusion.

With continuing growth (fig. 6*C*), the hinges of the host-rock flexure tightened and the central limb steepened, as at Mount Ellsworth and Mount Hillers. Radial and circumferential stretching of the overlying layers became more important, relative to bending. Radial dikes cut across steeply dipping sills on the flanks of the dome. At the edge of the dome, the cumulative effect of continued intrusion of satellitic sills and laccoliths was to incline the overburden over the length of the long outer limb. During the entire intrusive episode, brecciation and disruption of the host rock were limited to a thin zone at the immediate contact with the central intrusion.

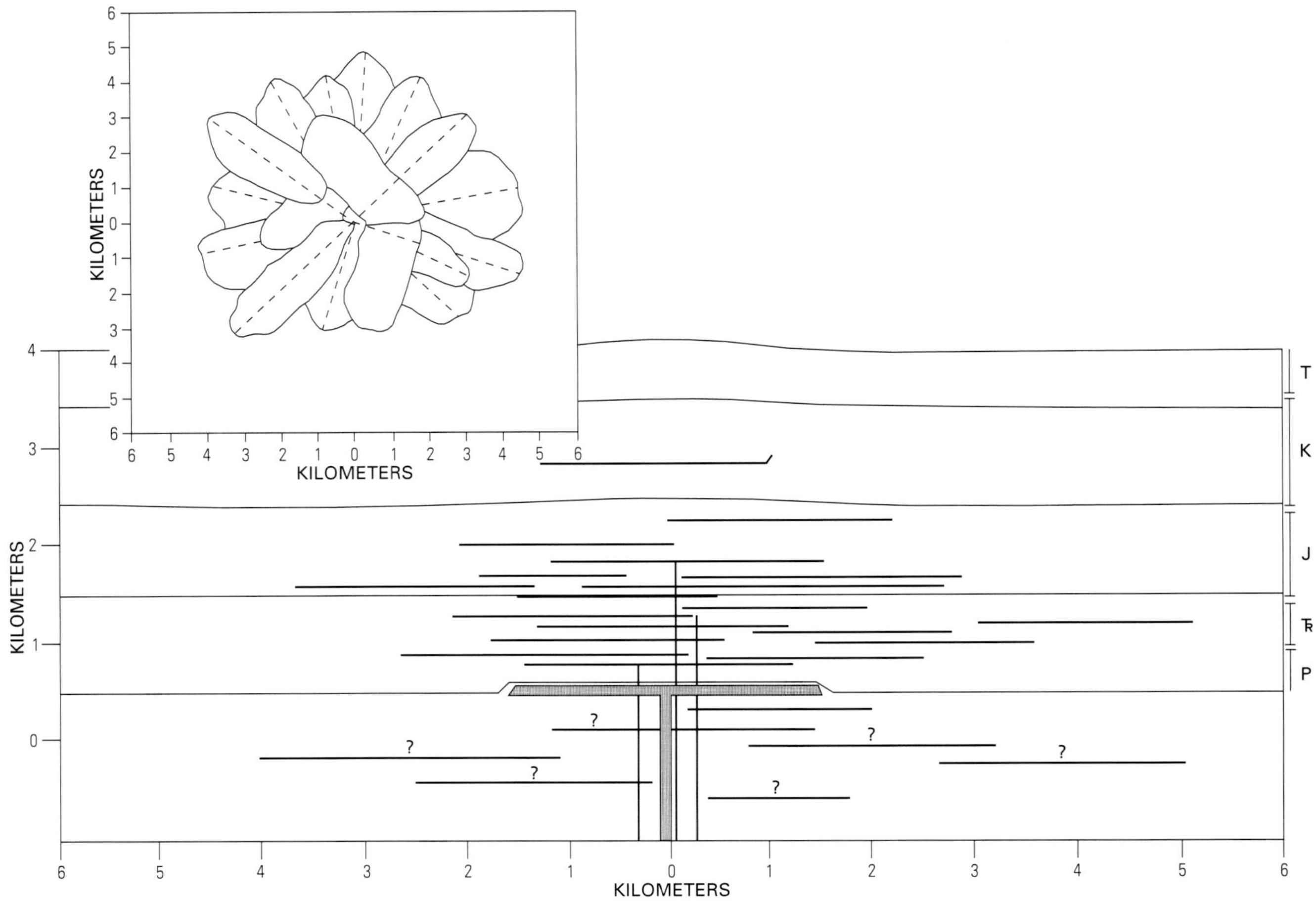

A

Figure 6 (above and facing page). Vertical cross sections showing states in growth of central intrusions and domes in the Henry Mountains. *A*, Emplacement of a stack of tongue-shaped sills and thin laccoliths fed by vertical dikes. Inset: plan view of early-formed intrusions, showing their tongue-like shape. The incipient major laccolith (stipple pattern) has a circular plan shape. *B*, Thickening of the major laccolith induces bedding-plane faulting, and the overlying intrusions are faulted. Peripheral dikes and faults form as lateral growth of the laccolith stops. Some sills and thin laccoliths intrude laterally under the peripheral limb of the dome. *C*, The major laccolith continues to thicken as the dome grows in amplitude. Beds steepen and stretch on flanks of the dome, numerous faults lift the roof rock, zone of peripheral intrusion enlarges, and radial dikes cut upward through overburden. P, T̄R, J, K, T: Permian, Triassic, Jurassic, Cretaceous, and Tertiary sedimentary host rocks. From Jackson and Pollard, 1988, figure 19.

REFERENCES CITED

Armstrong, R.L., 1969, K-Ar dating of laccolithic centers of the Colorado Plateau and vicinity: Geological Society of America Bulletin, v. 80, p. 2081–2086.

Delaney, P.T., and Pollard, D.D., 1981, Deformation of host rocks and flow of magma during growth of minette dikes and breccia-bearing intrusions near Ship Rock, New Mexico: U.S. Geological Survey Professional Paper 1201, 61 p.

Gilbert, G.K., 1877, Report on the geology of the Henry Mountains: U.S. Geographical and Geological Survey of the Rocky Mountain Region (Powell), 160 p.

Hunt, C.B., Averitt, Paul, and Miller, R.L., 1953, Geology and geography of the Henry Mountains region, Utah: U.S. Geological Survey Professional Paper 228, 234 p., 22 pl.

Irving, E., and Irving, G.A., 1982, Apparent polar wander paths Carboniferous through Cenozoic and the assembly of Gondwana: Geophysical Surveys, v. 5, p. 141–188.

Jackson, M.D., and Champion, D.E., 1987, Paleomagnetism of rotated sills at Mount Hillers, Henry Mountains, Utah—Implications for the early stages of growth of near surface igneous domes: Geological Society of America Abstracts with Programs, v. 19, no. 6, p. 391.

Jackson, M.D., and Pollard, D.D., 1988, The laccolith-stock controversy—New results from the southern Henry

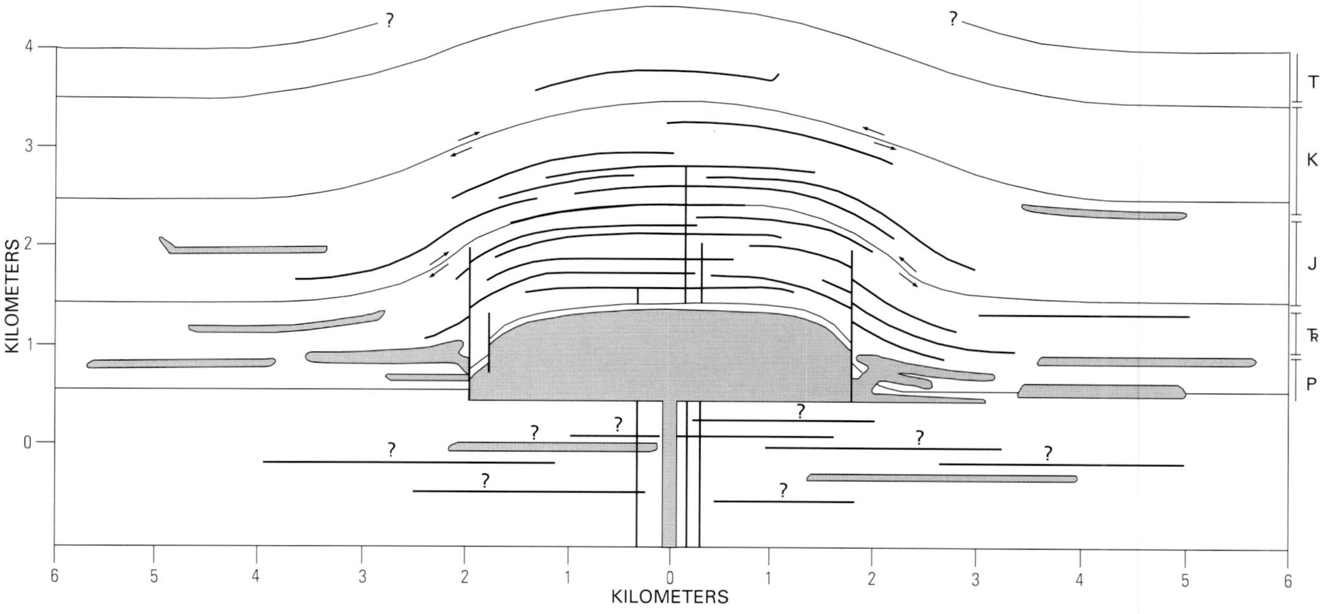

B

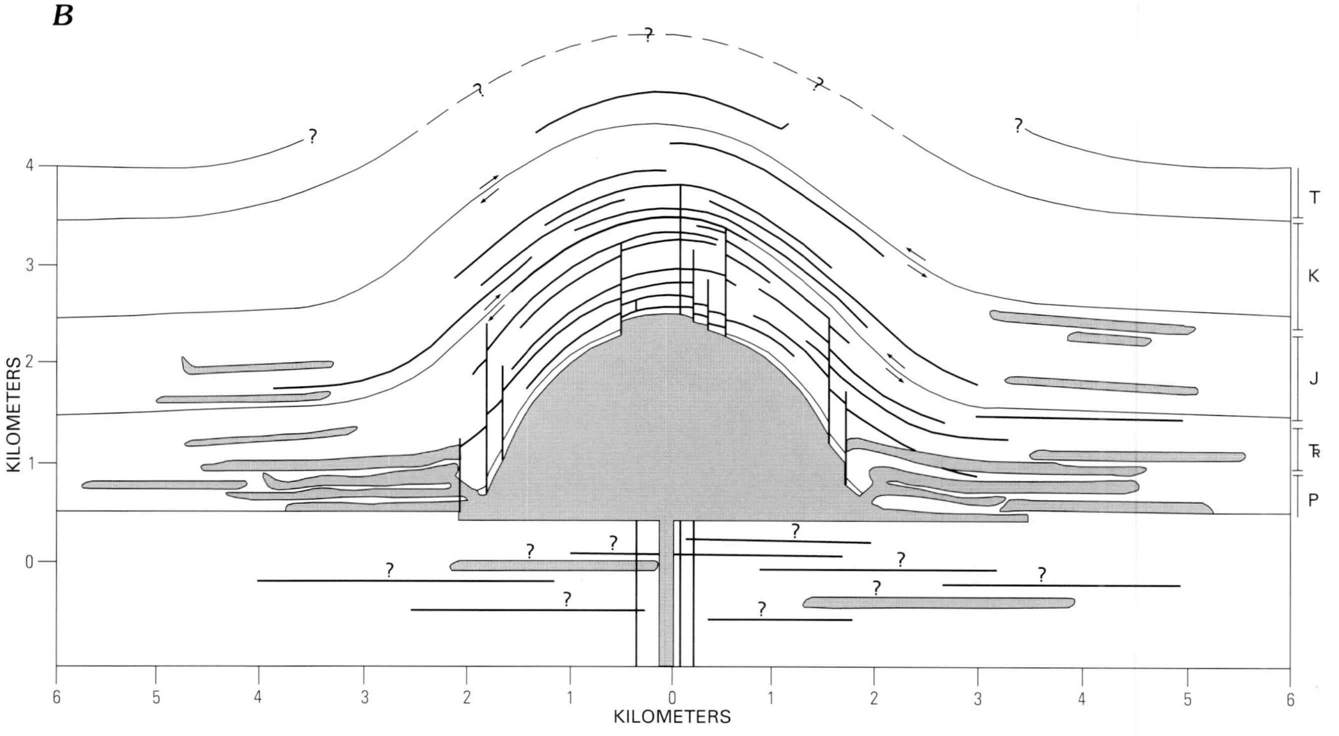

C

Mountains, Utah: Geological Society of America Bulletin, v. 100, p 117–139.

———1990, Flexure and faulting of sedimentary host rocks during growth of igneous domes, Henry Mountains, Utah: Journal of Structural Geology, v. 12, p. 185–206.

Sullivan, K.R., 1987, Isotopic ages of igneous intrusions in southeastern Utah—Relation to space-time-composition patterns of Cenozoic igneous activity in Nevada, Utah, and Colorado: Geological Society of America Abstracts with Programs, v. 19, no. 5, p. 337.

Geology of the Tertiary Intrusive Centers of the La Sal Mountains, Utah—Influence of Preexisting Structural Features on Emplacement and Morphology

By Michael L. Ross[1]

CONTENTS

FIGURES

---

[1] ERM—Southwest, Inc., 16300 Katy Freeway, Suite 300, Houston, TX 77094.

ABSTRACT

The results of geologic mapping and subsurface data interpretation, combined with previous regional gravity and magnetic surveys, define the structural setting and emplacement history of the late Oligocene intrusive centers of the La Sal Mountains, Utah. The La Sal Mountains contain three intrusive centers: northern, middle, and southern; they are located on a broad dome, which has about 600 meters of relief across a diameter of about 32 kilometers. The intrusions are estimated to have been emplaced at shallow levels ranging between 1.9 and 6.0 kilometers. The intrusions are holocrystalline and porphyritic and have a very fine to fine-grained groundmass. Each of the centers consists predominantly of hornblende plagioclase trachyte emplaced as laccoliths, plugs, sills, and dikes. The northern mountains intrusive center contains additional bodies of quartz plagioclase trachyte, peralkaline trachyte and rhyolite, and nosean trachyte that intrude the earlier hornblende plagoclase trachyte.

The northern and southern intrusive centers were emplaced into preexisting anticlines cored by salt diapirs. Geophysical and drill-core data suggest that older rocks beneath these salt diapirs are offset by northwest-striking, high-angle faults that influenced the location and development of the diapirs during the late Paleozoic and early Mesozoic. These northwest-striking en echelon faults may be connected by a northeast-striking fault or ramp-monocline structure. The northwest-striking faults and northeast-trending connecting structures form a structural boundary that separates a southern region of shallower level pre-salt rocks from a northern region of deeper level pre-salt rocks. The intrusions of the La Sal Mountains were emplaced along a kink in the structural boundary that separates these two parts of the late Paleozoic Paradox Basin. The ascending magmas for the La Sal Mountains intrusions appear to have exploited the preexisting zone of structural weakness in the upper crust during their ascent.

Most surface faults in the area around the La Sal Mountains postdate magma emplacement: the surface faults formed during late Tertiary to Quaternary collapse of the crests of the salt-cored anticlines. The collapse of the salt-cored anticlines was caused by dissolution of the salt and salt flowage.

INTRODUCTION

The La Sal Mountains of southeastern Utah are one of several mountain ranges in the central Colorado Plateau that contain hypabyssal intrusion-cored domes (fig. 1). In general, these intrusions have similar morphologies, lithologies, and chemical compositions (Eckel and others, 1949; Hunt and others, 1953; Hunt, 1958; Witkind, 1964; Ekren and Houser, 1965). A common form for these shallow intrusions is a laccolith, as initially described by Gilbert (1877) in the Henry Mountains. Therefore, the intrusive centers are commonly referred to as laccolithic centers. Previous workers (Kelley, 1955; Hunt, 1956; Shoemaker, 1956; Witkind, 1975; Warner, 1978) have speculated about whether structures represented by lineaments may have influenced the distribution and form of the conspicuous and prominent intrusive centers of the Colorado Plateau. The problem I will address in this report is: Did northwest-striking subsurface faults and associated salt-cored anticlines of the northern Paradox Basin control the location and form of the La Sal Mountains intrusive centers? If so, what is the geological and geophysical evidence to support this hypothesis?

Peale (1877; 1878) originally described the La Sal Mountains. Gould (1926) and Hunt (1958) mapped the mountains at a reconnaissance scale and studied the mineral deposits. Uranium and petroleum exploration during the 1950's and 1960's prompted additional geologic mapping (1:24,000 scale) in the middle and southern La Sal Mountains (Carter and Gualtieri, 1957, 1958; Weir and Puffet, 1960; Weir and others, 1960). Case and others (1963) conducted regional gravity and magnetic surveys in the area.

This report presents some of the results of recent (1988–92) 1:24,000-scale geologic mapping of the northern La Sal Mountains by the Utah Geological Survey and a synthesis of the previous investigations. The report will focus on the general structural geology of the La Sal Mountains intrusive centers and the possible influence of subsurface faults and salt-cored anticlines on the emplacement and morphology of the intrusions.

ACKNOWLEDGMENTS

This paper has benefited from discussions with Michael Shubat, Grant Willis, Steve Nelson, Hellmut Doelling, and Eugene Shoemaker. Jon King, Michael Shubat, Doug Sprinkel, and Grant Willis provided thorough and constructive reviews. My appreciation goes to Jules Friedman of the USGS for inviting me to participate in the symposium. This research is supported by the Utah Geological Survey as part of its 7.5-minute quadrangle mapping program.

REGIONAL GEOLOGIC SETTING

Geophysical studies suggest that the Colorado Plateau is underlain by 40- to 50-km-thick Precambrian cratonic crust (Thompson and Zoback, 1979; Allmendinger and others, 1987). Surface exposures of the Precambrian basement in the plateau region suggest that it consists of gneissic rocks of Early Proterozoic age (1,800 to 1,600 Ma) and granitoid rocks of Middle Proterozoic age (1,400 to 1,500 Ma) (Tweto, 1987; Bowring and Karlstrom, 1990; Case, 1991). East-northeast of the La Sal Mountains, on the

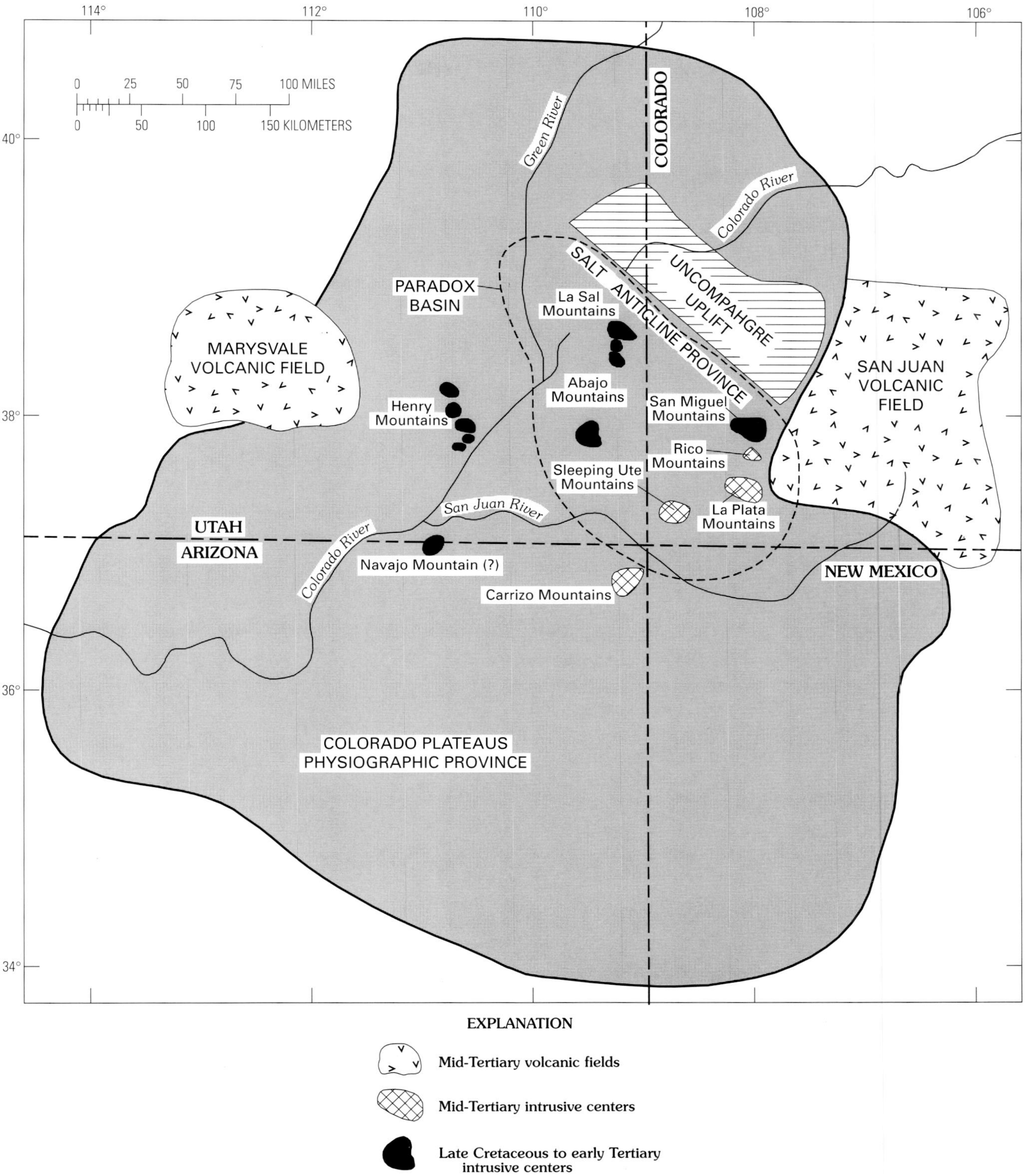

Figure 1. Selected magmatic and structural features of the Colorado Plateau (modified from Woodward-Clyde Consultants, 1983, figure 6–1, p. 136).

Uncompahgre Plateau, the basement rocks contain northwest-striking faults that show evidence of Proterozoic shearing (Case, 1991). Farther east, in the mountains of central Colorado, Proterozoic rocks contain both northwest-striking and northeast-striking fault zones that also show evidence of Proterozoic movement (Hedge and others, 1986; Tweto,

1987). In the southwestern part of the Colorado Plateau the basement rocks are faulted by northeast-striking Proterozoic shear zones (Bowring and Karlstrom, 1990). Subsurface data indicate that a 600-m-thick sequence of Cambrian to Mississippian epicontinental sedimentary rocks overlies the Proterozoic basement rocks of the central Colorado Plateau (Hintze, 1988).

During the late Paleozoic, regional tectonism formed the ancestral Rocky Mountains and associated basins (Woodward-Clyde Consultants, 1983). The Paradox Basin formed on the southwest side of the ancestral Uncompahgre Uplift during this tectonism (fig. 1). The basin is asymmetric, having its northwest-trending axis along its northeastern margin adjacent to the faulted ancestral uplift. Drilling and seismic data show that the uplift is bounded along the Paradox Basin by a high-angle reverse fault (Uncompahgre fault). The fault accommodated approximately 6,100 m of vertical offset and 9 km of horizontal offset during the late Paleozoic (Frahme and Vaughn, 1983; White and Jacobson, 1983; Potter and others, 1991). The amount of subsidence and sedimentation was greatest in the northeastern part of the basin adjacent to the uplift.

During the Middle Pennsylvanian, a sequence of cyclic evaporites (chiefly halite, gypsum, and anhydrite), fine-grained siliciclastics, and carbonates of the Paradox Formation were deposited in the basin. In the northeastern part of the basin, later flowage of the evaporites formed northwest-trending elongated salt diapirs, which may be referred to as salt walls (Jackson and Talbot, 1991). In the vicinity of the La Sal Mountains, the salt diapirs form three near-parallel rows: the Moab Valley–Spanish Valley–Pine Ridge salt diapir system, the Castle Valley–Paradox Valley salt diapir system, and the Salt Valley–Cache Valley–Fisher Valley–Sinbad Valley salt diapir system (fig. 2). Some early workers in the Paradox Basin suggested the existence of faults and (or) folds in the underlying rocks that controlled the locations and development of the salt diapirs (Harrison, 1927; Stokes, 1948). Data from drilling and geophysical surveys in the region strongly suggest the existence of high-angle, large-displacement, northwest-striking faults in the Pennsylvanian and older rocks below some of the salt diapirs (Shoemaker and others, 1958; Case and others, 1963; Baars, 1966; Woodward-Clyde Consultants, 1983). These faults strike subparallel to the Uncompahgre fault.

The growth of the Paradox Basin salt diapirs affected the deposition of synkinematic Upper Pennsylvanian to Upper Triassic sediments and caused considerable variation in the stratigraphic thickness of these units (Shoemaker and others, 1958; Cater, 1970; Doelling, 1988). Localized salt flowage along some of the diapirs continued until the Late Jurassic (Shoemaker and others, 1958; Cater, 1970; Tyler and Ethridge, 1983). Synkinematic Upper Pennsylvanian to Upper Jurassic formations are thin or absent over some salt diapirs, but are significantly thicker in the rim synclines that formed adjacent to the salt diapirs during salt flowage. The piercement, thinning, and folding of strata over the salt diapirs produced the salt-cored anticlines.

Northeast-southwest compressive stress of the Laramide orogeny affected the region during the Late Cretaceous to Eocene (McKnight, 1940; Cater, 1970; Baars and Stevenson, 1981; Heyman and others, 1986). Some structures were reactivated and formed large uplifts. The Uncompahgre Plateau, for instance, rose on the site of the Paleozoic Uncompahgre Uplift. North-northwest-striking high-angle faults and monoclines formed in association with these uplifts (Cater, 1970; Jamison and Stearns, 1982). North-northwest-trending gentle folds are also interpreted to have formed during this orogenic period (Cater, 1970; Johnson, 1985). Broad folds may have been superimposed on the preexisting salt-cored anticlines and adjacent rim synclines (fig. 2) (Cater, 1970; Doelling, 1988). The crustal structural fabric produced during these periods of deformation may have influenced the emplacement of the Colorado Plateau intrusive centers.

Even though the Colorado Plateau intrusive centers have structural and geochemical similarities, recent isotopic dating (Sullivan and others, 1991; Nelson, Heizler, and Davidson, 1992; Nelson, Davidson, and Sullivan, 1992) combined with previous studies (Armstrong, 1969; Cunningham and others, 1977) indicates that the centers form two temporally and spatially distinct belts (fig. 1). A northeast-southwest-trending belt of Late Cretaceous to early Tertiary intrusive centers extends from the Four Corners region into Colorado. An east-west-trending belt of late Oligocene to early Miocene magmatic centers extends across the Colorado Plateau from the Marysvale volcanic field to the San Juan volcanic field (fig. 1), including several intrusive centers (Best, 1988; Nelson, Davidson, and Sullivan, 1992). The La Sal Mountains magmatic activity occurred in the late Oligocene and is part of the east-west belt (Nelson, Davidson, and Sullivan, 1992).

The Colorado Plateau has undergone epeirogenic uplift during the late Cenozoic (Hunt, 1956; Lucchitta, 1979; Johnson and Finn, 1986), and regional erosion has been pronounced. However, the salt-cored anticlines remain distinct physiographic features because of late Cenozoic collapse along their crests. The collapse transformed most of them into steep-walled valleys, which are generally floored by Quaternary deposits (figs. 2 and 3). Collapse of the crests along marginal faults of the salt-cored anticlines was caused by salt dissolution or salt flowage.

Figure 2 (facing and following pages). Geologic and geographic features of the La Sal Mountains, Utah, and vicinity. *A*, Simplified geologic map, modified and reduced from the 1:250,000 map of Williams (1964). For the locations of geologic features not labeled, compare this to the geographic map on page 67, which is at the same scale. *B*, Notable geographic features in and near the La Sal Mountains, Utah.

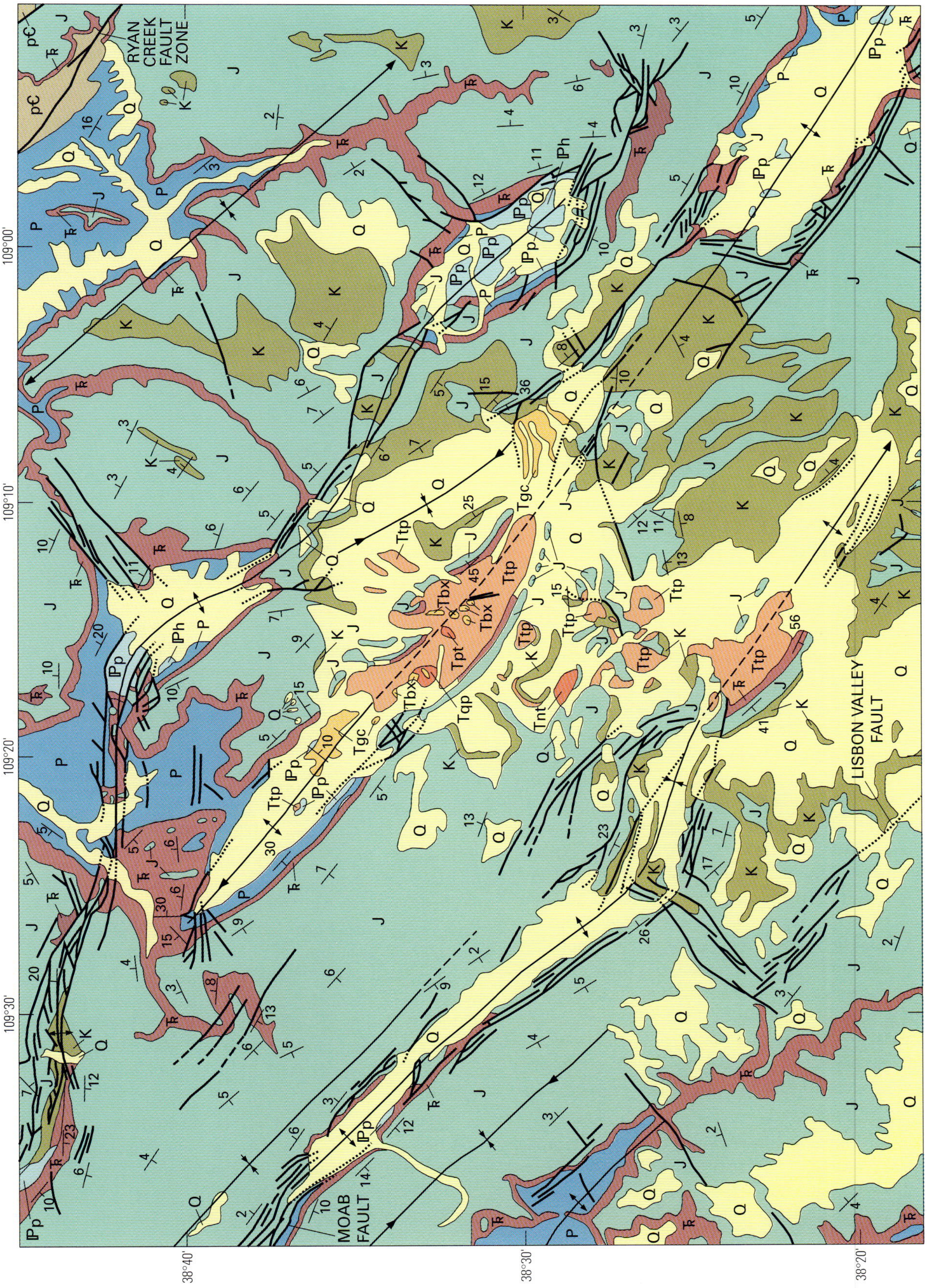
RYAN CREEK FAULT ZONE
MOAB FAULT
LISBON VALLEY FAULT
109°00'
109°10'
109°20'
109°30'
38°40'
38°30'
38°20'

A

EXPLANATION

- Q — **Quaternary deposits (includes some Pliocene fan gravels)**
- **Tertiary rocks**
  - Tgc — **Geyser Creek Fanglomerate (Pliocene?)**
  - Tnt — **Nosean trachyte** (youngest)
  - Tbx — **Breccia**
  - Tpt — **Peralkaline trachyte** (Oligocene)
  - Tqp — **Quartz plagioclase trachyte**
  - Ttp — **Hornblende plagioclase trachyte** (oldest)
- K — **Cretaceous strata**
- J — **Jurassic strata**
- TR — **Triassic strata**
- P — **Cutler Formation (Lower Permian)**
- ℙh — **Honaker Trail Formation (Upper and Middle Pennsylvanian)**
- ℙp — **Paradox Formation (Middle Pennsylvanian)**
- p€ — **Precambrian rocks**

- **Contact**
- **Fault**—Dashed where approximate; dotted where covered
- 23 — **Strike and dip**
- **Anticline axial trace**—Dashed where approximate or covered
- **Syncline axial trace**—Dashed where approximate or covered

0 5 10 15 20 KILOMETERS

GEOLOGY OF THE LA SAL MOUNTAINS

The La Sal Mountains consist of three distinct clusters of peaks separated by high passes: the northern, middle, and southern La Sal Mountains (fig. 4). Each of the mountain clusters is an intrusive center consisting of hypabyssal intrusions of trachyte and rhyolite porphyries emplaced as laccoliths, plugs, sills, and dikes.

The three intrusive centers in the La Sal Mountains intruded upper Paleozoic and Mesozoic sedimentary rocks and have a north-south alignment (fig. 2). As first recognized by Gould (1926), both the northern and southern mountains are cored by large elliptical igneous intrusions elongated northwest-southeast. The northern mountains are elongated along the Castle Valley–Paradox Valley salt-cored anticline system. The southern mountains are elongated along the Moab Valley–Spanish Valley–Pine Ridge salt-cored anticline system. The elongation of these intrusive centers suggests that shallow-level magma emplacement was influenced by the preexisting salt-cored anticlines. The intrusions of the middle mountains form several laccoliths, plugs, and sills in a circular cluster arrangement. These were emplaced into subhorizontal sedimentary rocks between the two salt-cored anticline systems.

The La Sal Mountains intrusive centers are on a broad dome that has approximately 600 m of relief across a diameter of about 32 km. Regional magnetic data suggest there is no large intrusion in the subsurface below the La Sal Mountains from which the intrusive centers were supplied. The magnetic low at the La Sal Mountains indicates that no such intrusion is present within 11–14 km beneath the mountains (Case and others, 1963). It is possible, though, that one could be present at greater depth.

Hunt (1958) interpreted each of the La Sal Mountain intrusive centers as consisting of a central stock surrounded by outward-radiating laccoliths. In his view, the stocks forced up the individual domes and the laccoliths were injected laterally from the stocks. This interpretation is identical to his conclusions on the physical injection of stocks and laccoliths for the Henry Mountains. The mechanism of emplacement of the Henry Mountains laccoliths is disputed. Research by Johnson and Pollard (1973), Pollard and Johnson (1973), Jackson and Pollard (1988a), and Corry (1988) in the Henry Mountains indicates that a large central laccolith, not a discordant stock, formed the intrusive domes of the Henry Mountains. The laccoliths were probably fed by a network of radiating dikes, but the data do not rule out the possibility of a stock at depth (Jackson and Pollard, 1988a). Jackson and Pollard (1988a,b), Hunt (1988), and Corry (1988) discuss the two opposing interpretations at length. No field evidence demonstrates convincingly that any of the La Sal Mountains centers are cored by discrete central stocks that have fed magma outward to radiating laccoliths. Surficial deposits, frost wedging, hydrothermal alteration, and poor bedrock exposures make detailed structural analysis of the La Sal Mountains intrusives problematic.

Igneous rocks of the La Sal Mountains are holocrystalline and porphyritic and have a very fine to fine-grained groundmass. Chemically, the igneous rocks are mildly subalkaline to alkaline and contain 59–71 percent SiO_2. Based on the Total Alkali-Silica classification of LeBas and others (1986), rocks of the La Sal Mountains consist of various types of trachyte and some rhyolite (Ross, 1992). Mapping (figs. 2 and 5) shows the following lithologic units in the La Sal

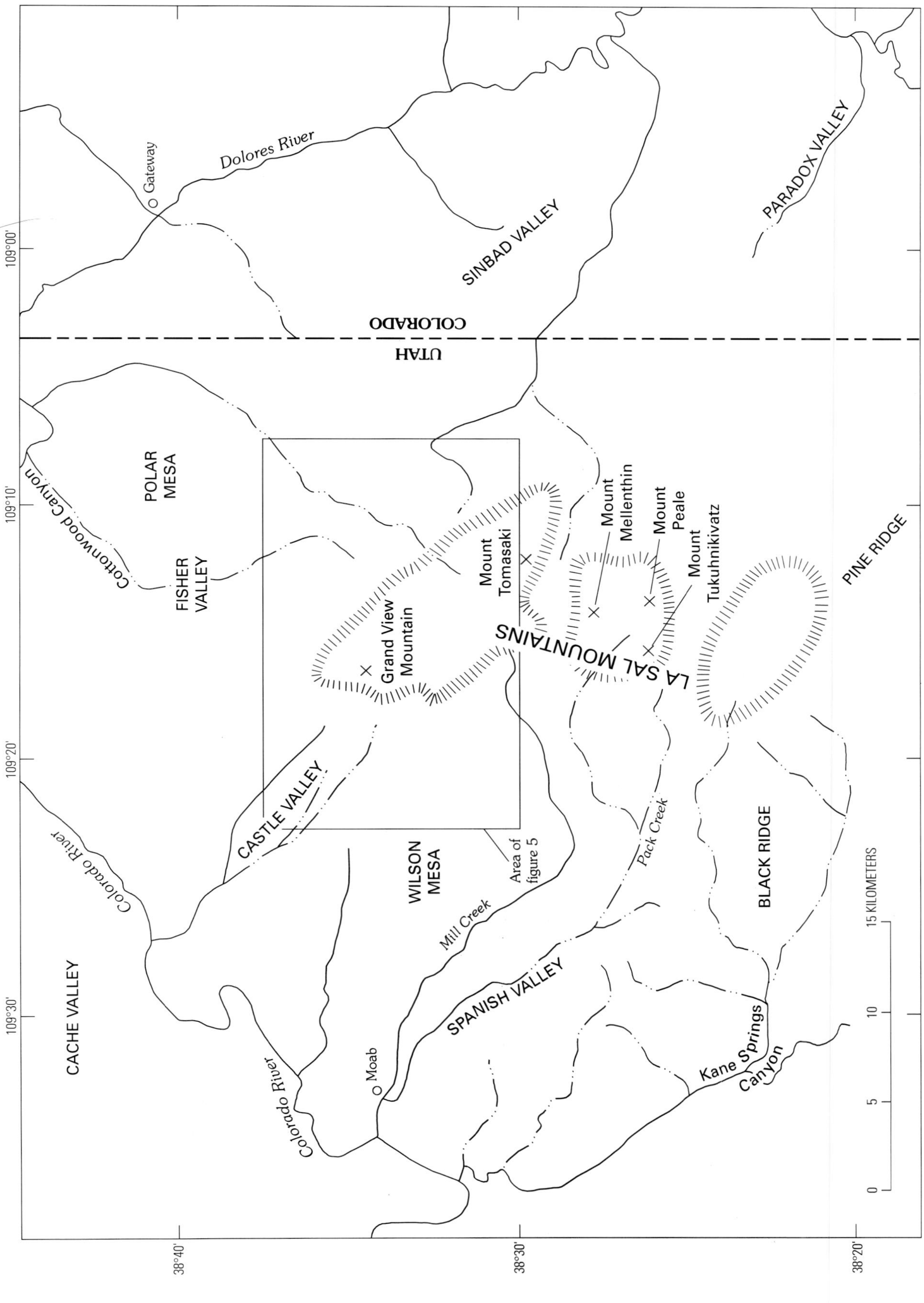

Dolores River
Gateway
SINBAD VALLEY
PARADOX VALLEY
COLORADO
UTAH
POLAR MESA
Cottonwood Canyon
FISHER VALLEY
Mount Tomasaki
Mount Mellenthin
Mount Peale
Mount Tukuhnikivatz
Grand View Mountain
LA SAL MOUNTAINS
PINE RIDGE
CASTLE VALLEY
Colorado River
WILSON MESA
Area of figure 5
Pack Creek
BLACK RIDGE
Mill Creek
CACHE VALLEY
SPANISH VALLEY
Moab
Kane Springs Canyon
15 KILOMETERS
0
5
10
15
109°00'
109°10'
109°20'
109°30'
38°40'
38°30'
38°20'

B

Figure 3. View to the southeast up Castle Valley, Utah, showing the northern La Sal Mountains. Note the high cliffs along the valley margins. Round Mountain is in the center of the valley (foreground).

Figure 4. The La Sal Mountains, Utah, as seen from Dead Horse Point State Park, about 40 km west of the mountains. Northern mountains are on the left, middle mountains are in the center, and southern mountains are on the right. Photograph by Craig Morgan, Utah Geological Survey.

Mountains (in order from oldest to youngest): hornblende plagioclase trachyte (Ttp), quartz plagioclase trachyte (Tqp), peralkaline trachyte (Tpt), peralkaline rhyolite (Trp), and nosean trachyte (Tnt).

Hornblende plagioclase trachyte is the predominant igneous rock type in the La Sal Mountains. It contains phenocrysts of plagioclase (20–50 percent) and hornblende (≤15 percent), ± clinopyroxene (≤5 percent). The plagioclase phenocrysts are subhedral to euhedral laths, some complexly zoned, ranging from 1 to 8 mm. The hornblende phenocrysts are subhedral to euhedral and generally range from 2 to 7 mm. The clinopyroxene forms equant microphenocrysts. Accessory minerals are apatite, opaque oxides, sphene, ± biotite, and ± zircon. The hornblende plagioclase trachyte also contains many conspicuous amphibolite xenoliths, and some of the hornblende crystals are xenocrysts derived from disintegration of those xenoliths (Nelson, Davidson, and Sullivan, 1992). Quartz plagioclase trachyte is largely similar to the hornblende plagioclase trachyte, except that its phenocryst assemblage includes ≤10 percent anhedral (resorbed?) quartz.

The peralkaline trachyte contains 20–70 percent phenocrysts of alkali feldspar, plagioclase, and microperthite, along with aegirine-augite (≤10 percent), ± hornblende (≤5 percent). The alkali feldspar forms large (0.5–2.0 cm) equidimensional euhedral crystals that are complexly zoned and twinned. Some crystals have plagioclase cores. Aegirine-augite forms euhedral microphenocrysts (≤3 mm) scattered in the groundmass, which consists of alkali feldspar and minor quartz, apatite, opaque oxides, and zircon. The texture of the peralkaline trachyte varies from seriate to porphyritic. Conspicuous protoclastic(?) and filter-pressing textures are locally present.

The peralkaline rhyolite contains 5–40 percent phenocrysts of subhedral to euhedral feldspar. Feldspar phenocrysts are commonly plagioclase, but some are complexly zoned, having plagioclase cores and alkali feldspar rims. Sanidine forms phenocrysts in certain rhyolite intrusions. Quartz (≤10 percent) and aegirine-augite (≤5 percent) may be phenocrysts. The groundmass is feldspar-rich and also includes some quartz, clinopyroxene, apatite, zircon, and opaque oxides.

The nosean trachyte is a distinct rock because it contains abundant large (0.5–5.0 cm) equidimensional multiphase feldspar megacrysts and glomerocrysts. The megacrysts are complexly zoned albite, perthite, and alkali feldspar. Besides the megacrysts, the rock contains 30 percent euhedral plagioclase and alkali feldspar (0.2–1.0 cm), 10 percent subhedral to euhedral nosean (≤2 mm), and 2–5 percent subhedral to euhedral aegirine-augite (≤4 mm) as phenocrysts. Microphenocrysts of anhedral melanite garnet, commonly in small clusters with aegirine-augite and opaque oxides, compose about 2 percent of the rock. Sphene, biotite, and opaque oxides are accessory minerals. The groundmass is feldspar-rich and trachytic.

Based on $^{40}Ar/^{39}Ar$, K/Ar, and fission-track geochronology, magmatic activity in the La Sal Mountains ranged in age from 25.1 to 27.9 Ma (Nelson, Heizler, and Davidson, 1992). This timing is contemporaneous with magmatic activity in the Henry and Abajo Mountains (Nelson, Davidson, and Sullivan, 1992).

FIELD RELATIONSHIPS OF THE LA SAL MOUNTAINS INTRUSIVE CENTERS

NORTHERN LA SAL MOUNTAINS INTRUSIVE CENTER

The northern La Sal Mountains intrusive center consists of a large composite pluton forming an elliptical structural dome, 13.7×4.8 km, surrounded by several smaller satellite laccoliths and sills (fig. 5). The main intrusive dome shows about 1,520 m of structural relief along the base of the Lower Jurassic Glen Canyon Group sandstones across the short axis of the dome (fig. 5, section *A–A′*). Preliminary restoration of the cross section *A–A′* prior to intrusion suggests about 300 m of relief was present on the base of the Glen Canyon Group over the preexisting salt-cored anticline. The salt diapirs in the core of the Castle Valley–Paradox Valley salt-cored anticlines have near-vertical margins and height-to-width ratios greater than one. The estimated height of the Castle Valley salt diapir is between 2,700 and 3,000 m. The estimated height of the Paradox Valley salt diapir is between 2,400 and 4,600 m (Shoemaker and others, 1958; Case and others, 1963; Cater, 1970).

Hornblende plagioclase trachyte is the oldest and most voluminous igneous rock type at the northern intrusive center. The exact shape and size of the hornblende plagioclase trachyte intrusion or intrusions that form the main pluton (within the preexisting salt-cored anticline) are poorly understood. Field observations and petrographic and geochemical analyses have not produced sufficient information to determine if the main pluton is formed by one or several intrusions. Even the pluton's shape and original size cannot be determined because its roof rocks have been removed by erosion, its floor is not exposed, and only a few outcrops show its marginal contacts. Though conclusive evidence is lacking, the main hornblende plagioclase trachyte pluton probably consists of several coalesced laccoliths that were emplaced in the upper part of the salt diapir below its contact with the overlying country rock.

Hunt (1958) used flow banding, linear hydrothermal alteration zones, sheeted joints, xenolithic blocks of contact-metamorphosed country rock, and the eroded shape of

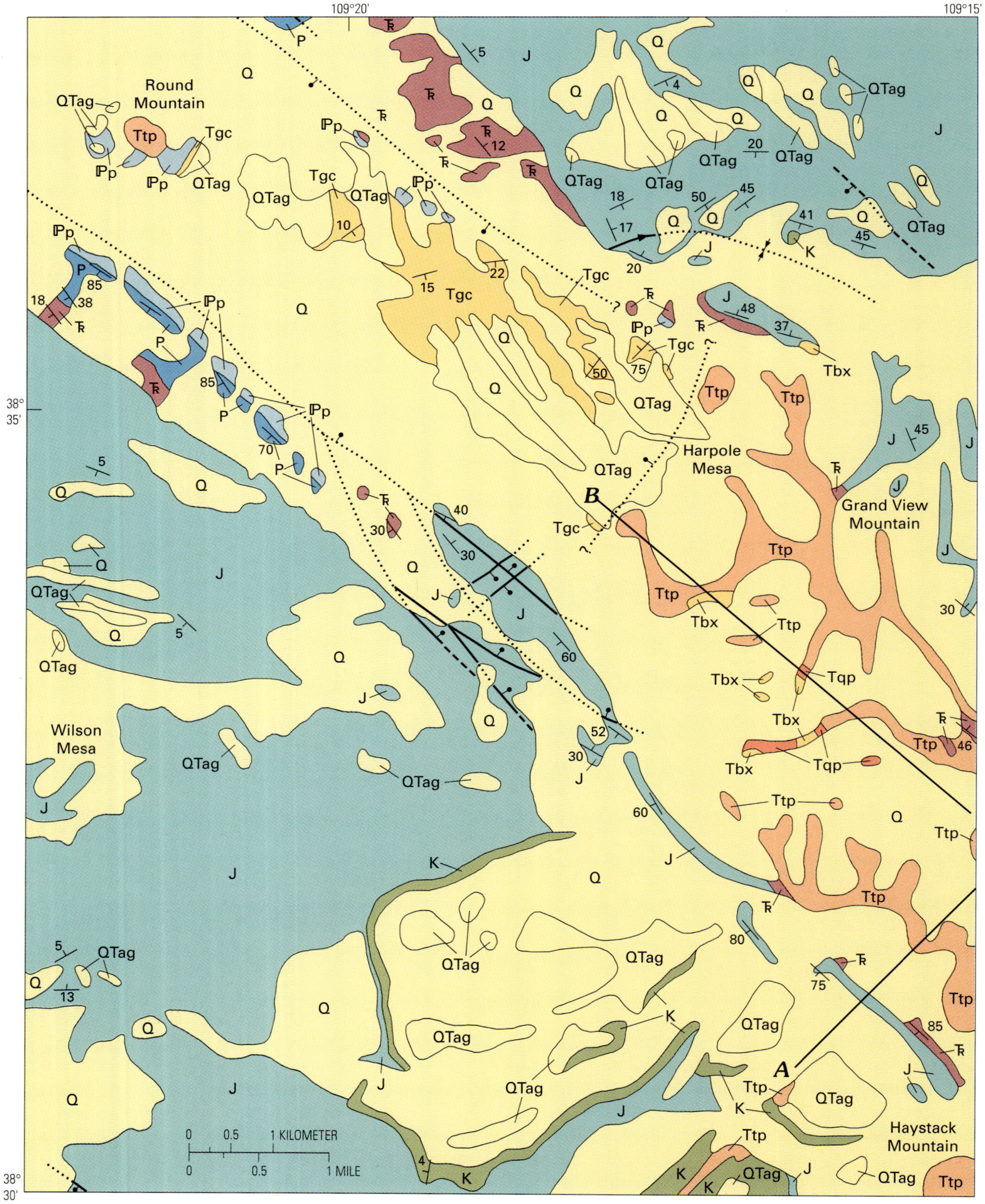

Figure 5 (above and following pages). Simplified geologic maps and cross sections of the northern La Sal Mountains, Utah. Maps above show the Warner Lake (left) and Mt. Waas (right) 7.5-minute quadrangles. See following pages for cross sections and explanations.

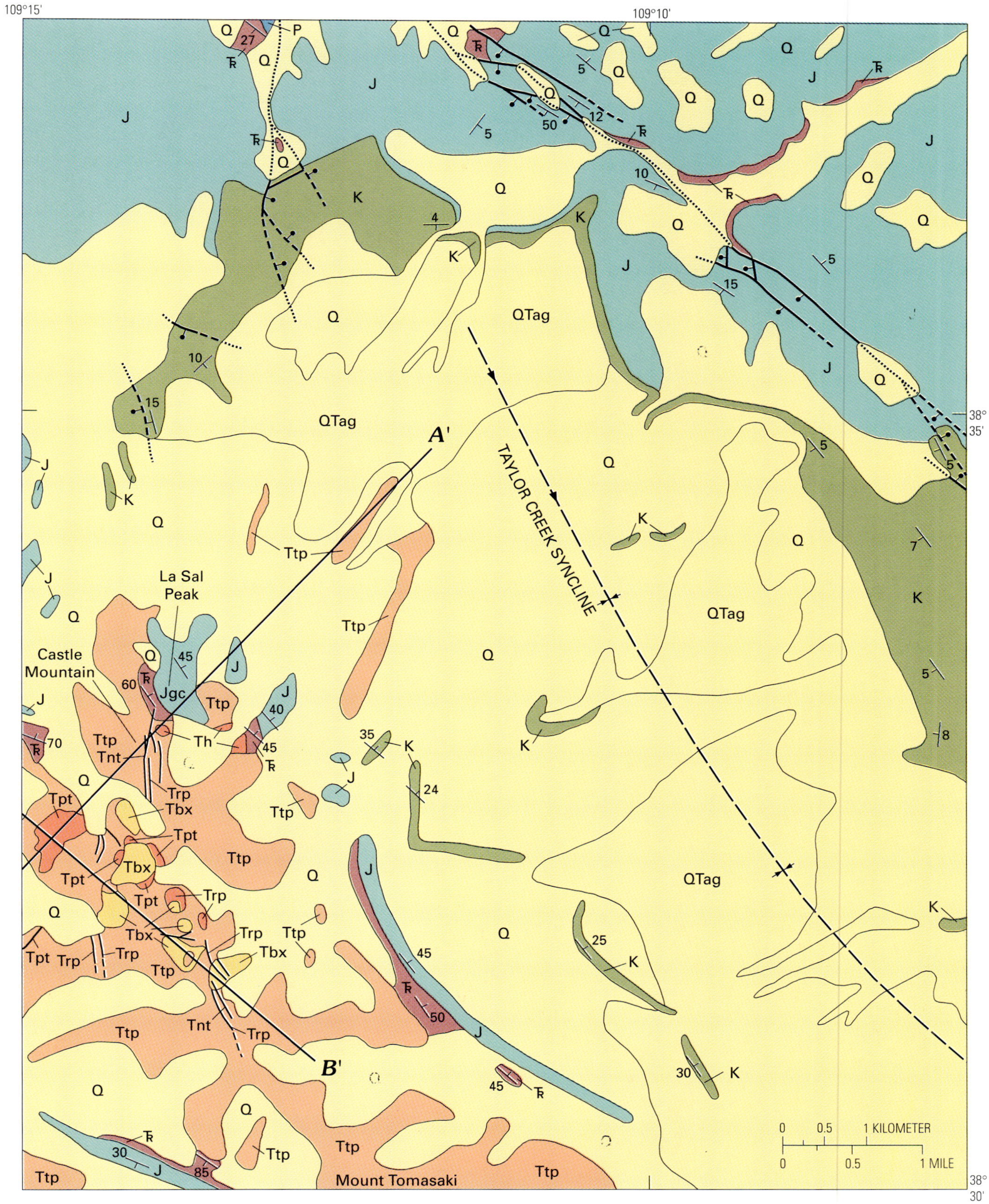
109°15'
109°10'
38° 35'
38° 30'
TAYLOR CREEK SYNCLINE
La Sal Peak
Castle Mountain
Mount Tomasaki
A'
B'
Q
QTag
Ttp
Tpt
Trp
Tbx
Tnt
Th
Jgc
J
K
P
Ꞧ
0 0.5 1 KILOMETER
0 0.5 1 MILE

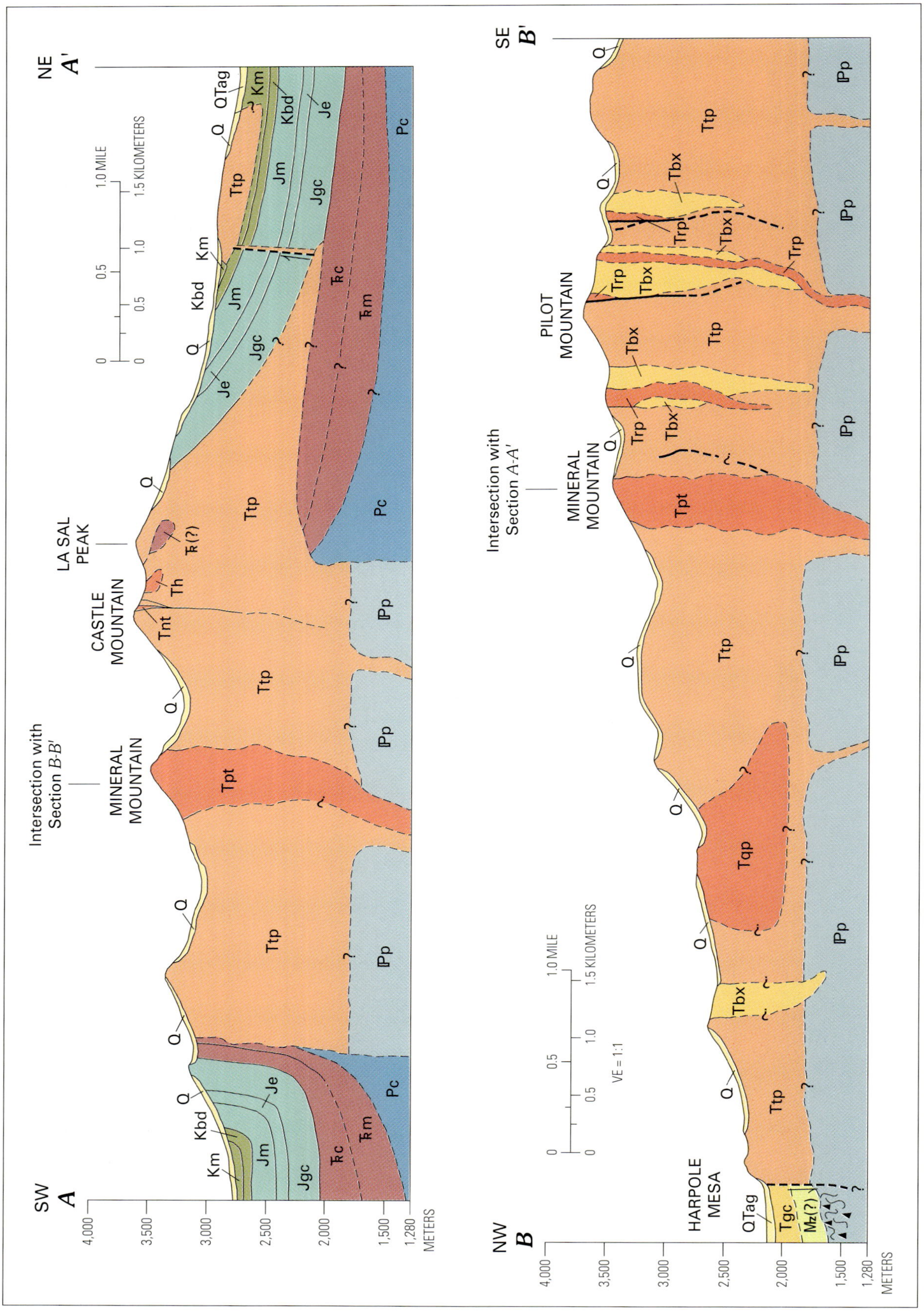

Figure 5 (above and facing page—Continued). Interpretive cross sections through the Warner Lake and Mt. Waas quadrangles. See maps on the preceding pages for lines of section. No vertical exaggeration.

EXPLANATION FOR MAPS AND CROSS SECTIONS

Q	**Quaternary deposits**	
QTag	**Pleistocene and Pliocene alluvial fan gravels**	
	Tertiary rocks	
Tgc	**Geyser Creek Fanglomerate (Pliocene?)**	
Th	**Hornfels**	youngest
Tnt	**Nosean trachyte**	
Trp	**Peralkaline rhyolite**	
Tbx	**Breccia**	(Oligocene)
Tpt	**Peralkaline trachyte**	
Tqp	**Quartz plagioclase trachyte**	
Ttp	**Hornblende plagioclase trachyte**	oldest
Mz	**Mesozoic rocks, undifferentiated**	
K	**Cretaceous strata**	
Km	**Mancos Shale (Upper Cretaceous)**	
Kbd	**Dakota Sandstone (Upper Cretaceous) and Burro Canyon Formation (Lower Cretaceous)**	
J	**Jurassic strata**	
Jm	**Morrison Formation (Upper Jurassic)**	
Je	**Entrada Sandstone (Middle Jurassic)**	
Jgc	**Glen Canyon Group (Lower Jurassic; includes Navajo Sandstone, Kayenta Formation, Wingate Sandstone)**	
TR	**Triassic strata**	
TRc	**Chinle Formation (Upper Triassic)**	
TRm	**Moenkopi Formation (Middle and Lower Triassic)**	
P	**Cutler Formation (Lower Permian)**	
IPp	**Paradox Formation (Middle Pennsylvanian)**	

Contact

Normal fault—Dashed where approximate; dotted where covered; bar and ball on downthrown side

23 **Strike and dip**

Anticline axial trace—Dashed where approximate or covered

Syncline axial trace—Dashed where approximate or covered

mountain peaks of igneous rocks to define the axial trends of "laccoliths that radiate from the central stock." However, in remapping the area I have found that (1) the hydrothermal alteration and fractures are peripheral to and related to the emplacement of later peralkaline trachyte and rhyolite intrusions and breccia pipes; (2) the metamorphosed xenolithic blocks are small isolated roof pendants that have no systematic distribution; and (3) the conical eroded peaks in the core of the northern mountains are not the tops of individual laccoliths.

On the southwest flank of the northern mountains dome, the sedimentary rocks are abruptly folded from a dip of 5° to dips of 60°–90° southwest, forming a northwest-trending monocline (fig. 5, section *A–A′*). Triassic strata are in near-vertical contact with the hornblende plagioclase trachyte pluton along the entire flank. A thin contact-metamorphic aureole of hornfels indicates minimal baking of the country rock along the contact. At several locations, Triassic strata adjacent to the pluton form thin breccia zones. This clast-supported breccia has well-indurated clasts in a matrix of calcite and crushed rock. Many of the frost-heaved Lower Jurassic Glen Canyon Group sandstone blocks that cover the large flatiron ridge of the monocline have slickenside surfaces and cataclastic shear bands. The breccia zones, slickensides, and shear bands suggest near-bedding-plane faulting and stretching of the sedimentary rocks as they were arched across the main igneous pluton. Similar features have been described on the flanks of the laccolithic domes in the southern Henry Mountains (Johnson and Pollard, 1973; Jackson and Pollard, 1988a).

Along most of the northeast flank of the northern mountains, Triassic and Jurassic strata dip about 45°–50° and 45°–60° northeast, respectively (fig. 5). The pluton–country rock contact along this flank appears to be nearly vertical or to dip slightly northeast. At La Sal Peak, the structure is complex because the hornblende plagioclase trachyte intrusion breached the flank of the anticline and was injected as much as 1.7 km into the flanking rocks (fig. 5, section *A–A′*). Exposures in this area (fig. 6) show that hornblende plagioclase trachyte cuts discordantly across the Triassic and Glen Canyon Group strata. Small dikes and sills penetrate several meters into the country rock beyond the main contact. Triassic strata are unmetamorphosed to variably baked, forming an irregular hornfels zone. Some areas of well-developed hornfels may be partially stoped blocks.

The elevation of the floor of the hornblende plagioclase trachyte pluton in the northern mountains is problematic. Case and others (1963) suggested, on the basis of gravity and magnetic data, that the northwest part of the main pluton (Grand View Mountain area) and the southeast part (Mount Tomasaki area) both extend outward over thickened masses of salt (fig. 5). By modeling the geophysical data, they suggested that the floor of the pluton is approximately 2,700 m above sea level at the northwest and southeast ends and drops to about 2,100 m near the center. Hunt (1958) estimated that the maximum thickness of the laccoliths in the northern mountains is about 600–900 m, indicating that the floor of the main pluton is at an elevation of 2,400–2,700 m. However, outcrops of hornblende plagioclase trachyte are present at elevations as

Figure 6. La Sal Peak (left) and Castle Mountain (right) in the northern La Sal Mountains, Utah. View is to the east from Horse Mountain. Castle Mountain and the flanks of La Sal Peak are composed of hornblende plagioclase trachyte (Ttp) and several dikes of nosean trachyte and peralkaline rhyolite (not labeled). La Sal Peak is capped by Mesozoic strata. Dashed lines mark approximate locations of contacts between the Triassic (₸) strata (Chinle and Moenkopi Formations), the Jurassic (Jgc) strata (Glen Canyon Group), and the intrusive rocks.

low as 2,100 m at the base of Grand View Mountain. The outcrop thickness of hornblende plagioclase trachyte at Grand View Mountain is about 1,000 m. At the southeast end of the main pluton, the lowest hornblende plagioclase trachyte outcrops are at about 2,700 m, but colluvium and glacial deposits cover bedrock below this elevation.

The best inference for the elevation of the pluton's floor may be one that derives from the observation that the underlying salt diapir does not appear to have dissolved. If it had, it would have caused collapse of the extended parts of the pluton. In the adjacent Castle Valley, late Cenozoic dissolution of the upper part of the salt diapir has lowered the upper surface of the diapir to elevations between 1,200 and 1,900 m for most of the valley (based on drilling records from water wells and on well cuttings and geophysical logs from two petroleum wells: Gold Bar Resources, Castle Valley #1 and Grand River Oil & Gas, Sid Pace #1, sec. 16, T. 25 S., R. 23 E.). An elevation of 2,400–2,700 m for the floor of the main pluton would be significantly higher (>500 m) than the level of collapse and the upper surface of the salt diapir in adjacent Castle Valley. If the pluton's floor were that much higher, then the salt dissolution in Castle Valley would have undermined the northwestern part of the pluton, and the igneous rocks there should show collapse structures, such as faults, folds, and open-fracture zones, like those in adjacent sedimentary strata that were subject to salt-dissolution-induced collapse. In fact, collapse appears to wrap around the intrusion at Grand View Mountain (fig. 5). These data suggest that the base of the pluton is at an elevation below the current level of salt dissolution. Thus, I estimate that the floor of the hornblende plagioclase trachyte pluton in the northern mountains is at an elevation of about 1,800 m (fig. 5, section *B–B'*), which is slightly lower than the lowest igneous outcrops at Grand View Mountain and the upper surface of the diapir in the upper part of Castle Valley. Assuming the pluton's contact with the underlying salt diapir is generally subhorizontal and yet locally complex, then its minimum thickness is about 1,800 m at the center of the northern mountains. At Grand View Mountain the pluton is estimated to be 1,400 m thick.

If this interpretation of the main hornblende plagioclase trachyte pluton is correct, then in cross section the pluton may resemble a cluster of coalesced mushrooms (laccoliths) with thin vertical feeder stalks rooted through the salt diapir (fig. 5, sections). As the laccoliths grew, the intervening evaporite rocks were pushed aside until the intrusions coalesced and the margins of the individual laccoliths became indistinguishable. A concentrated network of stalks would be located at the center of the northern mountains, where additional intrusions were emplaced.

Hornblende plagioclase trachyte samples from the northern, middle, and southern La Sal Mountains have similar weight-percent concentrations of Na_2O, K_2O, total alkalies, CaO, and Rb. These data suggest that during emplacement into the salt diapir and coalescence of the intrusions, the magma assimilated little or no material from the evaporite rocks.

Several satellite hornblende plagioclase trachyte laccoliths and sills are present around the northern La Sal Mountains dome. The satellite intrusions may connect at depth to the feeder system of the main pluton or may have their own feeder systems. Field observations and geophysical data are inconclusive about their relations to the main pluton. Round Mountain consists of an intrusion of hornblende plagioclase trachyte in contact with the contorted cap rock of the salt diapir in Castle Valley (fig. 5). Several small roof pendants of cap rock are preserved on its top. The base of the intrusion is not exposed and the marginal contacts are nearly vertical. Chill-zone rock at its margin locally shows randomly oriented slickensides, suggesting that the intrusion is fault bounded. Hunt (1958) and Corry (1988) interpreted the faults as having formed during emplacement. I believe the faulting postdates emplacement and occurred during salt dissolution of the diapir and collapse of cap rock and country rock around the rooted Round Mountain intrusion (fig. 3).

Northeast of the northern mountains dome are two small, poorly exposed laccoliths(?) of hornblende plagioclase trachyte that are elongated to the northeast (fig. 5, section *A–A'*). A thin sill probably connects them. These laccoliths(?) were emplaced in gently dipping Cretaceous strata (fig. 5).

To the southwest, the Haystack Mountain laccolith was emplaced in Upper Jurassic and Cretaceous strata at the lower hinge of the monocline that marks the steep southwest flank of the northern mountains dome (fig. 5). At the base of the mountain along its south side are scattered outcrops of strata that dip steeply away from the laccolith (Carter and Gualtieri, 1958; Weir and Puffet, 1960), suggesting that the floor of the laccolith is not exposed. Two thin sills are also present in the Upper Cretaceous strata in this area.

The later trachytes, rhyolites, and breccia pipes in the northern La Sal Mountains either intrude or are in gradational(?) contact with the hornblende plagioclase trachyte (fig. 5). Quartz plagioclase trachyte forms a bulbous or sill-like mass in gradational(?) contact with hornblende plagioclase trachyte. Peralkaline trachyte and peralkaline rhyolite intrude hornblende plagioclase trachyte as plugs and dikes and lie at the center of a synmagmatic hydrothermal alteration system that also includes calcite-quartz breccia pipes (Ross, 1992). The breccias found in the intrusive pipes range from crackle breccia to matrix-supported breccia. The matrix is predominantly calcite and includes subordinate amounts of quartz, crushed rock, and opaque grains. Quartz veins, stockworks, and pods are locally present. Hematite pseudomorphs after various sulfides are common. At the northwest end of the northern mountains, other discrete areas of similar breccias are present near the margins of the main pluton and adjacent to the quartz plagioclase trachyte body. In these breccias clast types are variable: some contain only fragments of either Triassic rock and (or) Glen Canyon Group sandstone, some have only fragments of hornblende plagioclase trachyte or quartz plagioclase trachyte, and at least one contains a mixture of both sedimentary and igneous rock fragments. The breccia formed at the sedimentary-igneous contact because of either forceful emplacement of magma or the release of volatile-rich fluids. The hydrothermal alteration mineral assemblages that formed during emplacement of the later intrusions and breccia pipes range from mainly argillic to propylitic to locally phyllic.

Several late-stage peralkaline rhyolite and nosean trachyte dikes, trending generally northward (N. 15° W. to N. 10° E.), cut most of the intrusive phases (fig. 5). Previous studies (Price and Henry, 1984; Best, 1988) have used the orientation of dikes in slightly older host rocks to determine the paleostress orientations at the time of dike emplacement. Following the assumptions used in these studies, the orientation of the late-stage rhyolite and nosean trachyte dikes suggests an overall east-west direction of horizontal least principal stress during their emplacement. The overall north-south alignment of the La Sal Mountains intrusive centers also supports an east-west direction of horizontal least principal stress during their emplacement in the late Oligocene. Best (1988) suggested that a northerly least-principal-stress orientation for the Western United States rotated to an east-northeast direction during the latest Oligocene to early Miocene. Even though the La Sal Mountains data set is small and only qualitative, the observations are worth noting.

Determining the depth of emplacement for the intrusions in the northern mountains is problematic due to episodic growth on the Castle Valley–Paradox Valley salt diapirs. (See Shoemaker and others, 1958; Cater, 1970; Doelling and Ross, 1993.) Surface geologic mapping and subsurface petroleum well data indicate that only the Middle to Upper Pennsylvanian, Permian, and Triassic strata show significant variations in stratigraphic thickness in the area around the northern La Sal Mountains, Castle Valley, and Fisher Valley (Shoemaker and others, 1958; Goydas, 1990; Doelling and Ross, 1993; H.H. Doelling, unpublished data; M.L. Ross, unpublished data). Jurassic and Cretaceous strata in the same area have relatively uniform stratigraphic thickness, with some local variation. Construction of preliminary restored cross sections across the salt diapir at Castle Valley and the northern La Sal Mountains (prior to emplacement of the intrusions) suggests that Upper Pennsylvanian and Permian strata were not preserved across the crest of the diapir. A relatively thin section of Triassic strata rested on the cap rock. Supporting this hypothesis is the fact that Triassic rocks are the oldest strata continuously exposed along the margins of the northern mountains dome.

Taking into account the effects of salt diapir movement, the reconstructed stratigraphic section of Lower Triassic (base of the Moenkopi Formation) to Upper Cretaceous (top of the Mancos Shale) strata (compiled from Hintze, 1988; Doelling and Ross, 1993; M.L. Ross, unpublished data; and Willis, 1991) is approximately 1.9 km thick near the La Sal Mountains. When the northern La Sal Mountains intrusions were emplaced, this section was probably overlain by a sequence comprising the Upper Cretaceous Mesaverde Group through the Eocene Green River Formation. This sequence would have added about 0.76 km to the section, based on its present thickness in the Book Cliffs, about 70 km north of the area (compiled from Willis, 1991; Willis, unpublished data; and Franczyk and others, 1992), and on thickness modifications suggested by G.C. Willis (oral commun., February 1993). According to these estimates, approximately 2.7 km of sedimentary rocks covered the crest of the salt diapir at the time the main intrusions and the Round Mountain intrusion were emplaced. The other satellite intrusions have an estimated depth of emplacement of about 1.9 km. These estimates assume a minimal amount of erosion from post-Green River Formation time to the late Oligocene.

SOUTHERN LA SAL MOUNTAINS INTRUSIVE CENTER

The southern La Sal Mountains intrusive center is structurally similar to that of the northern mountains in that it was emplaced into a salt-cored anticline (fig. 2). However, the intrusive dome is smaller (8.0 × 4.4 km), and hornblende plagioclase trachyte is the only igneous rock exposed.

Salt-dissolution-induced collapse of the Moab–Spanish Valley salt-cored anticline formed the northwest-trending Pack Creek syncline and many high-angle normal faults that terminate near the margin of the pluton. Wells drilled in the Pine Ridge salt-cored anticline southeast of the southern intrusive center (fig. 2) indicate that strata as young as the Chinle Formation locally rest on Paradox Formation cap rock (Hite and Lohman, 1973). The southwest margin of the dome is a steep (40°–85°) southwest-dipping monocline in Permian to Lower Cretaceous strata. The hornblende plagioclase trachyte pluton is in apparent discordant contact with the Lower Permian Cutler Formation along this side. (See Weir and Puffet, 1960, and Weir and others, 1960, for details.) Carter and Gualtieri (1958) and Hunt (1958) interpreted the intrusion to be in concordant contact with the Cutler along part of its northeast flank, forming a 45°–60° northeast-dipping monocline. Along the northeast flank the sedimentary rocks have been breached by several small sill-like intrusions that cut discordantly across the Upper Jurassic and Cretaceous strata (Hunt, 1958; Weir and Puffet, 1960). I estimate that the structural relief on the base of the Lower Jurassic Glen Canyon Group across the short axis of the southern mountains dome is 1,500–1,700 m. Taking into consideration the effects of salt diapir growth on the post-Paradox strata, the estimated depth of emplacement for the southern La Sal Mountains intrusions is between 2.60 and 2.74 km.

MIDDLE LA SAL MOUNTAINS INTRUSIVE CENTER

The middle La Sal Mountains consist of three prominent mountains: Mount Mellenthin, Mount Peale, and Mount Tukuhnikivatz, each underlain by hornblende plagioclase trachyte intrusions (fig. 2). Smaller satellite intrusions of hornblende plagioclase trachyte are adjacent to the three larger ones. Mount Mellenthin is capped by a laccolith emplaced in the Mancos Shale. The floor of the laccolith is about 60 m above the underlying Burro Canyon Formation and the Dakota Sandstone. Several irregular feeder dikes to the laccolith, striking N. 20°–25° E., are exposed in a glacial valley on the southwest side of the mountain (Gould, 1926; Hunt, 1958; Corry, 1988). Northeast of the mountain a smaller discordant intrusion was emplaced in the Upper Jurassic Morrison Formation (Carter and Gualtieri, 1958; Hunt, 1958).

Mount Peale consists of a large laccolith, which has faulted marginal contacts that are discordant with the country rock (Carter and Gualtieri, 1958). The level of emplacement for the Mount Peale laccolith is uncertain. Hunt (1958) showed the laccolith as being emplaced into the Morrison Formation; however, Carter and Gualtieri (1958) mapped the floor of the laccolith as being at the Kayenta Formation–Navajo Sandstone contact. They also mapped the Navajo Sandstone, Entrada Sandstone, and Morrison Formation in a large roof pendant.

The intrusions at Mount Tukuhnikivatz appear to have a complex morphology. The main intrusion, which forms the highest peak, cuts discordantly across Jurassic strata along its margins, and its floor is not exposed (Hunt, 1958; Weir and Puffet, 1960). Because of the lack of information on the base of this intrusion, its form has been interpreted as a plug (Gould, 1926), a bysmalith (Hunt, 1958), or a "punched laccolith" (Corry, 1988). On the northwest flank of the mountain are two intrusions previously described as laccoliths. Hunt (1958) and Weir and Puffet (1960) show the intrusions as having floors (basal contacts) that cut discordantly across northwest-dipping strata and extending laterally from the main Mount Tukuhnikivatz intrusion. (See cross section *B–B′* in Weir and Puffet, 1960.) As mentioned previously, recent work by Jackson and Pollard (1988a) in the Henry Mountains and by Corry (1988) indicate that the intrusive relations illustrated and discussed by Hunt (1958) and Weir and Puffet (1960) for the Mount Tukuhnikivatz intrusion(s) are unlikely and difficult to reconcile with current mechanical models. The level of emplacement of the Mount Tukuhnikivatz intrusion(s) is also uncertain. Hunt (1958) showed the intrusion(s) as being emplaced into the Morrison Formation. Weir and Puffet (1960) showed the intrusion(s) as being emplaced at different horizons within Jurassic strata.

An intrusion of nosean trachyte in the Morrison Formation is located in the northwest part of the middle mountains. The contact between the intrusion and the Morrison Formation is poorly exposed. The intrusion appears to be sill-like with a northwest dip subparallel to the dip of the Morrison Formation exposures nearby.

Exxon Corporation drilled the Gold Basin Unit #1 well (sec. 15, T. 27 S., R. 24 E.) near the center of the middle mountains (fig. 7). Geophysical and drilling logs indicate multiple levels of intrusions in the subsurface down to about 1,100 m below sea level. The lowest intrusions appear to be near the base of the Paradox Formation. The maximum thickness of an individual igneous rock horizon in the well is about 300 m. The intrusions in the middle mountains appear to have been emplaced at multiple stratigraphic horizons from the Paradox Formation to the Mancos Shale, resulting in a Christmas tree-like appearance in cross section. Using the data from the Exxon well and the structural relationships of Weir and Puffet (1960) the estimated overall range for depth of emplacement of the intrusion of the middle La Sal Mountains is 1.9–6.0 km.

STRUCTURAL CONTROL FOR THE LA SAL MOUNTAINS INTRUSIVE CENTERS

The northern and southern La Sal Mountains intrusive centers were emplaced into preexisting salt-cored anticlines. The salt-cored anticlines influenced the form of the main hornblende plagioclase trachyte plutons. Northwest-trending faults that offset the Paradox Formation and older rocks controlled the localization, linear form, and northwest-trending parallel belts of the salt diapirs that core the folds of the northern Paradox Basin. Northwest-trending high-angle faults, which displace Pennsylvanian and older rocks, have been identified in the subsurface beneath several of the salt-cored anticlines (Shoemaker and others, 1958; Case and others, 1963; Cater, 1970; Parker, 1981; Woodward-Clyde Consultants, 1983). Baars (1966) interpreted west-northwest-trending fault blocks of Proterozoic and Paleozoic rocks in the San Juan Mountains as southeastward continuations of the northwest-trending subsurface faults of the Paradox Basin. The west-northwest-trending basement faults in the San Juan Mountains show evidence for episodic movement from the Late Proterozoic to the Late Permian. The subsurface faults of the northern Paradox Basin are believed to have a similar history of repeated movements (Baars, 1966; Stevenson and Baars, 1987). Many northwest-striking and several east-west-striking faults cutting Proterozoic and Phanerozoic rocks have been mapped in the Uncompahgre Plateau and the eastern Paradox Basin region (Tweto, 1987; Case, 1991). The predominance of northwest-striking faults indicates a pervasive northwest-trending structural fabric for the region.

Geophysical data (Joesting and Byerly, 1958; Joesting and Case, 1960) and well data (Case and others, 1963)

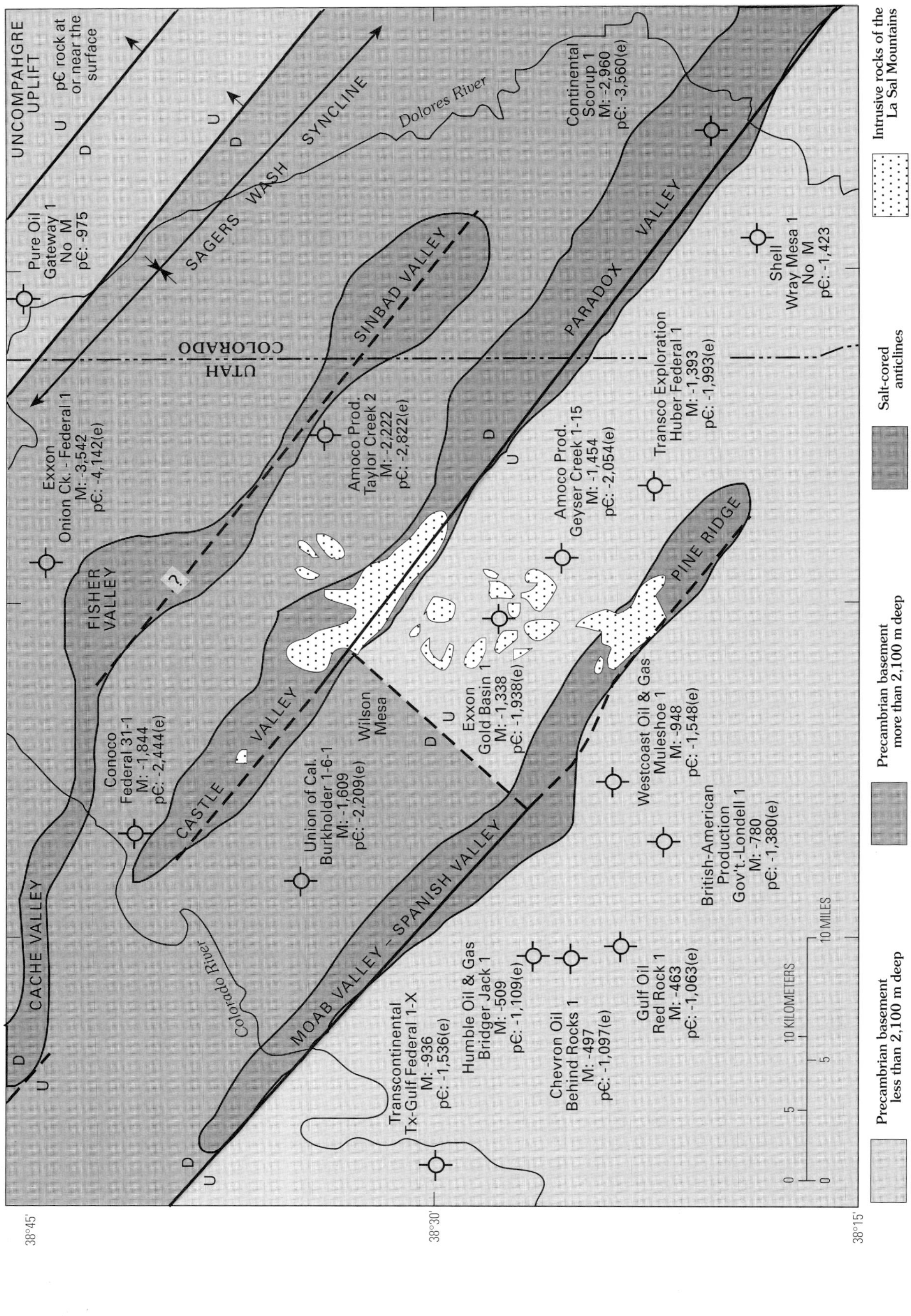

Figure 7. Structural and magmatic features in the La Sal Mountains region and the approximate locations of interpreted subsurface faults. Dashed lines are inferred faults. Subsurface faults on the southwest flank of the Uncompahgre Uplift are high-angle reverse faults. The locations of selected petroleum wells that penetrate pre-Paradox Formation rocks are shown. Subsurface elevations are given, in meters, for tops of the Mississippian (M) and Precambrian (p€) rocks. For wells not reaching the Precambrian, an estimated (e) thickness of 600 m (Hintze, 1988) was used for the Precambrian to Mississippian stratigraphic interval. Modified from Case and others (1963).

indicate that one of the Paradox Basin subsurface faults is along the southwest flank of the Paradox Valley salt-cored anticline (fig. 7). Steep northeastward gradients in both gravity and magnetic data (Paradox Valley regional gradient of Case and others, 1963) coincide with a vertical offset of about 1,800–2,100 m in the Proterozoic and Paleozoic rocks (Elston and others, 1962; Case and others, 1963). The zone of steep geophysical gradients extends northwestward to a point just beyond the northern La Sal Mountains. Figure 7 shows the estimated location and trend of the subsurface fault. The fault may continue northwestward along the southwest flank of Castle Valley, where the estimated vertical offset decreases to about 300 m (Doelling and Ross, 1993). Case and Joesting (1972) suggested that basement faults having as much as 300 m of displacement could exist in the salt-cored anticline region without causing an obvious gravity or magnetic anomaly.

Another pair of coincident northwest-trending gravity and magnetic gradients occurs subparallel to the southwest flank of the Moab–Spanish Valley salt-cored anticline (fig. 7) (Case and Joesting, 1972). Termed the Spanish Valley regional gradient, it trends west-northwest from Spanish Valley to the area between the Colorado and Green Rivers. Elevations of the tops of the Mississippian strata in the widely separated petroleum wells around Spanish Valley indicate an elevation difference of at least 1,100 m on pre-Paradox Formation rocks across the steep geophysical gradients, which most likely results from a subsurface fault or faults along the flank of the salt-cored anticline. However, the distance between wells is about 11.3 km, and a north-northeast dip of 6° on pre-Paradox Formation rocks could produce the same amount of elevation difference. If a fault does exist, geophysical data and well information suggest that displacement on the fault decreases gradually to the southeast of Spanish Valley.

At the southeast end of Spanish Valley the geophysical gradients change direction and trend northeast for 16–24 km along the west-northwest flank of the La Sal Mountains, transverse to the northwesterly regional gravity and magnetic trends. This segment of steep northeast-trending gradients (termed the Wilson Mesa gradient of Case and others, 1963) connects the Spanish Valley and Paradox Valley regional gradients. Both gravity and magnetic values decrease to the northwest along the Wilson Mesa gradient, and information from wells drilled since the geophysical study indicate a possible offset, down to the northwest, of about 300 m in the pre-Paradox Formation rocks across the Wilson Mesa gradient (fig. 7). The presence of the geophysical gradients and the apparent offset in pre-Paradox rocks suggest a buried fault, but the data are inconclusive. The distance between wells is large enough (22.5 km) that a northwest dip of about 2° in pre-Paradox Formation strata could also produce the same amount of elevation difference.

Case and others (1963) interpreted the Wilson Mesa gradient to be a major structural and lithologic discontinuity in the Proterozoic basement rocks. Other workers (Hite and Lohman, 1973; Hite, 1975) interpreted the coincidence of the Wilson Mesa gradient and the apparent left-lateral offset of both the Moab–Spanish Valley–Pine Ridge salt-cored anticline system and the Fisher Valley–Sinbad Valley salt-cored anticline system (fig. 2) as indications of a northeast-striking basement fault with left-lateral displacement. However, the Castle Valley–Paradox Valley salt-cored anticline system crosses the inferred basement fault with no offset, suggesting that the alignment of features is coincidental. Permian and younger strata in the area do not show any evidence for lateral offset. No large northeast-striking faults have been mapped in the Proterozoic basement or younger rocks on the Uncompahgre Plateau (Heyman, 1983; Heyman and others, 1986; Tweto, 1987; Case, 1991).

In summary, the coincidence of the geophysical gradients with displacement of the Paradox Formation and older rocks strongly supports the existence of a subsurface fault beneath Paradox Valley that extends to the northern La Sal Mountains (fig. 7). In the Moab–Spanish Valley area, the coincidence of steep geophysical gradients, a salt-cored anticline, and a large elevation difference for pre-Paradox rocks also suggests that a subsurface fault exists. Evidence for the existence of a subsurface fault coincident with the Wilson Mesa gradient is less conclusive. However, a fault or flexure in the pre-Paradox rocks along the Wilson Mesa geophysical gradient cannot be ruled out. Case and others (1963) suggested that (1) the subsurface faults are high-angle; (2) they separate a southern region of shallower, uniformly magnetized Proterozoic basement from a downthrown northern region of deeper, more heterogeneous Proterozoic basement; and (3) the geophysical gradients are manifestations of these faults. The sparse well data do support the correlation with geophysical gradients and the separation of a southern region of shallower Proterozoic basement from a northern region of deeper basement (fig. 7).

The locations of the La Sal Mountains intrusive centers along the trend of subsurface faults and possibly at the intersections of these faults suggest that the faults were avenues of weakness for the ascent of magma in the upper crust. This is especially true for the northern and southern centers.

SURFACE FAULTS IN THE LA SAL MOUNTAINS REGION

Most of the exposed faults around the La Sal Mountains strike northwesterly and formed by salt-dissolution-induced collapse after emplacement of the intrusions (fig. 2). The collapse-related faults are closely spaced, discontinuous to splaying, and generally short, and they form swarms or networks on or near the salt-cored anticlines. The fault networks form narrow, fault-block slivers that trend parallel to the length of the collapse valley. Most fault blocks are successively downdropped toward the valley. At several locations, seismic data show that the faults die out in the salt diapirs and do not displace pre-Paradox Formation strata. Previous studies (Colman, 1983; Harden and others, 1985; Oviatt, 1988; Goydas, 1990) and recent field mapping and geochronology (Doelling and Ross, 1993; Ross, unpublished data) suggest that salt-dissolution-induced collapse and faulting are Pliocene to Quaternary age for the Castle Valley, Moab–Spanish Valley, Fisher Valley, and Cache Valley salt-cored anticlines (fig. 2) and also for the Salt Valley salt-cored anticline (just northwest of area of fig. 2). Most collapse-related faults, like those at Moab and Spanish Valley, strike northwesterly. However, some collapse-related faults strike northeasterly, such as the Cottonwood graben at Fisher Valley (fig. 2). The amount of displacement on the graben boundary faults increases toward the Fisher Valley collapsed salt diapir, indicating that collapse of the diapir produced the graben (Goydas, 1990).

Some faults in the La Sal Mountains region are not related to salt tectonic movements. The Ryan Park fault zone, along the southwest flank of the Uncompahgre Plateau (northeast corner of the area of fig. 2), is part of a network of faults that formed during the Laramide orogeny and the late Cenozoic uplift of the Uncompahgre Plateau (Shoemaker, 1956; Cater, 1970; Heyman, 1983). The location and orientation of the Ryan Park fault zone were controlled by the late Paleozoic Uncompahgre fault(s), which formed during uplift of the ancestral Uncompahgre Uplift (Heyman and others, 1986).

Two northwest-striking normal faults near the La Sal Mountains, which are at the surface, offset strata across the crests of two salt-cored anticlines and are the result of regional extension and not late Cenozoic salt-dissolution-induced collapse. The Moab fault (western edge of the area of fig. 2) and the Lisbon Valley Fault (southern edge of the area of fig. 2) are both long (32–39 km) and have large displacements (800–1,800 m) (Parker, 1981; Doelling, 1988). Both have been interpreted as high-angle northeast-dipping listric faults that die out in the underlying salt diapirs. This interpretation was based on surface and well data for the Lisbon Valley fault (Parker, 1991) and on surface and seismic data for the Moab fault (D.M. Rawlins, Exxon Corp., oral commun., 1993). Both faults approximately overlie subsurface faults in pre-Paradox Formation rocks but are separated from them by the salt diapirs. In addition, both faults offset Upper Cretaceous and older strata, suggesting they are Tertiary age. McKnight (1940) interpreted the Moab fault to be a tectonic fault that formed in response to post-Laramide Tertiary extension. The Moab and Lisbon Valley faults may have formed in response to movement on the subsurface faults and salt diapirs during late Tertiary epeirogenic uplift of the Colorado Plateau. However, Hite and Lohman (1973) and Woodward-Clyde Consultants (1983) interpreted the Moab and Lisbon Valley faults as salt-dissolution-induced collapse structures because they believed the faults did not extend below the salt diapirs (now known to be the case).

In summary, most of the numerous surface faults around the La Sal Mountains intrusions postdate emplacement of the intrusions and represent shallow structural deformation resulting from salt dissolution or salt flowage of the diapirs.

CONCLUSIONS

No field evidence supports the hypothesis of Hunt (1958) that the individual La Sal Mountains intrusive centers consist of discrete central stocks surrounded by outward radiating laccoliths. Evidence once believed to define the central stock in the northern mountains is now interpreted to be the result of later emplacement of trachyte and rhyolite porphyry intrusions and breccia pipes. The main plutons in both the northern and southern La Sal Mountains probably consist of coalescent laccoliths of hornblende plagioclase trachyte. Intrusions in the middle La Sal Mountains were emplaced between salt-cored anticlines; surface structures and subsurface data suggest these laccoliths and sills are vertically stacked at various stratigraphic horizons from the Paradox Formation to the Mancos Shale.

Intrusions of the La Sal Mountains were probably emplaced at depths ranging between 1.9 and 6.0 km. These depths of emplacement are consistent with emplacement depths for the Henry and Abajo Mountains laccoliths (Witkind, 1964; Jackson and Pollard, 1988a).

The elliptical shape of the northern and southern La Sal Mountains intrusive centers and their positioning along two salt-cored anticline systems indicates that magma emplacement was controlled, in part, by the preexisting salt diapir and anticline. Geophysical data (Case and others, 1963) suggest that the extremities of the main pluton in the northern mountains overlie a thickened mass of salt. Field mapping in the northern mountains supports the geophysical data, indicating that thickened salt is present around the Grand View Mountain intrusion. The intersections between a possible concealed northeast-striking fault and the northwest-striking faults that underlie and controlled the salt-cored anticlines of the Castle Valley–Paradox Valley system and the Moab Valley–Spanish Valley–Pine Ridge system appear to have been magma conduits for the La Sal Mountains intrusions. The northeast-striking fault (or fold ramp) may be a connecting structure between the two northwest-striking en echelon faults. The La Sal Mountains intrusive centers are located along the structural boundary between deepest part of the Paradox Basin and slightly shallower area to the south.

Most surface faults in the area around the La Sal Mountains postdate magma emplacement and formed during late Tertiary to Quaternary salt-dissolution-induced collapse of the crests of the salt-cored anticlines.

REFERENCES CITED

Allmendinger, R.W., Hauge, T.A., Hauser, E.C., Potter, C.J., Klemperer, S.L., Nelson, K.D., Knuepfer, P.L.K., and Oliver, J., 1987, Overview of the COCORP 40°N transect, Western United States—The fabric of an orogenic belt: Geological Society of America Bulletin, v. 98, p. 308–319.

Armstrong, R.L., 1969, K-Ar dating of laccolithic centers of the Colorado Plateau and vicinity: Geological Society of America Bulletin, v. 80, p. 2081–2086.

Baars, D.L., 1966, Pre-Pennsylvanian paleotectonics—Key to basin evolution and petroleum occurrences in Paradox Basin, Utah and Colorado: American Association of Petroleum Geologists Bulletin, v. 50, p. 2082–2111.

Baars, D.L., and Stevenson, G.M., 1981, Tectonic evolution of the Paradox Basin, Utah and Colorado, *in* Wiegand, D.L., ed., Geology of the Paradox Basin: Denver, Colo., Rocky Mountain Association of Geologists, Field Conference, 1981, Guidebook, p. 23–31.

Best, M.G., 1988, Early Miocene change in direction of least principal stress, Southwestern United States—Conflicting inferences from dikes and metamorphic core-detachment fault terranes: Tectonics, v. 7, p. 249–259.

Bowring, S.A., and Karlstrom, K.E., 1990, Growth, stabilization, and reactivation of Proterozoic lithosphere in the southwestern United States: Geology, v. 18, p. 1203–1206.

Carter, W.D., and Gualtieri, J.L., 1957, Preliminary geologic map of the Mount Peale 1 SW quadrangle, San Juan County, Utah: U.S. Geological Survey Mineral Investigations Field Studies Map MF–124, scale 1:24,000.

———1958, Preliminary geologic map of the Mount Peale 1 NW quadrangle, San Juan County, Utah: U.S. Geological Survey Mineral Investigations Field Studies Map MF–140, scale 1:24,000.

Case, J.E., 1991, Geologic map of the northwestern part of the Uncompahgre Uplift, Grand County, Utah, and Mesa County, Colorado, with emphasis on Proterozoic rocks: U.S. Geological Survey Miscellaneous Investigations Series, Map I–2088, scale 1:24,000.

Case, J.E., and Joesting, H.R., 1972, Regional geophysical investigations in the central Colorado Plateau: U.S. Geological Survey Professional Paper 736, 31 p.

Case, J.E., Joesting, H.R., and Byerly, P.E., 1963, Regional geophysical investigations in the La Sal Mountains area, Utah and Colorado: U.S. Geological Survey Professional Paper 316–F, p. 91–116.

Cater, F.W., 1970, Geology of the salt anticline region in southwestern Colorado: U.S. Geological Survey Professional Paper 637, 80 p.

Colman, S.M., 1983, Influence of the Onion Creek salt diapir on the late Cenozoic history of Fisher Valley, southeastern Utah: Geology, v. 11, p. 240–243.

Corry, C.E., 1988, Laccoliths—Mechanics of emplacement and growth: Geological Society of America Special Paper 220, 110 p.

Cunningham, C.G., Naeser, C.W., and Marvin, R.F., 1977, New ages for intrusive rocks in the Colorado mineral belt: U.S. Geological Survey Open-File Report 77–573, 7 p.

Doelling, H.H., 1988, Geology of Salt Valley Anticline and Arches National Park, Grand County, Utah: Utah Geological Survey Bulletin 122, p. 1–58.

Doelling, H.H., and Ross, M.L., 1993, Geologic map of the Big Bend quadrangle, Grand County, Utah: Utah Geological Survey Open-File Report 285, 108 p., scale 1:24,000.

Eckel, E.B., Williams, J.S., and Galbraith, F.W., 1949, Geology and ore deposits of the La Plata district, Colorado, with sections by G.M. Schwartz, D.J. Varnes, and E.N. Goddard: U.S. Geological Survey Professional Paper 219, 179 p.

Ekren, E.B., and Houser, F.N., 1965, Geology and petrology of the Ute Mountains area, Colorado: U.S. Geological Survey Professional Paper 481, 74 p.

Elston, D.P., Shoemaker, E.M., and Landis, E.R., 1962, Uncompahgre front and salt anticline region of Paradox Basin, Colorado and Utah: American Association of Petroleum Geologists Bulletin, v. 46, p. 1857–1878.

Frahme, C.W., and Vaughn, E.B., 1983, Paleozoic geology and seismic stratigraphy of the northern Uncompahgre front, Grand County, Utah, *in* Lowell, J.D., and Gries, Robbie, eds., Rocky Mountain foreland basins and uplifts: Rocky Mountain Association of Geologists, Field Conference, Steamboat Springs, Colo., 1983, Guidebook, p. 201–211.

Franczyk, K.J., Fouch, T.D., Johnson, R.C., Molenaar, C.M., and Cobban, W.A., 1992, Cretaceous and Tertiary paleogeographic reconstructions for the Uinta–Piceance Basin study area, Colorado and Utah: U.S. Geological Survey Bulletin 1787–Q, 37 p.

Gilbert, G.K., 1877, Report on the geology of the Henry Mountains: U.S. Geographical and Geological Survey of the Rocky Mountains Region (Powell), 160 p.

Gould, L.M., 1926, The geology of the La Sal Mountains of Utah: Michigan Academy of Science Papers, v. 7, p. 55–106.

Goydas, M.J., 1990, Provisional geologic map of the Fisher Valley quadrangle, Grand County, Utah: Utah Geological Survey Open-File Report 167, 48 p., scale 1:24,000.

Harden, D.R., Biggar, N.E., and Gillam, M.L., 1985, Quaternary deposits and soils in and around Spanish Valley, Utah, *in* Weide, D.L., ed., Soils and Quaternary geology of the southwestern United States: Geological Society of America Special Paper 203, p. 43–64.

Harrison, T.S., 1927, Colorado-Utah salt domes: American Association of Petroleum Geologists Bulletin, v. 11, p. 111–133.

Hedge, C.E., Houston, R.S., Tweto, O.L., Peterman, Z.E., Harrison, J.E., and Reid, R.R., 1986, The Precambrian of the Rocky Mountain region: U.S. Geological Survey Professional Paper 1241–D, 17 p.

Heyman, O.G., 1983, Distribution and structural geometry of faults and folds along the northwestern Uncompahgre Uplift, western Colorado and eastern Utah, *in* Averett, W.R., ed., Northern Paradox Basin—Uncompahgre Uplift: Grand Junction, Colo., Grand Junction Geological Society, p. 45–57.

Heyman, O.G., Huntoon, P.W., and White-Heyman, M.A., 1986, Laramide deformation on the Uncompahgre Plateau—Geometry and mechanisms, *in* Stone, D.S., and Johnson, K.S., eds., New interpretations of northwest Colorado geology: Denver, Colo., Rocky Mountain Association of Geologists, p. 65–76.

Hintze, L.F., 1988, Geologic history of Utah: Brigham Young University Geology Studies Special Publication 7, 202 p.

Hite, R.J., 1975, An unusual northeast-trending fracture zone and its relation to basement wrench faulting in the northern Paradox Basin, Utah and Colorado, *in* Fassett, J.E., ed., Canyonlands country: Four Corners Geological Society, 8th Field Conference, 1975, Guidebook, p. 217–223.

Hite, R.J., and Lohman, S.W., 1973, Geologic appraisal of Paradox Basin salt deposits for waste emplacement: U.S. Geological Survey Open-File Report 73–114, 75 p.

Hunt, C.B., 1956, Cenozoic geology of the Colorado Plateau: U.S. Geological Survey Professional Paper 279, 99 p.

———1958, Structural and igneous geology of the La Sal Mountains, Utah: U.S. Geological Survey Professional Paper 294–I, p. 305–364.

———1988, Discussion *of* The laccolith-stock controversy—New results from the southern Henry Mountains, Utah [Jackson and Pollard, 1988a]: Geological Society of America Bulletin, v. 100, p. 1657–1658.

Hunt, C.B., Averitt, Paul, and Miller, R.L., 1953, Geology and geography of the Henry Mountains region, Utah: U.S. Geological Survey Professional Paper 228, 234 p.

Jackson, M.D., and Pollard, D.D., 1988a, The laccolith-stock controversy—New results from the southern Henry Mountains, Utah: Geological Society of America Bulletin, v. 100, p. 117–139.

———1988b, Reply *to* Discussion, by C.B. Hunt, *of* The laccolith-stock controversy—New results from the southern Henry Mountains, Utah [Jackson and Pollard, 1988a]: Geological Society of America Bulletin, v. 100, p. 1658–1659.

Jackson, M.P.A., and Talbot, C.J., 1991, A glossary of salt tectonics: Texas Bureau of Economic Geology Geological Circular 91–4, 44 p.

Jamison, W.R., and Stearns, D.W., 1982, Tectonic deformation of Wingate Sandstone, Colorado National Monument: American Association of Petroleum Geologists Bulletin, v. 66, p. 2584–2608.

Joesting, H.R., and Byerly, P.E., 1958, Regional geophysical investigations of the Uravan area, Colorado: U.S. Geological Survey Professional Paper 316–A, p. 1–17.

Joesting, H.R., and Case, J.E., 1960, Salt anticlines and deep-seated structures in the Paradox Basin, Colorado and Utah, *in* Short papers in the geological sciences: U.S. Geological Survey Professional Paper 400–B, p. B252–B256.

Johnson, A.M., and Pollard, D.D., 1973, Mechanics of growth of some laccolithic intrusions in the Henry Mountains, Utah—I, Field observations, Gilbert's model, physical properties, and flow of magma: Tectonophysics, v. 18, p. 261–309.

Johnson, R.C., 1985, Early Cenozoic history of the Uinta and Piceance Creek Basins, Utah and Colorado, with special reference to the development of Eocene Lake Uinta, *in* Flores, R.M., and Kaplan, S.S., eds., Cenozoic paleogeography of the west-central United States, Rocky Mountain Paleogeography Symposium 3: Denver, Colo., Rocky Mountain Section, Society of Economic Paleontologists and Mineralogists, p. 247–276.

Johnson, R.C., and Finn, T.M., 1986, Cretaceous through Holocene history of the Douglas Creek Arch, Colorado and Utah, *in* Stone, D.S., and Johnson, K.S., eds., New interpretations of northwest Colorado geology: Denver, Colo., Rocky Mountain Association of Geologists, p. 77–95.

Kelley, V.C., 1955, Regional tectonics of the Colorado Plateau and relationship to origin and distribution of uranium: University of New Mexico Publications in Geology, no. 5, 120 p.

Le Bas, M.J., Le Maitre, R.W., Streckeisen, A., and Zanettin, B.A., 1986, Chemical classification of volcanic rocks based on the total alkali-silica diagram: Journal of Petrology, v. 27, p. 745–750.

Lucchitta, Ivo, 1979, Late Cenozoic uplift of the southwestern Colorado Plateau and adjacent Colorado River region: Tectonophysics, v. 61, p. 63–95.

McKnight, E.T., 1940, Geology of area between Green and Colorado Rivers, Grand and San Juan Counties, Utah: U.S. Geological Survey Bulletin 908, 147 p.

Nelson, S.T., Davidson, J.P., and Sullivan, K.R., 1992, New age determinations of central Colorado Plateau laccoliths, Utah—Recognizing disturbed K-Ar systematics and re-evaluating tectonomagmatic relationships: Geological Society of America Bulletin, v. 104, p. 1547–1560.

Nelson, S.T., Heizler, M.T., and Davidson, J.P., 1992, New $^{40}Ar/^{39}Ar$ ages of intrusive rocks from the Henry and La Sal Mountains, Utah: Utah Geological Survey Miscellaneous Publication 92–2, 24 p.

Oviatt, C.G., 1988, Evidence for Quaternary deformation in the Salt Valley anticline, southeastern Utah: Utah Geological Survey Bulletin 122, p. 61–76.

Parker, J.M., 1981, Lisbon Field area, San Juan County, Utah, *in* Wiegand, D.L., ed., Geology of the Paradox Basin: Rocky Mountain Association of Petroleum Geologists, p. 89–101.

Peale, A.C., 1877, Report [on the Grand River district, Colo.]: U.S. Geological and Geographical Survey of the Territories (Hayden), 9th Annual Report, p. 31–101.

———1878, Geological report on the Grand River district: U.S. Geological and Geographical Survey of the Territories (Hayden), 10th Annual Report, p. 163–182.

Pollard, D.D., and Johnson, A.M., 1973, Mechanics of growth of some laccolithic intrusions in the Henry Mountains, Utah—II, Bending and failure of overburden layers and sill formation: Tectonophysics, v. 18, p. 311–354.

Potter, C.J., Tang, Rex, and Hainsworth, T.J., 1991, Late Paleozoic structure of the southern part of the Uinta Basin, Utah, from seismic reflection data: U.S. Geological Survey Bulletin 1787–V, 10 p.

Price, J.G., and Henry, C.D., 1984, Stress orientations during Oligocene volcanism in Trans-Pecos Texas—Timing the transition from Laramide compression to Basin and Range tension: Geology, v. 12, p. 238–241.

Ross, M.L., 1992, Geology and petrology of the Oligocene North Mountain intrusive center, La Sal Mountains, Utah: Geological Society of America Abstracts with Programs, v. 24, no. 6, p. 59.

Shoemaker, E.M., 1956, Structural features of the central Colorado Plateau and their relation to uranium deposits, *in* Page, L.R., Stocking, H.E., and Smith, H.B., compilers, Contributions to the geology of uranium and thorium ***: U.S. Geological Survey Professional Paper 300, p. 155–170.

Shoemaker, E.M., Case, J.E., and Elston, D.P., 1958, Salt anticlines of the Paradox Basin, *in* Sanborn, A.F., ed., Guidebook to the geology of the Paradox Basin: Intermountain Association of Petroleum Geologists, 9th Annual Field Conference, 1958, p. 39–59.

Stevenson, G.M., and Baars, D.L, 1987, The Paradox—A pull-apart basin of Pennsylvanian age, *in* Campbell, J.A., ed., Geology of Cataract Canyon and vicinity: Four Corners Geological Society, 10th Field Conference, 1987, Guidebook, p. 31–50.

Stokes, W.L., 1948, Geology of the Utah-Colorado salt dome region, with emphasis on Gypsum Valley, Utah: Utah Geological Society Guidebook no. 3, 50 p.

Sullivan, K.R., Kowallis, B.J., and Mehnert, H.H., 1991, Isotopic ages of igneous intrusions in southeastern Utah—Evidence for a mid-Cenozoic Reno-San Juan magmatic zone: Brigham Young University Geology Studies, v. 37, p. 139–144.

Thompson, G.A., and Zoback, M.L., 1979, Regional geophysics of the Colorado Plateau: Tectonophysics, v. 61, p. 149–181.

Tweto, O.L., 1987, Rock units of the Precambrian basement in Colorado: U.S. Geological Survey Professional Paper 1321–A, 54 p.

Tyler, Noel, and Ethridge, F.G., 1983, Depositional setting of the Salt Wash Member of the Morrison Formation, Southwest Colorado: Journal of Sedimentary Petrology, v. 53, p. 67–82.

Warner, L.A., 1978, The Colorado lineament—A middle Precambrian wrench fault system: Geological Society of America Bulletin, v. 89, p. 161–171.

Weir, G.W., Dodson, C.L., and Puffet, W.P., 1960, Preliminary geologic map and section of the Mount Peale 2 SE quadrangle, San Juan County, Utah: U.S. Geological Survey Mineral Investigations Field Studies Map MF–143, scale 1:24,000.

Weir, G.W., and Puffet, W.P., 1960, Preliminary geologic map and sections of the Mount Peale 2 NE quadrangle, San Juan County, Utah: U.S. Geological Survey Mineral Investigations Field Studies Map MF–141, scale 1:24,000.

White, M.A., and Jacobson, M.I., 1983, Structures associated with the southwest margin of the ancestral Uncompahgre Uplift, *in* Averett, W.R., ed., Northern Paradox Basin–Uncompahgre Uplift: Grand Junction, Colo., Grand Junction Geological Society, p. 33–39.

Williams, P.L., 1964, Geology, structure, and uranium deposits of the Moab quadrangle, Colorado and Utah: U.S. Geological Survey Miscellaneous Investigations Map I–360, scale 1:250,000.

Willis, G.C., 1991, Geologic map of the Harley Dome quadrangle, Grand County, Utah: Utah Geological Survey Open-File Report 216, 70 p., scale 1:24,000.

Witkind, I.J., 1964, Geology of the Abajo Mountains area, San Juan County, Utah: U.S. Geological Survey Professional Paper 453, 110 p.

———1975, The Abajo Mountains—An example of the laccolithic groups on the Colorado Plateau, *in* Fassett, J.E., ed., Canyonlands country: Four Corners Geological Society, 8th Field Conference, 1975, Guidebook, p. 245–252.

Woodward-Clyde Consultants, 1983, Overview of the regional geology of the Paradox Basin study area: Columbus, Ohio, Battelle Memorial Institute, Office of Nuclear Waste Isolation, Report ONWI–92, 433 p. Available from National Technical Information Service, Springfield, VA 22161.

The Petrogenesis of the Colorado Plateau Laccoliths and Their Relationship to Regional Magmatism

By Stephen T. Nelson[1] *and* Jon P. Davidson[2]

CONTENTS

FIGURES

TABLE

ABSTRACT

The petrogenetic processes that formed the Henry Mountains, Utah, may be the same as those responsible for other laccolithic intrusions in the Colorado Plateau, specifically the La Sal and Abajo Mountains. Each range consists of small separate intrusive centers where magma was emplaced into Phanerozoic sediments at shallow crustal levels. Two major rock suites, plagioclase-hornblende porphyry (95 volume percent) and syenite porphyry (5 volume percent), exist in both the Henry and La Sal Mountains, whereas plagioclase-hornblende porphyry alone is found in the Abajo Mountains. Plagioclase-hornblende porphyry evolved from mantle-derived magma, which was ponded in the deep crust and assimilated amphibolite crust during open-system differentiation before being emplaced at shallow crustal levels. Plagioclase-hornblende porphyry also shows isotopic provinciality at each intrusive center, which may, in part, reflect the isotopic diversity of the basement rocks.

Geochemically, the laccoliths and contemporaneous volcanic rocks outside the plateau appear to have strong affinities to arc rocks, although their volume is much less than that of a typical volcanic arc. We contend that the laccoliths are part of an east-west-oriented magmatic belt, itself a portion of a larger mid-Tertiary magmatic system in western North America, and that the minor volume of the laccoliths reflects the inability of large volumes of magma to penetrate the thick, strong, stable crust of the Colorado Plateau.

INTRODUCTION

The laccolithic intrusions of the Henry, La Sal, and Abajo Mountains of southwest Utah (fig. 1) represent much of the igneous activity of the Colorado Plateau interior during mid-Tertiary time. Understanding the origin of the laccoliths will help clarify the fundamental differences in contemporaneous magmatism on and off the plateau, crust-mantle dynamics involved in the formation of the magmas, and regional tectonomagmatic processes.

We review existing data and present new data for all three of the laccolithic ranges (fig. 1) and contemporaneous rocks in adjacent regions of the Western United States, to address the following topics:

1. We compare the geochemistry of the rocks from all three mountain ranges in order to assess their differences, their similarities, and their possible tectonic affiliation.

2. We outline the petrogenesis of the Henry Mountains laccoliths, which we propose as a model system to illustrate the interactions between mantle-derived melts and continental crust in the Colorado Plateau. A detailed study is given in Nelson and Davidson (1993).

3. We consider the regional relationships of the laccoliths to roughly contemporaneous and similarly evolved volcanic rocks in the vicinity of the Colorado Plateau in order to evaluate the origins of mid-Tertiary magmatism.

REGIONAL GEOLOGIC SETTING

Deformational and magmatic events have left the Colorado Plateau interior relatively unaffected during the entire Phanerozoic Eon (Allmendinger and others, 1987). However, the Colorado Plateau (fig. 1) is bounded by areas of intense Mesozoic and Cenozoic deformation and magmatism. Cretaceous to Eocene Laramide shortening may have been the result of shallow or flat subduction (Bird, 1988; Hamilton, 1988) and an accompanying magmatic lull in the region of the laccoliths and farther west (Armstrong and Ward, 1991). Mid-Tertiary andesitic to dacitic volcanism in the Reno-Marysvale, San Juan, and Mogollon-Datil belts produced great volumes ($\approx 5\times10^5$ km^3; Johnson, 1991) of ignimbrite and similar rocks, whereas contemporaneous (middle to late Oligocene) magmatism in the Henry, La Sal, and Abajo

---

[1]Depart of Geology, 673 WIDB, Brigham Young University, Provo, Utah 84602.

[2]Department of Earth and Space Sciences, University of California, Los Angeles, CA 90024–1567.

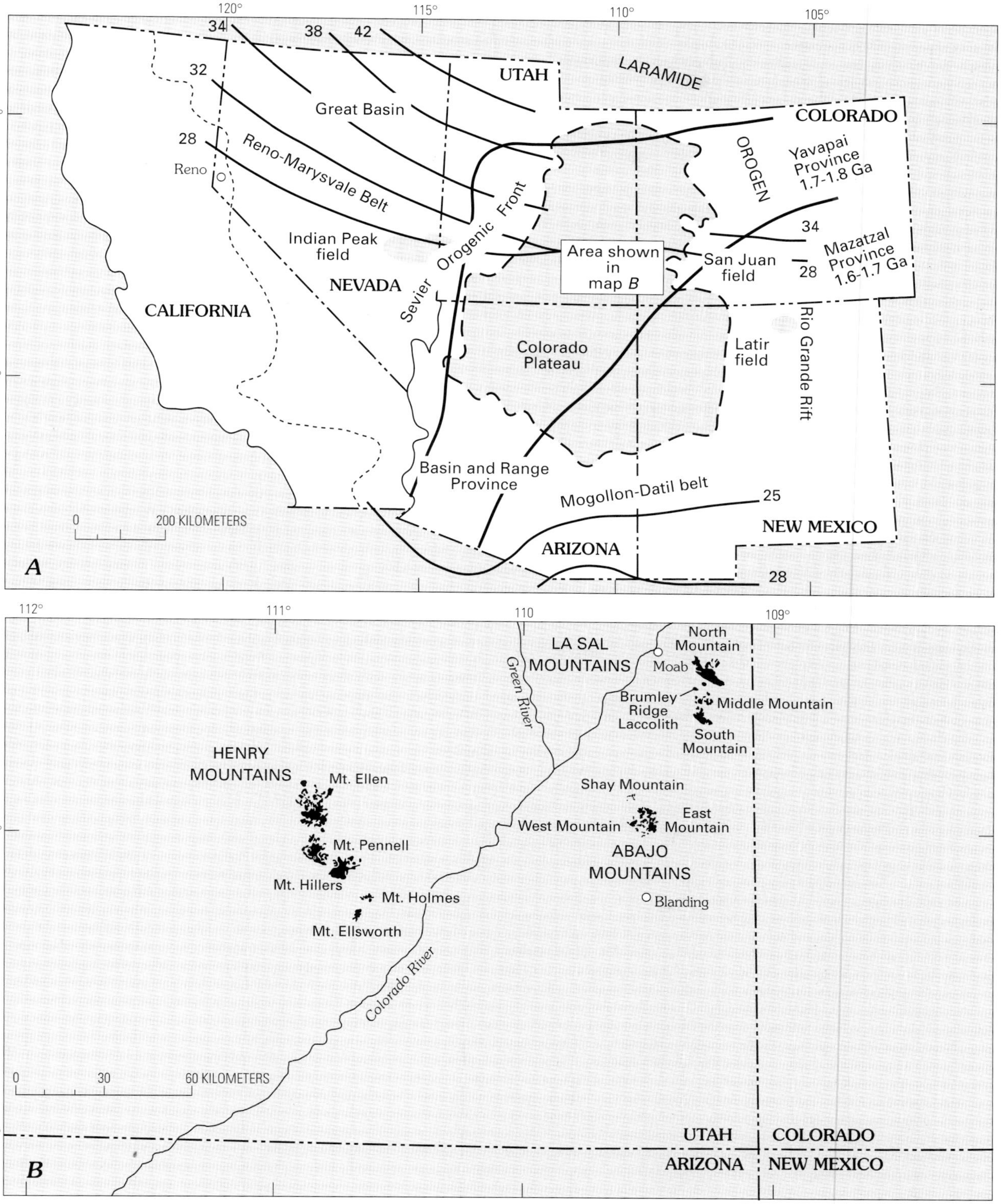

Figure 1. Locations of the Henry, La Sal, and Abajo Mountains, the Colorado Plateau, basement terrane boundaries, and other geographic and tectonomagmatic features of interest. *A*, General setting of the Southwestern United States: Numbers labeling hachured contours show the age (in millions of years) of the onset of mid-Tertiary magmatism as it swept through the Cordillera from the north and from the south (adapted from Cross and Pilger, 1978; Burke and McKee, 1979; and Glazner and Bartley, 1984). Proterozoic basement terrane boundaries from Bowring and Karlstrom (1990). *B*, Detail of the Henry, La Sal, and Abajo Mountains in the central part of the Colorado Plateau.

Mountains was volumetrically minor (69, 50, and 20 km^3, respectively) (Hunt, 1953; 1958; Witkind, 1964). The laccoliths may have been part of a large east-west-oriented late Oligocene magmatic belt extending from Reno, Nev., to the San Juan field in Colorado (Best, 1988; Sullivan and others, 1991; Nelson and others, 1992). This activity was followed by late Cenozoic (<17 Ma) basaltic magmatism around the margins of the Colorado Plateau. Crustal uplift and extension of terranes bordering the Colorado Plateau on the east (Rio Grande Rift) and on the west and south (Basin and Range province) were synchronous with the late Cenozoic volcanism. However, the interior of the Colorado Plateau has had little extension although it has been uplifted in Cenozoic time at rates exceeding those of the Basin and Range province (Lucchitta, 1979). To summarize, crustal deformation and magmatism in the Colorado Plateau have not been as intense as elsewhere in the Cordillera. The structural and rheological characteristics of the plateau may have inhibited large-scale upper-crustal magmatism and deformation.

COLORADO PLATEAU STRUCTURE, GEOPHYSICS, AND COMPOSITION

Geophysical data indicate that the Colorado Plateau is underlain by thick (45–50 km) stable crust (Thompson and Zoback, 1979; Allmendinger and others, 1986, 1987; and Beghoul and Barazangi, 1989) and is covered by a ≈6-km veneer of Late Proterozoic to Tertiary sedimentary rocks. Basement terranes range from about 1.6 to 1.9 Ga in age (Bennett and DePaolo, 1987; Karlstrom and others, 1987; Bowring and Karlstrom, 1990). Crystalline rocks beneath the laccolithic ranges formed a terrane termed the Yavapai province (fig. 1) (Condie, 1982; Karlstrom and others, 1987; Bowring and Karlstrom, 1990) along an east-northeast-trending convergent margin.

Some of the exposed Proterozoic basement rock to the east of the Colorado Plateau consists of metabasalt and other meta-igneous rocks (Knoper and Condie, 1988; Boardman and Condie, 1986), with mafic rocks making up as much as 80 percent of the assemblage in places (Robertson and Condie, 1989). The orientation of Proterozoic terrane boundaries and structural grain (fig. 1) indicates that the same mafic lithologies may form the basement to much of the Colorado Plateau. However, in New Mexico, much of the Proterozoic basement consists of granitoid rocks exposed in uplifts flanking the Rio Grande Rift (see Condie, 1978, for instance); therefore, much of the Mazatzal province (fig. 1) could be composed of silicic crust. Because quartz-rich crust is more easily strained than quartz-poor crust, one would expect deformation to be preferentially concentrated in quartz-rich regions during regional deformation. Thus, the lack of exposures of Proterozoic basement rock in the Colorado Plateau not only masks direct evidence of its composition, but could reflect a relatively mafic bulk composition and physical properties that distributed strain into surrounding terranes. Lower crustal P_n velocities of 6.8 km/s or greater in the Colorado Plateau (Smith and others, 1989; Wolf and Cipar, 1993) are appropriate for mafic rocks (Fountain and Christiansen, 1989). Velocities of 6.5–6.7 km/s or lower in the adjacent Basin and Range province (Smith and others, 1989; Wolf and Cipar, 1993) are more consistent with intermediate to silicic compositions (Fountain and Christiansen, 1989). More recently, Zandt and others (1995) measured poisson's ratio of the bulk crust and found values of 0.28–0.29 for the Colorado Plateau, consistent with an average mafic composition. In the Basin and Range province, values are about 0.20–0.25, indicating that the crust there is less mafic on average.

Some direct evidence indicates that the plateau is composed of mafic crust. Amphibolite inclusions, ranging from 1 to about 20 cm in diameter, are present in all intrusions and locally compose 1 percent of the laccoliths. They are probably from the Proterozoic Yavapai basement. Most are xenoliths of amphibolite-facies metabasalt, though some consist of hornblende gabbro (textural distinction). A statistical study of 200 randomly collected xenoliths from the Henry and La Sal Mountains showed that >95 percent of them were mafic (Hunt, 1953; 1958). McGetchin and Silver (1972) reported that >65 percent of crustal xenoliths in the Moses Rock dike, a mid-Tertiary diatreme in the Four Corners region of the Colorado Plateau, are basaltic. They estimated an average anhydrous composition for the crystalline portion of the plateau's crust that is surprisingly mafic (54 percent SiO_2, 8 percent MgO).

FIELD RELATIONS

The Henry, La Sal, and Abajo Mountains are cored by hypabyssal intrusions, with separate intrusive centers (5, 3, and 2, respectively) that were emplaced as multiple laccoliths, invading Mesozoic and upper Paleozoic sedimentary rocks. Structurally, individual mountains represent discrete intrusive loci composed of a central laccolith with radial sills and laccoliths and intervening sedimentary screens. Jackson and Pollard (1988) estimated the maximum depth of intrusion at 3 to 4 km (≈1 kbar) on the basis of the laccoliths' position within regional stratigraphic sequences. However, magmatic amphibole would be unstable at pressures less than 1 kbar, and yet breakdown textures are rare in the amphibole of the laccoliths. Thus the maximum intrusion depth given by Jackson and Pollard (1988) is close to the *minimum* depth based on petrographic considerations. One exception may be the middle La Sal Mountains, which show some breakdown of hornblende and are emplaced at a higher level in the stratigraphic section (Hunt, 1958; Michael Ross, Utah Geological Survey, oral commun., 1992).

The dominant rock type (95 volume percent) of the intrusions is plagioclase-hornblende porphyry (termed "diorite" by Hunt, 1953; 1958; Engel, 1959; Witkind, 1964; Irwin, 1973; and Hunt, 1988). It typically consists of 20–25 volume percent phenocrysts of plagioclase and about 10 volume percent hornblende in a fine-grained groundmass. The plagioclase is euhedral to subhedral, 0.5 to 5 mm in cross section, ranges from An_{30} to An_{45}, and shows complex zoning patterns. The hornblende is also euhedral to subhedral and is generally <1 mm in length, but can be as long as 5 mm. The remainder of the rock generally consists of an equigranular aphanitic groundmass of plagioclase, quartz, and alkali feldspar with trace quantities of apatite and sphene. Some plagioclase-hornblende porphyry, especially in the La Sal Mountains, contains clinopyroxene in subequal quantities to hornblende. At the Henry and La Sal Mountains, small volumes (5 volume percent) of fine-grained nepheline- to quartz-normative Na-rich syenite and rhyolite porphyries were emplaced as small stocks, laccoliths, and dikes that crosscut the plagioclase-hornblende porphyry. These rocks are described in detail in Nelson (1991) and are petrographically and geochemically diverse, especially in the La Sal Mountains. Syenite porphyry is restricted to Mount Pennell in the Henry Mountains and to North Mountain and the Brumley Ridge laccolith in the La Sal Mountains; it is apparently absent in the Abajo Mountains.

GEOCHEMICAL CHARACTERISTICS OF THE LACCOLITHS

In terms of major-element chemistry, the porphyries of the Abajo Mountains show considerable similarity to the plagioclase-hornblende porphyry of the Henry Mountains as determined from a small sample set (fig. 2*A*, *B*; table 1). The remaining compositional range shown in figure 2*B* represents data from Witkind (1964), who attributed at least some of the variation in total alkali contents to hydrothermal alteration. The trace-element systematics of the Abajo laccoliths are virtually identical to those of the plagioclase-hornblende porphyry of the Henry Mountains, which are summarized in figure 3*A*. Overall, the laccoliths have trace-element patterns similar to those of calc-alkaline basalt, although they have higher absolute elemental abundances except for titanium and phosphorus (fig. 3*B*). Fresh porphyry from the Abajo Mountains is, in general, somewhat more radiogenic than the plagioclase-hornblende porphyry of the Henry Mountains in terms of strontium isotopes (fig. 4). Given the similar age, tectonic and structural setting, and geochemical characteristics of the Henry and Abajo Mountains, it is our opinion that they reflect nearly identical igneous histories.

Despite locally pervasive alteration, the porphyries of the La Sal Mountains have primary geochemical differences that distinguish them from the Henry and Abajo Mountains. These rocks are also divided into a plagioclase-hornblende porphyry suite and a syenite-alkaline rhyolite suite (fig. 2*C*) on the basis of major-element criteria and

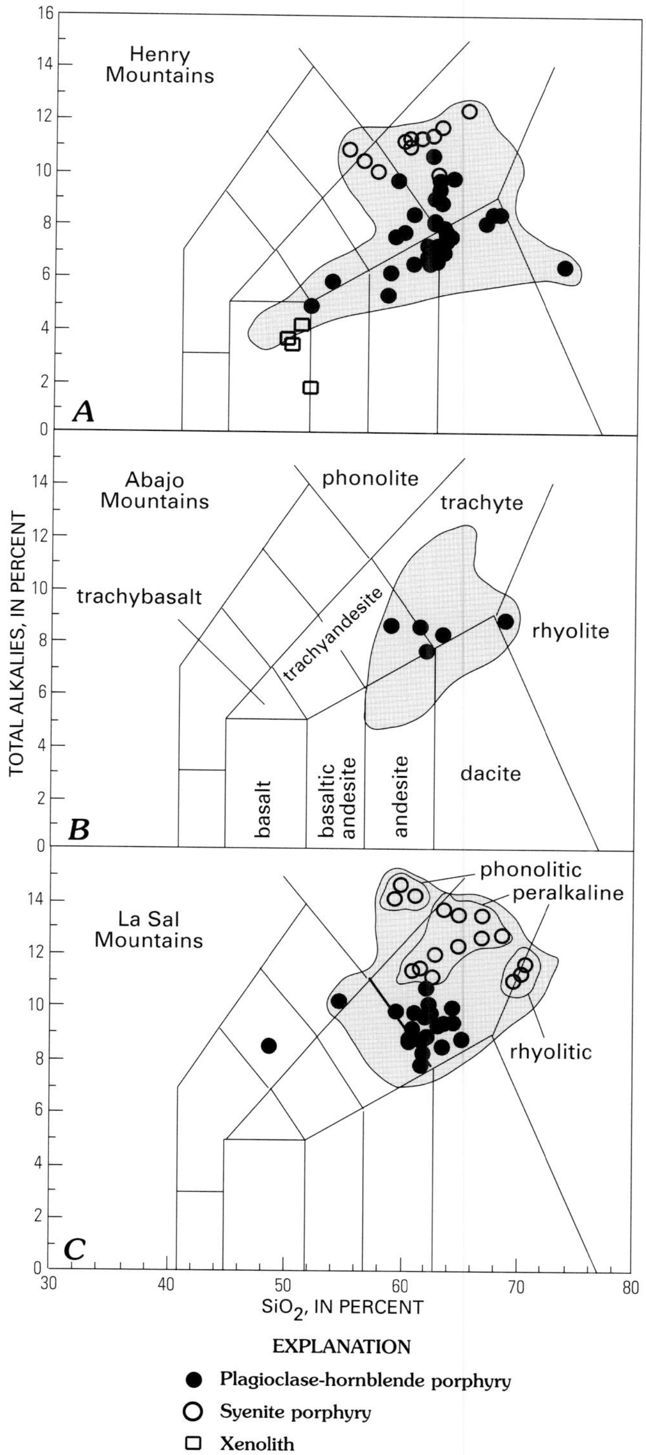

Figure 2. Total alkalies versus silica (Le Bas and others, 1986) for the Henry (*A*), Abajo (*B*), and La Sal (*C*) Mountains, Utah. Shaded fields represent the reported ranges of compositions from data in Nelson (1991), Hunt (1953; 1958), Engel (1959), Hunt (1988), and Witkind (1964).

petrographic characteristics. The plagioclase-hornblende porphyry is more alkali-rich than rocks of the Henry and

Table 1. Representative geochemical data and sample localities for rocks of the La Sal and Abajo Mountains, Utah.

[Analytical details are the same as reported by Nelson and Davidson (1993). LOI, loss on ignition; n.d., not determined]

	Syenite porphyry samples							Plagioclase-hornblende porphyry of La Sal Mountains			
Sample No.	CSTV–3KRS	MW–2	MW–6	MW–18	MW–17	LAS–1KRS	TUK–1	MW–3	MW–13	MW–14	WLKE–1
North latitude	109°16'34"	109°14'46"	109°14'48"	109°13'31"	109°13'29"	109°17'05"	109°15'57"	109°14'47"	109°14'05"	109°12'19"	109°15'55"
West longitude	38°32'56"	38°32'19"	38°32'17"	38°31'19"	38°31'19"	38°28'57"	38°27'23"	38°32'36"	38°32'34"	38°30'08"	38°30'21"
Major oxides in percent											
SiO_2	61.89	67.17	71.03	69.82	60.32	59.62	61.43	64.55	62.39	62.86	61.98
TiO_2	.55	.27	.20	.08	.16	.17	.15	.42	.52	.46	.50
Al_2O_3	16.89	16.86	12.55	15.01	19.54	19.08	19.70	17.24	16.90	17.39	17.76
FeO	3.75	2.06	2.87	1.14	1.82	1.71	1.67	3.71	4.03	3.77	4.01
MgO	1.01	.41	.23	.09	.10	.13	.16	1.23	1.47	.94	1.17
MnO	.14	.14	.14	.13	.13	.14	.14	.14	.14	.13	.16
CaO	2.78	.92	.41	.35	1.17	1.22	1.10	3.19	3.46	2.95	4.11
Na_2O	6.93	7.26	3.42	6.91	9.33	8.94	9.07	6.78	7.52	6.54	6.94
K_2O	4.68	5.56	8.29	4.28	5.46	5.34	5.33	2.91	3.42	3.30	2.99
P_2O_5	.19	.08	.08	.02	.03	.02	.02	.18	.24	.20	.21
LOI	1.18	.29	.06	1.38	1.09	3.40	1.60	.00	.59	.79	1.07
Total	99.80	100.94	99.20	99.18	99.12	99.75	100.35	100.17	100.44	99.12	100.69
Trace elements in parts per million (analyzed by X-ray fluorescence, except as noted) [1]											
Pb	7	14	13	82	34	36	35	10	13	10	26
Rb	128	135	246	69	149	138	135	49	80	83	76
Ba	772	556	183	724	211	424	361	757	969	719	837
Th	23	[2] 30.0	[2] 165.6	15	25	21	[2] 19.4	[2] 5.2	8	7	[2] 8.9
U	9	[2] 6.9	[2] 31.6	4	13	12	[2] 10.5	[2] 2.2	1	4	[2] 3.3
Nb	16	14	18	21	13	15	14	11	12	10	15
La	57	[2] 16.5	[2] 33.5	19	36	43	[2] 35.4	[2] 30.0	43	26	[2] 43.5
Sr	737	910	242	165	916	1416	1207	1043	1027	1197	1562
Zr	268	279	540	136	265	230	276	240	227	174	286
Y	31	12	18	7	10	12	12	21	21	18	21
Ni	4	n.d.	n.d.	n.d.	n.d.	n.d.	n.d.	2	2	3	1
Cr	10	7	7	5	5	8	7	8	8	6	8
Isotopic analyses (Pb ratios corrected for fractionation)											
$^{87}Sr/^{86}Sr \pm 2\sigma$	0.704938 ±0.000028	0.705061 ±0.000011	0.704806 ±0.000011	0.704574 ±0.000013	0.704293 ±0.000013	0.704226 ±0.000010	0.704325 ±0.000010	0.704174 ±0.000011	n.d.	0.704262 ±0.000010	0.704199 ±0.000010
$^{143}Nd/^{144}Nd \pm 2\sigma$	0.512525 ±0.000010	0.512427 ±0.000008	0.512450 ±0.000010	0.512443 ±0.000010	0.512489 ±0.000009	0.512490 ±0.000016	n.d.	n.d.	n.d.	0.512367 ±0.000008	0.512352 ±0.000016
εNd	−2.21	−4.12	−3.67	−3.80	−2.90	−2.89	n.d.	n.d.	n.d.	−5.29	−5.58
$^{206}Pb/^{204}Pb$	n.d.	19.411	n.d.	18.685	18.781	18.856	n.d.	n.d.	n.d.	18.574	n.d.
$^{207}Pb/^{204}Pb$	n.d.	15.650	n.d.	15.585	15.614	15.627	n.d.	n.d.	n.d.	15.582	n.d.
$^{208}Pb/^{204}Pb$	n.d.	38.790	n.d.	38.261	38.303	38.389	n.d.	n.d.	n.d.	37.971	n.d.

	Plagioclase-hornblende porphyry of La Sal Mountains						Plagioclase-hornblende porphyry of Abajo Mountains				
Sample No.	WLKE–3	WLKE–4	TUK–2	PEAL–2	TUK–5	LASE–1	MONT-1KRS	MONT-2KRS	MONT-3KRS	MONT-4KRS	LIN–1KRS
North latitude	109°21'44"	109°15'27"	109°16'15"	109°14'15"	109°15'10"	103°13'23"	109°27'09"	109°29'58"	109°27'09"	109°28'33"	109°30'12"
West longitude	38°36'46"	38°31'00"	38°31'36"	38°26'12"	38°24'27"	38°22'00"	37°48'11"	37°50'42"	37°51'08"	37°51'08"	37°49'34"
Major oxides in percent											
SiO_2	65.50	64.70	61.60	61.97	61.08	61.27	63.52	61.57	59.17	62.23	68.99
TiO_2	.40	.30	.49	.46	.58	.55	.43	.47	.54	.48	.20
Al_2O_3	17.06	16.06	17.88	17.88	17.70	17.50	17.47	17.62	17.12	17.51	15.38
FeO	2.33	2.90	3.81	3.98	5.06	4.68	3.70	3.82	4.48	3.92	1.75
MgO	1.42	.91	.99	.95	1.51	1.32	1.12	1.03	1.44	1.16	.52
MnO	.14	.13	.14	.10	.13	.13	.14	.16	.15	.14	.14
CaO	4.83	2.34	4.61	4.48	5.07	3.90	5.12	5.51	5.63	4.39	2.30
Na_2O	8.11	6.77	7.36	6.14	5.18	6.14	5.88	6.07	6.26	5.48	5.61
K_2O	.81	3.27	2.48	2.67	3.65	2.75	2.50	2.57	2.44	2.24	3.42
P_2O_5	.19	.15	.19	.20	.28	.27	.17	.17	.22	.18	.07
LOI	1.51	.63	1.31	.99	.67	1.00	.00	.64	4.30	.00	1.60
Total	102.11	98.01	100.67	99.61	100.90	99.50	100.05	99.63	101.75	97.73	99.98
Trace elements in parts per million (analyzed by X-ray fluorescence, except as noted) [1]											
Pb	7	8	13	13	13	7	16	15	13	11	39
Rb	20	58	44	54	92	41	42	51	45	42	64
Ba	274	940	780	773	996	921	767	964	721	724	958
Th	[2] 3.1	3	[2] 8.6	6	11	9	3	5	4	3	2
U	[2] 1.7	4	[2] 3.9	4	3	2	1	2	1	1	2
Nb	n.d.	10	n.d.	8	8	8	17	16	16	17	14
La	[2] 21.4	17	[2] 41.0	31	51	32	23	35	29	24	13
Sr	925	1035	1728	1393	1287	10004	827	964	864	1011	882
Zr	170	160	284	194	179	191	153	153	143	157	105
Y	14	15	24	24	26	27	30	26	35	30	18
Ni	0	5	1	4	6	4	1	n.d.	2	n.d.	n.d.
Cr	7	9	8	6	6	6	8	8	8	8	7
Isotopic analyses (Pb ratios corrected for fractionation)											
$^{87}Sr/^{86}Sr \pm 2\sigma$	0.703928 ±0.000010	n.d.	0.704426 ±0.000010	0.704443 ±0.000011	0.704810 ±0.000010	n.d.	0.705463 ±0.000010	0.705513 ±0.000010	n.d.	n.d.	0.705976 ±0.000010
$^{143}Nd/^{144}Nd \pm 2\sigma$	n.d.	n.d.	0.512419 ±0.000009	0.512466 ±0.000008	0.512436 ±0.000007	n.d.	0.512333 ±0.000026	0.512353 ±0.000007	n.d.	n.d.	0.512367 ±0.000007
εNd	n.d.	n.d.	–4.27	–3.36	–3.94	n.d.	–5.94	–5.55	n.d.	n.d.	–5.28
$^{206}Pb/^{204}Pb$	n.d.	18.430	n.d.	18.828	18.945	n.d.	18.309	n.d.	n.d.	n.d.	18.601
$^{207}Pb/^{204}Pb$	n.d.	15.557	n.d.	15.584	15.624	n.d.	15.533	n.d.	n.d.	n.d.	15.586
$^{208}Pb/^{204}Pb$	n.d.	37.721	n.d.	38.267	38.378	n.d.	37.772	n.d.	n.d.	n.d.	37.977

[1] X-ray fluorescence determinations below 10 ppm should be considered semiquantitative.
[2] Determined by instrumental neutron activation analysis (INAA).

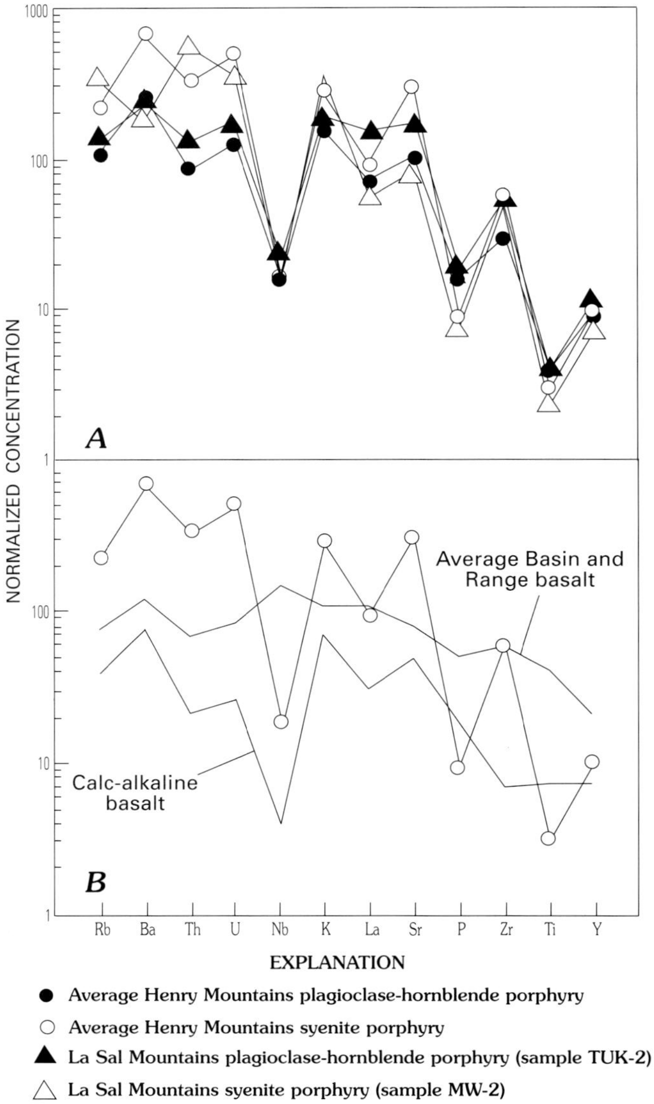

Figure 3. Relative trace-element abundances (normalized to primitive mantle, after Sun, 1980) of plagioclase-hornblende porphyry and syenite porphyry from the Henry and La Sal Mountains, Utah (*A*), compared to those of typical calc-alkaline basalts (Sun, 1980) and intraplate basalts of the Basin and Range province (Ormerod and others, 1988; Fitton and others, 1988) (*B*). Note the similarity of the porphyries to the calc-alkaline basalts and their dissimilarity to the Basin and Range basalts.

Abajo Mountains, and mostly ranges from 60 to 63 percent SiO_2. In fact some of the plagioclase-hornblende porphyry is sufficiently enriched in alkalies that it is nepheline-normative.

Syenitic to rhyolitic rocks of the La Sal Mountains are divided into subtypes on the basis of petrographic and geochemical criteria. A small group of phonolitic or feldspathoid- (nosean-) bearing syenite porphyry (fig. 2*C*) occurs in a large dike at North Mountain and in the Brumley Ridge laccolith (Hunt, 1958) of Middle Mountain. These nosean-bearing rocks are strongly undersaturated (nepheline ± nesosilicate normative). A second group consists of slightly nepheline- to quartz-normative and commonly peralkaline (acmite-normative) syenite porphyry, which is richer in silica and poorer in alkalies than the feldspathoid-bearing syenite (fig. 2*C*). A last group consists of quartz-phyric low-silica peralkaline rhyolite porphyry, which we interpret to be a differentiate of peralkaline syenite.

Trace-element patterns of the porphyries of the La Sal Mountains show some significant deviations from the patterns of the Henry and Abajo Mountains, although there is still a distinct relative depletion in Nb (figs. 3*A*, 5), suggesting an affinity to orogenic magmatism. $^{143}Nd/^{144}Nd$ and $^{87}Sr/^{86}Sr$ are more restricted in the La Sal porphyries (fig. 4).

The lack of trends in the La Sal Mountains major-element data set (fig. 6; table 1) and the distinct geochemical differences relative to the porphyries of the Henry and Abajo Mountains may be the result of some combination of major element mobility during alteration, different magma sources, and different petrogenetic processes. Because of a lack of correlation (fig. 6), the data are not amenable to extensive interpretation, and this lack of correlation extends to trace-element and isotope systematics as well. Therefore, petrogenetic processes in the Henry Mountains must serve as a general model for the La Sal Mountains, despite the geochemical differences.

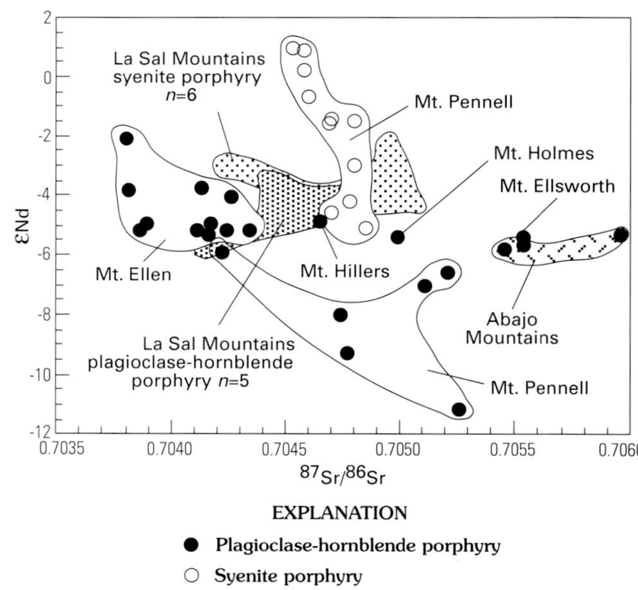

Figure 4. εNd versus $^{87}Sr/^{86}Sr$ in rocks of the Henry, La Sal, and Abajo Mountains, Utah. Note the isotopic provinciality of each of the intrusive centers of the Henry Mountains.

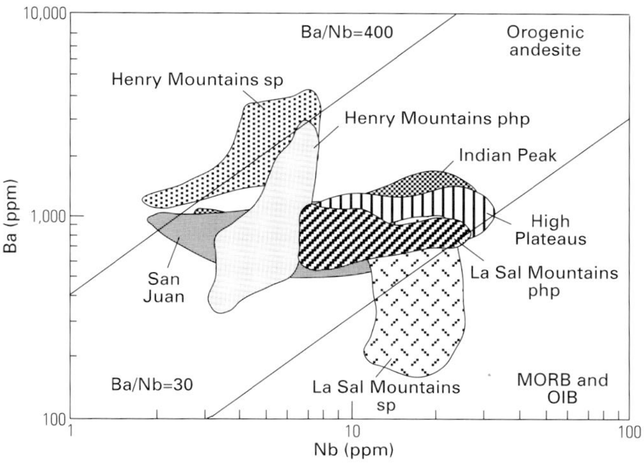

Figure 5. Barium versus niobium in syenite porphyry (sp) and plagioclase-hornblende porphyry (php) of the Henry and La Sal Mountains, Utah, and in contemporaneous magmatic rocks from the High Plateaus of Utah (Mattox, 1991), the San Juan field, Colorado (Colucci and others, 1991; Lipman and others, 1978), and the Indian Peak field, Nevada (Best and others, 1989). All of these show a general affinity to subduction-generated andesites. The field for orogenic andesite and the combined field for MORB (mid-oceanic-ridge basalt) and OIB (oceanic-island basalt) are modified from Gill (1981).

MAGMA CHEMISTRY AND PETROGENESIS IN THE HENRY MOUNTAINS

Fortunately, the major-element variations in the Henry Mountains are much more orderly and amenable to interpretation than those of the La Sal Mountains (fig. 6). As the syenite porphyry represents <5 percent of the volume of the Henry Mountains, we will focus on the origin and evolution of the plagioclase-hornblende porphyry, which, we suggest, will also serve as a model for the evolution of similar rocks in the other ranges. The following review is based on a detailed examination of the petrogenesis of both the syenite and the plagioclase-hornblende porphyry of the Henry Mountains, which can be found in Nelson and Davidson (1993).

Plagioclase-hornblende porphyry for each intrusive center has a distinct range of strontium, neodymium, and lead isotopic compositions. Meta-mafic rocks of the Yavapai and Mazatzal provinces and crustal xenoliths from the laccoliths show extreme variations, which are reflected in the isotopic diversity of the magmas (Nelson and Davidson, 1993; Nelson and DePaolo, 1984). The porphyries at Mount Ellen show systematic major-element, trace-element, and isotopic trends that allow the evaluation of petrogenetic processes. εNd is negatively correlated with rubidium among the plagioclase-hornblende porphyries at Mount Ellen, constraining the nature and extent of open-system behavior.

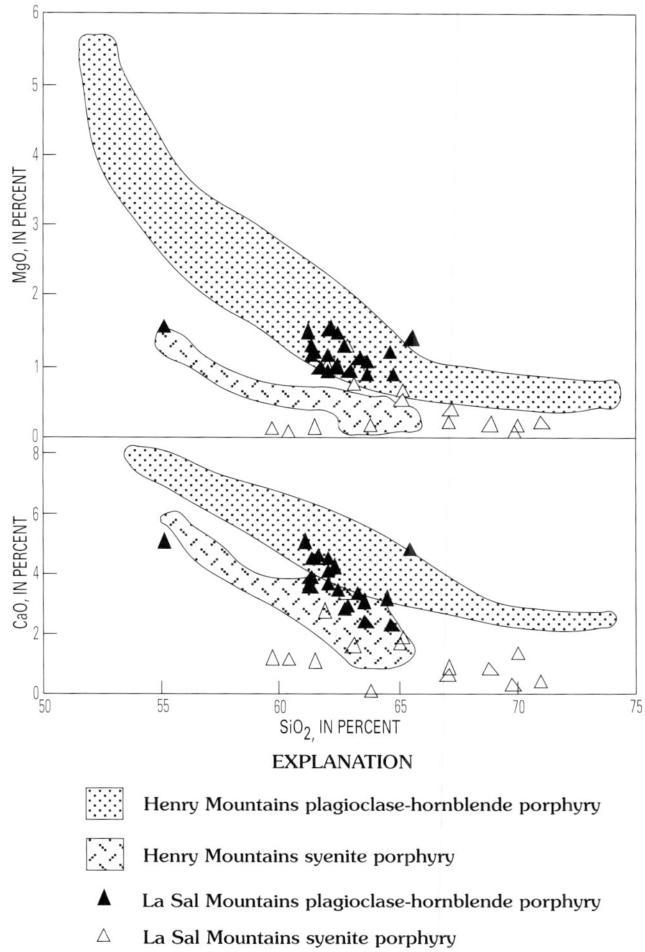

Figure 6. Harker variation diagrams comparing rocks of the Henry Mountains to those of the La Sal Mountains, Utah. The general lack of correlation in major-element data of the La Sal porphyries is also observed in their trace-element and isotope systematics.

AFC (assimilation and fractional crystallization; DePaolo, 1981) models indicate 40–45 percent crystallization and deep crustal magmatic evolution (rate of mass assimilation/mass fractionation (r) = 0.7–0.5; Nelson and Davidson, 1993).

Given the isotopic provinciality of the data set, heterogeneity of the crust, and the dominance of plagioclase-hornblende porphyry with 60–63 percent silica (fig. 2*A*), the following model is proposed. A flux of mantle-derived magmas impinged upon the base of the crust, and those magmas that penetrated into the crust ponded in deep-crustal magma reservoirs in which there was sufficient recharge or underplating for AFC to proceed with a high rate of mass assimilation to crystallization (r >0.5). Isotopic provinciality is attributed to heterogeneity in the net assimilant at each igneous center or mantle source region. The narrow range of SiO_2 concentrations and the general lack of mafic rocks in most laccoliths suggest that the magmas were able to rise to a shallow crustal level only when a critical density had been reached sufficient to overcome a strength or buoyancy barrier in the crust. Because magmas are so much more compressible than solid rock, the density contrast between magma and wall rock will be much smaller in the lower crust than at the surface, even though the magma is mafic. Based on a model by Herzberg (1987), a basaltic magma at 1,200°C and 10 kbar is only on the order of 0.15 g/cm^3 less dense than solid amphibolite, whereas the density contrast may double to 0.3 g/cm^3 at the surface. At 10 kbar and 850°C, a magma composition representative of the laccoliths (63 percent silica) is much more buoyant, having a density contrast of ≈0.35 g/cm^3. In this light, it may be perfectly reasonable to expect mafic magmas to pond deep in the crust, especially when they are surrounded by high-strength mafic wall rock.

The presence of isotopically primitive mafic rocks (in a relative sense) at Mount Ellen requires the source of the magmas to lie dominantly in the mantle. In addition, the trace-element systematics of both the plagioclase-hornblende porphyry, and especially the syenite porphyry, seem to require that apparent geochemical similarities to arc magmas (Nelson and Davidson, 1993) were probably inherited from the mantle rather than by contamination from continental crust.

REGIONAL TECTONOMAGMATIC CONSIDERATIONS

There is little disagreement that mid-Tertiary magmatism in the laccoliths and other areas of the Western United States, such as the Great Basin and the San Juan field, is fundamentally basaltic. In the rhyolitic magma, however, the primary tie to basaltic magma may be through (1) fractional crystallization accompanied by assimilation of crust or (2) crustal melting due to underplating of basalts. An additional consideration is the mechanism that triggers the influx of magma that drives silicic magmatism. As an example, the flux of basaltic magma might result from subduction, or it may have begun at "passive" hot spots in response to crustal extension, mantle upwelling, and decompression melting. Combining the two end members for the origin of mantle-derived magmas with both end members for the petrogenesis of silicic magmas yields a variety of tectonomagmatic scenarios for their origin. However, we present evidence that suggests that regional magmatism, including the laccoliths, resulted from subduction-related processes, and that the range of observed compositions (basaltic andesite to rhyolite) represents open-system evolution of mantle-derived magmas. In order to resolve differences between the models we review some of the major-element,

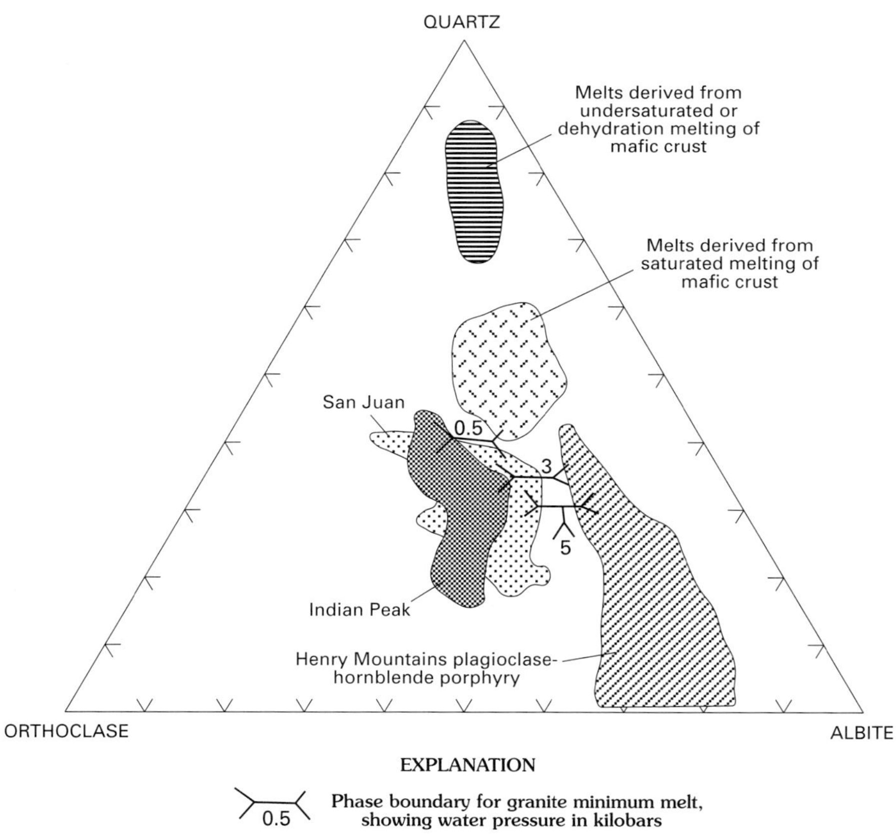

Figure 7. Ternary diagram plotting normative compositions of Henry Mountain laccoliths; of ash-flow sheets of the Indian Peak (Best and others, 1989) and San Juan (Lipman and others, 1978) fields; of melts derived from mafic crust in water-saturated (Helz, 1976), undersaturated (Allen and Boettcher, 1978), and dehydrated (Rushmer, 1991) conditions; and of granite minimum melts (labeled phase boundaries at 0.5, 3, and 5 kbar water pressure). Note that the laccoliths and ash-flow sheets do not correspond well to either type of crustal melt.

trace-element, and isotopic characteristics of regional mid-Tertiary (≈32–24 Ma) magmatism together with probable changes in the plate tectonic configuration of the Western United States during that period.

First, the major-element and isotopic characteristics of regional mid-Tertiary silicic magmas are not consistent with a crustal-melting model. In most instances, crustal anatexis ought to produce melts that are either substantially more silica-rich or more quartz-normative than the dacitic-rhyodacitic melts of the large ash-flow sheets of the ignimbrite flareup (Best and others, 1989; Lipman and others, 1978). In the melting of felsic crust, minimum melts (fig. 7) ought to predominate, unless the degree of melting is sufficient to exhaust quartz or one of the feldspars. However, few silicic ash-flow sheets have minimum-melt compositions (fig. 7), whereas most are displaced from thermal minima at all pressures. Thus, the dacite to low-silica rhyolites of many mid-Tertiary caldera complexes of the Western United States contain a significant fraction of mafic minerals (Best and others, 1989) and are not minimum melts because they are neither liquids derived from felsic crust, nor sufficiently evolved via fractional crystallization. Also, the plagioclase-hornblende porphyry of the Henry Mountains and the regional ash-flow sheets do not have appropriate compositions to be melts derived from mafic crust (fig. 7), although experimental data suggest that such melts may exhibit a range of compositions depending on the starting material and conditions of melting. Melting of mafic crust (fig. 7) generally produces liquids that are highly quartz normative (tonalites, also containing a significant anorthite component) unless large degrees of melting occur (Helz, 1976; Allen and Boettcher, 1978; Beard and Lofgren, 1991; Rushmer, 1991). In addition, these caldera complexes are surrounded by and intercalated with andesites and basaltic andesites (Best and others, 1989; Lipman and others, 1978). We find no compelling reason to suppose that the silicic ash flows are crustal melts rather than more evolved components of the same mantle-derived system as the andesites.

Based upon an exhaustive review of available isotopic data for the Western Cordillera, Johnson (1991) and Perry and others (1993) concluded that commonly 50 percent or more of the mass of the silicic ash flows originated as mantle-derived basalt that evolved via fractional crystallization or AFC processes. Therefore, we give further consideration to the isotopic character of the laccoliths and related rocks in the region, to assess the relative contributions of crustal and mantle sources (fig. 8). We have calculated ranges of expected isotopic compositions for 1,800-Ma crust that originated from either depleted or undepleted mantle and has evolved from Rb/Sr and Sm/Nd ratios matching those reported by Weaver and Tarney (1984) and Taylor and McLennan (1985) for average continental crust. The purpose of this exercise is not to infer the isotopic composition of the crust that has already been demonstrated to be heterogeneous (Nelson and Davidson, 1993), but to illustrate that relatively

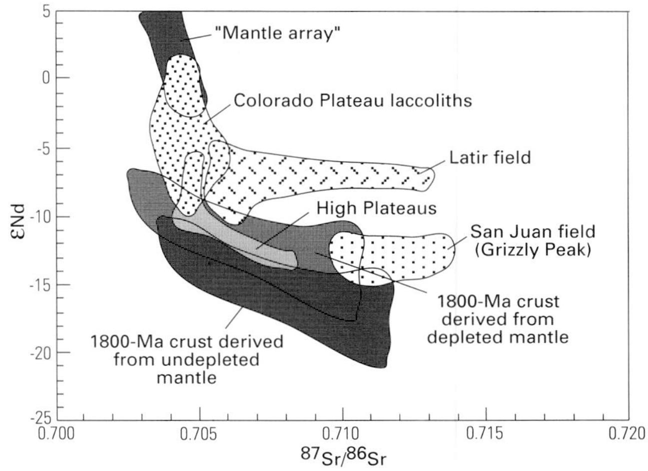

Figure 8. εNd versus $^{87}Sr/^{86}Sr$ in Colorado Plateau laccoliths, in rocks of the High Plateaus of Utah (S.R. Mattox, unpub. data, 1991), and in ignimbrites of the San Juan field, Colorado (Grizzly Peak; Johnson and Fridrich, 1990), and the Latir field, New Mexico (Johnson and others, 1990). Also shown are the hypothetical ranges of 1,800-Ma crust that had initial isotopic ratios derived from depleted and undepleted mantle and has evolved to the Rb/Sr and Sm/Nd ratios of average continental crust reported by Weaver and Tarney (1984) and Taylor and McLennan (1985).

mafic rocks (regional basaltic andesites) and silicic rocks (laccoliths and regional ash-flow sheets) may contain a substantial amount of mantle-derived material even if they are contaminated to "crust-like" isotopic compositions.

The Grizzly Peak Tuff (fig. 8) has isotopic characteristics similar to our calculated crustal values. Johnson and Fridrich (1990) note that the primitive end member of this zoned ash-flow sheet is somewhat mafic (57 percent SiO_2), and would have required an unrealistically large degree of melting of mafic crust (≈60 percent) to produce the observed SiO_2 concentration. Such a melt should also have concentrations of other major elements quite different from those of the observed magma. Furthermore, although the lavas of the High Plateaus (fig. 8) also resemble our calculated crust, they are too mafic (50–62 percent SiO_2) to be crustal melts. Clearly, "crustal" isotopic signatures alone may be misleading in assigning crustal anatexis origins even for silicic ignimbrites and laccoliths. The laccoliths lie between mantle and crustal isotopic end members (fig. 8), and the petrogenetic model for the plagioclase-hornblende porphyry described earlier in this paper may represent a link between the two reservoirs.

Many studies have suggested that subduction beneath the Western United States was shallow or flat during the Laramide orogeny but then steepened in mid-Tertiary time. (See, for example, Bird, 1988; Severinghaus and Atwater, 1990; Armstrong and Ward, 1991; and Best and Christiansen, 1991.) Models of passive extension, however, do not account for the influence of the subducted plate that must have existed beneath the Western United States during

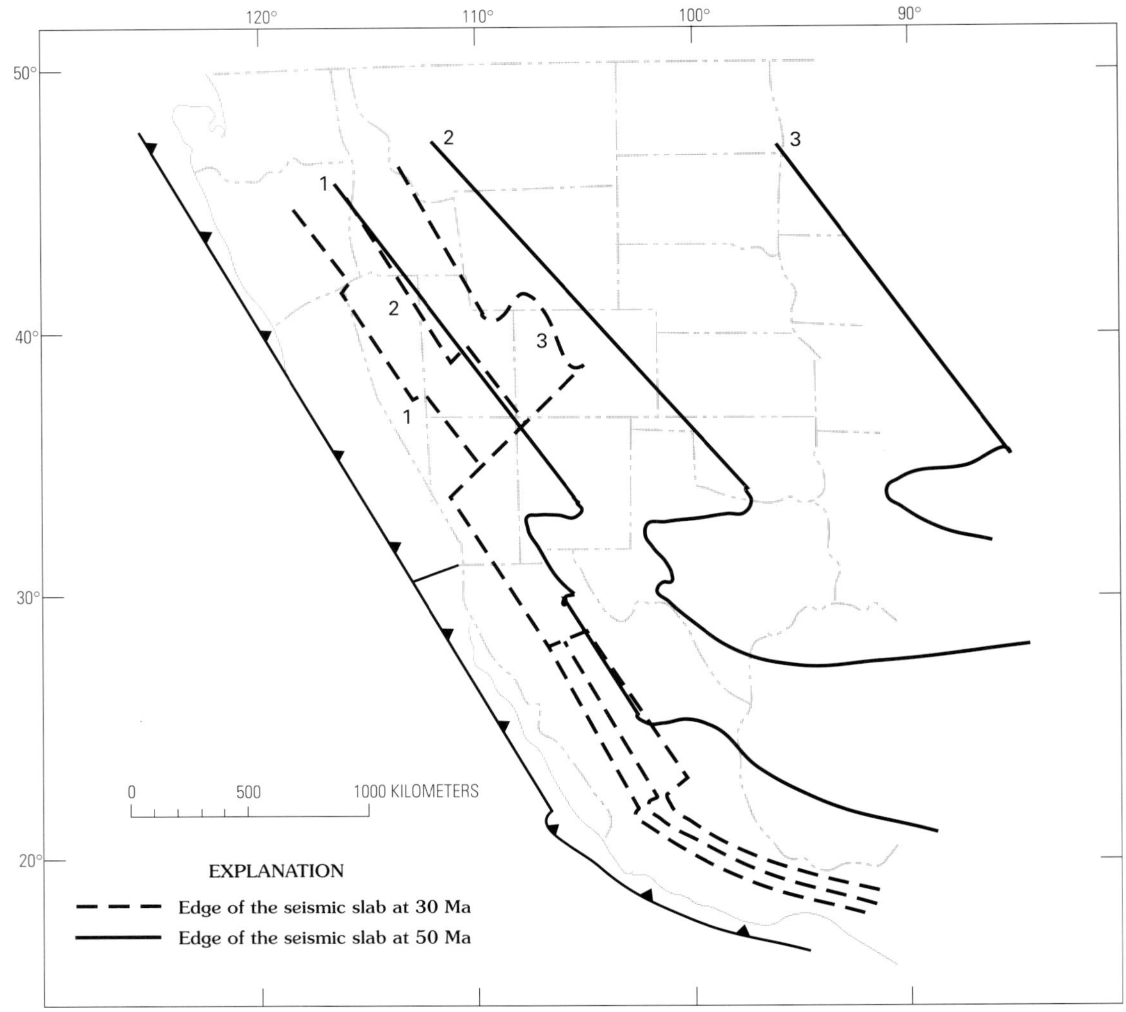

Figure 9. Plate-tectonic reconstruction of the Western United States during Tertiary time at 50 and 30 Ma. Numbered contours represent estimates for the inland limit of the seismically active slab, based on three different assumptions about how long after subduction a slab of a given age and thickness remains distinct from the overlying asthenosphere. (Estimate 1 is most conservative.) After Severinghaus and Atwater (1990).

mid-Tertiary time. Recent plate-tectonic reconstructions permit the existence of a seismically active subducted slab far inland from the paleo-plate margin—as far east as the San Juan field—as the subduction angle steepened following Laramide compression (fig. 9). Even in the most conservative case, the aseismic extensions of the slab may have contributed to petrogenesis. Therefore, it is reasonable to conclude that subducted oceanic lithosphere could have exerted primary control on the composition, distribution, and timing of magmatism after the Laramide orogeny.

Available data from the High Plateaus (Mattox, 1991), San Juan field (Colucci and others, 1991), Indian Peak caldera complex (Best and others, 1989) and the laccoliths, plotted in figure 5, show large ratios of LILE (large-ion lithophile elements, such as Ba) to HFSE (high-field-strength elements, such as Nb). These high ratios suggest the influence of subducted lithosphere; they are characteristic of arc magmas and are believed to result either from LILE-rich slab-derived fluids that infiltrate the overlying mantle wedge (Tatsumi and others, 1986) or from extensive interactions between basaltic melts and the mantle (Kelemen and others, 1990). In theory, any mantle-derived magma could become depleted in HFSE according to the model of Kelemen and others (1990). However, they note that this process is important only in arc settings. Regardless of how the trace-element signatures (fig. 3) were acquired, they seem to be a fundamental characteristic of magmas that are at least partly derived from subducted lithosphere. We recognize that high LILE/HFSE ratios could be stored in mantle lithosphere affected by subduction earlier in its history, as has been inferred by the geochemistry of post-subduction Cenozoic basalts of the Western United States

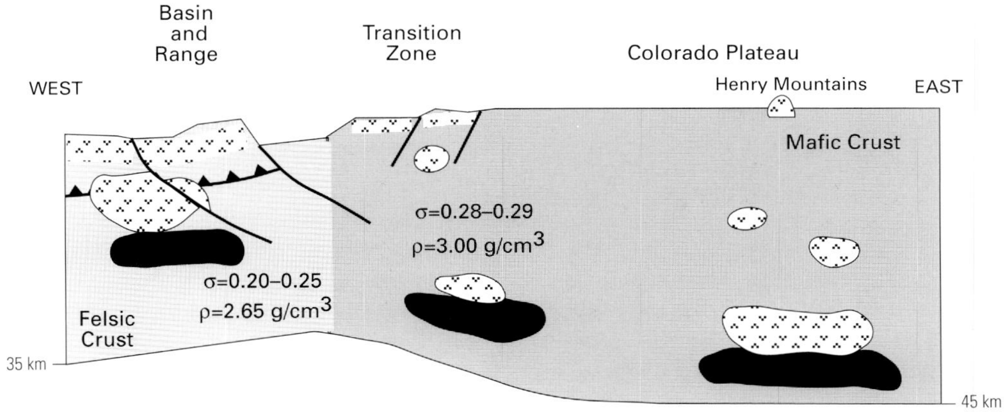

EXPLANATION

Mafic magma

Evolved magma

Felsic crust of the Basin and Range province

Mafic crust of the Colorado Plateau

Figure 10. Cartoon diagram illustrating differences in crustal structure, composition, and magma volumes across the Basin and Range province, the transition zone, and the Colorado Plateau interior. Crustal differences between the two provinces are indicated by differences in poisson's ration (σ) of the bulk crust (Zandt and others, 1995), and by differences in density (ρ). Crustal thicknesses after Allmendinger and others (1987).

(Kempton and others, 1991). However, we favor the interpretation that magmatism with an arc-like geochemical signature was genetically linked to active subduction beneath North America. Based upon our observations, therefore, we present a model for the tectonic setting of the laccoliths and related rock bodies throughout the region that is consistent with both their geochemistry and their temporal and spatial distributions.

We have reviewed geologic observations which suggest that the crust of the Colorado Plateau may be more mafic than that of surrounding regions. Meta-mafic rocks are stronger than their quartz-rich counterparts (Hacker and Christie, 1990), and therefore the difference in rheological properties between the Colorado Plateau and surrounding regions may be explained by compositional differences. The contrasting physical properties and tectonomagmatic history of the Colorado Plateau and surrounding regions are illustrated in a hypothetical cross section in figure 10.

Best and Christiansen (1991) and Severinghaus and Atwater (1990) suggested that the southwestward sweep of magmatism in the Western United States (fig. 1) was in response to the shortening of the subducting slab (fig. 9) as it adjusted to decreased convergence rates during the Tertiary Period. This resulted in the ingress of hot asthenosphere above the subducted plate, and magmatism was initiated as LILE-rich fluids invaded the growing mantle wedge and reduced solidus temperatures below ambient thermal conditions. The flux of slab-derived fluids accounts for the high Ba/Nb ratios observed throughout the Western United States (fig. 3).

Although the Colorado Plateau laccoliths seem related to the contemporaneous magmatic centers that surround the plateau (Nelson and others, 1992), the volume of magma reaching upper crustal levels at those other centers is several orders of magnitude greater. Best and Christiansen (1991) suggested that Mesozoic shortening surrounding the Colorado Plateau (fig. 10) preconditioned the crust in those regions for magmatic activity. Although shortening will initially depress isotherms, a greater net heat production in thickened radiogenic crust will result in subsequent warming. Overthickening of the crust may also produce gravitational instabilities and crustal extension, further warming the crust. The nearly identical normative (fig. 7) and trace-element (fig. 5) compositions, ages (fig. 1), and volumes of the ash-flow sheets of the Indian Peak and San Juan fields suggest that they were produced in a similar fashion. However, beneath the Colorado Plateau, the presence of substantial mantle-derived magma beneath a thick, unwarmed, and undeformed crust may have contributed to the difference in major-element chemistry (fig. 7) and the large contrast in volume between the Colorado Plateau intrusions and contemporary volcanic fields to the west and east (fig. 10). The model we describe is a consequence of the physical characteristics of the Colorado Plateau and the change in plate motions between Laramide and post-Laramide time. It explains the timing and distribution of magmatism in the

Western United States, the subduction chemistry of the magmas, and the differences in igneous volumes between the Colorado Plateau and surrounding regions.

CONCLUSIONS

Plagioclase-hornblende and minor syenite porphyries of the Henry, La Sal, and Abajo Mountains record petrogenetic processes in an unusual geologic setting, the Colorado Plateau interior. Although fractional crystallization could explain the major-element variations and many trace-element variations, radiogenic isotope systematics require open-system interaction of arc-like mantle-derived magmas with mafic but isotopically heterogeneous Proterozoic crust. Plagioclase-hornblende porphyry evolved via assimilation and fractional crystallization in deep-crustal magma chambers until it reached a critical density at which it could overcome the strength of the wall rock and rise to shallow crustal levels. In addition, the plagioclase-hornblende porphyry shows isotopic provinciality, indicating that rocks from each intrusive center were derived from, or interacted with, distinct reservoirs.

Both the plagioclase-hornblende porphyry and the syenite porphyry magma series have trace-element characteristics that indicate that the mantle source was enriched in large-ion lithophile elements relative to high-field-strength elements. This may be a primary source characteristic of the laccoliths and of contemporaneous andesites of the High Plateaus, Great Basin, and San Juan field.

We interpret the magmatism at these centers to be a consequence of subduction and changing plate motions rather than a result of passive hot spot activity. Although the Henry Mountains were far removed from the location of the Tertiary paleo-trench, recent work (Nelson and others, 1992; Nelson and Davidson, 1993) indicates that the Henry, La Sal, and Abajo Mountains were part of an east-west-oriented segment of a late Oligocene arc that extended from the vicinity of Reno, Nev., to the San Juan volcanic field of Colorado. This segment, in turn, was part of a much larger contemporaneous system that extended from Canada to southern Mexico. The arc-like signature of the Colorado Plateau laccoliths and other Oligocene magmatic rocks in the region supports this interpretation. The relatively small volume of the laccoliths suggests that the unusual physical properties of the Colorado Plateau inhibited large volumes of magma from reaching shallow crustal levels.

ACKNOWLEDGMENTS

The U.S. Geological Survey's symposium on the Colorado Plateau laccoliths expanded our view of the regional geology and gave us new problems to consider regarding their petrogenesis. For this we thank the Survey and symposium participants. We especially thank Mike Ross of the Utah Geological Survey for his insights and assistance in the field. The manuscript benefited from reviews by Brad Singer and Allen Glazner. This work was supported by National Science Foundation grant EAR 8917192 to Jon Davidson.

REFERENCES CITED

Allen, J.C., and Boettcher, A.L., 1978, Amphiboles in andesite and basalt—II, Stability as a function of P–T–f_{H_2O}–f_{O_2}: American Mineralogist, v. 63, p. 1074–1087.

Allmendinger, R.W., Farmer, Harlow, Hauser, E.C., Sharp, James, von Tish, Douglas, Oliver, Jack, and Kaufman, Sidney, 1986, Phanerozoic tectonics of the Basin and Range–Colorado Plateau transition from COCORP data and geologic data, *in* Barazangi, Muawia, and Brown, L.D., eds., Reflection seismology—The continental crust: American Geophysical Union Geodynamics Series, v. 14, p. 257–268.

Allmendinger, R.W., Hauge, T.A., Hauser, E.C., Potter, C.J., Klemperer, S.L., Nelson, K.D., Kneupfer, P.L.K., and Oliver, J., 1987, Overview of the COCORP 40°N transect, Western United States—The fabric of an orogenic belt: Geological Society of America Bulletin, v. 98, p. 308–319.

Armstrong, R.L., and Ward, P.L., 1991, Evolving geographic patterns of Cenozoic magmatism in the North American Cordillera—The temporal and spatial association of magmatism and metamorphic core complexes: Journal of Geophysical Research, ser. B, v. 96, p. 13201–13224.

Beard, J.S., and Lofgren, G.E., 1991, Dehydration melting and water-saturated melting of basaltic and andesitic greenstones and amphibolites: Journal of Petrology, v. 32, p. 365–401.

Beghoul, Noureddine, and Barazangi, Muawia, 1989, Mapping high P_n velocity beneath the Colorado Plateau constrains uplift models: Journal of Geophysical Research, ser. B, v. 94, p. 7083–7104.

Bennett, V.C., and DePaolo, D.J., 1987, Proterozoic crustal history of the western United States as determined by neodymium isotopic mapping: Geological Society of America Bulletin, v. 99, p. 674–685.

Best, M.G., 1988, Early Miocene change in direction of least principal stress, Southwestern United States—Conflicting inferences from dikes and metamorphic core-detachment fault terranes: Tectonics, v. 7, p. 249–259.

Best, M.G., and Christiansen, E.H., 1991, Limited extension during peak Tertiary volcanism, Great Basin of Nevada and Utah: Journal of Geophysical Research, ser. B, v. 96, p. 13509–13528.

Best, M.G., Christiansen, E.H., and Blank, R.H., Jr., 1989, Oligocene caldera complex and calc-alkaline tuffs and lavas of the Indian Peak volcanic field, Nevada and Utah: Geological Society of America Bulletin, v. 101, p. 1076–1090.

Bird, Peter, 1988, Formation of the Rocky Mountains, western United States—A continuum computer model: Science, v. 239, p. 1501–1507.

Boardman, S.J., and Condie, K.C., 1986, Early Proterozoic bimodal volcanic rocks in central Colorado, U.S.A., part II—Geochemistry, petrogenesis and tectonic setting: Precambrian Research, v. 34, p. 37–68.

Bowring, S.A., and Karlstrom, K.E., 1990, Growth, stabilization, and reactivation of Proterozoic lithosphere in the Southwestern United States: Geology, v. 18, p. 1203–1206.

Burke, D.B., and McKee, E.H., 1979, Mid-Cenozoic volcano-tectonic troughs in central Nevada: Geological Society of America Bulletin, v. 90, p. 181–184.

Colucci, M.T., Dungan, M.A., Ferguson, K.M., Lipman, P.W., and Moorbath, Stephen, 1991, Precaldera lavas of the southeast San Juan volcanic field—Parent magmas and crustal interactions: Journal of Geophysical Research, ser. B, v. 96, p. 13413–13434.

Condie, K.C., 1978, Geochemistry of Proterozoic granitic plutons from New Mexico, U.S.A.: Chemical Geology, v. 21, p. 131–149.

———1982, Plate tectonics model for Proterozoic continental accretion in the Southwestern United States: Geology, v. 10, p. 37–42.

Cross, T.A., and Pilger, R.H., 1978, Constraints on absolute motion and plate interaction inferred from Cenozoic igneous activity in the Western United States: American Journal of Science, v. 278, p. 865–902.

DePaolo, D.J., 1981, Trace element and isotopic effects of combined wallrock assimilation and fractional crystallization: Earth and Planetary Science Letters, v. 53, p. 189–202.

Engel, Celeste, 1959, Igneous rocks and constituent hornblendes of the Henry Mountains, Utah: Geological Society of America Bulletin, v. 70, p. 961–980.

Fitton, J.G., James, Dodie, Kempton, P.D., Ormerod, D.S., and Leeman, W.P., 1988, The role of lithospheric mantle in the generation of late Cenozoic basic magmas in the Western United States, *in* Menzies, M.A., and Cox, K.G., eds, Oceanic and continental lithosphere—Similarities and differences: Journal of Petrology, 1988, p. 331–349.

Fountain, D.M., and Christiansen, N.I., 1989, Composition of the continental crust and upper mantle—A review, *in* Pakiser, L.C., and Mooney, W.D., eds., Geophysical framework of the continental United States: Geological Society of America Memoir 172, p. 711–742.

Gill, J.B., 1981, Orogenic andesites and plate tectonics: New York, Springer Verlag, 390 p.

Glazner, A.F., and Bartley, J.M., 1984, Timing and tectonic setting of Tertiary low-angle normal faulting and associated magmatism in the Southwestern United States: Tectonics, v. 3, p. 385–396.

Hacker, B.R., and Christie, J.M., 1990, Brittle/ductile and plastic/cataclastic transitions in experimentally deformed and metamorphosed amphibolite, *in* Duba, A.G., Durham, W.B., Handin, J.W., and Wang, H.F., eds., The brittle-ductile transition in rocks: Geophysical Monograph 56, p. 127–147.

Hamilton, W.B., 1988, Laramide crustal shortening, *in* Schmidt, C.J., and Perry, W.J., eds., Interaction of the Rocky Mountain foreland and the Cordilleran thrust belt: Geological Society of America Memoir 171, p. 27–39.

Helz, R.T., 1976, Phase relations of basalts in their melting ranges at P_{H_2O} = 5 kb—Part II, Melt compositions: Journal of Petrology, v. 17, p. 139–193.

Herzberg, C.T., 1987, Magma density at high pressure—Part 2, A test of the olivine flotation hypothesis, *in* Mysen, B.O., ed., Magmatic processes—Physicochemical principles: Geochemical Society Special Publication 1, p. 47–58.

Hunt, C.B., 1953, Geology and geography of the Henry Mountains region, Utah: U.S. Geological Survey Professional Paper 228, 234 p.

———1958, Structural and igneous geology of the La Sal Mountains, Utah: U.S. Geological Survey Professional Paper 294–I, p. 305–394.

Hunt, G.L., 1988, Petrology of the Mt. Pennell central stock, Henry Mountains, Utah: Brigham Young University Geology Studies, v. 35, p. 81–100.

Irwin, T.D., 1973, The petrologic evolution of the North Mountains stock, La Sal Mountains, Utah: Tucson, Ariz., University of Arizona M.S. thesis, 71 p.

Jackson, M.D., and Pollard, D.D., 1988, The laccolith-stock controversy—New results from the southern Henry Mountains, Utah: Geological Society of America Bulletin, v. 100, p. 117–139.

Johnson, C.M., 1991, Large-scale crust formation and lithosphere modification beneath middle to late Cenozoic calderas and volcanic fields, western North America: Journal of Geophysical Research, ser. B, v. 96, p. 13485–13507.

Johnson, C.M., and Fridrich, C.J., 1990, Non-monotonic chemical and O, Sr, Nd, and Pb isotope zonations and heterogeneities in the mafic- to silicic-composition magma chamber of the Grizzly Peak Tuff, Colorado: Contributions to Mineralogy and Petrology, v. 105, p. 677–690.

Johnson, C.M., Lipman, P.W., and Czamanske, G.K., 1990, H, O, Sr, Nd, and Pb isotope geochemistry of the Latir volcanic field and cogenetic intrusions, New Mexico, and relations between evolution of a continental magmatic center and modifications of the lithosphere: Contributions to Mineralogy and Petrology, v. 104, p. 99–124.

Karlstrom, K.E., Bowring, S.A., and Conway, C.M., 1987, Tectonic significance of an Early Proterozoic two-province boundary in central Arizona: Geological Society of America Bulletin, v. 99, p. 529–538.

Kelemen, P.B., Johnson, K.T.M., Kinzler, R.J., and Irving, A.J., 1990, High-field-strength element depletions in arc basalts due to mantle-magma interaction: Nature, v. 345, p. 521–524.

Kempton, P.D., Fitton, J.G., Hawkesworth, C.J., and Ormerod, D.S., 1991, Isotopic and trace element constraints on the composition and evolution of the lithosphere beneath the southwestern United States: Journal of Geophysical Research, ser. B, v. 96, p. 13713–13735.

Knoper, M.W., and Condie, K.C., 1988, Geochemistry and petrogenesis of early Proterozoic amphibolites, west-central Colorado: Chemical Geology, v. 67, p. 209–225.

Le Bas, M.J., Le Maitre, R.W., Streckeisen, A., and Zanettin, B.A., 1986, A chemical classification of volcanic rocks based on the total alkali-silica diagram: Journal of Petrology, v. 27, p. 745–750.

Lipman, P.W., Doe, B.R., Hedge, C.E., and Steven, T.A., 1978, Petrologic evolution of the San Juan volcanic field, southwestern Colorado—Pb and Sr isotope evidence: Geological Society of America Bulletin, v. 89, p. 59–82.

Lipman, P.W., and Glazner, A.F., 1991, Introduction to middle Tertiary Cordilleran volcanism—Magma sources and relation to regional tectonics: Journal of Geophysical Research, ser. B, v. 96, p. 13193–13199.

Lucchitta, Ivo, 1979, Late Cenozoic uplift of the southwestern Colorado Plateau and adjacent lower Colorado River region: Tectonophysics, v. 61, p. 63–95.

Mattox, S.R., 1991, Petrology, age, geochemistry and correlation of the Tertiary volcanic rocks of the Awapa Plateau, Garfield, Piute, and Wayne Counties, Utah: Utah Geological Survey Miscellaneous Publication 91–5, 46 p.

McGetchin, T.R., and Silver, L.T., 1972, A crustal-upper-mantle model for the Colorado Plateau based on observations of crystalline rock fragments in the Moses Rock dike: Journal of Geophysical Research, v. 77, p. 7022–7037.

McKay, G.A., 1989, Partitioning of rare earth elements between major silicate minerals and basaltic melts, *in* Lipin, B.R., and McKay, G.A., eds. Geochemistry and mineralogy of the rare earth elements: Mineralogical Society of America Reviews in Mineralogy, v. 21, p. 45–78.

Nelson, B.K., and DePaolo, D.J., 1984, 1700-Myr greenstone volcanic successions in southwestern North America and isotopic evolution of Proterozoic mantle: Nature, v. 312, p. 143–146.

———1985, Rapid production of continental crust 1.7 to 1.9 b.y. ago—Nd isotopic evidence from the basement of the North American mid-continent: Geological Society of America Bulletin, v. 96, p. 746–754.

Nelson, S.T., 1991, Mid-Tertiary magmatism of the Colorado Plateau—The Henry and La Sal Mountains, Utah: Los Angeles, University of California Ph. D. thesis, 318 p.

Nelson, S.T., and Davidson, J.P., 1993, Interactions between mantle-derived magmas and mafic crust, Henry Mountains, Utah: Journal of Geophysical Research, ser. B, v. 98, p. 1837–1852.

Nelson, S.T., Davidson, J.P., and Sullivan, K.R., 1992, New age determinations of central Colorado Plateau laccoliths, Utah—Recognizing disturbed K-Ar systematics and re-evaluating tectonomagmatic relationships: Geological Society of America Bulletin, v. 104, p. 1547–1560.

O'Hara, M.H., 1968, The bearing of phase equilibria studies in synthetic and natural systems on the origin and evolution of basic and ultrabasic rocks: Earth Science Reviews, v. 4, p. 69–133.

Ormerod, D.S., Hawkesworth, C.J., Rogers, N.W., Leeman, W.P., and Menzies, M.A., 1988, Tectonic and magmatic transitions in the western Great Basin, USA: Nature, v. 333, p. 349–353.

Perry, F.V., DePaolo, D.J., and Baldridge, W.S., 1993, Neodymium isotopic evidence for decreasing crustal contributions to Cenozoic ignimbrites of the Western United States—Implications for the thermal evolution of the Cordilleran crust: Geological Society of America Bulletin, v. 105, p. 872–882.

Robertson, J.M., and Condie, K.C., 1989, Geology and geochemistry of early Proterozoic volcanic and subvolcanic rocks of the Pecos greenstone belt, Sangre de Cristo Mountains, New Mexico, *in* Grambling, J.A., and Tewksbury, B.J., eds., Proterozoic geology of the Southern Rocky Mountains: Geological Society of America Special Paper 235, p. 119–146.

Rushmer, Tracy, 1991, Partial melting of two amphibolites—Contrasting experimental results under fluid absent conditions: Contributions to Mineralogy and Petrology, v. 107, p. 41–59.

Severinghaus, Jeff, and Atwater, Tanya, 1990, Cenozoic geometry and thermal state of the subducting slabs beneath Western North America, *in* Wernicke, B.P., ed., Basin and Range extensional tectonics near the latitude of Las Vegas, Nevada: Geological Society of America Memoir 176, p. 1–22.

Smith, R.B., Nagy, W.C., Julander, K.A., Viveros, J.J., Barker, C.A., and Gants, D.G., 1989, Geophysical and tectonic framework of the eastern Basin and Range–Colorado Plateau–Rocky Mountain transition, *in* Pakiser, L.C., and Mooney, W.D., eds., Geophysical framework of the continental United States: Geological Society of America Memoir 172, p. 205–233.

Sullivan, K.R., Kowallis, B.J., and Mehnert, H.H., 1991, Isotopic ages of igneous intrusions in southeastern Utah—Evidence for a mid-Cenozoic Reno–San Juan magmatic zone: Brigham Young University Geology Studies, v. 37, p. 139–144.

Sun, S.S., 1980, Lead isotopic study of young volcanic rocks from mid-ocean ridges, ocean islands, and island arcs: Philosophical Transactions of the Royal Society of London, v. A27, p. 409–445.

Tatsumi, Y., Hamilton, D.L., and Nesbitt, R.W., 1986, Chemical characteristics of fluid phase released from a subducted lithosphere and origin of arc magmas—Evidence from high-pressure experiments and natural rocks: Journal of Volcanology and Geothermal Research, v. 29, p. 293–309.

Taylor, S.R., and McLennan, S.M., 1985, The continental crust—Its composition and evolution—An examination of the geochemical record preserved in sedimentary rocks: Oxford, Blackwell Scientific, 312 p.

Thompson, G.A., and Zoback, M.L., 1979, Regional geophysics of the Colorado Plateau: Tectonophysics, v. 61, p. 149–181.

Weaver, B.L., and Tarney, John, 1984, Empirical approach to estimating the composition of the continental crust: Nature, v. 310, p. 575–577.

Witkind, I.J., 1964, Geology of the Abajo Mountains area, San Juan County, Utah: U.S. Geological Survey Professional Paper 453, 110 p.

Wolf, L.W., and Cipar, J.J., 1993, Through thick and thin—A new model for the Colorado Plateau from seismic refraction data from Pacific to Arizona crustal experiment: Journal of Geophysical Research, ser. B, v. 98, p. 19881–19894.

Zandt, G., Myers, S.C., and Wallace, T.C., 1995, Crust and mantle structure across the Basin and Range–Colorado Plateau boundary at 37° N. latitude and implications for Cenozoic extensional mechanism: Journal of Geophysical Research, ser. B, v. 100, p. 10529–10548.

Relation Between Middle Tertiary Dike Intrusion, Regional Joint Formation, and Crustal Extension in the Southeastern Paradox Basin, Colorado

By Marilyn A. Grout[1] *and* Earl R. Verbeek[2]

CONTENTS

FIGURES

[1]Present address: Marilyn Grout, Consulting Geologist, 5407 S.W. Viewpoint Terrace, Portland, OR 97201.

[2]Present address: Earl Verbeek, Geologist, Bureau of Land Management, Bldg. 102, Military Circle, P.O. Box 911, Tonopah, NV 89049-0911.

ABSTRACT

Outcrop relations of a middle Tertiary monzonite dike near Rico, Colorado, suggest that the dike records the early phases of a period of regional crustal extension that affected much of the northern and central Colorado Plateau. The dike, about equidistant between the Rico and La Plata Mountains laccolithic centers in the southeastern part of the Paradox Basin, intruded Lower Permian strata that previously had been broken by vertical joints of north-northwest strike. The joints are members of a regionally extensive set of probable Permian age. Evidence that the dike intruded preexisting joints, rather than hydraulic fractures formed during the intrusion process, includes the restriction of the joints to the most well cemented beds of the host rock, their regular spacing, and the presence of predike, recrystallized calcite fillings. Two postdike regional joint sets also are present in the same area.

The laccoliths and associated dikes of the Paradox Basin are broadly contemporaneous with the eruption of voluminous volcanic rocks along the edges of the Colorado Plateau to the west (the Marysvale volcanic field in the High Plateaus transition zone of southwestern Utah) and to the east (the San Juan volcanic field of southwestern Colorado). All are probable manifestations of the same period of regional east-northeast crustal extension whose effects, as expressed by igneous activity, were much more subdued on the Colorado Plateau than along its margins. The monzonite dike near Rico, taken within the context of the regional fracture history of the Paradox Basin, is one of the earliest expressions of this extension within the basin. Continued extension resulted in formation of a regional set of joints with a median strike of N. 29° W., almost exactly parallel both to the dike and to the older, more weakly expressed set of joints that it intruded.

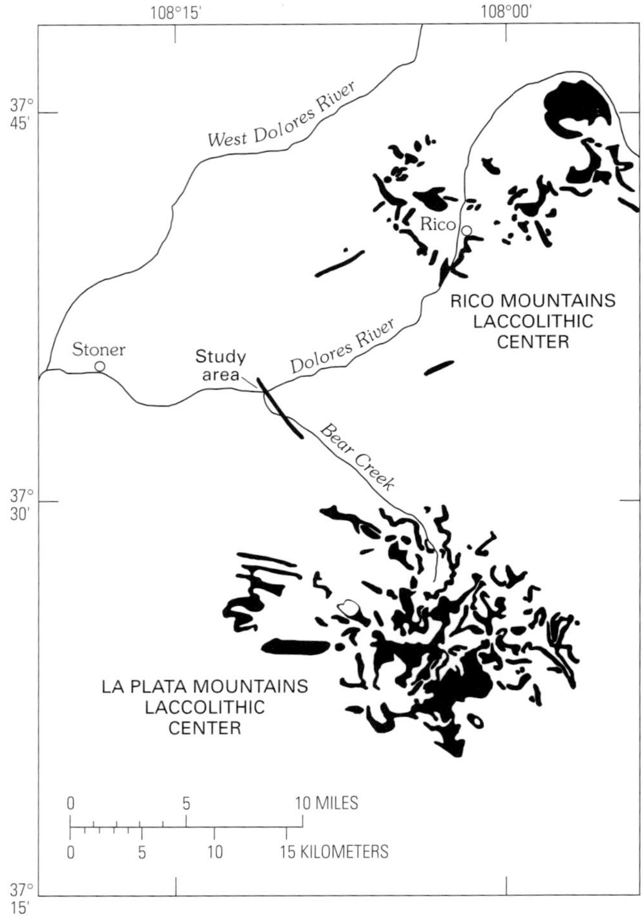

Figure 1. Location of the Rico and La Plata Mountains laccolithic centers and of the studied dike in the southeastern Paradox Basin, Colo. Because parts of the centers are still masked by cover rocks, they appear as discontinuous masses within roughly circular areas. Modified from Haynes and others (1972) and Steven and others (1974).

INTRODUCTION

This report discusses the correlation between timing of dike intrusion and regional joint formation as it may relate to crustal extension following Laramide compressive events in the Paradox Basin of southeastern Utah and southwestern Colorado. Conclusions presented herein are based on structural relations documented along a well-exposed monzonite dike in the southeastern part of the Paradox Basin, nearly equidistant from the Rico and La Plata Mountains laccolithic centers (fig. 1). The dike, of middle Tertiary age, intruded Permian through at least Lower Cretaceous sedimentary strata (Haynes and others, 1972). Field evidence at the locality studied, immediately north of the Dolores River about 18 km southwest of Rico, Colo. (fig. 1), suggests that the dike intruded preexisting joints in sandstone beds of the Lower Permian Cutler Formation. Outcrop relations of this dike are here discussed in relation to new age determinations on nearby laccolithic rocks (Nelson and others, 1992, and this volume; Cunningham and others, 1994); to recent work on the fracture and tectonic history of Paleozoic through Cretaceous rocks in nearby parts of the Paradox Basin (Grout and Verbeek, this volume); and to recent studies of plutonism, volcanism, and extension farther west (Rowley and others, this volume).

JOINT SYSTEMS IN THE PARADOX BASIN

The history of joint formation in the Paradox Basin was determined from such information as joint orientation, dimensions, surface structures, spacing, and shape, as well as abutting relations, mineral fillings, geographic and stratigraphic distribution, and relation to other structures.

Taken collectively, these characteristics uniquely define each joint set in the basin and form the basis for correlating sets from one outcrop to another, for determining the regional sequence of joint-set formation, and for relating the fracture history to the regional tectonic history. Care in interpreting relative ages of sets is one of the most critical aspects of such studies because joints of similar orientation may form more than once in a given region, as they have in the Paradox Basin. For discussion of field methods used and their general applications the reader is referred to Grout and Verbeek (1983) and Verbeek and Grout (1991).

Three joint systems, each comprising multiple sets of joints, have been delineated to date in the Paradox Basin (Grout and Verbeek, this volume). Regional joint-set correlations and the documented stratigraphic ranges of each set show that the three systems originated in Carboniferous, Permian, and middle Tertiary (Miocene?) time, respectively. The outcrop studied for this report contains one set of joints of Permian age and two sets of middle Tertiary age.

STRUCTURAL RELATIONS BETWEEN OTHER DIKES AND JOINTS

Structural relations between joints and various types of dikes in other areas near the Paradox Basin have been documented for Tertiary mafic igneous dikes in southern Utah by Delaney and others (1986), for middle Tertiary hydrocarbon (gilsonite) dikes in eastern Utah by Verbeek and Grout (1992, 1993), and briefly for tuffaceous clastic dikes of similar age in northwestern Colorado by Grout and Verbeek (1982). Joint-dike relations in these areas fall into three general categories, summarized below:

1. *Dike intrudes unbroken rock.* A dike in unbroken rock must have created its own conduit. The dike accomplishes this through a hydraulic fracture mechanism wherein magmatic pressure exceeds the minimum horizontal compressive stress in the vicinity of the dike. Subsidiary joints parallel to the dike, and within a narrow zone adjacent to it, commonly are created during this process (Delaney and others, 1986; Verbeek and Grout, 1992, 1993). Dike intrusion by hydraulic fracture is indicated where (a) local dike-parallel joints decrease in abundance away from the dike, typically to near zero within distances of 15–150 m for igneous dikes (Delaney and others, 1986) and 6–12 m for hydrocarbon dikes (Verbeek and Grout, 1992; 1993); (b) the fractures forming the dike walls are the largest joints in the rock (Verbeek and Grout, 1992; 1993); (c) joints in the host rock, excluding the local dike-related joints mentioned above, are not parallel to the dike; and (d) abutting relations among joints of the different sets show that the dike-parallel joints are the oldest fractures in the rock.

2. *Dike intrudes jointed rock.* Magma intruding jointed rock can exploit existing fractures as conduits. Only limited additional breakage of the host rock is needed to link dilated joints and so form a continuous dike, parts of whose walls are the faces of preexisting joints (Grout and Verbeek, 1982; Delaney and others, 1986). The intruding magma can dilate any fracture for which magmatic pressure equals or exceeds the normal component of compressive stress across the fracture walls. Thus, intrusion is not limited to joints that are exactly perpendicular to the minimum horizontal compressive stress but can include joints that strike to within about 30° of the perpendicular direction (Delaney and others, 1986). It follows that dikes intruded into jointed rock may not be reliable indicators of exact paleostress orientations.

Local dike-parallel joints similar to those formed during intrusion of dikes into unbroken rock can form also where magma invades previously jointed rock, but the spacing of the dike-parallel joints increases markedly with distance from the contact, and generally none are present beyond a few tens of meters from the dike (Delaney and others, 1986). Their size can also be stunted because some of the induced tensile stress ahead of the tip of the advancing dike is relieved by dilation of joints already present rather than by increasing the surface area of the new joints.

3. *Dike intrusion during regional jointing.* Dike-parallel joints in some areas initially seem to show a puzzling distribution: generally they are restricted to narrow zones adjacent to the dike, but in rare beds they are of regional extent, even where magmatic intrusion through a hydraulic fracture mechanism is strongly indicated by the field evidence. Delaney and others (1986) suggested that the stratigraphic discontinuity of these regional dike-parallel joints shows that they could not have resulted from intrusion of the dike. Their calculations of the widths of zones of induced tensile stresses adjacent to advancing dikes led to the same conclusion for all reasonable combinations of rock mechanical properties and magmatic pressures. Verbeek and Grout (1992, 1993) noted that all of the regional joints that parallel the large hydrocarbon dikes examined by them are in very thin, laterally persistent, exceptionally brittle beds—those most susceptible to extensional failure at low strains. They interpreted these joints as products of the earliest stages of tectonic extension in the area, when the dikes also formed. The dikes are hydraulic fractures that created their own path. The joints, however, are due to direct layer-parallel extension.

STRUCTURE AND AGE OF THE DIKE

The monzonite dike studied for this report is subvertical and is discontinuously exposed for about 6 km along its length; it strikes N. 28° W. in its northernmost part but curves to N. 42° W. near its southeast end (fig. 1). The dike is almost equidistant from the Rico and La Plata Mountains

Figure 2. Contact between Tertiary monzonite dike and jointed Cutler Formation (Lower Permian) host rocks in vertical roadcut. Contact is vertical and planar along a north-northwest-striking joint in host rock but otherwise irregular and curving in unjointed parts of the host rock. Hammer is on contact between one joint and the dike. View north-northwest.

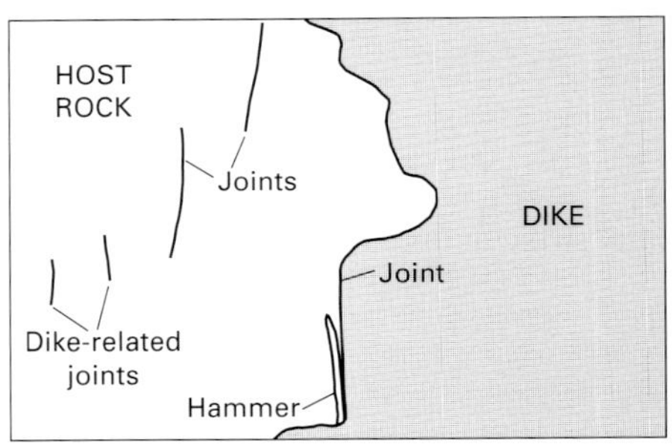

laccolithic centers, and thus is on the outermost fringes of each. Rocks of the two centers are similar in composition (Haynes and others, 1972). The dike, where studied immediately north of the Dolores River (fig. 1), intruded essentially horizontal sedimentary strata of Permian age and is 14–15 m thick. In vertical profile its walls are irregular, but locally, especially within the finer grained layers of the host rock, the dike is bounded by planar fractures of north-northwest strike (fig. 2). Field evidence discussed below suggests that the dike intruded preexisting joints of Permian age and subsequently was cut by joints of a later regional set of middle Tertiary age.

Although the dike has not been dated, samples from the nearby Rico and La Plata Mountains laccolithic centers yield Late Cretaceous to Paleocene K/Ar ages of 70–60 Ma (Mutschler and others, this volume) and Late Cretaceous to Miocene ages of 78–20 Ma (Cunningham and others, 1994). Igneous rocks from some of the other laccolithic centers in and near the western part of the Paradox Basin yielded similar K/Ar ages. Most intrusions from these centers were emplaced in Permian and Triassic strata, but some intruded the Upper Cretaceous Mancos Shale (Haynes and others, 1972; Steven and others, 1974). Younger rocks have been eroded from the immediate area. Radiometric ages of these and other central Colorado Plateau laccoliths, however, are now being reevaluated in light of recently developed $^{40}Ar/^{39}Ar$ methods that consistently yield the younger age determinations of 32–23 Ma (Oligocene/Miocene) for samples from the La Sal, Abajo, and Henry Mountains complexes in southeastern Utah. The dike in the study area therefore may also be middle Tertiary in age. Nelson and others (1992, and this volume) provide a review of the geochronology and its regional implications.

JOINTS IN THE LOWER PERMIAN HOST ROCKS

The Lower Permian host rocks cut by the dike consist of interlayered, locally crossbedded, silty, very fine grained to medium-grained sandstone and interlayered shale beds of the Cutler Formation. The finer grained layers are reddish brown and 12–80 cm thick, whereas the coarser grained layers are reddish purple and 10–100 cm thick. These rocks are cut by three sets of joints that are prominent in some strata but weakly developed or absent in others. The two older sets generally are most abundant in the finer grained layers and the youngest set in the coarser grained layers. Characteristics of the joints of each set are summarized below.

FIRST-FORMED SET

Joints of the first-formed set in the host rocks are vertical and have a median strike of N. 29° W. Among 15 of the most planar joints of this set selected for measurement (the most planar joints are the best indicators of paleostress directions), 14 of them are within 8° of the median strike. Most of these joints can be traced vertically for at least 85–190 cm through several sandstone and siltstone layers, but their full heights could not be determined because of soil cover; the highest observed was slightly more than 3 m. The joints strike at high angles to the steep outcrop face and thus have little of their lengths exposed; their true lengths probably are much greater than the exposed dimensions of 1.5 m or less. Spacings range from 2–8 cm locally to as much as 180 cm, but spacings of 30–90 cm are typical of much of the outcrop. The joints are planar or nearly so; plumose structure and arrest lines on their surfaces are indicative of failure by extension. The joints terminate laterally as hairline cracks and abut no other fractures, as they were the earliest set of fractures to form in the rock. Translucent, white to gray, granular calcite in seams 1.5–4 mm thick is preserved in some of these joints. Some of the calcite grains are as much as 2 cm long within the plane of the joint, possibly indicating thermal recrystallization. The walls of some of these joints adjacent to the dike are partially baked and blackened.

The large size, planarity, and low strike dispersion of the joints in this set are properties common to early sets of joints in sedimentary rocks because such joints develop unimpeded by the presence of other fractures.

SECOND-FORMED SET

The joints of the second-formed set in the Lower Permian host rocks are subvertical and have a median strike of N. 72° W. They are sparsely distributed in the host rock: spacings of 3–4 m are common in beds of fine-grained reddish-brown sandstone, but the set is absent from the interbedded siltstones. None of these joints were found in contact with the dike. The joints are 1.5–2.0 m high, but their true lengths could not be determined, as all extend into unexposed rock; exposed lengths of 2 m or less are common. Unlike joints of the earlier set, most joints of the second-formed set are subplanar rather than planar. Plumose structure and arrest lines on their surfaces show that they are extension joints. Some of them contain white calcite in thin seams less than one-third millimeter thick. Terminations of these joints against the north-northwest-striking joints of the earlier set amply establish the relative age of the two sets.

THIRD-FORMED SET

Joints of the youngest set to cut the Lower Permian host rocks are prominent in the coarser grained purple-weathering sandstone but are sparse in the reddish-brown beds of finer grain size. These joints are subvertical and have a median strike of N. 51° E. Overall they are about as abundant as joints of the oldest set but are much smaller in both dimensions; exposed heights of only 12–80 cm and exposed lengths of 80–150 cm are typical. Probably these values are close to the true joint dimensions, for the tops, bottoms, and one or both ends of a sufficient number of their surfaces are exposed that one can gain a good impression of their full size. Commonly they are spaced 16–40 cm apart; the observed range is 5–120 cm. Most of these joints are subplanar and curve broadly along strike and dip; few approximate a plane. The same array of surface structures as was found on joints of the other two sets establishes them as extension joints. A few of them are filled with white fibrous calcite, unlike the coarse granular calcite of the oldest set; the calcite fill is 4 mm or less thick, and the fibers are oriented approximately perpendicular to the joint surfaces. On the northeast side of the dike, joints of this set in the host rock locally are darkly stained by manganese or iron oxides.

Some of these joints cut across mineral-filled joints of the two older sets, and others terminate against the older joints. A few appear to have originated at a point on the surface of one of the older joints and grown outward into the rock. All these relations establish that the northeast-striking joints are the youngest of the three sets in the outcrop.

LOCAL DIKE-PARALLEL JOINTS

Small, north-northwest-striking joints that cut the host rock within 2 m of the southwestern dike wall probably are members of a weakly formed zone of dike-parallel joints that are true products of intrusion. They differ from the older tectonic joints of similar orientation in their smaller size, spatial relation to the dike, and lack of mineral fillings. Analogous joints were not found adjacent to the northeast dike wall. As noted above, prominent zones of dike-parallel joints commonly form near dikes that invade newly created hydraulic fractures, but such joints can also form in modest abundance where magma invades fractures already present, as at the Dolores River locality. The influence of the preexisting joints at this locality is clearly seen in the weak definition of the dike-parallel joint zone: the dike-parallel joints are sparse and affected only a small volume of rock. Similar observations have been made in other areas where magma has invaded preexisting fractures in the rock (Delaney and others, 1986; Rogers and Bird, 1987; Verbeek and Grout, 1992).

JOINTS IN THE MONZONITE DIKE

Subvertical joints in the monzonite dike strike from north-northeast to east-northeast and generally are much more irregular in shape and longer than joints in the host rock. These joints form two sets, one of which most likely is a set of cooling joints perpendicular to the dike wall; the other correlates with the youngest set of joints in the host rocks. The orientation distributions of the two sets overlap, but their respective joint planes are sufficiently different in overall style that the sets may be distinguished.

COOLING JOINTS

Cooling joints in the dike terminate laterally against the contact with the host sandstone and are more irregular in shape within the interior of the dike than near its margins. Spacings are variable and range from 3 to 24 cm. Most of these joints are exposed for 0.5–2 m both vertically and laterally. Their true heights are unknown, for most are covered, but true lengths could be observed for many of the joints, and they were not greatly different from the exposed lengths of the remaining joints. The fractures are subplanar and generally broadly sinuous along strike, but locally some are planar and contain plumose structure and a twist-hackle fringe.

YOUNGER JOINTS

The most planar, subvertical joints in the dike have a median strike of N. 57° E., but their strike variation is about 50°, more than double that of the corresponding joint set in the host rock. The wide strike dispersion likely reflects the irregular nature of the local stress field due to the presence of abundant, irregular, preexisting discontinuities (the cooling joints); old joints in the adjacent host rock were much more widely spaced. Many of the N. 57° E.-striking joints in the dike contain only thin films of white calcite, but calcite on one is as much as 1 mm thick and is clearly fibrous, similar to that in joints of the same set in the host rock. Numerous joints of this set cut across the dike-wall contacts into the host rock.

Subhorizontal fractures, also perpendicular to the dike wall, are present in the dike rock but were not measured. They either formed upon cooling of the dike or, more likely, are sheeting joints formed during exhumation. Secondary malachite on one such fracture near the northeast contact of the dike with the host rock probably reflects local migration of copper originally introduced during intrusion.

REGIONAL CORRELATION OF JOINT SETS

All three sets of joints in the Lower Permian host rocks are of regional extent; we have documented them at many localities throughout the central Paradox Basin, from the Green River in Utah to the eastern margin of the basin in southwestern Colorado. Joints of the oldest set near the dike (PX_2 set of Grout and Verbeck, this volume) are members of the second of three regional sets of probable Permian age, as they have been found to date only in Lower Permian and older rocks and not in the Lower Triassic and younger rocks above. Some of these subvertical joints were intruded by the dike, whose walls are planar along these joints but irregular between them (fig. 2). Joints of the second- and third-formed sets in the host rocks are members of the third (PX_6) and fourth (PX_7) sets, respectively, of six regional sets of Tertiary age in the basin (Grout and Verbeek, this volume). Since joints of these two sets postdate the dike, their ages are constrained by the time of intrusion. Assuming the dike to be of similar age to the newly dated laccolithic centers west of the study area (Nelson and others, 1992, and this volume), the age of the Tertiary joints in the dike would thus be conclusively established as no older than 32 Ma. This suggested age accords well with our field experience: the joints of both sets have been traced eastward into volcanic units dated as Miocene in the San Juan volcanic field along the eastern margin of the basin (Grout and Verbeek, this volume).

TERTIARY JOINT SET PARALLEL TO DIKE BUT NOT PRESENT HERE

Joints of the oldest known regional Tertiary set in the Paradox Basin (excepting an older set of more local extent along the northern and eastern margins of the basin) strike N. 22°–34° W. (Grout and Verbeek, this volume), about parallel to the Permian set discussed above in relation to the dike. These Tertiary joints form the third most prevalent set (PX_4) in the basin, especially in Mesozoic rocks. Wherever they are present in Paleozoic rocks they either originate at, terminate against, or cut across the older, mineral-filled, Permian joints; their relatively young age is thus well established by field evidence at numerous localities.

Though these NNW.-striking Tertiary joints are prominent on a regional basis, and have been found at nearly one-third of the 494 localities studied, they are absent from the outcrop of the dike. Their absence is due to the prior formation of joints of nearly identical orientation—the Permian PX_2 joints discussed above, which were the first to form in the Lower Permian beds at this particular locality. Any new increments of extensional strain in these beds likely were accommodated by dilation of the Permian joints rather than by formation of new fractures. In other areas, however, where older fractures favorably oriented for reactivation were absent, the Tertiary joints formed in abundance. Suppression of joint-set formation by a preexisting set of broadly similar orientation is an expected and well-known effect in many localities (Verbeek and Grout, 1984; Grout and Verbeek, 1992; Throckmorton and Verbeek, 1995).

Below we suggest that the dike is an early product of a prolonged period of regional crustal extension and that the prominent Tertiary set of NNW.-striking PX_4 joints is a slightly later manifestation of the same extension. We further suggest, tentatively, that this extension affected not only the Paradox Basin but also much of the northern Colorado Plateau and is related to the extensional volcanism discussed by Rowley and others (this volume) in the Marysvale volcanic field farther west.

DISCUSSION

DIKE INTRUSION OF PREVIOUSLY JOINTED ROCK

Evidence that the N. 29° W.-striking PX_2 set of joints formed prior to intrusion rather than as a result of it is compelling. The relatively regular spacing of these joints, their stratigraphic discontinuity, and the evidence of predike calcite fillings are properties much more in keeping with regional joints than with hydraulic fractures formed during intrusion. So too is the overall dike morphology, wherein planar dike segments in some beds are linked by irregular, curving segments in others (fig. 2). We interpret these observations as evidence of intrusion into a heterogeneous sequence of rock, one in which the most well cemented beds had already been jointed but the more weakly cemented beds had not. The intruded joints correspond to a well-documented set of probable Permian age that is present in Paleozoic rocks over an extensive area across the central Paradox Basin, from the Green River in Utah to the eastern margin of the basin in Colorado (Grout and Verbeek, this volume). Joints of this set have a regional median strike of N. 28° W. and are present at more than one-third of the Paleozoic outcrops studied.

DIKE INTRUSION DURING CRUSTAL EXTENSION

We suggest that the monzonite dike near Rico and the laccolithic complexes nearby intruded the crust during a time of renewed ENE.-WSW. crustal extension in this part of the Western United States during Oligocene/Miocene time. Beginning in the middle Cenozoic, calc-alkaline magmatism became widespread in the Western United States as far east as Colorado (Lipman and others, 1972). Laccolithic intrusive centers formed more abundantly in the Paradox Basin than in other areas of the Colorado Plateau (see Kelley and Clinton, 1960), but they are minor features compared to the large volcanic fields both west and east of the Paradox Basin, on the edges of the Colorado Plateau (Rowley and others, this volume; Mutschler and others, this volume).

Volcanic rocks of the Marysvale field, west of the Paradox Basin in southwestern Utah, were erupted at 34(?)–21 Ma during a period of regional crustal extension. The extension direction is regarded by Rowley and others (this volume) to have been east-northeast, consistent with the Cenozoic stress history presented by Zoback and others (1981). The volcanic and underlying igneous intrusive rocks of the Marysvale field define two igneous belts, the Pioche-Marysvale and Delamar–Iron Springs belts, both of which trend east-northeast. The parallelism of this trend to the postulated extension direction led Rowley and others (this volume) to interpret the igneous belts as continental analogs of deep-seated oceanic transform faults which localized plutonism and volcanism over significant periods of time. On the opposite side of the Colorado Plateau, volcanic rocks of the San Juan Mountains erupted about 35–30 Ma from scattered central volcanoes and are overlain by widespread, voluminous ash-flow sheets from calderas dated at 32–23.1 Ma (Lipman, 1989). The times of the eruptions in both fields are similar and correspond to the 32.3–22.6 Ma dates suggested by the new $^{40}Ar/^{39}Ar$ age determinations for emplacement of three of the laccolithic centers in the Paradox Basin (Nelson and others, 1992, and this volume), and presumably also for the dike studied. Magmatism thus was broadly contemporaneous within a span of about 10 million years during Oligocene/Miocene time for a roughly east-trending belt that spans the entire width of the central Colorado Plateau, from the Marysvale field on the High Plateaus to the west, through the Paradox Basin, to the San Juan field on the east (Nelson and others, 1992). The interplay between crustal extension, magmatism, and regional joint-set formation as outlined in this report had not previously been recognized in the Paradox Basin. We stress, however, that further work on reconciling the broad spread of age determinations on igneous rocks in the region is needed before these relations can be more fully understood.

CENOZOIC STRESS-FIELD ROTATION ON THE COLORADO PLATEAU

Evidence from regional joint-set formation in the Paradox Basin strongly suggests that the dike discussed in this report is an early product of a prolonged period of Tertiary regional extension during which the stress field rotated counterclockwise with time. The Tertiary paleostress history of the Colorado Plateau, however, remains controversial. The present-day state of stress has been documented in several reports (for example, Zoback and Zoback, 1980; Wong and Humphrey, 1989), and aspects of the paleostress field during a given epoch in others (for example, Zoback and others, 1981, 1994). The horizontal component of Tertiary stress-field rotation, however, has variously been reported as (1) clockwise, based on relative ages of joint sets at 200 outcrops scattered across the plateau (Bergerat and others, 1992); (2) unchanged through time, based on interpretation of surface and subsurface fracture data on the northeastern margin of the plateau (Lorenz and Finley, 1991); and (3) counterclockwise, based on joint-set chronology at more than 1,650 outcrops in the northern part of the plateau (Verbeek and Grout, 1986, 1993, and this volume; Grout and Verbeek, 1992, and this volume).

In the central Paradox Basin, the sequence in which regional sets of joints have formed in Cretaceous and younger rocks furnishes a clear record of counterclockwise stress rotation since the cessation of Laramide compressive events (see Grout and Verbeek, this volume). Also in the basin, but farther north, a similar counterclockwise stress rotation has just been suggested from joint-chronology mapping on the flanks of the Salt Valley Anticline (Cruikshank and Aydin, 1995). The Tertiary fracture history in these areas is similar to that documented in the Piceance and Uinta Basins still farther north by Verbeek and Grout (1986, 1993, and this volume) and Grout and Verbeek (1992). We thus suggest that a period of regional crustal extension and counterclockwise stress-field rotation affected the central and northern Colorado Plateau over an area of at least 80,000 km^2. Additional products of this extension, listed here in order from oldest to youngest, include (1) a set of NNW.-striking joints over the entire region (this is the set related to the dike and associated crustal extension discussed in this report); (2) NW.-striking gilsonite (hydrocarbon) dikes in the eastern Uinta Basin (Verbeek and Grout, 1992, 1993); (3) a strongly expressed regional set of NW.- to WNW.-striking joints; (4) minor, WNW.-striking normal faults, many of them corresponding to zones of reactivated joints, particularly in the Piceance Basin; (5) a set of E.-striking joints, common in the Paradox Basin but more sparse farther north; and (6) a strongly expressed set of ENE.-striking joints over the entire region.

CONCLUSIONS

The monzonite dike in the Dolores River area near Rico intruded Lower Permian strata during middle Tertiary time. Where existing joints in the host rock were favorably oriented, as were the N. 29° W.-striking PX_2 joints of the oldest (Permian) set known in the local area, the northern segment of the dike followed those joints. Younger, middle Tertiary PX_4 joints of similar orientation are not present near the dike because the older joints suppressed their formation, but elsewhere, especially in younger rocks, the Tertiary joints are fairly abundant. Both the dike and the regional set of Tertiary joints parallel to it are related products of post-Laramide, regional, ENE.-WSW. crustal extension.

The central Paradox Basin, in the central part of the Colorado Plateau, lies between two major Oligocene-Miocene eruptive centers—the Marysvale volcanic field (34?–21 Ma) to the west and the San Juan volcanic field (32–23.1 Ma) to the east. Times of eruption in these volcanic centers are similar to those suggested by new 32.3–22.6 Ma age determinations for the Paradox laccolithic complexes, and presumably also the dike near Rico. Tracking of calc-alkaline magmatic eruptive events in the Marysvale field led Rowley and others (this volume) to link the aligned ENE.-WSW. eruptions and the direction of major crustal extension to shallow oceanic plate subduction beneath the Colorado Plateau. These relations suggest that volcanism was relatively synchronous with crustal extension of similar direction within a broad belt across the central Colorado Plateau between the two volcanic fields. On the plateau, however, this extension was manifested in a much more subdued manner than in the volcanic fields: only scattered laccolithic centers and minor aligned dikes are present in the Paradox Basin. The major tectonic expression of Oligocene-Miocene extension in this area is a pervasive, regional set of NNW.-striking joints. These structures are the first expression of regional extension in the central Paradox Basin following the cessation of Laramide compressive events earlier in the Tertiary.

ACKNOWLEDGMENTS

The U.S. Geological Survey's Evolution of Sedimentary Basins and National Geologic Mapping programs provided funding for this study. We thank D.L. Brevik for assistance in the field and G.J. Calderone and C.K. Throckmorton for valiant reviews of earlier versions of this paper. P.D. Rowley commented on later versions.

REFERENCES CITED

Bergerat, Françoise, Bouroz-Weil, Catherine, and Angelier, Jacques, 1991, Paleostresses inferred from macrofractures, Colorado Plateau, western U.S.A.: Tectonophysics, v. 206, p. 219–243.

Cunningham, C.G., Naeser, C.W., Marvin, R.F., Luedke, R.G., and Wallace, A.R., 1994, Ages of selected intrusive rocks and associated ore deposits in the Colorado mineral belt: U.S. Geological Survey Bulletin 2109, 31 p.

Cruikshank, K.M., and Aydin, Atilla, 1995, Unweaving the joints in Entrada Sandstone, Arches National Park, Utah, U.S.A.: Journal of Structural Geology, v. 17, no. 3, p. 409–422.

Delaney, P.T., Pollard, D.D., Ziony, J.L., and McKee, E.H., 1986, Field relations between dikes and joints—Emplacement processes and paleostress analysis: Journal of Geophysical Research, v. 91, no. 85, p. 4920–4938.

Grout, M.A., Abrams, G.A., Tang, R.L., Hainsworth, T.J., and Verbeek, E.R., 1991, Late Laramide thrust-related and evaporite-domed anticlines in the southern Piceance Basin, northeastern Colorado Plateau: American Association of Petroleum Geologists Bulletin, v. 75, no. 2, p. 205–218.

Grout, M.A., and Verbeek, E.R., 1982, Intrusion of clastic dikes during diagenesis and fracture of oil shales: Geological Society of America Abstracts with Programs, v. 14, no. 7, p. 503.

———1983, Field studies of joints—Insufficiencies and solutions, with examples from the Piceance Creek basin, Colorado, *in* Gary, J.H., ed., 16th Oil Shale Symposium, Proceedings: Golden, Colo., Colorado School of Mines Press, p. 68–80.

———1985, Fracture history of the Plateau Creek and adjacent Colorado River valleys, southern Piceance Basin—Implications for predicting joint patterns at depth: U.S. Geological Survey Open–File Report 85–744, 17 p.

———1992, Fracture history of the Divide Creek and Wolf Creek anticlines and its relation to Laramide basin-margin tectonism, southern Piceance Basin, northwestern Colorado: U.S. Geological Survey Bulletin 1787–Z, p. Z1–Z32.

———1993, Joint-history summary and orientation data for Upper Cretaceous sandstones, Rawlins and Rock Springs uplifts, Washakie Basin, southern Wyoming: U.S. Geological Survey Open-File Report 92–388, 32 p.

Haynes, D.D., Vogel, J.D., and Wyant, D.G., 1972, Geology, structure, and uranium deposits of the Cortez quadrangle, Colorado and Utah: U.S. Geological Survey Miscellaneous Geologic Investigations Map I–629, 2 sheets, scale 1:250,000.

Lipman, P.W., ed., 1989, Oligocene-Miocene San Juan volcanic field, Colorado, IAVCEI May field trip guide, excursion 16B: New Mexico Bureau of Mines and Mineral Resources Memoir 46, p. 303–380.

Lipman, P.W., Prostka, H.J., and Christiansen, R.L., 1972, Cenozoic volcanism and plate-tectonic evolution of the Western United States—I, Early and middle Cenozoic: Royal Society of London Philosophical Transactions, ser. A, v. 271, p. 217–248.

Lorenz, J.C., and Finley, S.J., 1991, Regional fractures II—Fracturing of Mesaverde reservoirs in the Piceance Basin, Colorado: American Association of Petroleum Geologists Bulletin, v. 75, no. 11, p. 1738–1757.

Nelson, S.T., Davidson, J.P., and Sullivan, K.R., 1992, New age determinations of central Colorado Plateau laccoliths, Utah—Recognizing disturbed K-Ar systematics and re-evaluating tectonomagmatic relationships: Geological Society of America Bulletin, v. 104, no. 12, p. 1537–1560.

Rogers, R.D., and Bird, D.K., 1987, Fracture propagation associated with dike emplacement at the Skaergaard intrusion, East Greenland: Journal of Structural Geology, v. 9, no. 1, p. 71–86.

Steven, T.A., Lipman, P.W., Hail W.J., Jr., Barker, Fred, and Luedke, R.G., 1974, Geologic map of the Durango quadrangle, southwestern Colorado: U.S. Geological Survey Miscellaneous Investigations Series Map I–764, scale 1:250,000.

Throckmorton, C.K., and Verbeek, E.R., 1995, Joint networks in the Tiva Canyon and Topopah Spring Tuffs of the Paintbrush Group, southwestern Nevada: U.S. Geological Survey Open-File Report 95–2, 182 p.

Verbeek, E.R., and Grout, M.A., 1983, Fracture history of the northern Piceance Creek basin, northwestern Colorado, *in* Gary, J.H., ed., 16th Oil Shale Symposium, Proceedings: Golden, Colo., Colorado School of Mines Press, p. 26–44.

———1984, Fracture studies in Cretaceous and Paleocene strata in and around the Piceance Basin, Colorado—Preliminary results and their bearing on a fracture-controlled natural-gas reservoir at the MWX site: U.S. Geological Survey Open-File Report 84–156, 30 p.

———1986, Cenozoic stress rotation, northeastern Colorado Plateau [abs.], *in* Stone, D.S., and Johnson, K.S., eds., New interpretations of northwest Colorado geology: Denver, Colo., Rocky Mountain Association of Geologists, p. 97.

———1991, Some general properties of joints and joint networks in horizontally layered sedimentary and volcanic rocks—An overview, *in* Gomberg, Joan, ed., U.S. Geological Survey Committee for the Advancement of Science in the Yucca Mountain Project Symposium on "Fractures, Hydrology, and Yucca Mountain"—Abstracts and summary: U.S. Geological Survey Open-File Report 91–125, p. 10–12.

———1992, Structural evolution of gilsonite dikes, eastern Uinta Basin, Utah, *in* Fouch, T.D., Nuccio, V.F., and Chidsey, T.C., Jr., eds., Hydrocarbon and mineral resources of the Uinta Basin, Utah and Colorado: Utah Geological Association Guidebook 20, p. 237–255.

———1993, Geometry and structural evolution of gilsonite dikes in the eastern Uinta Basin, Utah: U.S. Geological Survey Bulletin 1787–HH, p. HH1–HH42.

Verbeek, E.R., Grout, M.A., and Van Gosen, B.S., 1988, Structural evolution of a Grand Canyon breccia pipe—The Ridenour copper-vanadium-uranium mine, Hualapai Indian Reservation, Coconino County, Arizona: U.S. Geological Survey Open-File Report 88–006, 75 p.

Wong, I.G., and Humphrey, J.R., 1989, Contemporary seismicity, faulting and the state of stress in the Colorado Plateau: Geological Society of America Bulletin, v. 101, no. 9, p. 1127–1146.

Zoback, M.L., Anderson, R.E., and Thompson, G.A., 1981, Cainozoic evolution of the state of stress and style of tectonism in the

Basin and Range province of the Western United States: Royal Society of London Philosophical Transactions, v. A300, p. 407–434.

Zoback, M.L., McKee, E.H., Blakely, R.J., and Thompson, G.A., 1994, The northern Nevada rift—Regional tectono-magmatic relations and middle Miocene stress direction: Geological Society of America Bulletin, v. 106, no. 3, p. 371–382.

Zoback, M.L., and Zoback, M.D., 1980, State of stress in the conterminous United States: Journal of Geophysical Research, v. 85, no. B11, p. 6113–6156.

Relation Between Basement Structures and Fracture Systems in Cover Rocks, Northeastern and Southwestern Colorado Plateau

By Earl R. Verbeek[1] *and* Marilyn A. Grout[2]

CONTENTS

FIGURES

ABSTRACT

The degree to which the structural geometry of Precambrian crystalline basement rocks beneath the Colorado Plateau influenced fracture development in overlying sedimentary rocks was assessed for three areas: the Hualapai Indian Reservation (Hualapai and western Coconino Plateaus) bordering the Grand Canyon in northwestern Arizona, the southern Marble Plateau in north-central Arizona, and the Piceance Basin along the northeastern edge of the Colorado Plateau in western Colorado. Depths to basement rock range from 460 meters on parts of the Hualapai Plateau to at least 7,900 meters in the deepest portions of the Piceance Basin. The fracture system in all three areas includes local fracture zones related to movement along basement structures as well as regional sets of extension joints that developed independently of basement control.

Differential strain due to reactivation of basement fault zones is expressed in overlying rocks in a variety of ways: on the Hualapai Reservation, as zones of closely spaced joints and well-developed karst features in Mississippian limestones above high-angle fault zones in the basement rocks; on the Marble Plateau, as 0.5–1.0 kilometer-wide belts of minor faults in Permian limestones, again above high-angle basement fault zones; and in the Piceance Basin, as a 25-kilometer-wide, 135-kilometer-long zone of joints in Cretaceous and Paleocene rocks above a basement-involved thrust fault. Common to all these basement-related fracture sets is their local rather than regional extent and their position above a known or inferred basement fault zone in deeper rocks.

In addition to fracture zones related to basement structures, strata in all three areas contain multiple sets of regionally pervasive joints, which resulted from post-Laramide tectonic extension and decreasing burial depths due to regional uplift and erosion. These sets are present over vast areas and in most places dominate the fracture network of the sedimentary rocks. Though continuous upward propagation of preexisting joint networks in the basement rocks has been suggested for their development, particularly for the Arizona examples, the regional joint sets have many properties incompatible with any such mechanism, including their orientations, stratigraphic distribution, mineralization histories, and sequence of formation. Orientations of these joints are unrelated to basement structure and instead reflect regional stress trajectories in the sedimentary cover during sequential episodes of failure, each of which affected areas of thousands to tens of thousands of square kilometers. Much of the fracture system as we see it today in exposed rocks of the Colorado Plateau is a comparatively young element (commonly Miocene or younger) of that region's complex geologic history.

INTRODUCTION

We studied relations between basement structures in Precambrian crystalline rocks and fracture systems in overlying sedimentary rocks in three parts of the Colorado Plateau (fig. 1) and, subsequently, in the Paradox Basin (Grout and Verbeek, this volume). Areas and stratigraphic units discussed here include (1) the Hualapai Indian Reservation (Hualapai and western Coconino Plateaus) in northwestern Arizona, where Upper Mississippian strata of the Redwall Limestone are exposed over a large area; (2) the southern Marble Plateau of north-central Arizona, capped by Permian strata of the Kaibab Limestone and locally by Lower to Middle Triassic rocks of the Moenkopi and Chinle Formations; and (3) the Piceance Basin of northwestern Colorado, where Tertiary strata of the Wasatch, Green River, and Uinta Formations overlie Upper Cretaceous strata of the Mesaverde Group. Units exposed on the Marble Plateau are comparable in age to those exposed in the Paradox Basin, but those on the Hualapai Reservation and in the Piceance Basin are mostly older and younger, respectively. Depths to crystalline basement in the three areas range from 460 m to at least 7,900 m, providing excellent opportunity to assess the influence of basement structure on surface fracture systems as a function of depth to basement. Fracture systems in all three areas are complex and contain not only local fracture sets possibly related to movements along basement structures but also multiple regional sets demonstrably unrelated to them. The results of these basin studies have helped guide our interpretation of the relationship between basement structure and fracture systems in the Paradox Basin (Grout and Verbeek, two reports in this volume).

Our methods for investigating basement-cover fracture relations inevitably differed from one region to another depending on the relative availability of detailed geologic, geophysical, and outcrop fracture data. The fragmentary nature of structural knowledge in each region necessitated more than the usual amount of caution in interpreting the structural record, as illustrated by our first example.

HUALAPAI INDIAN RESERVATION, NORTHWESTERN ARIZONA

The Hualapai Reservation in northwestern Arizona (fig. 1) is capped by nearly flatlying sedimentary rocks of Mississippian through Triassic age. The Mississippian rocks are extensively exposed on the Hualapai Plateau in the

[1]Present address: Earl R. Verbeek, Geologist, Bureau of Land Management, Bldg. 102, Military Circle, P.O. Box 911, Tonopah, NV 89049-0911.

[2]Present address: Marilyn A. Grout, Consulting Geologist, 5407 S.W. Viewpoint Terrace, Portland, OR 97201.

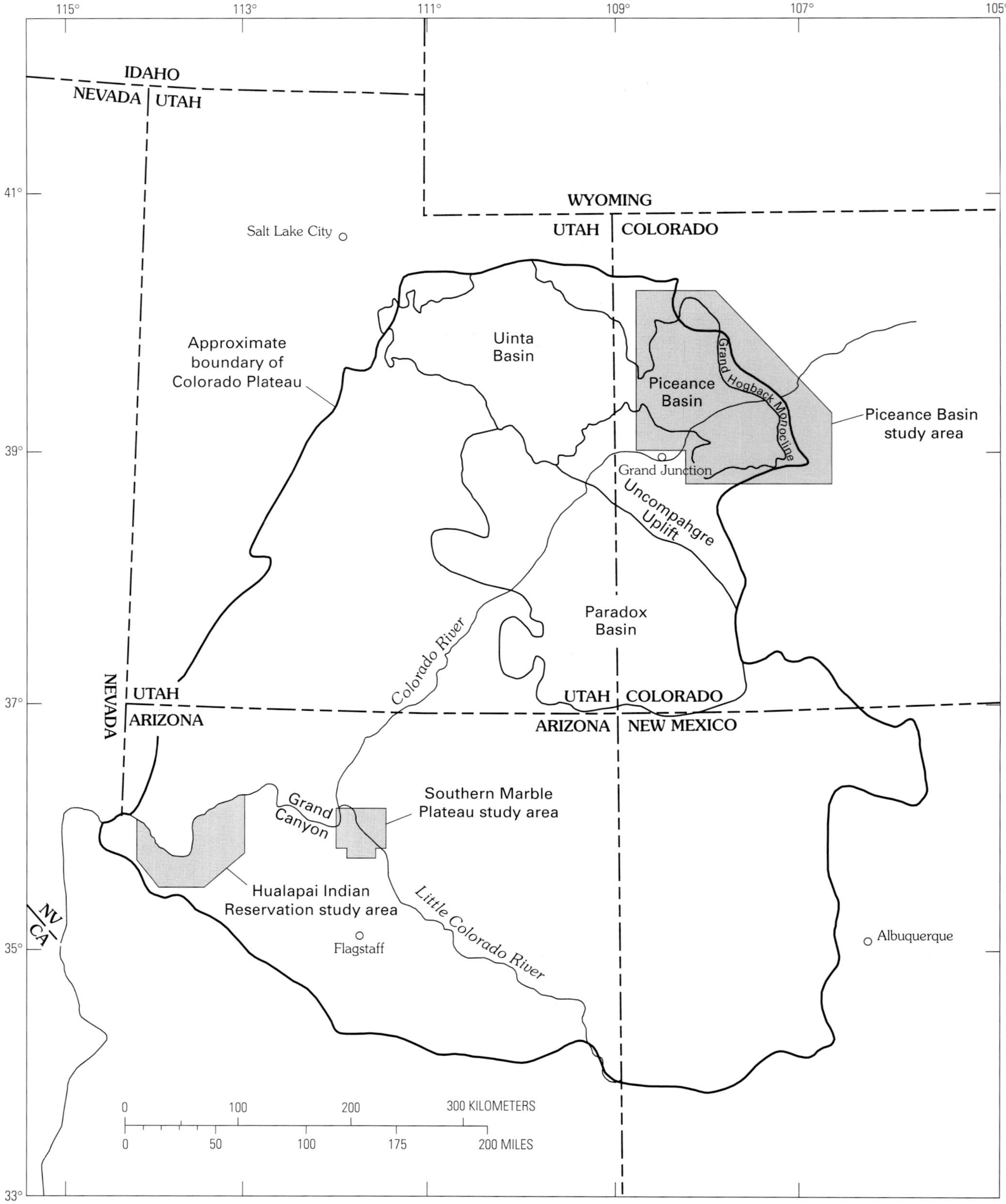

Figure 1. Location of the Colorado Plateau, selected Tertiary sedimentary basins, and three study areas discussed in this report.

western half of the reservation; the adjacent Coconino Plateau to the east is capped mostly by strata of Permian age. Older rocks are well exposed within the Grand Canyon to the north, where the Colorado River has cut deeply into the Paleozoic succession and locally into the crystalline Proterozoic basement rocks beneath. The region offers an ideal opportunity to trace the influence of basement structure on the development of fracture networks in overlying strata.

New attention was focused on the Hualapai Reservation in the 1980's during geologic studies of uranium-mineralized breccia pipes of the Grand Canyon region. The breccia pipes (fig. 2) are solution-collapse features that originated from extensive cavern systems in the Upper Mississippian Redwall Limestone (Wenrich, 1985) and that stoped upward through overlying rocks for vertical distances of 200–920 m. Today, more than 2,000 confirmed and suspected pipes are known from the Grand Canyon region, about 900 of them on the Hualapai Reservation alone (Billingsley and others, 1986, in press; Wenrich and others, 1995a,b). The existence in parts of the region of conspicuous linear belts of pipes led to repeated suggestions that pipe positions were influenced by underlying structure, possibly of basement origin (Sutphin and others, 1983; Sutphin and Wenrich, 1983; Sutphin, 1986). The nature and extent of that influence, however, remained conjectural. Accordingly, we and our colleagues began reconnaissance work in 1986 to document the fracture system in exposed units on the Hualapai Reservation and to test the possible relations between fracture-system evolution, cavern development in the Redwall Limestone, and underlying basement structures (Roller, 1987, 1989; Sutphin and Wenrich, 1988; Verbeek and others, 1988; Wenrich and others, 1989).

TECTONIC OVERVIEW

The earliest Paleozoic sediments of the Hualapai Reservation were deposited upon a Proterozoic basement complex that had already been highly metamorphosed, intruded, extensively and recurrently faulted, and deeply eroded. Proterozoic faults striking northwest through north to northeast and dipping steeply westward are abundant in the basement rocks (Billingsley and others, in press), whereas those striking within 40° of east-west are decidedly less common. Similar faults are known from throughout the Grand Canyon region (Sears, 1973; Huntoon, 1974; Shoemaker and others, 1978). Offsets on the largest faults were several hundred meters or more and were dominantly normal (Billingsley and others, in press), although right-lateral and reverse movements on some faults of north and northeast strike have also been documented in areas nearby (Shoemaker and others, 1978).

The whole of the Paleozoic Era, and much of the Cenozoic Era as well, was a time of net regional subsidence, accumulation of 2,500–4,000 m of sedimentary strata, and relative tectonic quiescence. The term *relative*, however, is used advisedly, for at least some of the ancient Proterozoic faults of the Grand Canyon region were mildly reactivated during this period (Huntoon, 1974), and minor episodes of uplift and emergence have been documented from the sedimentary record. As discussed below, such movements, though slight, nevertheless loom large in the interpretation of the region's fracture history. Among the most important events in the present context was a Late Mississippian period of regional uplift and erosion, which signaled a major change in the geologic development of the Grand Canyon region (McKee, 1979).

The Laramide orogeny of Late Cretaceous through Eocene time was the most important Phanerozoic tectonic event to have affected the Grand Canyon region, though its effects there were slight compared to most other areas of the Laramide orogenic belt. Crustal shortening during Laramide compression, with maximum horizontal compressive stress directed approximately N. 70° E. (Reches, 1978), reactivated many of the steeply west dipping basement faults as high-angle reverse faults. Slip was thus in the opposite sense from that which had occurred in Proterozoic time, and amounts of offset generally were smaller (Shoemaker and others, 1978). The overlying Paleozoic strata failed along reverse faults above the preexisting basement faults and, at higher levels, were flexed to form a series of east-facing monoclines. The structure of a typical Grand Canyon monocline, which passes upward from a high-angle reverse fault to a tight, steep monoclinal flexure and thence into a broad, gentle fold nearer the surface, was reviewed by Huntoon (1981, 1993) and described in detail for one monocline by Reches (1978). The numerous monoclines of the Grand Canyon region are that region's most prominent structural features and provide one means by which the position of major reactivated basement faults can be traced far beyond the area of exposed Proterozoic rocks. Dating the monoclines more closely than "Laramide" in the broadest sense of that term is made difficult by the absence from most of the region of rocks younger than Triassic but older than Miocene.

The late Tertiary Period in the Grand Canyon region was a time of dominantly west directed crustal extension and normal faulting, related by most authors to the inception of the Basin and Range orogeny farther west. During this time many of the ancient basement faults were reactivated once more, as were their upward extensions in Paleozoic rocks; also many new faults were created. Slip on preexisting faults in the Paleozoic rocks was opposite in sense to that during Laramide compression. A common result was faulted monoclines, the monoclines facing generally east but the faults in many places having the west side downthrown. Some of the principal late Tertiary faults on the Hualapai Reservation are coincident with early Tertiary monoclines for many kilometers and extend beyond them for considerable distances, providing another means by which the positions of the underlying basement faults can be traced.

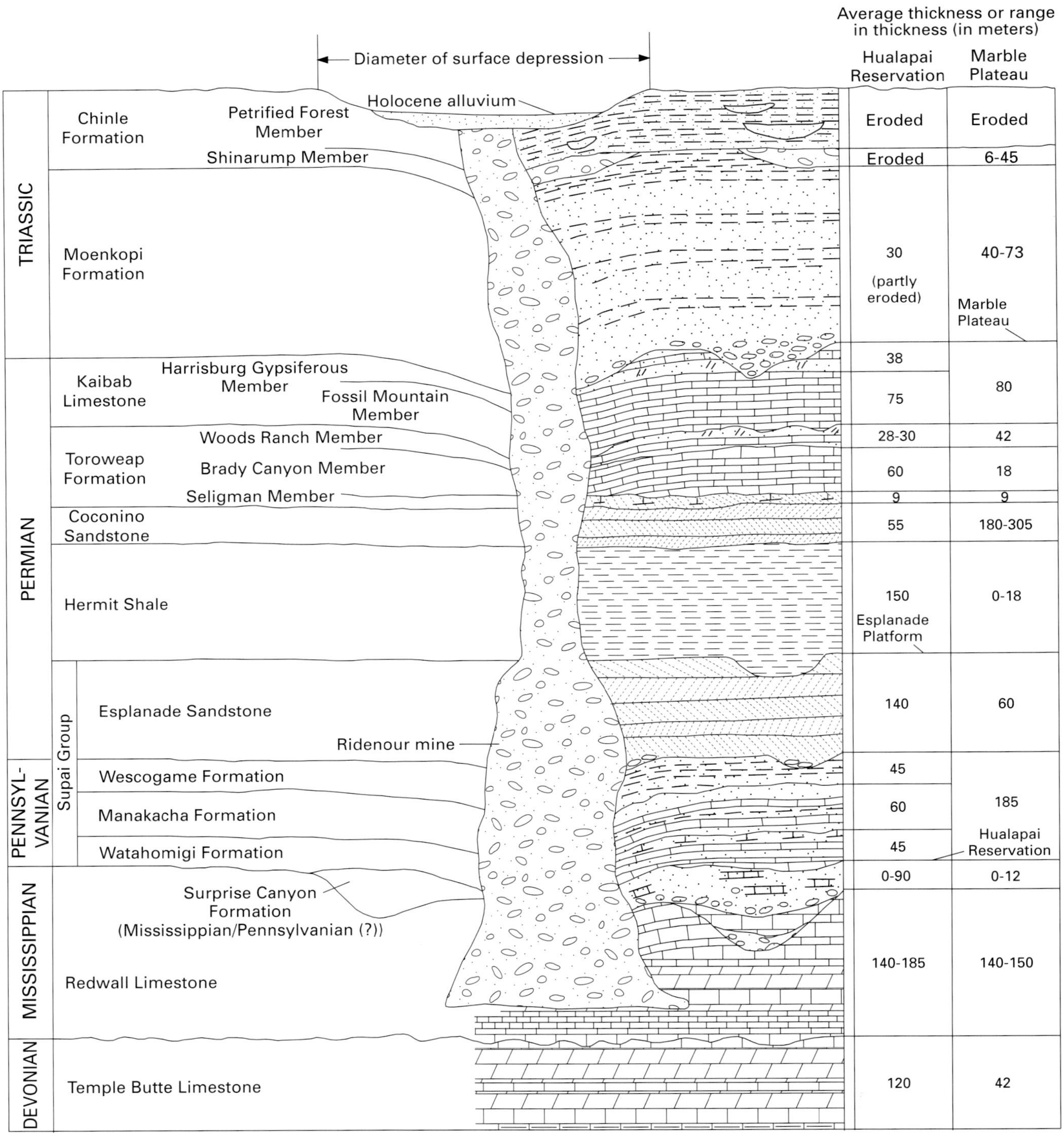

Figure 2. Generalized vertical section through a solution-collapse breccia pipe of the Grand Canyon region. Thicknesses and lithologies of the stratigraphic units shown are typical of those on and near the Hualapai and Marble Plateaus. Figure modified from Van Gosen and Wenrich (1988); thickness data from Wenrich and others (1995a,b), Billingsley and others (1985), and G.H. Billingsley (oral commun., 1994).

The Cenozoic Era in the Grand Canyon region, commencing with the Laramide tectonic movements and continuing into the period of late Tertiary normal faulting, was a time of regional uplift and erosion on a grand scale. Vertical uplift ultimately totaled 3.2–4.8 km and resulted in the removal of at least 1.5 km of rock—nearly all of the post-Paleozoic strata—from the Hualapai lands (Wenrich and others, 1995b). More than any other event, rapid Cenozoic uplift and erosional unloading had a pronounced effect on the fracture history of exposed rocks in the Grand Canyon region and beyond. That topic is discussed at some length later. We first discuss, for the Hualapai Reservation, the possible relation of selected elements of the fracture network to basement structure.

BASEMENT STRUCTURE BENEATH HUALAPAI RESERVATION

SOURCES OF EVIDENCE

Outlines of basement blocks and locations of major fault zones beneath the Hualapai Reservation and adjacent areas of the Grand Canyon region have been inferred from several lines of evidence, the first two already noted in the preceding section.

1. Major exposed faults, many repeatedly active over time. Some of the faults can be traced directly into exposed basement rocks and shown to be of Precambrian ancestry (Huntoon, 1974, 1993). Others of similar dimension and orientation but exposed only in younger rocks are assumed to have a similar history.

2. Monoclines, widely regarded as the surface expressions of deep Precambrian fault zones that were reactivated during the Laramide and that define boundaries between major basement blocks (Davis, 1978). The structure of several monoclines transected by the Grand Canyon can be followed in continuous outcrop from a broad fold in Paleozoic sedimentary rocks to a high-angle fault zone in the Precambrian crystalline rocks, thus establishing the basement-cover relationship directly (Lucchitta, 1974; Huntoon, 1974, 1993; Huntoon and Sears, 1975). A similar relationship is presumed for numerous other monoclines exposed on plateau surfaces flanking the Grand Canyon (Davis, 1978).

3. Geophysical data, chiefly linear aeromagnetic and gravity anomalies, that presumably reflect either differential elevation of the basement blocks to either side (Shoemaker and others, 1978) or different rock types juxtaposed along a basement fault zone. The use of geophysical data as a guide to basement structure is especially effective in the Grand Canyon region because the sedimentary cover rocks are very weakly magnetic and in most places are less than 2 km thick (Shoemaker and others, 1978).

4. Aligned volcanic features, including not only individual vents (McLain, 1965) but also apparent alignments of major eruptive centers over distances of 65–175 km (Eastwood, 1974; Shoemaker and others, 1978).

Many individual basement fault zones are reflected along different parts of their length by one or more of the types of features listed, often in combination. The coincidence of gravity and aeromagnetic anomalies with each other and with exposed portions of major fault zones, for example, led Shoemaker and others (1978) to conclude that the geophysical anomalies are reliable indicators of underlying basement structure. Similarly, volcanic vents aligned with the on-strike extensions of known fault zones provide good evidence that the fault zones persist beneath the volcanic rocks. Thus, a combination of geologic and geophysical data can be used to trace the buried extensions of known fault zones far beyond their limits of exposure and across plateau surfaces of only modest topographic relief.

INTERPRETED BASEMENT STRUCTURE

A generalized, regional interpretation of the major basement fault zones beneath much of northern Arizona, as depicted by Shoemaker and others (1978), is shown in figure 3. Three major trends are apparent, with approximate average directions of N. 40° E. (Sinyala, Bright Angel, and Mesa Butte systems), N. 40° W. (Chino Valley, Cataract Creek, Kaibab, and Mormon Ridges systems), and N. 5° E. (Toroweap and Oak Creek Canyon systems). The recent 1:48,000 geologic maps of Wenrich and others (1995a,b) and Billingsley and others (1986, in press) show that all three trends persist into the area of the Hualapai Reservation (fig. 4), though gradations from one to another are apparent as well. The northeast trend of the Sinyala system, for example, is defined by the Meriwhitica and Peach Springs monoclines; the Grand Wash, Separation, Lava, and Sinyala faults; and by parts of the sinuous Hurricane, Toroweap, and Lone Mountain monoclines. The Aubrey monocline, northern segment of the Hurricane monocline, and Toroweap fault define the north trend of the Toroweap system in the eastern part of the reservation. Farther west,

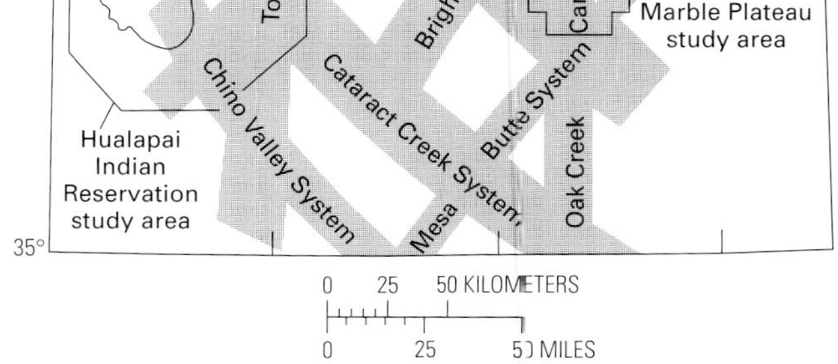

Figure 3. Generalized basement fault systems of northwestern and north-central Arizona, from Shoemaker and others (1978).

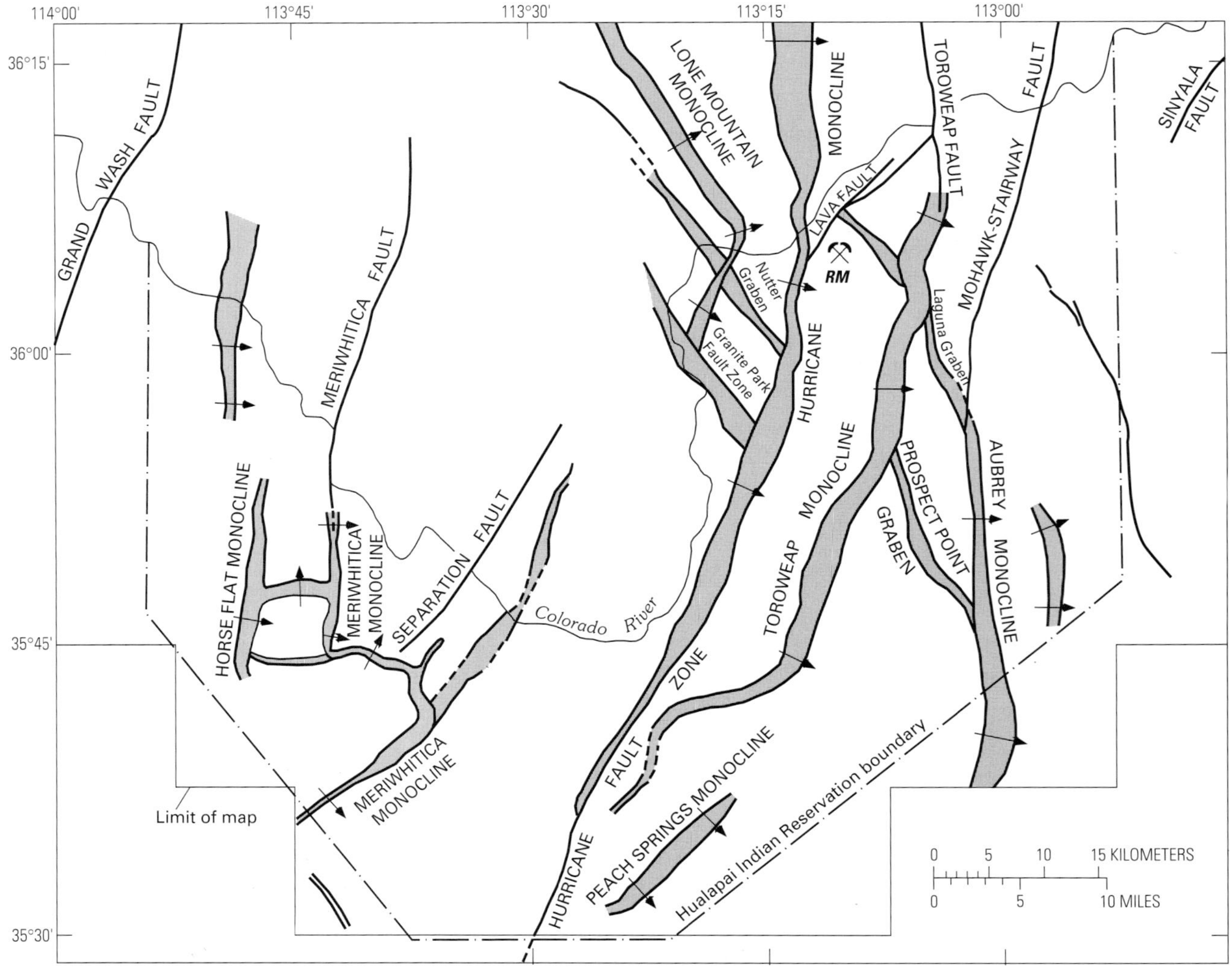

Figure 4. Major basement-related structures on the Hualapai Indian Reservation, as interpreted from the geologic maps of Wenrich and others (1995a,b) and Billingsley and others (1986; in press). Arrows indicate facing direction of monoclines. ***RM***, Ridenour mine.

the Horse Flat monocline, part of the Meriwhitica monocline, and an unnamed monocline to the north are part of the same trend. The northwest trend is reflected mostly in smaller features such as the northern segment of the Lone Mountain monocline, the Granite Park fault zone, and the Nutter, Laguna, and Prospect Point grabens; all are part of the Cataract Creek system. Of the three major trends, their relative prominence as judged from surface structures on the Hualapai Reservation is northeast (strongest), north (intermediate), and northwest (weakest, but locally conspicuous).

EARLY FRACTURE SETS IN THE REDWALL LIMESTONE

The Redwall Limestone, well known to legions of hikers and rafters of the Grand Canyon for its tendency to form imposing cliffs, is a fine-grained, thickly bedded limestone and dolomite unit commonly 135–180 m thick (Billingsley and others, 1986, in press; Wenrich and others, 1995a). The importance of this unit in the present context derives from the possibility that reactivation of basement faults influenced the distribution of fractures in the brittle carbonate rock and thus also influenced early cavern development and the stoping of breccia pipes. The depth from the base of the Redwall to the underlying Precambrian metamorphic basement rocks is 460–610 m on the Hualapai lands (Billingsley and others, 1986, in press; Wenrich and others, 1995a).

The Redwall Limestone, the upper part of which was deposited in Late Mississippian time, was uplifted shortly thereafter and exposed to subaerial weathering and erosion. The effects of this event included the development of at least two sets of joints, erosional incision of the exposed limestone surface, extensive cave development, and minor reactivation of basement faults. Late Mississippian clastic sediments (Surprise Canyon Formation, fig. 2), flushed into and preserved within the cave system, show that uplift and karst development began before the close of the Mississippian Period (Billingsley and Beus, 1985). The erosion surface subsequently was buried during deposition of the Watahomigi Formation (Pennsylvanian) and overlying units.

Joints of the two earliest sets within the Redwall Limestone have median strikes of N. 50° E. and N. 51° W. (figs. 5 and 6; Roller, 1987, 1989), similar to two of the three regional trends of basement fault zones defined by Shoemaker and others (1978) from geologic and geophysical data. Both joint sets are absent from the Watahomigi Formation and younger units, suggesting that they formed during the same period of Late Mississippian uplift that resulted in karstification. A possible scenario suggested by Wenrich and others (1989, 1995b) is that (a) minor reactivation of high-angle basement faults during uplift resulted in low-amplitude flexing of the overlying strata and attendant formation of joints along those flexures; (b) the joints formed in greatest abundance within the flexed zones, where extensional strains presumably were greatest, and in lesser abundance within the interfault areas; and (c) caverns in the Redwall Limestone developed preferentially within the zones of most-fractured rock. If this scenario is correct, the distribution of solution-collapse breccia pipes on the Hualapai Reservation was controlled at least in part by basement structure (Wenrich and others, 1995b; Billingsley and others, in press). As supporting evidence we note the following:

1. Field evidence for minor post-Redwall, pre-Supai Group movement on several ancient faults of the Grand Canyon region was noted by McKee and Gutschick (1969), Huntoon (1970), and Huntoon and Sears (1975). Similar movements may well have occurred on other basement faults beneath the Hualapai Reservation.

2. Joints of the two earliest sets in the Redwall Limestone tend to be unusually abundant near breccia pipes, locally to such an extent that the limestone looks like rubble (Roller, 1989). Roller (1989, p. 31) suggested as one likely explanation that the closely spaced early fractures "localized and concentrated fluid flow, which initiated cavern formation***."

3. Pipes that stoped through structurally intact, unjointed rock of the Supai Group above the Redwall Limestone are bordered by a well-defined zone of "ring fractures" that dip outward from the pipe and that formed during stoping (Verbeek and others, 1988). No such ring fractures were found by Roller (1989) adjacent to pipes within the Redwall, suggesting that the limestone was already well fractured when stoping of the pipes commenced.

4. Many breccia pipes mapped by Billingsley and others (1986) in the Blue Mountain area of the southeastern Hualapai Reservation are elongated in a N. 50°–60° E. direction. The joint-set maps of Roller (1989, fig. 7) show that her F_1 (oldest) set, with a median strike of N. 50° E., is exceptionally prominent in this area. The observed pipe asymmetry was attributed by Verbeek and others (1988) to stoping of the pipes through prejointed rock.

5. The distribution of pipes mapped by Billingsley and others (1986, in press) and Wenrich and others (1995a,b) on the Hualapai Reservation shows several prominent northeast-trending alignments. In a recent, informal test, all but one of seven geologists working independently with maps showing pipe distribution but no other information recognized the northeast trend (K.J. Wenrich, oral commun., 1993). Fourteen alignments of varying prominence, all trending N. 46°–48° E., have been mapped by Wenrich and others (1995b).

6. Mapping by Wenrich and Sutphin (1989) of breccia pipes within one large Redwall cave disclosed five pipes within cave passages parallel to the two early joint sets. Nearly rectilinear cave passages of different directions—passages inferred by Wenrich and Sutphin to reflect dissolution along younger joint sets—contain no pipes.

The above relations seemingly imply a close link between Late Mississippian regional uplift, the formation of early joint sets in the Redwall Limestone, cavern development, and the distribution of solution-collapse breccia pipes. The link between all of these and basement structure, however, is less clear: the principal evidence for it is the *approximate* parallelism between median strikes of Roller's (1987, 1989) early joint sets and the generalized basement fault trends defined earlier by Shoemaker and others (1978) for the entire Grand Canyon region. A more rigorous appraisal of the degree of parallelism is given in figure 7, in which basement trends for the Hualapai lands specifically (top, from fig. 4) are compared to strike-frequency distributions of the F_1 and F_2 joint sets (bottom, from Roller's data). Some obvious observations: (1) A clear distinction between the north trend of the Toroweap system (fig. 3) and northeast trend of the Sinyala system is not evident at this scale; (2) the earliest (F_1) joint set, with common strikes of N. 35°–60° E., corresponds only to a weak maximum in the basement-trend data; (3) the F_2 joint peak, with common strikes of N. 40°–60° W., is offset nearly 20° from the N. 25°–40° W peak in the basement-trend data; and (4) neither joint set parallels the prominent, broad basement trend between N. 10° W. and N. 40° E. Viewed in this manner, the notion that basement structure influenced early joint formation in the Redwall Limestone seems considerably less appealing.

Assessing the degree of parallelism still more closely, by comparing joint orientations in specific areas to the trends of individual basement structures nearby, is limited by the irregular distribution of data from the Haulapai Reservation. The joint measurements of Roller (1987, 1989) cluster into four areas; between them, no information is available. The F_1 joints in two of those areas (fig. 5)

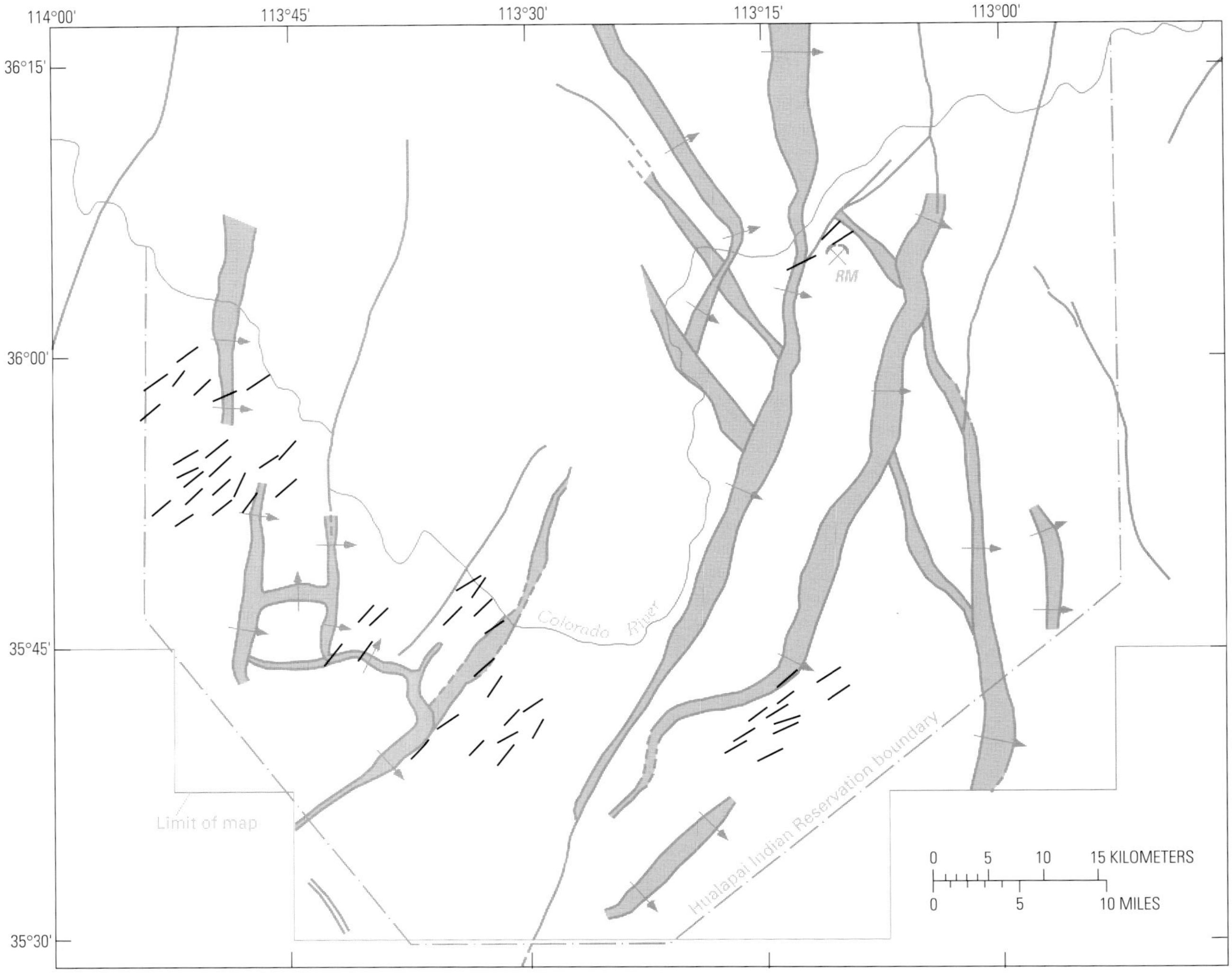

Figure 5. Average orientations of joints of the F_1 regional set of Roller (1987, 1989) in the Redwall Limestone in relation to interpreted basement-related structures (from fig. 4) on the Hualapai Indian Reservation.

parallel nearby basement structures closely and in a third show tolerably good agreement. The fourth area contains no known northeast-trending basement structures for comparison. Similarly, joints of the F_2 set (fig. 6) strike subparallel to a nearby basement structure of northwest trend in one area near the Ridenour mine, but the other three areas lack any basis for comparison. One might thus argue that the imperfect correspondence shown in figure 7 between joint strikes and basement structures reflects only the fact that the two were measured in largely different places. Nonetheless, other properties of the same joint sets are fully consistent with an origin unrelated to basement structure: the joints of both sets undeniably are widely distributed, both near and far from known basement structures of like trend (figs. 5 and 6), and neither set shows any tendency to curve in response to the sinuosity of individual monoclines or basement-related faults. Thus, though few would argue the strong influence of basement structure on the Cenozoic *fault* pattern of the Hualapai Reservation, firm evidence of the possible influence of basement structure on *joint* formation in the Redwall Limestone remains elusive.

We emphasize that much of the evidence (both pro and con) discussed here is circumstantial and based on reconnaissance data. No one has yet shown how the fracture history of the Redwall Limestone compares with that of any unit below: from the crystalline basement through the entire lower Paleozoic succession, the nature of the fracture network and of vertical variations within it remains largely unknown. Specific mechanism(s) by which rejuvenated basement faults in the Grand Canyon region could have influenced rock failure in the Redwall strata more than 460 m above have been discussed only in broad, qualitative terms, and much of the field evidence required to address the topic is lacking. That the hypothesis of basement control can seem alternately appealing or unconvincing, depending on how one looks at the evidence, underscores the need for care—and a generous measure of skepticism—in any study of this type.

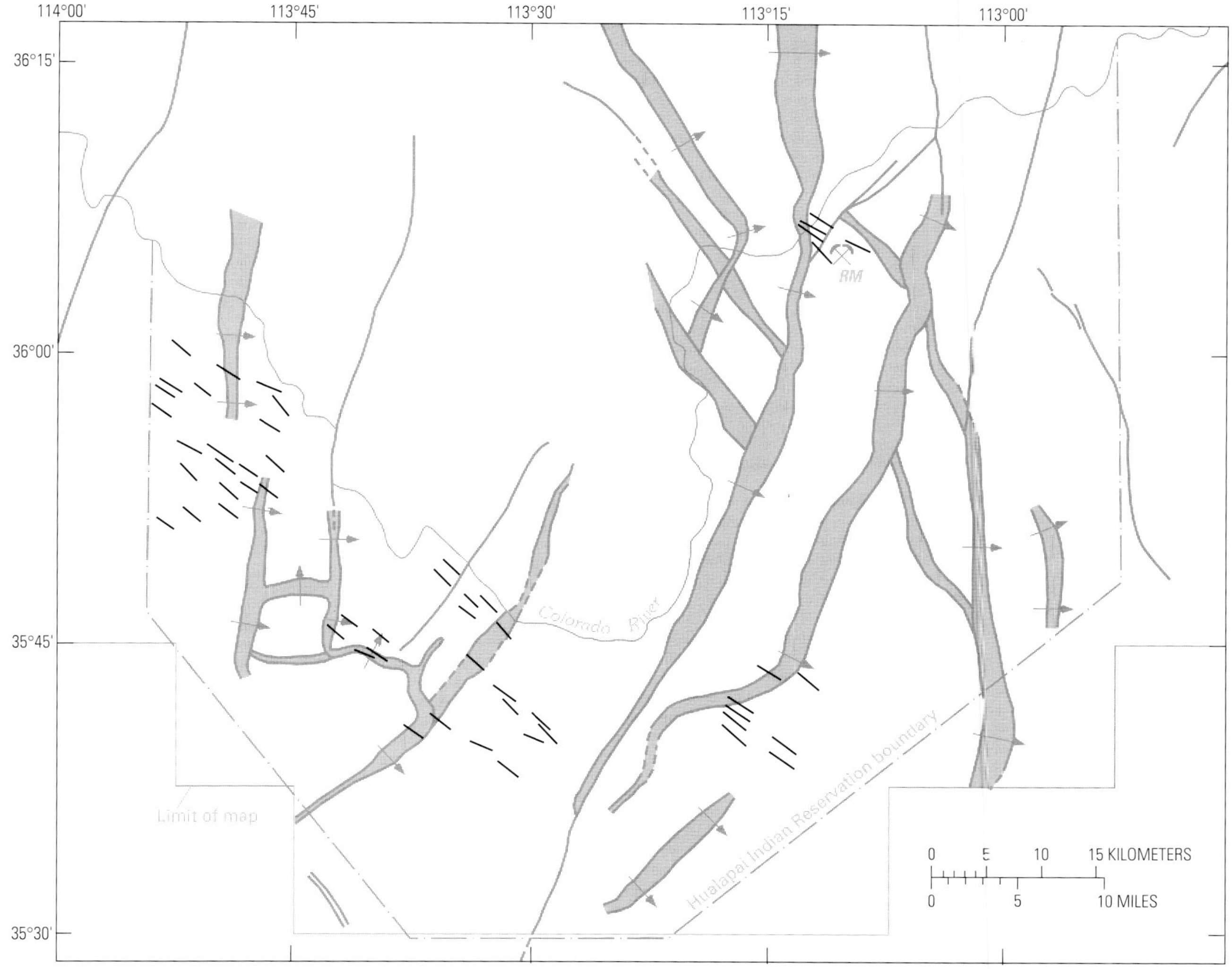

Figure 6. Average orientations of joints of the F_2 regional set of Roller (1987, 1989) in the Redwall Limestone in relation to interpreted basement-related structures (from fig. 4) on the Hualapai Indian Reservation.

LATER FRACTURE SETS

At least six later sets of joints are present both in the Redwall Limestone and in overlying strata, greatly adding to the complexity of the regional fracture network (Roller, 1987, 1989). Apparent counterparts to most of these sets occur not only in the post-Redwall Paleozoic strata but also in (1) Tertiary basalts capping erosional remnants of the Supai Group, (2) Tertiary gravels filling ancient stream valleys, and (3) the Miocene Peach Springs Tuff (18 Ma). Roller's work thus suggests that much of the fracture network is post-Laramide. The possible relation of any of these young fracture sets to reactivated basement structures has not yet been addressed; again, existing data are of a reconnaissance nature, and much remains to be learned. The suggestive evidence for a geologically young, post-Laramide, rapidly evolving regional fracture system is nonetheless a recurring theme of Colorado Plateau geology, as discussed later in this report.

SOUTHERN MARBLE PLATEAU, NORTH-CENTRAL ARIZONA

The southern Marble Plateau northeast of Flagstaff, Ariz. (fig. 1), is an elongate, northwest-trending crustal block of Precambrian metamorphic rocks (unexposed) capped by 950–1,200 m of dominantly flat-lying Paleozoic and lower Mesozoic sedimentary rocks. The study area as outlined in figure 1 includes the southern half of this plateau and, along its western margin, the easternmost parts of the adjacent (and topographically higher) Coconino Plateau. Much of the study area has been stripped by erosion to the top of the Kaibab Limestone, a resistant unit of Permian age (fig. 2), although sandstone of the Moenkopi Formation (Middle? and Lower Triassic) and mudstone of the Chinle Formation (Upper Triassic) are preserved as isolated buttes and small mesas in some places. Local relief generally is 100 m or less except near the Colorado and Little Colorado Rivers, where canyon floors lie 370–460 m below the plateau surface.

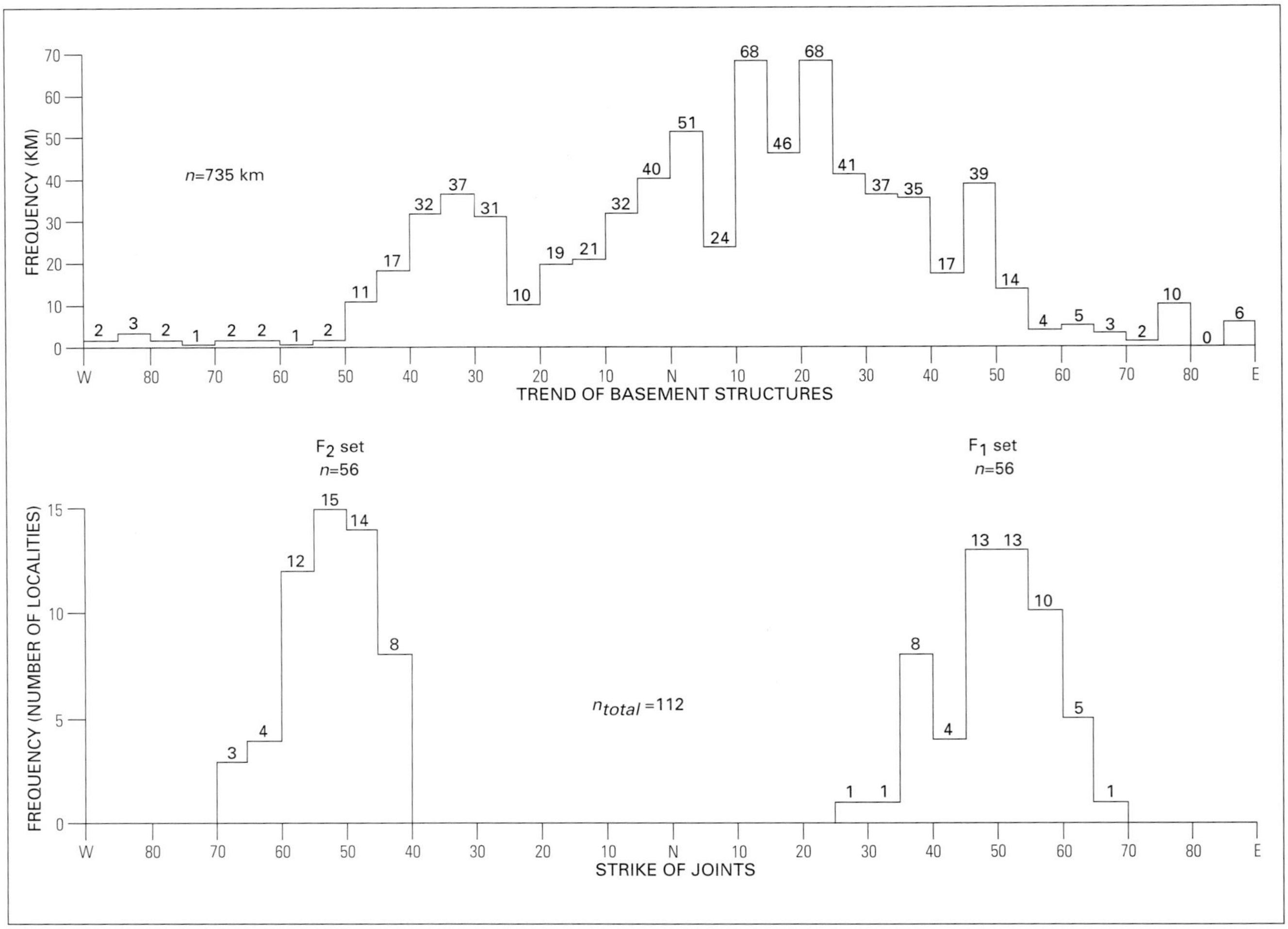

Figure 7. Histograms comparing trends of major basement structures on the Hualapai Plateau (top, from map of fig. 4) to strikes of joints of the F_1 and F_2 sets of Roller (1987, 1989) (bottom, from figs. 5 and 6). In order to show basement trends for the Hualapai lands specifically, we divided the 735 km of inferred basement structures shown in figure 4 into small (average 2.85 km) linear segments, measured the length and orientation of each of the 258 segments so defined, and plotted the length-weighted frequency distribution of figure 7*A*. Below this, in figure 7*B*, are shown strike frequency distributions of the F_1 and F_2 joint sets derived from the data of Roller (1987, 1989).

The impetus for fracture studies on the Marble Plateau, like that on the Hualapai Indian Reservation 170 km farther west, stemmed from USGS work on uranium-mineralized breccia pipes during the 1980's. Breccia pipes on the Marble Plateau are well exposed and were first described in detail by Sutphin (1986). Recent fracture work includes a field study of joint networks at 18 localities (Sutphin, 1986), detailed photogeologic mapping of fracture traces across large tracts of exposed bedrock (E.R. Verbeek, unpub. data, 1980–1987), and delineation of faults during geologic quadrangle mapping (Billingsley and others, 1985).

TECTONIC OVERVIEW

Much of what is known of the geologic evolution of the Hualapai lands of the western Grand Canyon region applies as well to the Marble Plateau. The same formations underlie both areas (fig. 2), albeit with some notable thickness and facies changes, and the structural inventory is virtually identical. Principal differences are that structures on the Marble Plateau are exposed at a higher stratigraphic level than those on the Hualapai lands, and the largest Cenozoic faults on the Marble Plateau are of much lesser displacement than their counterparts farther west.

The most prominent structures on and near the southern Marble Plateau are breccia pipes, monoclines, and normal faults. The breccia pipes of this area were extensively investigated by Sutphin (1986), who mapped 90 of them and interpreted them as solution-collapse features related to Mississippian-age caverns in the Redwall Limestone. The common presence of pipes in Late Triassic strata of the Chinle Formation, the youngest bedrock unit preserved in the area, shows that the upper parts of some pipes formed 100 m.y. or more after cavern formation began. Some of the pipes show elevated gamma radiation counts (Sutphin, 1986; Sutphin and Wenrich, 1988), and one of them, the Riverview pipe south of the study area (fig. 8, lower right), was mined for uranium (Chenoweth and Blakemore, 1961).

Lengthy, sinuous monoclines are the dominant structures on and near the Marble Plateau and are the main expression of Laramide crustal compression in the area. Several monoclines—notably the Grandview, East Kaibab, Coconino Point, Black Point, and Echo Cliffs monoclines (fig. 8)—exert a profound influence on the topography of the region. The 500-m elevation difference between the Marble and Coconino Plateaus in the south-central part of the study area, for example, is a direct reflection of structural relief across the Coconino Point monocline. The major, northwest-trending monoclines of the region all face northeast, a feature interpreted to reflect reverse movement on reactivated Proterozoic basement faults of steep southwest dip (Reches, 1978; Davis, 1978).

Minor normal faults are abundant on the Marble Plateau and conspicuous on aerial photographs. Some of the faults are coincident with monoclines or lie along their on-strike projections and reflect post-Laramide normal movement on the same basement faults that earlier had been reactivated in a reverse sense. Most of the other faults strike within 30° of due north and probably are products of crustal extension related to basin-range extensional tectonism farther west. Dating the onset of normal faulting on the Marble Plateau is difficult owing to insufficient stratigraphic control, but the abundance of faults in late Paleozoic and Triassic rocks, contrasted with their paucity in Pliocene to Pleistocene volcanic rocks immediately to the south, shows that much of the faulting occurred before 6 Ma (Babenroth and Strahler, 1945). In a few places, however, faulted lava flows (Babenroth and Strahler, 1945; Barnes, 1974), terrace gravels (Reiche, 1937), and debris fans (Holm, 1987) show that normal faulting continued into the Quaternary Period, and contemporary seismicity (Sturgul and Irwin, 1971; Wong and Humphrey, 1989) suggests that it continues still.

Late Cenozoic regional uplift resulted in erosion of nearly all post-Paleozoic strata from much of the Marble Plateau. The drainage net that developed on the exhumed Kaibab surface is of probable Miocene age (G.H. Billingsley, oral commun., 1989) and is incised in the areas of maximum uplift (Barnes, 1987). Normal faults have disrupted the original drainage in many places.

BASEMENT STRUCTURE BENEATH SOUTHERN MARBLE PLATEAU

Interpretation of basement structure beneath the southern Marble Plateau is drawn from some of the same sources of evidence already discussed for the Hualapai lands. Laramide monoclines outlining the edges of crustal blocks are particularly abundant on and near the Marble Plateau, and parts of several fault zones have a strong geophysical expression (Shoemaker and others, 1978). Only one monocline, however, has been sufficiently dissected by erosion that its deep structure and underlying fault zone, well exposed within the gorge of the Colorado River and one of its tributaries, have been studied in detail (Reches, 1978). Moreover, major fault zones similar to the ancient and repeatedly active Hurricane and Toroweap faults farther west (fig. 4) are missing from the Marble Plateau area; in their place are faults of similar style but much lesser displacement. Greater depth to basement (915–1,220 m), shallower erosional incision, and consequent lack of basement exposure make interpretation of basement structure more inferential on the Marble Plateau than for areas farther west.

Parts of the Marble Plateau overlie the zone of intersection of three major basement fault systems (fig. 3) with trends similar to those across other parts of northern Arizona. Individual surface structures that define these trends on and near the southern Marble Plateau are shown in figure 8. The prominent northwest trend of the Kaibab fault system is expressed at the surface principally by monoclines, notably the East Kaibab and Blue Springs monoclines within the study area, and the Black Point monocline farther south. The equally prominent northeast basement trend is defined by the southern segment of the Coconino Point monocline, the Additional Hill monocline nearby, and by at least six belts of minor faults, including topographically expressed grabens; collectively these structures mark the northeasternmost extent of the Mesa Butte fault system of Shoemaker and others (1978). The north trend of the Oak Creek Canyon fault system is more weakly defined than the other two and within the study area is expressed only by the easternmost portion of the Coconino Point monocline and by several lengthy segments of the Snake graben. Farther south, however, this fault system gains in prominence and coincides with a magnetic anomaly that marks its signature in basement rocks (Shoemaker and others, 1978).

The interpretive map of figure 8 combines information from existing geologic maps with results from more recent structural work. Some of the features shown on that map, particularly those of northeast trend as noted above, correspond to belts of minor faults. Also present (fig. 9) are apparent alignments of solution-collapse breccia pipes analogous to the pipes already discussed for the Hualapai Indian Reservation. The possible relation of both kinds of features to basement structure is discussed below.

ALIGNMENTS OF BRECCIA PIPES ON SOUTHERN MARBLE PLATEAU

Numerous breccia pipes on the Marble Plateau appear to be aligned within northeast- and northwest-trending belts interpreted by Sutphin and Wenrich (1983, 1988), Sutphin

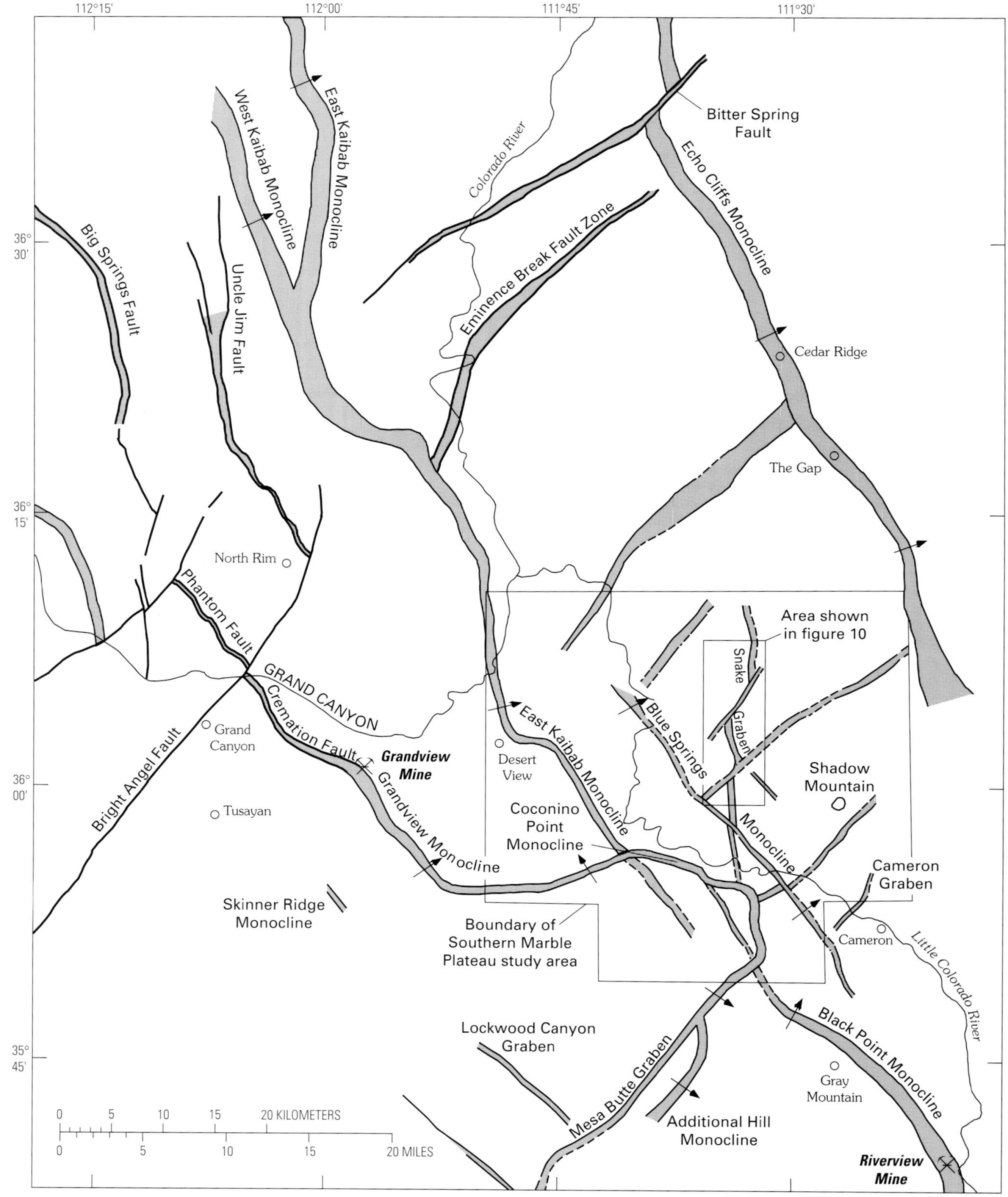

Figure 8. Map showing interpreted basement-related structures in the eastern Grand Canyon region. Box outlines Marble Plateau study area. Positions of the features shown were compiled from the geologic maps of Akers and others (1962), Huntoon and others (1976), Haynes and Hackman (1978), Ulrich and others (1984), and Billingsley and others (1985), plus the photogeologic fracture-trace map of Verbeek (unpub. data, 1981–1987). Arrows indicate facing direction of monoclines. Unlabeled features are zones of unnamed minor faults.

(1986), and Wenrich and others (1989) as evidence of basement influence on pipe position. The map of figure 9, modified from that of Sutphin and Wenrich (1988), shows nine such alignments, labeled A–I in approximate decreasing order of believability. Seven of the alignments parallel known structures of the Mesa Butte fault system; the other two parallel monoclines of the Kaibab system. One line of 19 pipes, labeled A on the map, extends N. 45° W. for 27 km and coincides throughout its length with the Blue Springs monocline. A second alignment (E) of similar trend, 18 km long and including 15 pipes, lies between and parallel to two monoclines and probably overlies a buried basement fault zone that has no other known expression in the surface rocks. A third alignment (B, fig. 9) trends N. 40° E. and lies wholly within one of the northeast-trending fracture zones shown in figure 8. That breccia-pipe alignments, belts of minor faults, and monoclines are mutually parallel, and in some places spatially coincident, reinforces the view that all are manifestations of underlying basement structure.

SURFACE FRACTURE SYSTEM OF THE SOUTHERN MARBLE PLATEAU

Photogeologic mapping of the surface fracture network of the southern Marble Plateau (E.R. Verbeek, unpub. data, 1980–1987) was done at 10× magnification on 1:50,000 black-and-white vertical aerial photographs of good to excellent resolution. The visibility of many fracture traces from the air is enhanced by the sparsity of soil and vegetation cover and, in carbonate rocks and calcite-cemented sandstones, by solution-widening of fracture openings. A close correspondence between fracture traces mapped from the photographs and those measured at the outcrop was demonstrated by the field work of Sutphin (1986). Fracture sets visible on the aerial photographs are broadly divisible into two classes: (1) fractures within rectilinear to gently curved zones, between which fractures of the same orientation are sparse or absent, and (2) areally pervasive sets present over much of the plateau.

ZONED FRACTURE SETS

Straight to gently curved belts of fractures 0.5–1.0 km wide and 5–25 km long are locally conspicuous elements of the surface fracture network of the Marble Plateau. More than a dozen such belts have been identified. The individual fractures (or narrow fracture zones) within the belts generally have traces 0.3–1.5 km long, and many are small faults with throws of only a few meters or less. Fractures between the zones are generally shorter, more pervasively distributed, and, with the exception of a few north-trending zones, of different orientation. Few of the fracture belts are portrayed as such on conventional geologic maps, and most remained unrecognized until recent photogeologic mapping of the plateau. Six of the fracture belts trend N. 40°–60° E., four others N. 30°–45° W., and several more nearly due north. These are the same directions identified by Shoemaker and others (1978) from independent data (monoclines, geophysical anomalies, exposed fault zones, aligned volcanic features) as the principal basement trends in this and adjacent areas (fig. 3). The coincidence in trend suggests to us that these fracture belts, like the larger Hurricane and Toroweap fault zones to the west, are the surface expression of underlying high-angle basement fault zones; hence they are included in figure 8. Along these basement faults, however, movement since the end of the Paleozoic resulted only in minor faulting of the overlying strata and was of insufficient magnitude to produce either monoclines or large offsets in the Permian and Triassic surface rocks.

Some of the most interesting fracture zones on the Marble Plateau are those that collectively define the Snake graben, a lengthy (>20 km), topographically expressed fault trough that bisects the study area from north to south (fig. 8). The Snake graben, sinuous in plan view (fig. 10*A*), is composed of several north-trending fracture zones offset from one another in a dextral sense (fig. 10*B*), presumably by slip along some of the northeast-trending fault zones mentioned above (fig. 10*C*). Inasmuch as no evidence exists of substantial strike-slip movement in the exposed Paleozoic rocks, the apparent offsets of 1.5–5 km presumably reflect displacement of basement fault blocks in Precambrian time. This interpretation agrees with the findings of Sears (1973) and Shoemaker and others (1978), who discussed evidence of 1,300–1,600 Ma dextral offsets along northeast-trending basement fault zones in the Grand Canyon region. Similar dextral offsets of 2–5 km occur where the Phantom and Cremation faults are intersected by the Bright Angel fault, west of the study area (fig. 8), and where the Bitter Spring fault crosses the Echo Cliffs monocline north of the study area.

AREALLY PERVASIVE FRACTURE SETS

Combined field and photogeologic work suggests that at least five areally pervasive sets of joints are present in the Permian and Triassic rocks capping the Marble Plateau. The joints of all five sets are vertical, or nearly so, and have the following general properties:

Average Strike	Photogeologic Expression
N. 10°–15° E.	Strong to moderate
N. 00°–05° W.	Strong to moderate
N. 15°–25° W.	Strong to moderate
N. 60°–70° E.	Weak
N. 70°–80° W.	Weak

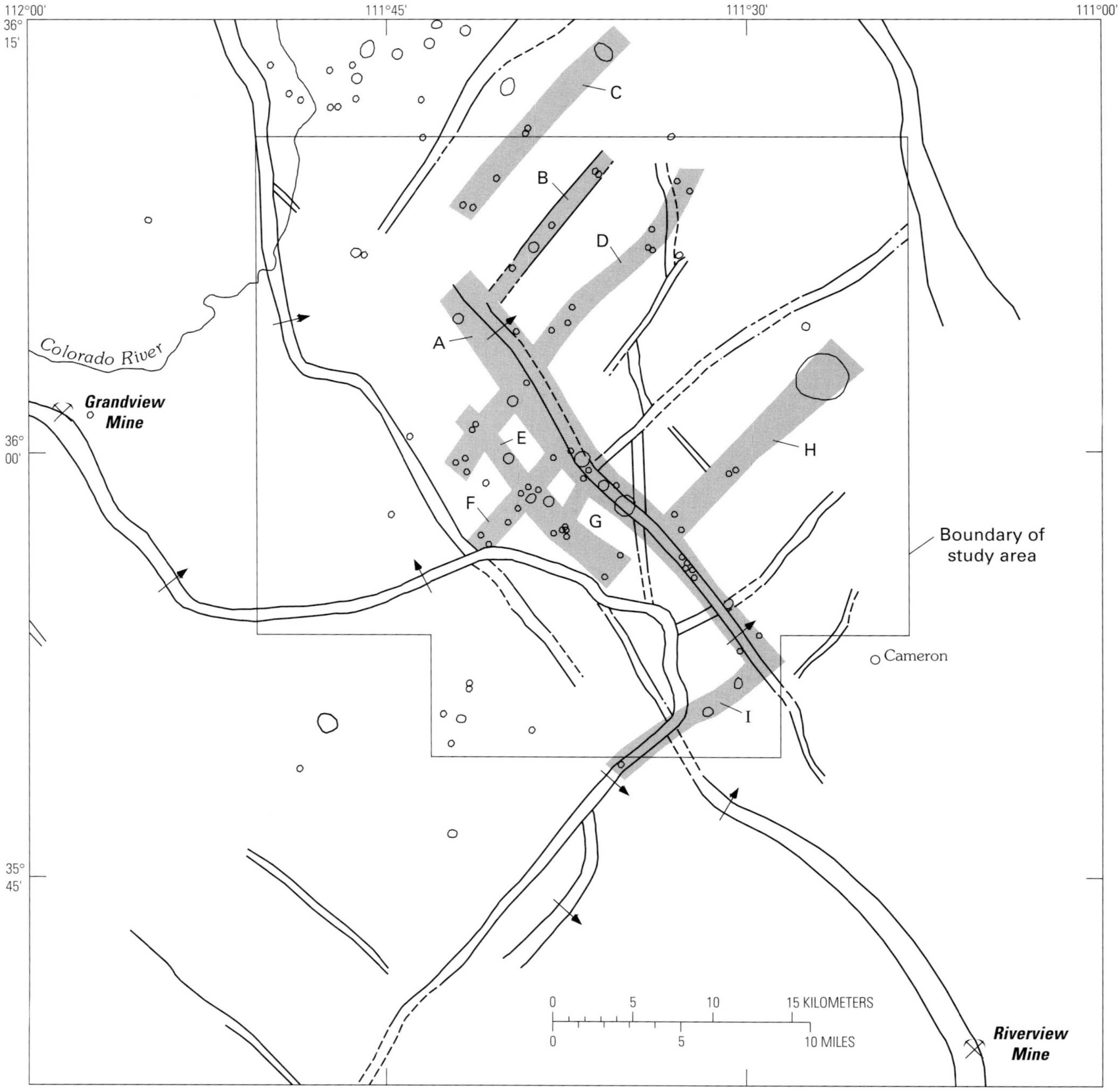

Figure 9. Map showing interpreted alignments of solution-collapse breccia pipes on the southern Marble Plateau in relation to basement-related monoclines and fault zones (from fig. 8). Modified from Sutphin and Wenrich (1988). A–I, explained in text.

Many joints of the first three sets, nearly all of which strike within 30° of due north, were reactivated as small normal faults during post-Laramide regional extension. The Marble Plateau thus exhibits a pronounced northerly structural "grain" of elongate fault blocks, with each fault corresponding to a narrow zone of faulted joints (fig. 11). Most such faults are highly visible on aerial photographs as long, low scarps across the landscape and are readily mapped; between them, the unfaulted joints of the same sets exhibit much shorter traces. Evidence that the faulting is of post-Laramide age, and that it is related to the onset of basin-range tectonism farther west, is based partly on analogy to similar but larger faults of the Grand Canyon region (Lucchitta, 1974; Huntoon, 1974) and partly on new information presented in a later section of this report. The west-northwest- and east-northeast-striking joints of the other two sets, in contrast, are oriented at low angles to the regional extension direction (approximately east-west) and thus generally were not reactivated. Their traces on aerial photographs are invariably short and their photogeologic expression subdued.

No evidence exists at present that any of the five areally pervasive joint sets are genetically related to basement structure. Their distribution shows no obvious relation to known or inferred basement fault zones, and the two major basement-fault trends (N. 40° W., N. 50°E.) have no counterpart among common strike directions of the joints. The parallelism between the third, subordinate basement trend and the many north-striking fractures at the surface probably is fortuitous; the broad, nonzoned distribution of the joints again is inconsistent with basement-fault reactivation and instead is a product of regional crustal extension. At present, the only north-trending element of the fracture network that we can confidently relate to basement structure is the Snake graben (fig. 10), by far the longest north-trending graben on the Marble Plateau. The position of this graben, in line with the north-trending portion of the Coconino Point monocline (fig. 8), suggests that both features are surface expressions of the same underlying fault zone.

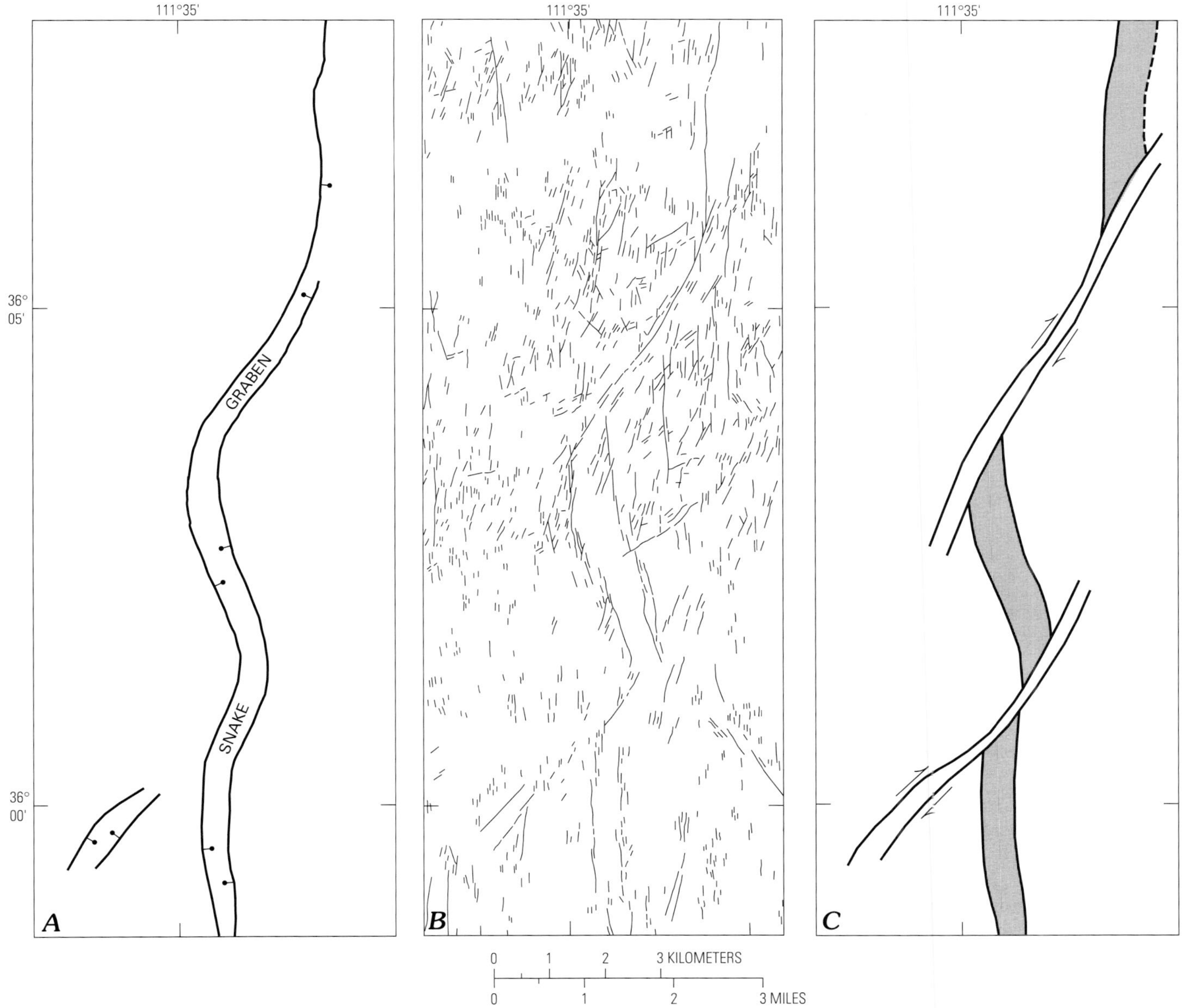

Figure 10. Three portrayals of the structure of the Snake graben. See figure 8 for location. *A*, Continuously curved walls as portrayed on published geologic map (Billingsley and others, 1985). *B*, Actual fracture structure as mapped from aerial photographs (E.R. Verbeek, unpub. data, 1981–1987). *C*, Interpretation of underlying basement structure. Bar and ball, downthrown sides of normal faults bounding graben.

The sequence of formation and absolute ages of the regional joint sets of the Marble Plateau have not yet been established with certainty, but available evidence suggests that most or all of the sets are post-Laramide. Along the northwest-trending Black Point monocline south of Cameron, for example, joints of the N. 15°–25° W. set dip within 5° of vertical on both the horizontal and tilted limbs of the fold (fig. 12), showing that this joint set—the oldest set present in these particular rocks—was superimposed on a preexisting Laramide structure. Two additional sets of joints in weakly cemented volcaniclastic sandstones, dated by Damon and others (1974) at less than 700,000 years B.P., provide additional evidence of geologically young jointing in this part of Arizona (fig. 13). Though much remains to be done in studying fracture evolution in this region, one plausible interpretation consistent with known facts is that (1) the three major sets that strike within 30° of due north are products of regional basin-range extension, the differing orientations of the joints reflecting noncoaxial extension over time, and (2) continuing crustal extension resulted in minor faulting by dominantly dip-slip movement along preexisting joints. The dogleg bends and discontinuous nature of many of the minor grabens that offset the Kaibab surface (fig. 11) are a natural consequence of faulting by reactivation of multiple, preexisting fracture sets. The other two sets, whose short joints strike at high angles to the northerly structural grain of the surface rocks, probably reflect near-surface stress-relief jointing upon progressive reduction of confining pressure by erosion. Similar small joints are common to many areas of uplifted, flat-lying sedimentary rocks; the mechanism of their formation was discussed recently by Gross (1993).

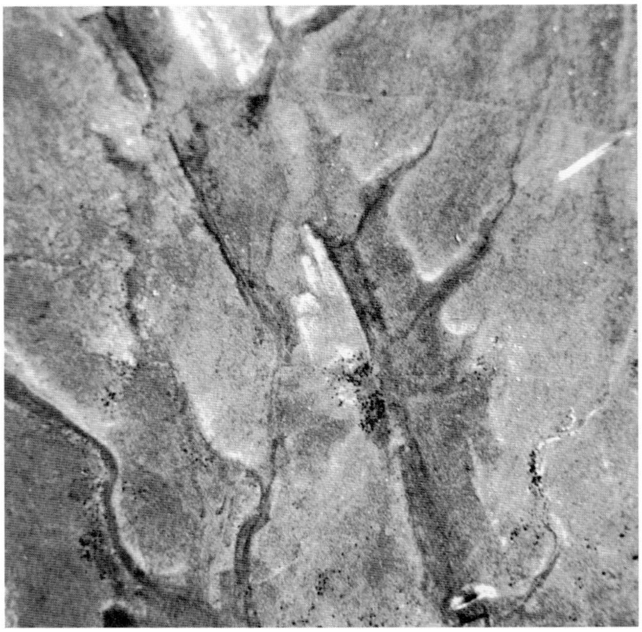

Figure 11. Vertical aerial photograph showing discontinuous graben typical of those found on and near the Marble Plateau. The serrated walls and discontinuous fault trough are the result of reactivation of preexisting vertical joints. Width of horizontal field of view approximately 4 km.

SUMMARY OF SURFACE FRACTURE NETWORK

The fracture network in exposed rocks of the Marble Plateau has two major components: (1) basement-related fractures that owe their origin to episodic reactivation of Precambrian fault zones, and (2) shallow, dominantly post-Laramide, high-angle normal faults and joints resulting from regional extension related to basin-range extensional tectonism in areas to the west. The major basement-controlled fracture zones trend, on average, about N. 40° W. and N. 50° E., directions that fortuitously are poorly represented among the post-Laramide fractures. The basement-related structures generally fall within gently sinuous zones that are expressed at the surface as prominent monoclines, belts of minor faults, and chains of volcanoes. The post-Laramide joints and normal faults, in contrast, are more widespread and dominate the fracture network at the surface. The different styles of expression and different trends between these two major groups of fractures are the principal means, exclusive of geophysical methods, by which buried basement fault zones can be recognized in the region.

PICEANCE BASIN, NORTHWESTERN COLORADO

The Piceance Basin of northwestern Colorado lies along the northeastern edge of the Colorado Plateau (figs. 1 and 14) and is one of a series of intermontane basins that developed during Laramide orogenesis by segmentation of the Late Cretaceous seaway that once stretched north-to-south across North America. The basin is separated from the adjacent Rocky Mountains to the east by the Grand Hogback monocline, from the Uinta Basin on the west by the Douglas Creek arch, and from the Paradox Basin on the south by the Uncompahgre uplift. Nearly flat-lying upper Paleocene and Eocene sedimentary rocks of the Wasatch, Green River, and Uinta Formations (fig. 15) are exposed within the Piceance Basin, and Upper Cretaceous (prebasin) rocks of the Mesaverde Group crop out over large areas along the basin margins. The nearest exposures of Precambrian basement rocks lie east of the basin (principally in Glenwood Canyon, where the Colorado River has cut deeply into the uplifted White River block) and in small areas of the Uncompahgre uplift bordering the basin on the southwest. Depths to crystalline basement within the basin interior range from about 5,000 m to more than 7,900 m.

Figure 12. Vertical joints (N. 15°–25° W.) in tilted beds of the Moenkopi Formation along the Black Point monocline at the Riverview mine. Approximate height of outcrop is 5 m. See figure 8 (lower right) for location.

Figure 13. Members of two well-developed sets of vertical joints in weakly cemented volcaniclastic sandstones near the Riverview mine; see figure 8 (lower right) for location. Large fracture in foreground strikes about N. 20° W.; sunlit fracture above canteen is member of second set striking about N. 70° E.

TECTONIC OVERVIEW

Seismic reflection lines across the east-central part of the Piceance Basin and basin margin (Waechter and Johnson, 1986; Grout and others, 1991) reveal evidence of two episodes of basement-involved deformation. The earliest is recorded by northwest-trending high-angle faults that penetrate crystalline basement and persist upward into Pennsylvanian rocks (fig. 16). These faults were active during Middle Pennsylvanian time and controlled facies patterns within the evaporitic rocks that formed during that period (Dodge and Bartleson, 1986; Johnson and others, 1988), when hypersaline deposits (halite and gypsum) accumulated within subsiding grabens, and penesaline and clastic sediments were deposited in adjacent areas. The faults subsequently were buried beneath 5.5–6.1 km of overlying sediment and have no expression in Tertiary rocks at the surface, either directly or through components of the fracture network.

The second episode of basement-involved deformation took place during the Laramide Orogeny when a large, basement-cored block (Perry and others, 1988) advanced southwestward beneath the eastern part of the basin along a 135-km-long front. As thrusting proceeded, strata above the thrust block were uplifted and tilted basinward to form the Grand Hogback monocline, which now marks the boundary between the Piceance Basin on the Colorado Plateau to the west and the structurally higher White River uplift of the Rocky Mountains province to the east. The dogleg trace of the monocline in map view (fig. 14) suggests that this great fold developed through reactivation of one or more preexisting basement fault zones. From its northern end, the monocline trends approximately S. 5° E. for 45 km, bends abruptly to a S. 70° E. trend for 50 km, and then bends abruptly once more to a S. 10° E. trend for an additional 40 km before dying out. MacQuown (1945) and Stone (1969) speculated that the northern and southern legs were once continuous and that their current positions reflect ancient sinistral slip along the middle segment, whose trend parallels the regional schistosity of the basement rocks. Strata along the steep limb dip from 30° to slightly overturned. Involvement of the middle Eocene Green River Formation in this tilting suggests that much of the fold development is late Eocene or younger. Much of the thrust-induced strain was accommodated in this large fold, but strata basinward of the thrust block were shortened slightly along a series of imbricate splay faults that mark the leading edge of a décollement within the mechanically weak Pennsylvanian evaporitic rocks (Grout and others, 1991; fig. 16). At and near the thrust front, gas-producing intrabasin folds formed by tectonic repetition of Middle Pennsylvanian through Upper Cretaceous strata (for example, the Divide Creek anticline) and flowage of Middle Pennsylvanian salt (the Wolf Creek anticline); details of their geologic evolution are given in Grout and others (1991), Gunneson and others (1995), and Hoak and Klawitter (1996). Additional consequences of the same general deformation include the local development of several basement-related fracture sets, as described in a following section.

BASEMENT STRUCTURE BENEATH THE PICEANCE BASIN

Several factors inhibit confident interpretation of basement structure beneath the Piceance Basin. First is the large depth to basement, more than 5,000 m for much of the basin interior. Second, most of the natural resources for which the region has been explored (chiefly oil shale, petroleum, and natural gas) are located within the upper 2,500 m of the sedimentary section—thus, with few exceptions, the deepest boreholes penetrate only into Permian and Pennsylvanian sedimentary rocks. Seismic lines depicting deep structure are likewise few in the public domain. Interpretation of basement structure beneath this part of northwestern Colorado, then, rests chiefly on aeromagnetic and gravity data and on extrapolation from more deeply exposed areas farther southwest (Uncompahgre uplift, Paradox Basin) and east (Rocky Mountains).

In a general way the pattern of northwest- and northeast-trending basement fault zones discussed earlier for the Grand Canyon region is considered by many geologists to be characteristic also of areas farther northeast, including the Paradox Basin, the Uncompahgre uplift, and at least part of the Piceance Basin. The existence of most of these features has been inferred from geophysical data rather than surface evidence, and their character and geologic history remain topics of lively debate. Discussion is well beyond the scope of this paper, but see, for example, Case (1966), Case and Joesting (1972), Hite (1975), Friedman and Simpson (1980), Friedman and others (1994), and Johnson (1983). The interpretive map of basement faults in this last report includes the southwestern part of the area shown in figure 14, along the border between the Uncompahgre uplift and the Piceance Basin. Dominant basement trends in this area, as in the Paradox Basin to the south, are approximately N. 50° W. and N. 45° E.

Surface structural evidence for the northeast trend within the Paradox and Piceance Basins is sparse, but the northwest trend is reflected in monoclines and faults similar to those along the same trend (Shoemaker and others, 1978) in the Grand Canyon region. Along the southwestern margin of the Piceance Basin, west of Grand Junction (fig. 14), principal components of this trend include the Devils Canyons, Lizard Canyon, Fruita Canyon, and Ladder Creek monoclines and associated Redlands and Kodels Canyon faults (Williams, 1964; Cashion, 1973; Lohman, 1981). Collectively these features separate Jurassic and younger rocks of the Piceance Basin on the northeast from older rocks, including Proterozoic schists and gneisses, of the Uncompahgre uplift on the southwest. On the opposite side

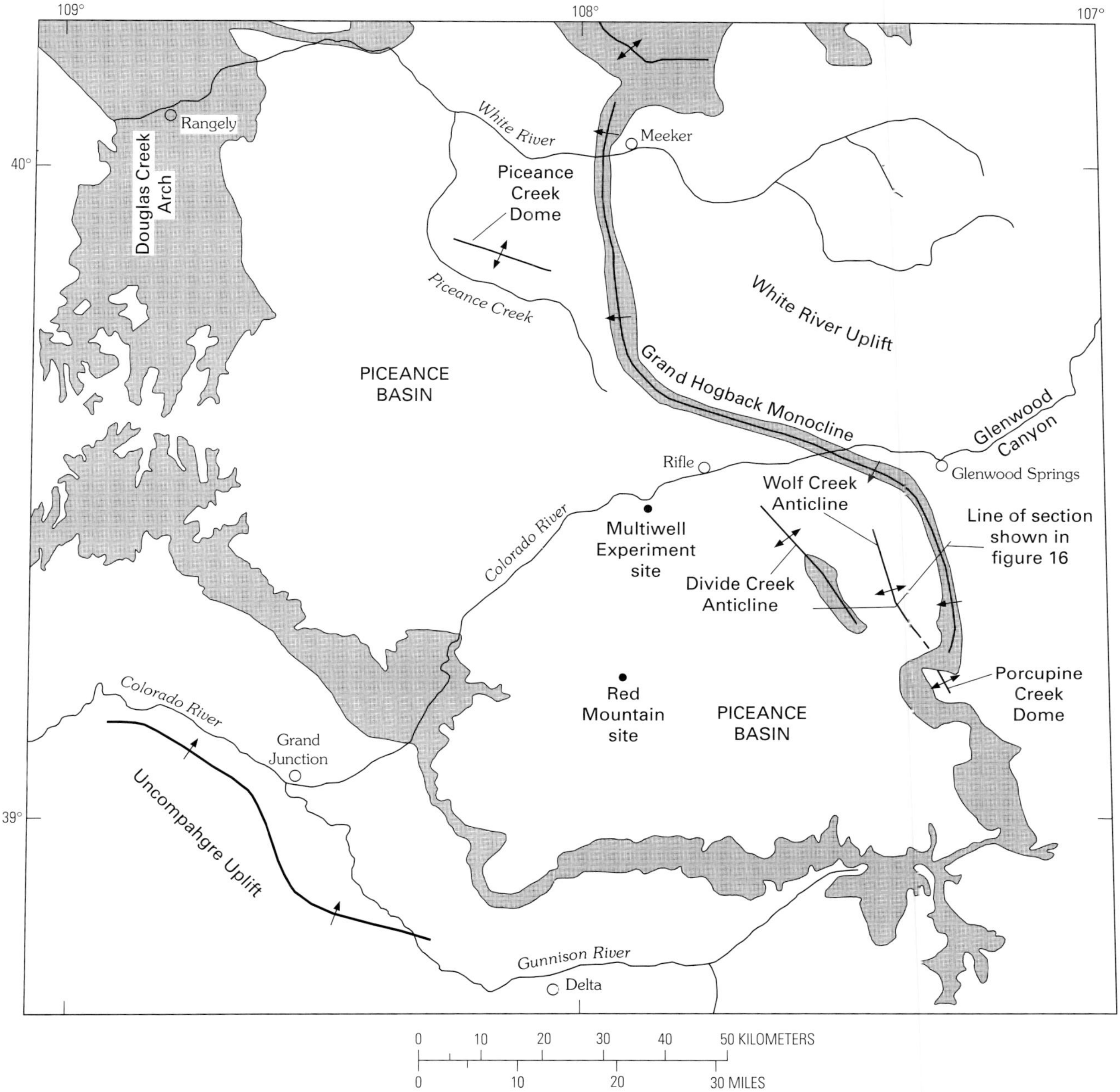

Figure 14. Major structural features in and near the Piceance Basin and outcrop belt of the Upper Cretaceous Mesaverde Group (shaded) along the basin margins. Profile line of cross section shown in figure 16 also shown. Arrows indicate facing direction of monoclines. Grand Hogback monocline marks the boundary between the Colorado Plateaus province on the west and the Rocky Mountain province on the east. Trace of monocline southwest of Grand Junction (from Williams, 1964; Cashion, 1973; and Lohman, 1981) represents zone of contiguous basement-related features along northern edge of Uncompahgre uplift, including Devils Canyon, Lizard Canyon, Fruita Canyon, and Ladder Creek monoclines and associated Redlands and Kodels Canyon faults.

of the basin, the aforementioned Grand Hogback monocline separates the Cenozoic basin rocks from older rocks, including Proterozoic crystalline rocks, of the White River uplift to the northeast. As interpreted by Davis (1978, his fig. 7), the Grand Hogback represents the middle section of a lengthy (220 km) basement fault zone of overall N. 35° W. trend. Between these basin-margin structures, the basement-penetrating faults beneath the Piceance Basin, shown on deep seismic lines by Waechter and Johnson (1986) and Grout and others (1991), are additional elements of the regional northwest trend. As noted above, during Pennsylvanian time these faults controlled the subsidence of elongate troughs within

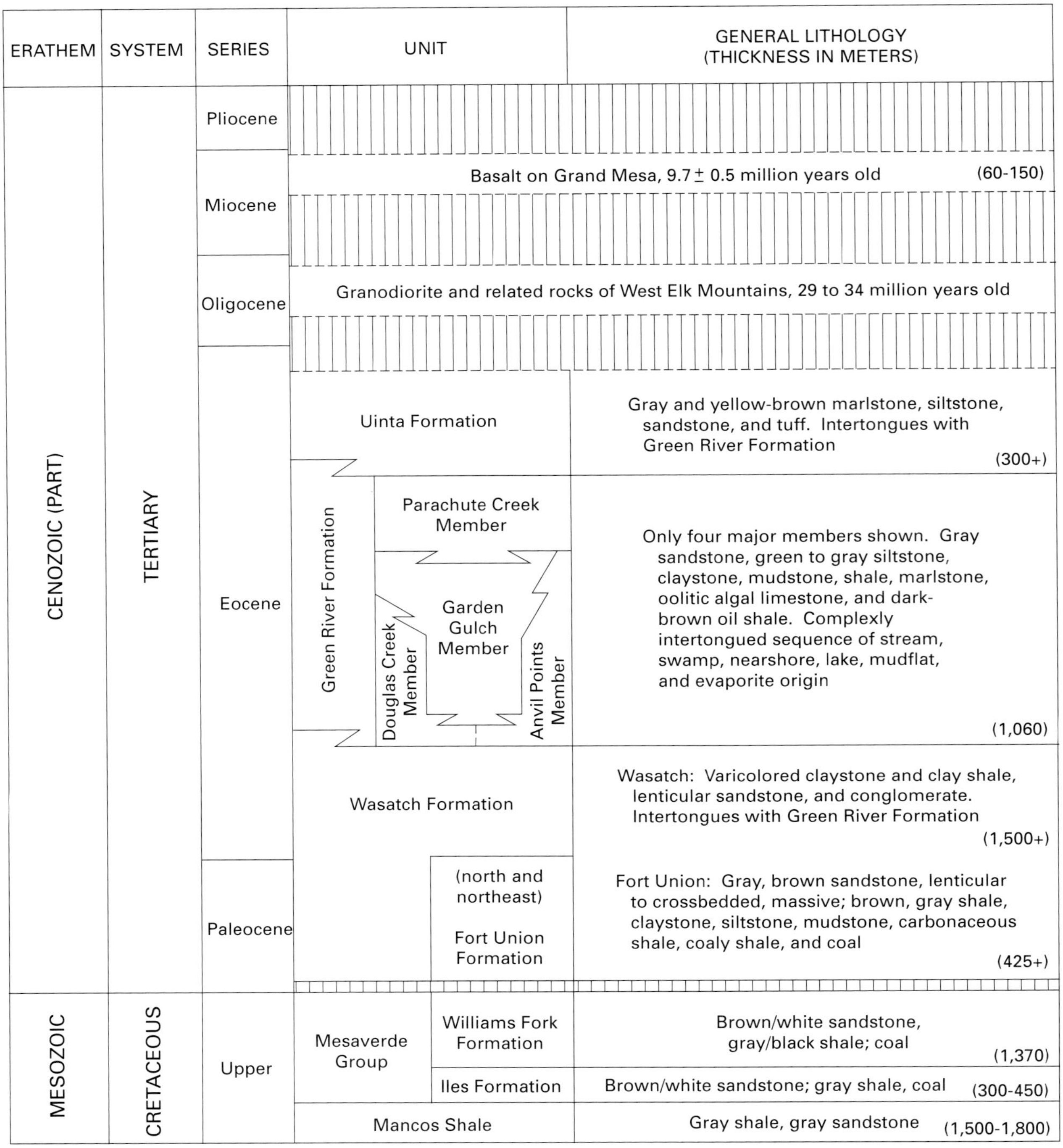

Figure 15. Stratigraphic column of Upper Cretaceous and Tertiary rocks in and bordering the Piceance Basin north of the Colorado River, from MacLachlan (1987) and MacLachlan and Welder (1987).

which evaporite sediments accumulated. Similar fault-bounded troughs farther east were inferred from stratigraphic relations by Dodge and Bartleson (1986).

For many of these structures their early history is poorly known, but some, like the basin-margin fault zones discussed previously, are reactivated elements of an older, probably Precambrian, fault pattern. The discussion below centers on the Grand Hogback monocline and associated structures because the record of late basement-related fracturing is clearest in this area.

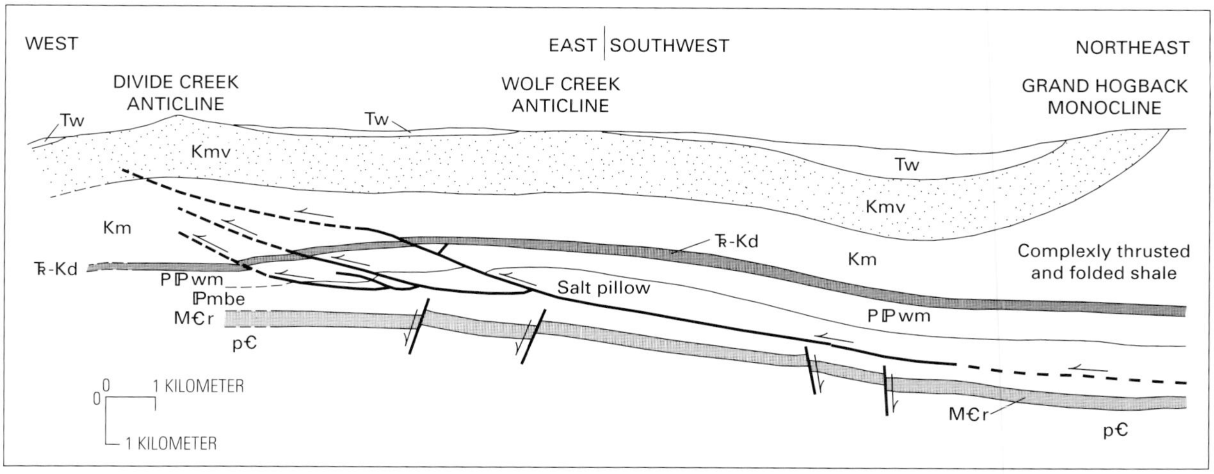

Figure 16. Structure beneath Divide Creek and Wolf Creek anticlines near the eastern margin of the Piceance Basin, from Grout and others (1991). Location of section shown in figure 14. Geology shown is combined interpretation from seismic, gravity, and drill-hole data. Geologic units: Tw, Tertiary Wasatch Formation; Kmv, Upper Cretaceous Mesaverde Group; Km, Upper Cretaceous Mancos Shale; TR-Kd, Triassic strata through Upper Cretaceous Dakota Sandstone, undivided; PIPwm, Lower Permian to Middle Pennsylvanian Weber Sandstone and Maroon Formation; IPmbe, Middle Pennsylvanian Minturn and Belden Formations, including Eagle Valley evaporite sequence; M€r, Mississippian through Cambrian rocks, undivided; p€, Precambrian metamorphic basement rocks.

BASEMENT-RELATED FRACTURE SETS ALONG THE GRAND HOGBACK MONOCLINE

The oldest fracture sets known within the Upper Cretaceous rocks bordering the Piceance Basin are those along the Grand Hogback monocline. Two joint sets, both possibly related to basement tectonism, dominate this early fracture system; a third set, weakly developed and little studied, will be considered no further here. Collectively these joint sets constitute the *Hogback system* as described in Verbeek and Grout (1984a, 1984b). Surface structures on joints of all three sets show that they are extension fractures.

Joints of the two dominant sets form a rectangular network of fractures everywhere perpendicular to bedding regardless of present bed orientation, which ranges widely both in strike (from northeast through north to west-northwest) and dip (from 30° through vertical to slightly overturned). The joints thus formed when the beds were nearly horizontal and were tilted with those beds to new attitudes as the Grand Hogback developed. Bed-parallel slickenside striations, common on the joints of both sets, record minor shear adjustments during tilting. Restored, pretilt strikes for the older set (fig. 17) range from west-northwest through west along the entire length of the monocline; those for the younger set (fig. 18) range from northeast through north. Average strikes are about N. 80° W. and N. 10° E., respectively. Joints of the older set are abundant within sandstones of the Late Cretaceous Mesaverde Group but penetrate no higher stratigraphically than the lower part of the Paleocene to early Eocene Wasatch Formation. Joints of the younger set, in contrast, are present higher in the Wasatch Formation, but they too are missing from the uppermost Wasatch beds and from the overlying middle to late Eocene beds of the Green River and Uinta Formations. The stratigraphic evidence thus suggests an age of late Paleocene for the older set and latest Paleocene to early Eocene for the younger set.

Interpretation of both sets as basement-related rests on two principal lines of evidence: (1) their restricted areal distribution, as documented both in outcrop and in oriented core from two well sites in the basin (Multiwell Experiment site and Red Mountain site in fig. 14), and (2) the stratigraphic record of early movements in the area of the future Grand Hogback monocline. Joints of the older set are abundant along the entire 135-km-long outcrop belt of the Mesaverde Group along the Grand Hogback (Verbeek and Grout, 1984a) and are present also in Mesaverde strata beneath the Multiwell Experiment (MWX) site 16 km into the basin (Lorenz and Finley, 1987; Finley and Lorenz, 1989; Lorenz and others, 1989), but they are absent from the Divide Creek anticline (Grout and Verbeek, 1992), whose crest lies 15–19 km from the monocline. They are similarly absent from the Red Mountain (RM) well site, 43 km into the basin (Seccombe and Decker, 1986), and from the extensive outcrop belt of the Mesaverde Group (fig. 14) along the southern and western margins of the basin (Verbeek and Grout, 1984b; Grout and Verbeek, 1985). The available evidence thus suggests that these joints exist only in and near the Grand Hogback. Joints of the younger set are similarly restricted: they are abundant only along and near the Grand Hogback and are sparsely present on the Divide Creek fold (Grout and Verbeek, 1992); they have been found nowhere else. That two prominent joint sets traceable for more than 130 km along the length of a monocline should die out so dramatically within 25 km away from it suggests some causative link between all these structures.

The nature of that link now seems more clear from new stratigraphic evidence. Eastward thinning of the Paleocene section (R.C. Johnson, USGS, oral commun., 1991), in response to broad warping along the trace of the future Grand Hogback, likely records the earliest stages of Laramide reactivation of the underlying basement fault zone. Along the northern end of the monocline, for example, the combined Fort Union and Wasatch Formations thin eastward from about 1,550 m to 1,060 m over a lateral distance of only 5.5 km as the monocline is approached (Izett and others, 1985). These early, premonocline movements probably record the initial stages of Laramide compression along the eastern margin of the basin (Grout and Verbeek, 1992). The oldest of the two prominent joint sets along the Grand Hogback (fig. 17) probably resulted from this compression; if so, the early compressive movements were directed west-northwest to west, parallel to the strike of the older joints, and only later were directed more southwesterly.

Joints of the younger set (fig. 18), unlike the older joints, are present only in structurally elevated areas: they are found along the monocline itself, and on the Divide Creek anticline, but not at depth beneath the MWX site. This distribution suggests that they developed upon uplift as a set of stress-release joints nearly at right angles to those of the earlier set; their consistent orientation perpendicular to bedding shows that bed dips at the time were still quite low. In our interpretation (Grout and Verbeek, 1992), they formed during the early (late Paleocene–early Eocene) stages of fold growth, as strata along the trace of the nascent monocline were being warped upward over the advancing thrust wedge at depth, and as splay faults had just started to develop beneath the Divide Creek anticline.

LOCAL FRACTURE SETS ON DIVIDE CREEK ANTICLINE

The Divide Creek anticline near the eastern margin of the Piceance Basin (fig. 14) is a northwest-trending intrabasin fold approximately 35 km long and 15 km wide; limb dips at the surface are 15° or less (Grout and others, 1991). Two local fracture sets on this anticline are basement-related, but only in the indirect sense that they formed near the leading edge of a thrust system that involved basement rocks farther east. The joints of both sets strike N. 28°–55° W., about parallel to the axial trace of the fold (fig. 19), and dip in opposite directions to intersect bedding at angles of 60°–70°, thereby dividing the beds into rhomboidal blocks. Joints of the two sets are unequal in size: on both fold limbs the largest fractures generally dip toward the axial trace of the fold and the smallest dip away, implying that the anticline had already started to form when jointing occurred so that structural position on the fold determined which set would grow to larger size. Moreover, the set dipping toward the fold axis has the shallower dip, by amounts of 2°–11°, suggesting that folding continued after jointing so that joints inclined toward the fold axis were rotated to shallower dips and their counterparts to steeper ones. Joint development *during* growth of the Divide Creek anticline thus best explains the observed geometry (Grout and Verbeek, 1992). Abutting relations between coexisting joint sets confirm that these joints are younger than the N. 10° E.-striking premonocline joints discussed above, which formed while bed dips were still nearly horizontal.

LATER REGIONAL FRACTURE SETS

The regional fracture network of the Tertiary surface rocks in the Piceance Basin consists of five sets of vertical extension joints collectively termed the *Piceance system* (Verbeek and Grout, 1984a; Grout and Verbeek, 1985). Orientations of joints of the first four sets are shown in figure 20. Joints of the fifth and youngest set (not shown) form only a minor component of the fracture network; they are parallel to the present-day direction of maximum horizontal compressive stress (Bredehoeft and others, 1976; Zoback and Zoback, 1980; Wong and Humphrey, 1989) and are present only in near-surface rocks. All five sets are of regional extent and have been traced throughout the whole of the Piceance Basin and far beyond, into prebasin Paleocene and Cretaceous rocks to the south and southwest. The same five sets are present in the neighboring Uinta Basin farther west (Verbeek and Grout, 1992, 1993), across the Douglas Creek arch between the two basins, and probably within the Paradox Basin of southwestern Colorado and southeastern Utah (Grout and Verbeek, this volume). The geographic limits of each set remain only partially defined.

The area *dominated* by joints of the Piceance system is at least 25,000 km^2 and includes much of the northern Colorado Plateau north of lat 39° N. Within this vast region the joints of all five sets show broad variations in relative abundance, especially among the three oldest sets. Joints of the F_1 set, for example, are the most strongly expressed joints in scattered parts of the Piceance Basin north of the Colorado River but are sparse elsewhere; in the Uinta Basin they are uncommon in the eastern part of the basin but are superbly developed farther west (M.A. Grout and E.R. Verbeek, unpub. data, 1992). Joints of the F_2 and F_3 sets show comparable geographic variations in prominence, each being the dominant set in parts of both basins and subordinate to one of the other sets elsewhere. These areal changes in prominence among the joint sets of the Piceance

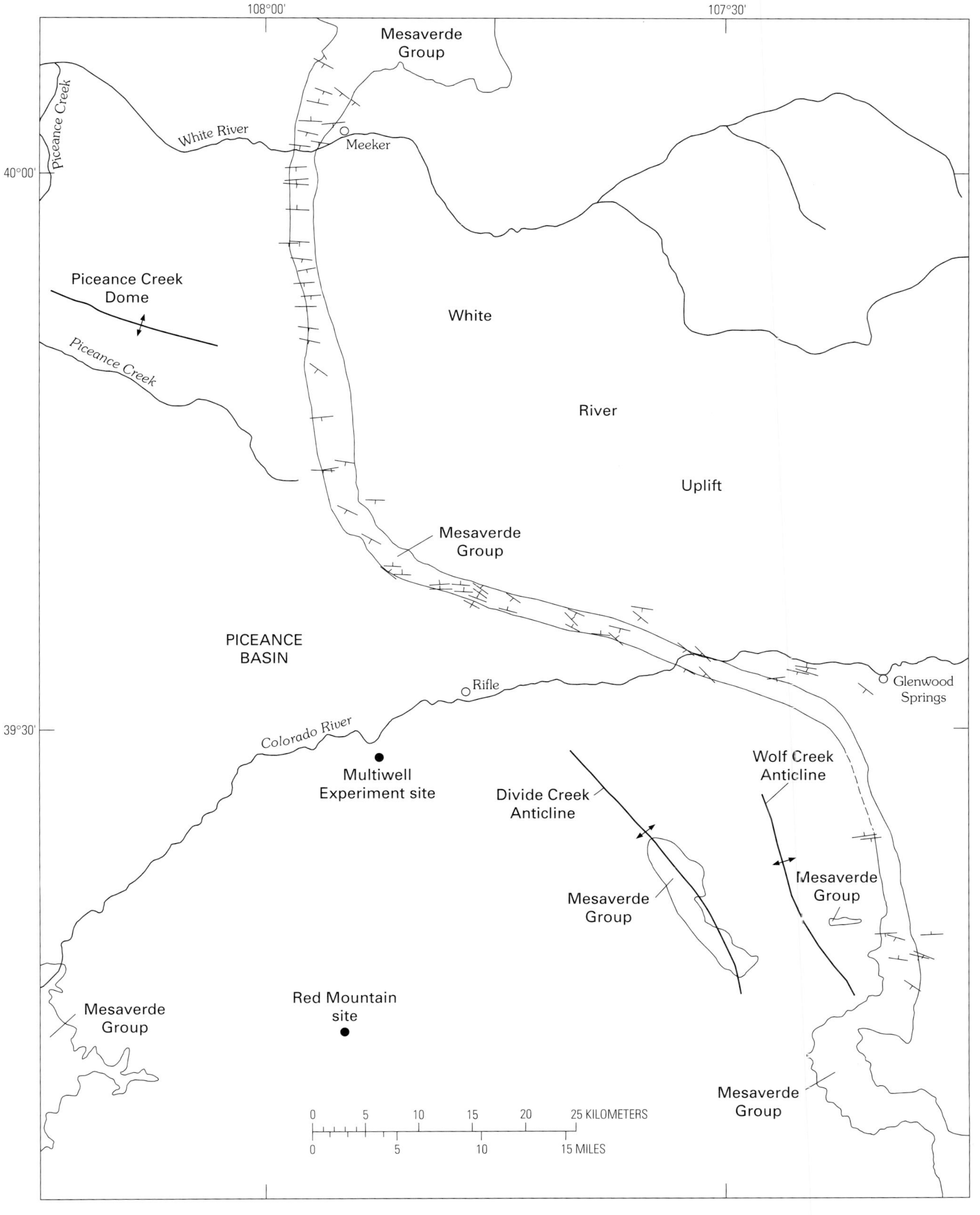

Figure 17. Reconstructed (bed-horizontal) orientations of the older of two prominent sets of joints in beds of the Upper Cretaceous Mesaverde Group and lowermost Wasatch Formation (Paleocene) along the Grand Hogback monocline.

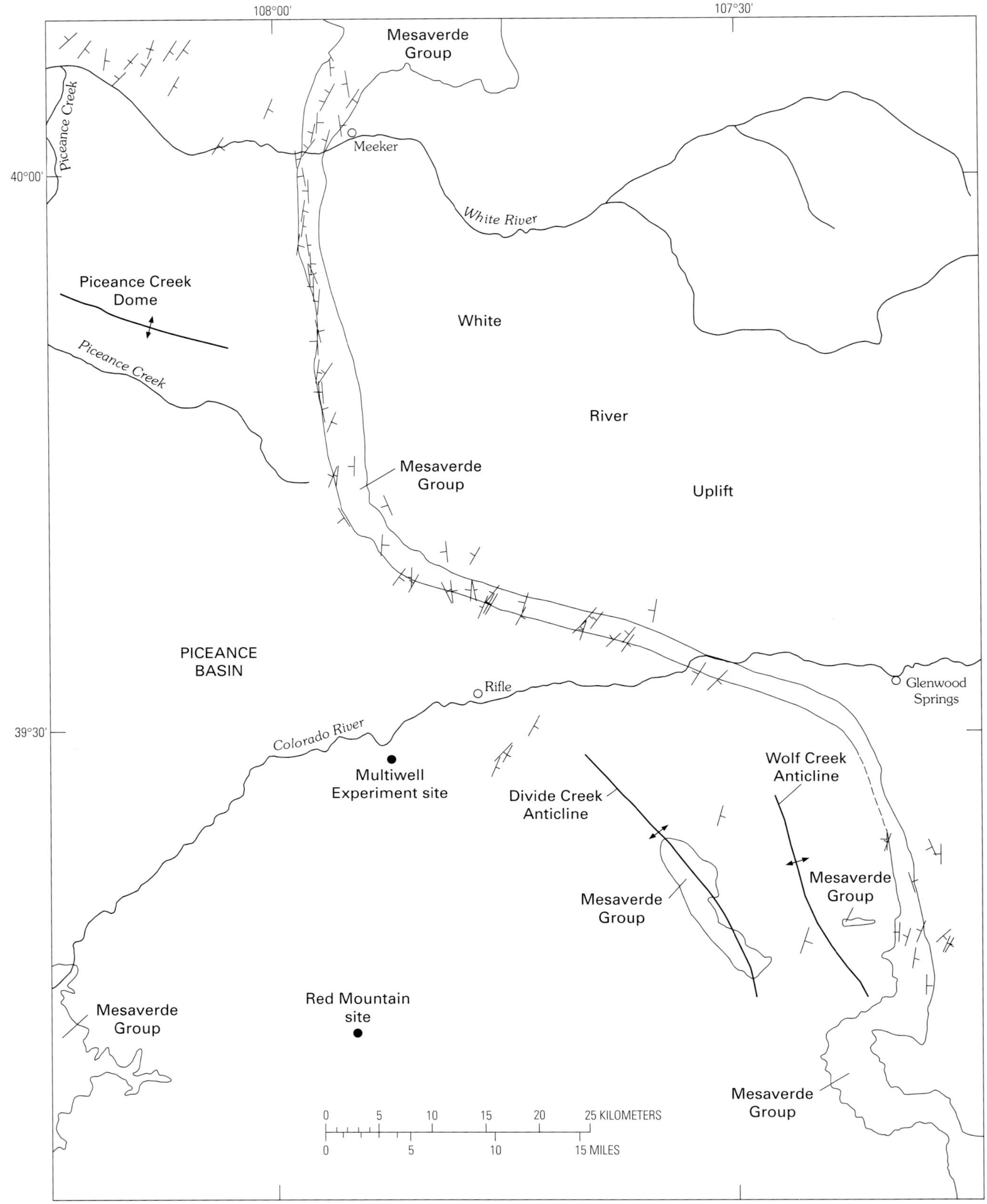

Figure 18. Reconstructed (bed-horizontal) orientations of the younger of two prominent sets of joints in beds of the Upper Cretaceous Mesaverde Group and Paleocene to lower Eocene Wasatch Formation along and near the Grand Hogback monocline.

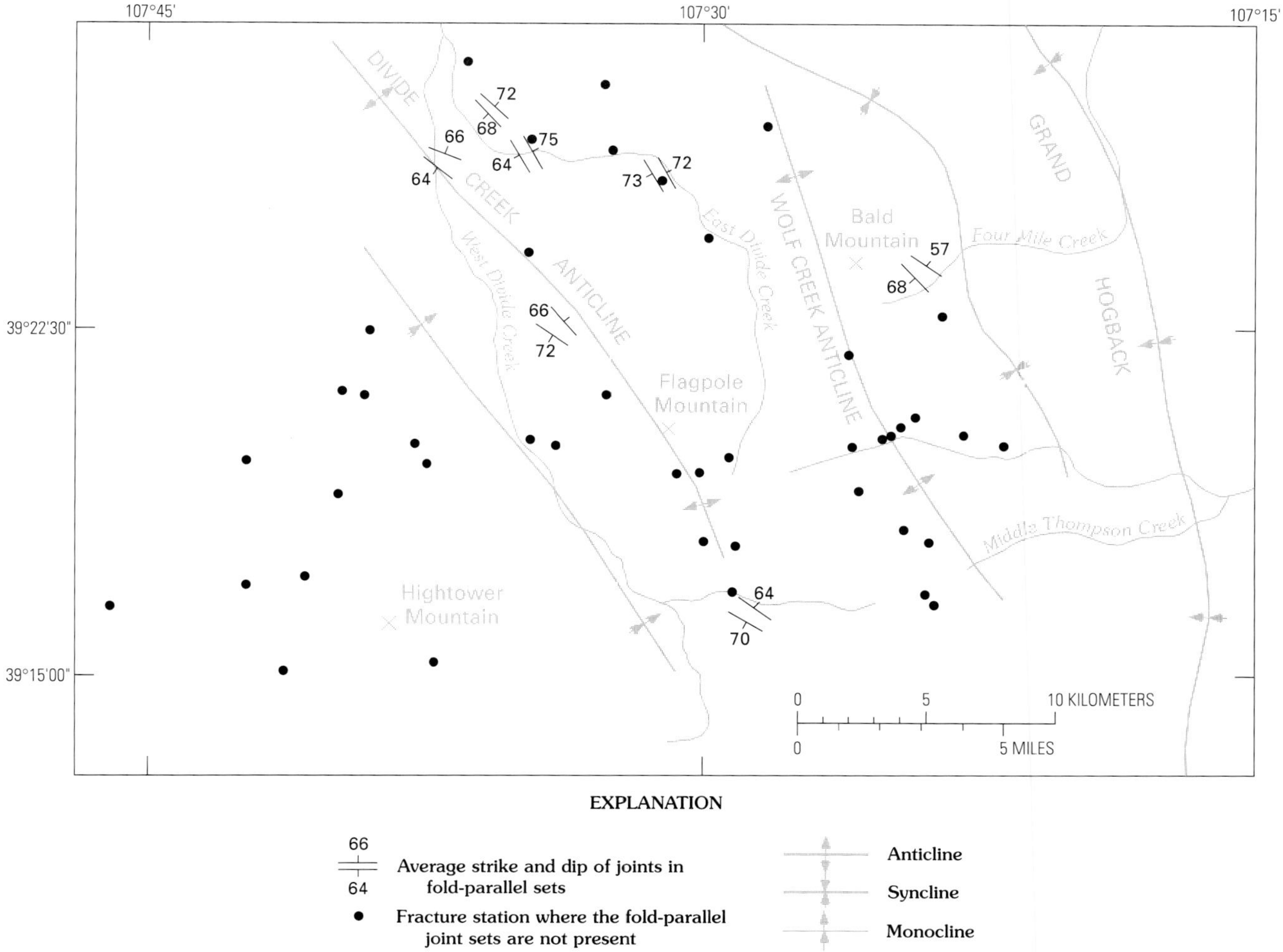

Figure 19. Orientations of two sets of fold-parallel joints on the Divide Creek and Wolf Creek anticlines near the eastern margin of the Piceance Basin, from Grout and Verbeek (1992).

system are broad and gradual, unlike the abrupt changes described earlier for the older, thrust-related fracture sets. We have noted no relation between degree of joint-set prominence and proximity to any recognized basement-related structure for any of the five sets.

We emphasize here the young age of the Piceance system of joints. Within the Piceance Basin, for example, minor calcite-cemented thrust faults (fig. 21) created during Laramide compression along the Grand Hogback are cut through by joints of the F_2 set. Kink folds within the finely laminated Green River beds also formed during the Laramide compressive events, and they everywhere predate all joints present. Farther west, in the Uinta Basin, joints of the earliest (F_1) set of the Piceance system have been traced upward through the Green River and Uinta Formations into the youngest dated beds (about 32–34 Ma; Bryant and others, 1989) of the Oligocene Duchesne River Formation (M.A. Grout and E.R. Verbeek, unpub. data, 1994). The abundant evidence for a post-Laramide (Oligocene and younger) age for all joint sets of the Piceance system explains their lack of spatial correlation to exposed structures; all five sets were superimposed on structures already present.

SUMMARY OF SURFACE FRACTURE NETWORK

Two of the oldest fracture sets in Cretaceous and Paleocene rocks along the eastern margin of the Piceance Basin are spatially coincident with the leading edge of a large, basement-cored crustal block that was thrust westward into the basin during Laramide time. The principal surface expression of this crustal boundary is the Grand Hogback monocline; along and west of this fold the fractures developed within a sinuous belt at least 135 km long but less than 25 km wide. Farther west, and within a much smaller area, two additional sets of fractures formed parallel to the axial

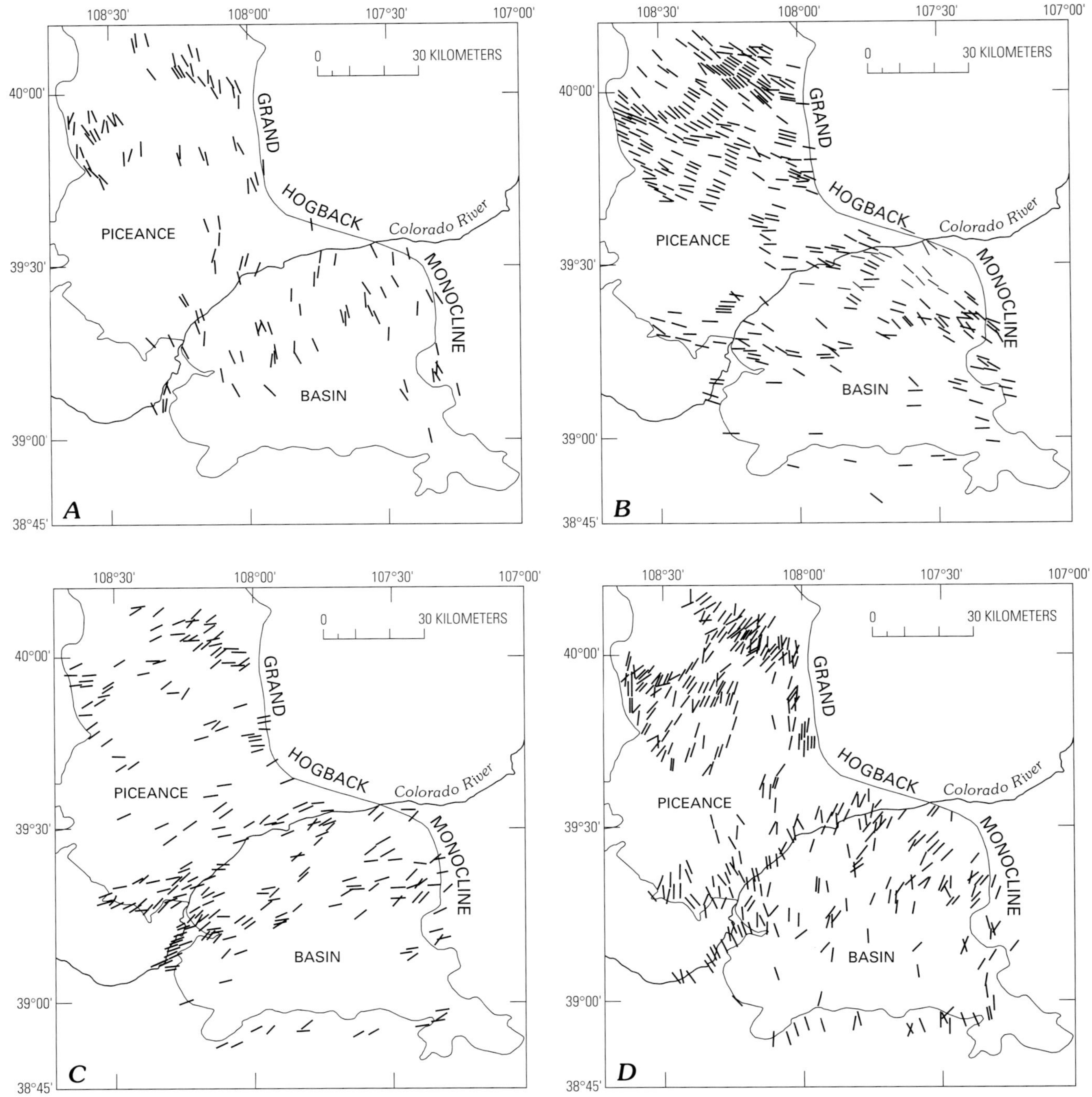

Figure 20. Distribution and orientations of four regional sets of post-Laramide joints in Tertiary rocks of the Piceance Basin, from Grout and Verbeek (1992). *A*, F_1 set; *B*, F_2 set; *C*, F_3 set; *D*, F_4 set.

trace of an intrabasin anticline that developed above a series of imbricate splay faults that sole into a décollement related to the same thrust system. These two fracture sets are related to basement structure only in an indirect sense; both are located about 15–25 km west of the inferred position of the thrusted basement block. Though basement structure exerted a seemingly clear influence over the formation and distribution of all four fracture sets discussed here, it did so through mechanisms much different from those discussed previously for the Grand Canyon region.

Younger sets of fractures, five in all, are present within an area far greater than that occupied by the Piceance Basin and are dominantly of post-Laramide age. In their distribution and expression we find no relation to basement structure and conclude that they were imposed largely on structures already present.

Figure 21. Small thrust fault in thin coal bed of Wasatch Formation, central Piceance Basin. Geologic pick for scale.

DISCUSSION

EFFECT OF DEPTH AND GEOLOGIC HISTORY ON BASEMENT-COVER FRACTURE RELATIONS

The degree to which basement structure influences fracture evolution in overlying rocks is related in large part to nature of the basement fault zones, their history of reactivation, and depth to basement. We comment briefly on the first topic before turning to the other two.

The gross pattern of known and inferred basement fault zones is broadly similar in all three areas discussed in this report. The northeast basement-fault trend is particularly well defined and areally extensive, as shown for the Grand Canyon region through the work of Shoemaker and others (1978), Huntoon (1974, 1993), and Sears (1973); for the Paradox Basin by Case and Joesting (1972), Friedman and Simpson (1980), and Friedman and others (1994); and for the Uncompahgre uplift and southern Piceance Basin by Case (1966) and Johnson (1983). Case and Joesting (1972, p. 2) believed that it continues "from the Grand Canyon area, Arizona, through the Rocky Mountains of northern Colorado and southern Wyoming, and into the High Plains of Wyoming." The northwest trend is likewise well expressed in all three areas, as shown in figures 4, 8, and 14 and discussed by most of the same authors just cited. The relative strength of these trends differs in different parts of the Colorado Plateau, but both are evident in many areas, as on the Marble Plateau (fig. 8). A subordinate north trend, locally prominent in the Grand Canyon region (Shoemaker and others, 1978) and the Paradox Basin (Case and Joesting, 1972), seems to weaken farther north but nevertheless finds expression in the north and south segments of the Grand Hogback monocline and, probably, the Douglas Creek arch between the Piceance and Uinta Basins (fig. 14). To a first approximation, then, we will assume that observed differences in basement-related fracture patterns from one area to another are due mostly to effects other than regional differences in basement fault pattern.

The major faults and faulted monoclines of the Hualapai Indian Reservation, in the western Grand Canyon region, offer particularly clear (and long recognized) examples of basement influence on local fracture of overlying sedimentary rocks. Among the three areas discussed in this report, it is here that the sedimentary veneer is thinnest (625–770 m over much of the area), the effects of Laramide compression on monocline development clearest, and Cenozoic reactivation of basement faults greatest. During the latter event many of the Laramide monoclines were faulted as the basement faults beneath them were reactivated in a normal sense. Spacings of 10–20 km between basement-related faults and faulted monoclines are typical (fig. 4), and several faults in the region display normal offsets exceeding 100 m. Farther east, on the Marble Plateau, faulted monoclines and basement-related faults are no less abundant (fig. 8), but few of the faults show normal offsets exceeding several tens of meters. Depth to basement in this area is slightly greater (950–1,200 m) than that on the Hualapai Reservation, but the greatest influence on structural differences between the two areas undoubtedly is the rapid eastward decline in basin-range-related crustal extension. Several previously unrecognized belts of fractures that we regard as the surface manifestation of reactivated basement structures beneath the Marble Plateau consist largely of faults with offsets too small to merit portrayal on conventional geologic maps.

The Piceance Basin differs markedly from these two areas in its much greater depth to basement (5,000–7,900 m) and in its comparative freedom from the effects of basin-range crustal extension. Both effects contribute to the general lack of expression of basement-related structures within the basin interior. The numerous basement-penetrating faults shown on seismic lines in Waechter and Johnson (1986) and Grout and others (1991), for example, show the persistence of the northwest basement-fault trend into this region but have no expression at the surface in Tertiary rocks. Similarly, the northeast basement-fault trend, so well defined both northeast and southwest of the Piceance Basin, has no known expression among fractures within it; the only fractures corresponding to that direction are late cross joints of the F_4 regional set (fig. 20). Basement-related fractures on this part of the Colorado Plateau are instead confined almost exclusively to the basin margins, and they are related to Laramide crustal compression rather than post-Laramide extension.

FRACTURE NETWORKS AND TECTONIC HEREDITY

The degree to which ancient structures in Precambrian crystalline basement rocks have influenced or controlled the development of fractures in overlying sedimentary strata has been long debated. Opinions in the literature have been widely divergent for decades, and remain so. Nowhere is this more apparent than in the proceedings volumes for conferences on "The New Basement Tectonics," the first of which was held in 1974 (Nickelsen, 1975; Hodgson and others, 1976) and the ninth in 1990 (Rickard and others, 1992). The problem is best considered in two parts: local formation of zoned fracture sets related to specific basement structures, and regional formation of areally pervasive fracture sets.

LOCAL FORMATION OF ZONED FRACTURE SETS

Evidence of basement involvement in local fracture development, typically as belts of fractures in overlying strata, is uncontested in such places as the Grand Canyon, where the deep structure of many faults can be studied directly in outcrop. Multiple episodes of movement, often in opposing senses, can be documented for faults exposed over thousands of feet in vertical section and traceable in continuity from Mesozoic and Paleozoic sedimentary rocks into the crystalline schists below. Huntoon (1974), for example, presented evidence for seven episodes of movement on the Hurricane fault from the Precambrian to the Quaternary. Evidence of basement-related fracturing of cover rocks is likewise strong in some areas where basement rocks are not exposed but the underlying structure is well documented through seismic evidence or drilling. Reactivation of basement structures is expressed at the surface most often as faults, but sets of extension joints can form during the same movements; the thrust-related joint sets of the Piceance Basin (Grout and Verbeek, 1992) are one example, and the early joint sets of the Redwall Limestone described by Roller (1987, 1989) may be another. In all cases, however, it should be recognized that the formation of zoned fracture sets on the Colorado Plateau is related to reactivation of individual, and generally large, basement faults.

REGIONAL FORMATION OF AREALLY PERVASIVE FRACTURE SETS

The case for basement-related fracturing on a finer scale is not nearly as convincing, despite the many papers devoted to the subject. The notion that ancient fracture networks in basement rocks are reflected in regionally widespread joint sets of the cover rocks found many adherents from the 1950's through the 1970's, a period which coincided with widespread use of aerial photographs to interpret regional fracture patterns. Satellite images, beginning with the first ERTS images in the late 1960's and high-resolution Skylab photographs shortly thereafter, quickly were put to similar use. A process of continuous fracture propagation—inheritance of entire fracture networks from underlying units—was envisioned by many investigators of this period (Blanchet, 1957; Mollard, 1957; Haman, 1961; Hodgson, 1961a, b; Gol'braikh and others, 1968; Rumsey, 1971; Burford and Dixon, 1977, 1978). Suggested causes of upward fracture propagation included earthquakes (fracture induced by transient shock waves), earth tides (fracture due to low-amplitude cyclic stress, a kind of geologic fatigue failure), and tectonic compression. The renewal of fracture networks over time, if true, meant that joint patterns in surface rocks might reflect nothing of the stress fields that affected those rocks (Hodgson, 1961b; Gol'braikh and others, 1968), thereby complicating attempts to decipher regional paleostress histories. Rumsey (1971) introduced a further complication by suggesting that *additional* fracture sets, unrelated to those already present, could form at any time and at any stratigraphic level due to tectonic forces; if so, he maintained, it might be impossible to deduce from surface evidence in which horizon each fracture set originated, or which set is genetically related to any other.

EVIDENCE AGAINST UPWARD PROPAGATION OF REGIONAL JOINT SETS

SOME COMMON PROBLEMS

Common problems with the concept of fracture inheritance, as listed by Engelder (1982), include (1) the puzzling selectivity of the process, wherein some basement fracture sets have apparent counterparts in overlying rocks but others do not; (2) the existence of additional sets in the cover rocks with orientations different from those in the jointed basement; and (3) the existence of unjointed strata overlain by beds that are jointed. Examples of all three types of problems can be drawn from the areas discussed in this report. The mechanism of joint inheritance cannot explain, for example, why the two oldest joint sets in the Redwall Limestone on the Hualapai Plateau disappear abruptly at the erosional unconformity that truncates this unit. Nor can it explain why the two most prominent joint sets in Cretaceous strata along the Grand Hogback die out upward in Paleocene strata and are replaced in younger beds by five sets that have no counterparts in the older rocks, or why Cretaceous beds containing abundant joints are interlayered with other beds, typically weakly cemented sandstones, that remained unjointed. More telling evidence, however, comes from the relative-age relations and mineralization histories of coexisting joint sets. Consistent abutting relations among joint sets at more than 1,100 localities in and near the Piceance and Uinta Basins show that all of the joint sets discussed above—the three sets along the Grand Hogback, the fold-related sets on the Divide Creek anticline, and the five regional sets in the basin rocks—formed in a well-defined sequence, one set after the other. Moreover, joints of the F_2 regional set (fig. 20), to select one prominent example, commonly are coated with calcite of two and locally three generations, whereas only a single generation of calcite was deposited in joints of the later F_4 set, and joints of the F_5 set everywhere are unmineralized. There is no suggestion in any of these relations of upward propagation of a fracture network from complexly jointed older beds into younger ones; instead we see the clear record of discrete fracture events over time, each one adding another joint set to an evolving and increasingly complex network.

JOINT DEVELOPMENT IN A ROTATIONAL STRESS FIELD

The formation of joints during periods of stress-field rotation furnishes additional evidence incompatible with the hypothesis of upward propagation of regional fracture sets. Lacustrine strata of the Green River Formation in the Piceance and Uinta Basins provide an unusually clear example of this effect. Green River marlstone of nearly zero organic content is a compact, well-cemented rock that, under conditions corresponding to known burial depths of this unit, is mechanically brittle. With increasing organic content the marlstones grade into oil shales, the change in lithology being reflected in a marked increase in ductility of the rock. The richest oil-shale beds contain more than 40 percent organic matter by weight (Tisot and Murphy, 1962) and are some of the most ductile rocks known (Tisot and Sohns, 1971). As these beds experienced layer-parallel stretching during post-Laramide tectonic extension, the most brittle beds failed first and increasingly less brittle (more ductile) beds failed progressively later. The record of progressive failure and stress rotation is clear at many localities where beds of differing organic content are interlayered at the outcrop scale. Figure 22 shows three examples from widely separated areas, where median strikes of the F_2 regional joint set show a marked counterclockwise shift from bed to bed as a function of increasing organic content. The oil shales preserve a complete record of the transition between the F_2 and F_3 episodes of fracture and show that jointing occurred in different beds at different times as the regional stress field rotated counterclockwise at least 40°. The close tie between joint orientation and lithology at the outcrop scale is incompatible with any mechanism of fracture propagation upward from deeper, previously jointed rocks.

TIMING OF JOINT DEVELOPMENT

The concept of continuous upward propagation of fracture networks from jointed basement implies that most rock units should fracture not long after deposition, as soon as cementation renders them capable of brittle failure (Hodgson, 1961). Reports of systematic joint sets in geologically young deposits (Gilbert, 1882; Burford and Dixon, 1977, 1978; see also fig. 13) lend credence to the hypothesis, but other evidence shows that rock units can, and commonly do, persist tens of millions of years in an unfractured state. We briefly summarize that evidence here for the three areas discussed in this report.

Hualapai Plateau.—Lower Permian strata of the Esplanade Sandstone near the Ridenour mine breccia pipe (fig. 4) are cut by five well-developed sets of subvertical extension joints (Verbeek and others, 1988). The pipe is well exposed at the surface, in a tributary gorge to the Grand Canyon, and also underground, in historic mine workings for the extraction of copper, uranium, and vanadium ores. Most of the ore minerals were precipitated within an annular zone of outward-dipping ring fractures that formed during the stoping process. Verbeek and others (1988) presented eight independent lines of evidence showing that the pipe stoped upward through the Esplanade Sandstone and was mineralized before the first set of regional joints had formed in the host rocks. These include the observations that all five joint sets are present in the cemented pipe breccia, that unmineralized joints terminate against mineralized ring fractures, and that solution pockets are common along the ring fractures but do not occur along the joints, whose surfaces

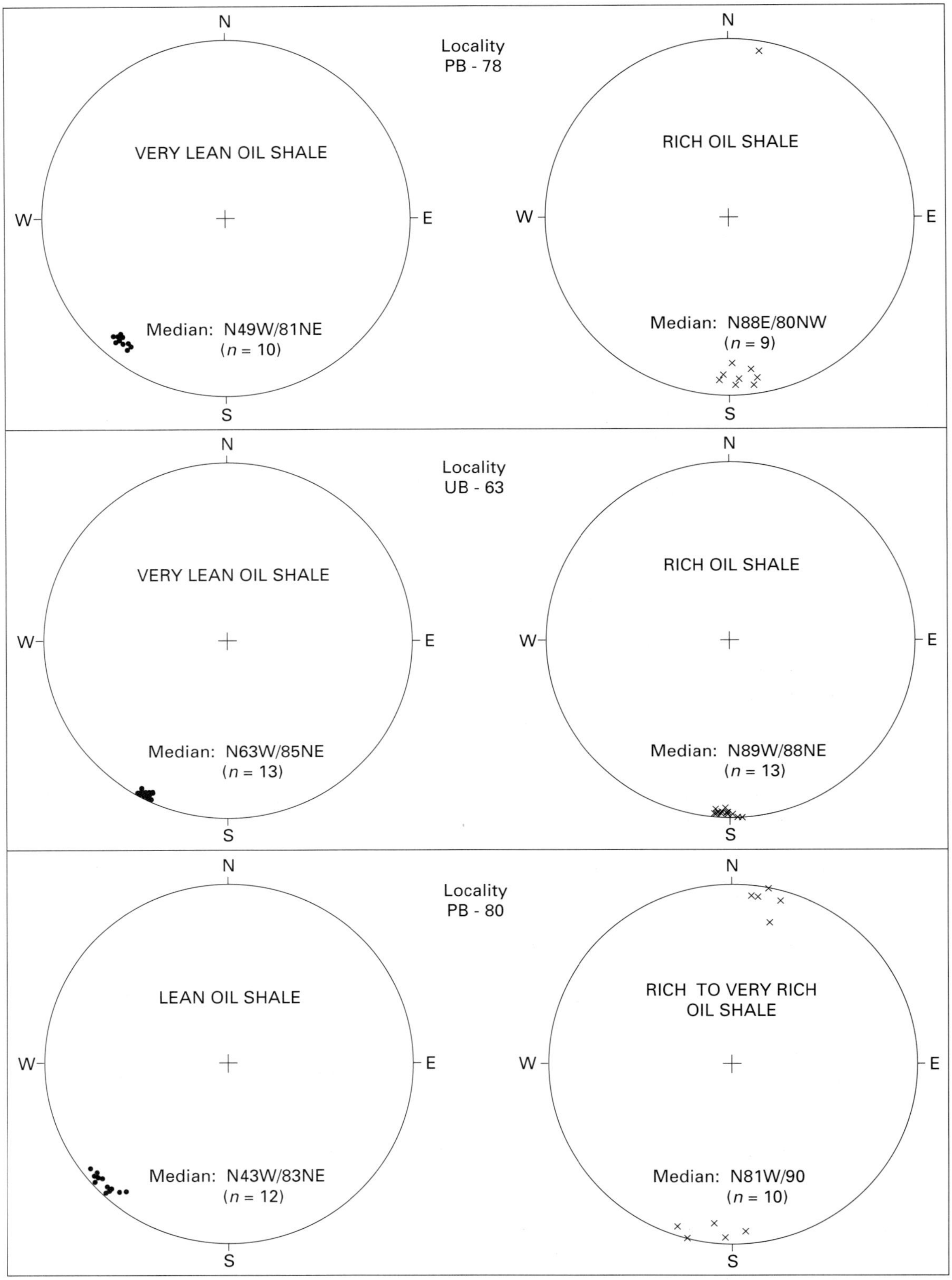

Figure 22. Paired lower-hemisphere equal-area projections of poles to F_2 joints, showing their difference in orientation in lean oil shale (left) versus rich oil shale (right) in the same outcrops, for three localities in the Piceance and Uinta Basins. Localities PB-78 and PB-80 are near the junction of Piceance Creek and the White River in the north-central part of the Piceance Basin, in the NW$^1/_4$SW$^1/_4$ sec. 11, T. 1 N., R. 97 W., Rio Blanco County, Colo. Locality UB-63 is about 70 km farther west, on the eastern flank of the Uinta Basin in the NE$^1/_4$NE$^1/_4$ sec. 20, T. 10 S., R. 25 E., Uintah County, Utah.

preserve fine details of their original morphology. If primary mineralization of the Ridenour pipe occurred at 200–220 Ma (Triassic), as suggested by U-Pb age determinations on other pipes nearby (Ludwig and Simmons, 1992), then at least 60 m.y. elapsed between deposition of the Esplanade Sandstone in Permian time and the development of the first joint set within that unit. Joints of similar orientation in overlying strata, including Tertiary basalts and conglomerates, suggest further that most or all of the fracture sets may be younger than 18 Ma (Roller, 1989) and thus that the Esplanade Sandstone may have lain unfractured for fully 250 m.y. or more.

Southern Marble Plateau.—One of the strongest pieces of evidence for long-delayed jointing in this area was mentioned previously: that joints of the oldest set in lower Triassic strata are nearly vertical on both the horizontal and tilted limbs of the Laramide Black Point monocline (fig. 12). We note also that remnant channels of a dendritic drainage system incised into the Permian Kaibab Limestone on the Marble Plateau, and especially on the adjacent Coconino Plateau to the west, show no obvious evidence from aerial photographs as having developed in jointed rock. The drainage system is of probable Miocene age (G.H. Billingsley, USGS, oral commun., 1989). Though little work has been done on the joint network of this large region, the available evidence suggests that most or all joint sets in the uppermost Paleozoic and younger rocks are of post-Laramide age. The joint set shown in figure 12, for example, is the oldest one present in the unit pictured but is at least 150 m.y. younger than the rocks that contain it.

Piceance Basin.—Upper Cretaceous beds of the Mesaverde Group and underlying Mancos Shale along the south and southwest margins of the Piceance Basin are cut by five sets of joints that have been traced in stratigraphic and areal continuity into late Eocene beds of the basin interior and into Oligocene beds of the Uinta Basin farther west. No older joint sets are present in the southern Piceance Basin, even in rock types normally considered especially susceptible to early fracture; coal beds in the lower part of the Mesaverde Group, for example, contain the same joint sets as those in associated and overlying clastic rocks (Grout, 1991). Strata of the lower Mesaverde Group were deposited 75–73 Ma and the Mancos Shale somewhat earlier, but the joints within these beds are younger than 32–34 Ma (Bryant and others, 1989), the age of the youngest dated beds within which we have documented them (M.A. Grout and E.R. Verbeek, unpub. data, 1994). The evidence thus points to at least a 40-m.y. hiatus between deposition of the Late Cretaceous beds and the development of the first joint set within them. Earlier fracture of the Mesaverde Group occurred only within a narrow (<25 km wide) belt of rock along and near the thrust front that marks the eastern edge of the basin (Grout and Verbeek, 1992).

Summary.—The evidence in all three areas for discrete episodes of regional jointing, preceded by geologically long periods during which no jointing occurred, is incompatible with basement control of surface-fracture networks as commonly envisioned. We conclude that if the basement rocks in any way influenced the development of regional joint sets in overlying sedimentary strata, they did not do so through any mechanism of continuous upward propagation of fracture sets.

GEOLOGIC SIGNIFICANCE OF REGIONAL JOINT SETS ON THE COLORADO PLATEAU

In all three areas studied on the Colorado Plateau, the existence of sedimentary units that remained unjointed for 40–250 m.y. after deposition suggests that conditions favorable for extension fracture did not exist for geologically long periods of time. The history of post-Redwall strata in the Grand Canyon region, following the latest Mississippian period of uplift, incision, and jointing, furnishes the clearest example. Preserved thicknesses of post-Redwall, Paleozoic and Mesozoic strata in this region, from the base of the Supai Group to the highest preserved remnants of the Chinle Formation, range from about 2,300 m on the Hualapai Plateau to 2,700 m on the Marble Plateau (fig. 2). As noted above, at least some of these rock units remained unjointed until the Tertiary Period. The inferred sequence of deformation, based on field relations in the Little Colorado River valley (E.R. Verbeek and M.A. Grout, unpub. data, 1986), is (1) formation of monoclines and attendant local thrust faults during Laramide crustal compression, (2) formation of multiple sets of vertical, dominantly post-Laramide extension joints, and (3) local reactivation of joints to form numerous minor normal faults with vertical walls during post-Laramide crustal extension. Evidence that regional jointing of the rocks occurred *after* the Laramide compressive events suggests that burial depths during monoclinal folding were still too great, and fluid pressures too low, for extension fractures to form. This interpretation in turn implies that erosion had not yet appreciably reduced lithostatic load. Moreover, much of the post-Redwall succession consists of sandstones and carbonate rocks poor in organic material. Had more organic material been present, its maturation might have produced sufficient fluid pressure to promote extensile failure at considerable depth. The paucity of organic matter, however, meant that jointing of these rocks was not possible until their return to conditions of low confining pressure as overlying strata were stripped by erosion. Maximum burial depths of these rocks remain poorly constrained because whatever younger deposits that may once have existed above them are not preserved, but removal of about 2.7–4.5 km of overburden seems assured (Dumitru and others, 1994).

A characteristic common to all three areas discussed here, and perhaps to most of the Colorado Plateau, is the abundance of geologically young fracture sets, many of them post-Laramide in age. The Tertiary Period in all three areas was a time of strong regional uplift and tectonic extension, both of which created conditions conducive to repeated extensile failure of the upper crust. Regional uplift promotes jointing by extensile failure in at least three ways:

1. Removal of overburden through erosion. Reduction in the vertical component of stress (lithostatic load) results in a corresponding decrease in horizontal stress by an amount proportional to the elastic properties of the rock (Price, 1966; Engelder, 1982).

2. Cooling and consequent lateral contraction of the rock, which further relieves horizontal stresses (Voight and St. Pierre, 1974; Haxby and Turcotte, 1976).

3. Layer-parallel extension resulting from the increasing radius of curvature of strata being uplifted (Price, 1966, 1974; Haxby and Turcotte, 1976).

In addition, active tectonic extension can further greatly increase the potential for jointing through progressive decrease in the regional component of horizontal stress in the direction of extension. Given these considerations, the evidence for repeated Tertiary jointing on the Hualapai Plateau as reported by Roller (1987, 1989) is not at all surprising; the region has been uplifted 3–5 km since the end of the Cretaceous Period (Wenrich and others, 1995b) and has been mildly extended during basin-range tectonism from Miocene time onwards (Dickinson, 1981). Similarly, on the Marble Plateau, Cenozoic erosion has stripped most post-Paleozoic strata from the area to leave only erosional remnants of Triassic rocks atop a vast, nearly bare surface of Permian limestone, and the area contains numerous minor normal faults that resulted from late Tertiary, nearly east-west extension. On the northeastern part of the Colorado Plateau, the estimated uplift of the Piceance Basin since deposition of the Uinta Formation in late Eocene time is about 2.8 km (Lorenz, 1985). Mild but prolonged, post-Laramide northeast-southwest tectonic extension of much of the northeastern Colorado Plateau is reflected in a regional set of joints (F_2) and minor normal faults within an area far larger than the Piceance Basin itself; to date we have mapped them over an area of more than 80,000 km^2 (Verbeek and Grout, 1992; Grout and Verbeek, this volume, p. 163). In all three areas discussed in this report, post-Laramide tectonic extension and decreasing burial depths resulted in a complex record of repeated episodes of jointing during the Tertiary. In all three areas, too, the resultant joint sets are present throughout an appreciable thickness of rocks of diverse geologic age, thereby creating a false appearance of upward propagation of fracture networks.

CONCLUSIONS

Fracture networks in Paleozoic and younger sedimentary rocks in all three areas studied include some elements related to movements along major basement faults. Common properties of these basement-related fractures are their local, rather than regional, extent and their occurrence in zones more-or-less directly above known basement structures. Offset of basement fault blocks during episodic reactivation of the basement structures, either in compression or extension, was expressed in the cover rocks by zones of shear failure (minor faults) in some areas and by zones of extensile failure in others, depending principally upon lithostatic load (depth of fracture) and fluid pressure at the time of fracture. Examples include the local thrust faults associated with monoclinal folding during Laramide compression in the Grand Canyon region, the elongate belts of minor faults due to vertical offsets along basement faults during post-Laramide extension of the same region, and the <25 km-wide zone of extension joints resulting from Laramide compression near the leading edge of a basement-involved thrust zone in the Piceance Basin. Fracture orientations in such basement-involved zones may be unreliable indicators of regional stress fields because of local perturbations of the stress field in the vicinity of the basement structures. All known and suspected basement-related fracture zones so far investigated by us have resulted from discrete episodes of movement; we have found no evidence for continuous upward propagation of basement fracture zones through any mechanism.

Joint sets of regional, rather than local, extent are common elements of Colorado Plateau geology and within vast areas are the dominant components of the overall fracture network. Numerous properties of the regional joint sets, including their orientations, stratigraphic range, spatial distribution, mineralization history, and consistent sequence of formation, are incompatible with upward propagation of fracture sets from deeper rocks. Each set is instead the preserved record of a discrete episode of failure that affected broad areas of thousands to tens of thousands of square kilometers; the joint sets reflect no direct influence of basement structures. Orientations of these joints reflect regional stress trajectories at the time of fracture and provide a useful means of reconstructing the paleostress history of the rocks in which they occur.

We conclude that the influence of basement rocks on fracture development in overlying sedimentary rocks on the Colorado Plateau is more limited than commonly envisioned. Individual basement structures, when reactivated, have influenced fracturing in overlying rocks and resulted in localized failure within discrete belts of rock, but nowhere during the course of our work have we found evidence for the upward propagation of entire joint networks.

UNRESOLVED STRUCTURAL PROBLEMS

An obvious limitation of any interpretive study such as this one is the fragmentary nature of the data on which it is based. In the literature for the Grand Canyon region, for example, is an astonishing dearth of actual data and specific field observations on how the regional joint network as we see it today evolved through time. Some of the papers cited here signal a good beginning upon which to build, but nowhere is there a comprehensive account of the geometry, let alone the genesis, of the joint network of this classic region. Conversely, basic elements of the basement structure are fairly well known through field study of exposed Precambrian rocks in the Grand Canyon and through geophysical studies on the flanking plateaus. Tracing of exposed basement fault zones upward through the sedimentary section long ago established the critical link between these deep structures and their surface expression in rocks hundreds of meters above, enabling geologists to trace the lateral extensions of some basement faults for scores of kilometers from the nearest exposure of Precambrian rock.

The state of knowledge on basement structure and fracture evolution in the northeastern part of the Colorado Plateau, in western Colorado and eastern Utah, is nearly the opposite. Only recently have studies clarified the true nature of the basement fault zone beneath the Grand Hogback monocline, the great fold that defines the boundary between this part of the Colorado Plateau and the adjacent Rocky Mountains. Knowledge of basement structure deep beneath the sedimentary basins to the west remains sketchy at best. The deepest boreholes in the Piceance Basin penetrate only into Pennsylvanian sedimentary rocks, and over large areas the basement remains hidden beneath more than 5,000 m of sedimentary rock. The fracture history of Paleozoic rocks in the region likewise remains almost wholly unknown, in part because these rocks crop out over such a small proportion of the total area. The joint network in Upper Cretaceous and younger rocks, in contrast, has been documented at more than 1,100 localities throughout an area of more than 25,000 km^2, and its evolution is now known in considerable detail.

The mechanisms by which reactivation of basement structures resulted in diverse types of fractures in overlying sedimentary rocks remain imperfectly understood in all three areas discussed here. In the Grand Canyon region, for example, how has strain been partitioned in cover rocks above a growing step in the basement? What governs the relation between displacement at basement level and width of an induced fracture zone at various levels above? In the Piceance Basin, the two sets of inclined joints on the Divide Creek anticline are especially intriguing in that no close analog to them seems to exist among popular models of fold-related fracturing (Hancock, 1985). Modeling of stress orientations during fold growth probably will be required to understand their genesis.

The basement-cover relations discussed here thus remain incompletely understood, but for different reasons in different regions. Geophysical study of deep basement structure, still an evolving art, is likewise an expensive one; and the study of fracture evolution in sedimentary cover rocks at the surface, if done properly, is tedious at best. Still, enough seems known to support the basic tenets of this paper: that individual basement fault zones, when reactivated, have resulted in local and in some cases repeated fracture of the cover rocks; that regional joint systems reflect the stress conditions under which they formed and did not propagate upward from preexisting basement networks; and that much of the exposed joint network as it exists today on the Colorado Plateau is a comparatively young element of that region's complex geologic history.

ACKNOWLEDGMENTS

This report is an outgrowth of several different projects funded by various agencies during the period 1978–1995. Funding by the U.S. Geological Survey was provided through the following programs: National Uranium Resource Evaluation (NURE), Oil Shale, National Geologic Mapping, Coal, Construction Hazards, and Evolution of Sedimentary Basins. Additional funds were provided by the Hualapai Indian Tribe through the Bureau of Indian Affairs for work in the Grand Canyon region; and by the U.S. Department of Energy, through their Tight Gas Sands Program, for work in the Piceance Basin.

The impetus for this report came from Gene Shoemaker, not only through his studies of basement-cover fracture relations in the Grand Canyon region, but also through his encouragement, during the "Utah laccoliths" conference, that our results be published. Field work and conversations with numerous colleagues contributed in no small measure to our understanding of Colorado Plateau geology and to the evolution of our ideas. Those deserving of special thanks include Karen Wenrich, Hoyt Sutphin, Brad Van Gosen, Pete Rowley, and Ivo and Barbara Lucchitta—fine geologists all. Thanks also to Brad Van Gosen for his thorough review; to the numerous members of the National Association of Geology Teachers and assorted volunteers who assisted us in the field; and to our long-suffering editors, Jules Friedman and Curt Huffman, for their patience, encouragement, and thumb screws.

The Central Publications Group of the U.S. Geological Survey performed their usual miracles in turning a rough manuscript and hand-drawn illustrations into a finished product. Pete Martin edited many an awkward phrase into readable English. Gayle Dumonceaux used skill and patience in drafting the illustrations, some of them twice after revised copies went astray.

REFERENCES CITED

Akers, J.P., Irwin, J.H., Stevens, P.R., and McClymonds, N.E., 1962, Geology of the Cameron quadrangle, Arizona: U.S. Geological Survey Geologic Quadrangle Map GQ–162, 1 sheet, scale 1:62,500.

Babenroth, D.L., and Strahler, A.N., 1945, Geomorphology and structure of the East Kaibab monocline, Arizona and Utah: Geological Society of America Bulletin, v. 56, no. 2, p. 107–150.

Barnes, C.W., 1974, Interference and gravity tectonics in the Gray Mountain area, Arizona, *in* Karlstrom, T.N.V., Swann, G.A., and Eastwood, R.L., eds., Geology of Northern Arizona, Part I—Regional studies: Geological Society of America, Rocky Mountain Section meeting [guidebook], p. 442–453.

———1987, Geology of the Gray Mountain area, Arizona: Geological Society of America Centennial Field Guide, Rocky Mountain Section, p. 379–384.

Billingsley, G.H., Barnes, C.W., and Ulrich, G.E., 1985, Geologic map of the Coconino Point and Grandview Point Quadrangles, Coconino County, Arizona: U.S. Geological Survey Miscellaneous Investigations Series Map 1644, scale 1:62,500.

Billingsley, G.H., and Beus, S.S., 1985, The Surprise Canyon Formation—An Upper Mississippian and Lower Pennsylvanian(?) rock unit in the Grand Canyon, Arizona: U.S. Geological Survey Bulletin 1605–A, p. A27–A33.

Billingsley, G.H., Wenrich, K.J., and Huntoon, P.W., 1986, Breccia pipe and geologic map of the southeastern Hualapai Indian Reservation and vicinity, Arizona: U.S. Geological Survey Open-File Report 86-458-B, 26 p., scale 1:48,000, 2 plates.

Billingsley, G.H., Wenrich, K.J., Huntoon, P.W., and Young, R.A., in press, Breccia pipe and geologic map of the southwestern Hualapai Indian Reservation and vicinity, Arizona: U.S. Geological Survey Miscellaneous Investigations Map I–2554, 33 p., scale 1:48,000, 2 plates.

Blanchet, P.H., 1957, Development of fracture analysis as exploration method: American Association of Petroleum Geologists Bulletin, v. 41, no. 8, p. 1748–1759.

Bredehoeft, J.D., Wolff, R.G., Keys, W.S., and Shuter, Eugene, 1976, Hydraulic fracturing to determine the regional in situ stress field, Piceance Basin, Colorado: Geological Society of America Bulletin, v. 87, no. 2, p. 250–258.

Bryant, Bruce, Naeser, C.W., Marvin, R.F., and Mehnert, H.H., 1989, Upper Cretaceous and Paleogene sedimentary rocks and isotopic ages of Paleogene tuffs, Uinta Basin, Utah: U.S. Geological Survey Bulletin 1787–J, p. J1–J22.

Burford, A.E., and Dixon, J.M., 1977, Systematic fracturing in clay and its relation to bedrock joints and drainage segments: Geological Society of America Abstracts with Programs, v. 9, no. 7, p. 916.

———1978, Systematic fracturing in young clay of the Cuyahoga River valley, Ohio, and its relation to bedrock jointing and drainage segments: Géotechnique, v. 28, p. 201–206.

Case, J.E., 1966, Geophysical anomalies over Precambrian rocks, northwestern Uncompahgre Plateau, Utah and Colorado: American Association of Petroleum Geologists Bulletin, v. 50, no. 7, p. 1423–1443.

Case, J.E., and Joesting, H.R., 1972, Regional geophysical investigations in the central Colorado Plateau: U.S. Geological Survey Professional Paper 736, 31 p.

Cashion, W.B., 1973, Geologic and structure map of the Grand Junction Quadrangle, Colorado and Utah: U.S. Geological Survey Miscellaneous Investigations Series Map I–736, scale 1:250,000.

Chenoweth, W.L., and Blakemore, P.P., 1961, The Riverview mine, Coconino County, Arizona: Plateau, v. 33, no. 4, p. 112–114.

Damon, P.E., Shafiqullah, Muhammad, and Leventhal, J.S., 1974, K-Ar chronology for the San Francisco volcanic field and rate of erosion of the Little Colorado River, *in* Karlstrom, T.N.V., Swann, G.A., and Eastwood, R.L., eds., Geology of northern Arizona, with notes on archaeology and paleoclimate—Part I, Regional studies: Geological Society of America, Rocky Mountain Section Meeting, Flagstaff, Ariz., April 27–30, 1974 (Guidebook for Field Trips, No. 27), p. 221–235.

Davis, G.H., 1978, Monocline fold pattern of the Colorado Plateau, *in* Matthews, Vincent, III, ed., Laramide folding associated with basement block faulting in the western United States: Geological Society of America Memoir 151, p. 215–233.

Dickinson, W.R., 1981, Plate tectonic evolution of the southern Cordillera, *in* Dickinson, W.R., and Payne, W.D., eds., Relations of tectonics to ore deposits in the southern Cordillera: Arizona Geological Society Digest, v. 14, p. 135–172.

Dodge, C.N., and Bartleson, B.L., 1986, The Eagle basin—A new exploration frontier, *in* Stone, D.S., ed., New interpretations of northwest Colorado geology: Denver, Colo., Rocky Mountain Association of Geologists, p. 113–121.

Dumitru, T.A., Duddy, I.R., and Green, P.F., 1994, Mesozoic-Cenozoic burial, uplift, and erosion history of the west-central Colorado Plateau: Geology, v. 22, no. 6, p. 499–502.

Eastwood, R.L., 1974, Cenozoic volcanism and tectonism of the southern Marble Plateau, *in* Karlstrom, T.N.V., Swann, G.A., and Eastwood, R.L., eds., Geology of northern Arizona, with notes on archaeology and paleoclimate—Part I, Regional studies: Geological Society of America, Rocky Mountain Section Meeting, Flagstaff, Ariz., April 27–30, 1974 (Guidebook for Field Trips, No. 27), p. 236–256.

Engelder, Terry, 1982, Is there a genetic relationship between selected regional joints and contemporary stress within the lithosphere of North America?: Tectonics, v. 1, no. 2, p. 161–177.

Finley, S.J., and Lorenz, J.C., 1989, Characterization and significance of natural fractures in Mesaverde reservoirs at the Multiwell Experiment site: Society of Petroleum Engineers Joint Rocky Mountain Regional/Low Permeability Reservoir Symposium and Exhibition, Denver, SPE Paper 19007, p. 721–728.

Friedman, J.D., Case, J.E., and Simpson, S.L., 1994, Tectonic trends of the northern part of the Paradox Basin, southeastern Utah and southwestern Colorado, as derived from Landsat multispectral scanner imaging and geophysical and geologic mapping, chap. C *of* Evolution of sedimentary basins—Paradox Basin: U.S. Geological Survey Bulletin 2000–C, p. C1–C30.

Friedman, J.D., and Simpson, S.L., 1980, Lineaments and structure of the northern Paradox Basin, Colorado and Utah: U.S. Geological Survey Miscellaneous Field Studies Map MF–1221, scale 1:250,000.

Gilbert, G.K., 1882, Post-glacial joints: American Journal of Science, Third series, v. 23, no. 133, p. 25–27.

Gol'braikh, I.G., Zabaluyev, V.V., Lastochkin, A.N., Mirkin, G.R., and Reinin, I.V., 1968, Morfostrukurnyye metody izucheniya tektoniki zakrytykh platformennykh neftegazonosnykh oblastey [Morphostructural methods for the tectonic study of covered platform oil and gas regions]: Leningrad, Izdatel'stvo

Nedra, 151 p. Partial English translation [1974] by M.H. Podwysocki (NASA Goddard Space Flight Center, Greenbelt, MD 20771), 136 p.

Gross, M.R., 1993, The origin and spacing of cross joints—Examples from the Monterey Formation, Santa Barbara coastline, California: Journal of Structural Geology, v. 15, no. 6, p. 737–751.

Grout, M.A., 1991, Cleats in coal beds of the southern Piceance Basin, Colorado—Correlation with regional and local fracture sets in associated clastic rocks, *in* Schwochow, S.D., Murray, D.K., and Fahy, M.F., eds., Coalbed methane of western North America: Denver, Colo., Rocky Mountain Association of Geologists, p. 35–47.

Grout, M.A., Abrams, G.A., Tang, R.L., Hainsworth, T.J., and Verbeek, E.R., 1991, Late Laramide thrust-related and evaporite-domed anticlines in southern Piceance Basin, northeastern Colorado Plateau: American Association of Petroleum Geologists Bulletin, v. 75, no. 2, p. 205–218.

Grout, M.A., and Verbeek, E.R., 1985, Fracture history of the Plateau Creek and adjacent Colorado River valleys, southern Piceance Basin—Implications for predicting joint patterns at depth: U.S. Geological Survey Open-File Report 85-744, 17 p.

———1992, Fracture history of the Divide Creek and Wolf Creek anticlines and its relation to Laramide basin-margin tectonism, southern Piceance Basin, northwestern Colorado: U.S. Geological Survey Bulletin 1787–Z, 32 p.

Gunneson, B.G., Wilson, M.S., and Labo, J.A., 1995, Divide Creek Anticline—A decapitated pop-up structure with two detachment zones: Rocky Mountain Association of Geologists, High-Definition Seismic Guidebook, p. 31–45.

Haman, P.J., 1961, Lineament analysis on aerial photographs exemplified in the North Sturgeon Lake area, Alberta: West Canadian Research Publications of Geology and Related Sciences, series 2, no. 1, 20 p.

Hancock, P.L., 1985, Brittle microtectonics—Principles and practice: Journal of Structural Geology, v. 7, nos. 3–4, p. 437–457.

Haxby, W.F., and Turcotte, D.L., 1976, Stress induced by the addition or removal of overburden and associated thermal effects: Geology, v. 4, p. 181–185.

Haynes, D.D., and Hackman, R.J., 1978, Geology, structure, and uranium deposits of the Marble Canyon 1°×2° quadrangle, Arizona: U.S. Geological Survey Miscellaneous Investigations Series Map I–1003, 2 sheets, scale 1:250,000.

Hite, R.J., 1975, An unusual northeast-trending fracture zone and its relation to basement wrench faulting in the northern Paradox Basin, *in* Fassett, J.E., ed., Canyonlands country guidebook: Four Corners Geological Society Field Conference, 8th, p. 217–223.

Hoak, T.E., and Klawitter, A.L., 1996, Structural compartmentalization in a decapitated anticline—the example of the Divide Creek fractured reservoir, Piceance Basin [abs.]: American Association of Petroleum Geologists, Annual meeting, San Diego, California, Poster Session abstract, p. A66.

Hodgson, R.A., 1961a, Regional study of jointing in Comb Ridge–Navajo Mountain area, Arizona and Utah: American Association of Petroleum Geologists Bulletin, v. 45, no. 1, p. 1–38.

———1961b, Reconnaissance of jointing in Bright Angel area, Grand Canyon, Arizona: American Association of Petroleum Geologists Bulletin, v. 45, no. 1, p. 95–97.

Hodgson, R.A., Gay, S.P., Jr., and Benjamins, J.Y., eds., 1976, First international conference on the new basement tectonics, Salt Lake City, Utah, June 3–7, 1974, Proceedings: Utah Geological Association Publication 5, 636 p.

Holm, R.F., 1987, Geomorphic evidence for youngest dated faults in north-central Arizona: The Mountain Geologist, v. 24, no. 1, p. 19–25.

Huntoon, P.W., 1970, Hydro-mechanics of the groundwater system in the southern portion of the Kaibab Plateau, Arizona: Tucson, Ariz., University of Arizona Ph. D. dissertation, 251 p.

———1974, The post-Paleozoic structural geology of the eastern Grand Canyon, Arizona, *in* Breed, W.J., and Roat, E.C., eds., Geology of the Grand Canyon: Flagstaff, Ariz., Museum of Northern Arizona, p. 82–115.

———1993, Influence of inherited Precambrian basement structure on the localization and form of Laramide monoclines, Grand Canyon, Arizona, *in* Schmidt, C.J., Chase, R.B., and Erslev, E.A., eds., Laramide basement deformation in the Rocky Mountain foreland of the western United States: Geological Society of America Special Paper 280, p. 243–256.

Huntoon, P.W., Billingsley, G.H., Breed, W.J., Sears, J.W., Ford, T.D., Clark, M.D., Babcock, R.S., and Brown, E.H., 1976, Geologic map of the Grand Canyon National Park, Arizona: Grand Canyon, Ariz., Grand Canyon Natural History Association, 1 sheet, scale 1:62,500.

Huntoon, P.W., and Sears, J.W., 1975, Bright Angel and Eminence faults, eastern Grand Canyon, Arizona: Geological Society of America Bulletin, v. 86, p. 465–472.

Izett, G.A., Honey, J.G., and Brownfield, M.E., 1985, Geologic map of the Citadel Plateau quadrangle, Moffat County, Colorado: U.S. Geological Survey Miscellaneous Investigations Series Map I–1532, 1 sheet, scale 1:48,000.

Johnson, S.Y., Schenk, C.J., and Karachewski, J.A., 1988, Pennsylvanian and Permian depositional systems and cycles in the Eagle basin, northwest Colorado, *in* Holden, G.S., ed., Geological Society of America field trip guidebook, 1988—Centennial meeting, Denver, Colorado: Golden, Colo., Colorado School of Mines Professional Contributions 12, p. 156–174.

Johnson, V.C., 1983, Preliminary aeromagnetic interpretation of the Uncompahgre Uplift and Paradox Basin, west-central Colorado and east-central Utah, *in* Averett, W.R., ed., Northern Paradox Basin–Uncompahgre Uplift field trip: Grand Junction, Colo., Grand Junction Geological Society, p. 67–70.

Lohman, S.W., 1981, The geologic story of Colorado National Monument: U.S. Geological Survey Bulletin 1508, 142 p.

Lorenz, J.C., 1985, Tectonic and stress histories of the Piceance Creek basin and the MWX site, from 75 million years ago to the present: Albuquerque, N.Mex., Sandia National Laboratories, Report SAND84-2603, 50 p.

Lorenz, J.C., and Finley, S.J., 1987, Differences in fracture characteristics and related production of natural gas in different zones of the Mesaverde Formation, northwestern Colorado: Society of Petroleum Engineers Annual Technical Conference and Exhibition, 62nd, Dallas, Sept. 27–30, 1987, SPE Paper 16809, p. 589–596.

Lorenz, J.C., Warpinski, N.R., Branagan, Paul, and Sattler, A.R., 1989, Fracture characteristics and reservoir behavior of stress-sensitive fracture systems in flat-lying lenticular formations: Journal of Petroleum Technology, v. 4, no. 1, p. 11–16.

Lucchitta, Ivo, 1974, Structural evolution of northwest Arizona and its relation to adjacent Basin and Range province structures, *in* Karlstrom, T.N.V., Swann, G.A., and Eastwood, R.L., eds., Geology of northern Arizona, with notes on archaeology and paleoclimate—Part I, Regional studies: Geological Society of America, Rocky Mountain Section Meeting, Flagstaff, Ariz., April 27–30, 1974 (Guidebook for Field Trips, No. 27), p. 336–354.

Ludwig, K.R., and Simmons, K.R., 1992, U-Pb dating of uranium deposits in collapse breccia pipes of the Grand Canyon region: Economic Geology, v. 87, p. 1747–1765.

MacLachlan, M.E., 1987, General geology of the Piceance Basin, *in* Taylor, O.J., compiler, Oil shale, water resources, and valuable minerals of the Piceance Basin, Colorado—The challenge and choices of development: U.S. Geological Survey Professional Paper 1310, p. 7–15.

MacLachlan, M.E., and Welder, F.E., 1987, Paleozoic and Mesozoic formations and their potential as ground-water reservoirs, *in* Taylor, O.J., compiler, Oil shale, water resources, and valuable minerals of the Piceance Basin, Colorado—The challenge and choices of development: U.S. Geological Survey Professional Paper 1310, p. 95–106.

MacQuown, W.C., Jr., 1945, Structure of the White River Plateau near Glenwood Springs, Colorado: Geological Society of America Bulletin, v. 56, p. 877–892.

McKee, E.D., 1979, Arizona, *in* Craig, L.C., and Connor, C.W., coordinators, Paleotectonic investigations of the Mississippian System in the United States—Part 1, Introduction and regional analyses of the Mississippian System: U.S. Geological Survey Professional Paper 1010–L, p. 197–207.

McKee, E.D., and Gutschick, R.C., 1969, History of the Redwall Limestone of northern Arizona: Geological Society of America Memoir 114, 692 p.

McLain, J.P., 1965, Photographic analysis of lineaments in the San Francisco volcanic field, Coconino and Yavapai Counties, Arizona: Phoenix, Ariz., University of Arizona M.S. thesis, 31 p.

Mollard, J.D., 1957, Aerial mosaics reveal fracture patterns on surface materials in southern Saskatchewan and Manitoba: Oil in Canada, v. 9, no. 40, p. 26–50.

Nickelsen, R.P., 1975, "New basement tectonics" evaluated in Salt Lake City: Geotimes, v. 20, no. 10, p. 16–17.

Perry, W.J., Jr., Grout, M.A., Tang, R.L., and Hainsworth, T.J., 1988, Wedge model for late Laramide basement-involved thrusting, Grand Hogback monocline and White River uplift, western Colorado: Geological Society of America Abstracts with Programs, v. 20, no. 7, p. A384–A385.

Price, N.J., 1966, Fault and joint development in brittle and semi-brittle rock: Oxford and New York, Pergamon Press, 176 p.

———1974, The development of stress systems and fracture patterns in undeformed sediments, *in* Advances in rock mechanics—Proceedings of the Third Congress of the International Society for Rock Mechanics, Denver, Colo., Sept. 1–7, 1974: Washington, National Academy of Sciences, v. 1, pt. A, p. 487–496.

Reches, Ze'ev, 1978, Development of monoclines—Part 1, Structure of the Palisades Creek branch of the East Kaibab monocline, Grand Canyon, Arizona, *in* Matthews, Vincent, III, ed., Laramide folding associated with basement block faulting in the western United States: Geological Society of America Memoir 151, p. 235–271.

Rickard, M.J., Harrington, H.J., and Williams, P.R., eds., 1992, Basement tectonics 9—Australia and other regions: Ninth International Conference on Basement Tectonics, Proceedings, Canberra, Australia, July 2–6, 1990, 272 p.

Roller, J.A., 1987, Fracture history of the Redwall Limestone and lower Supai Group, western Hualapai Indian Reservation, northwestern Arizona: U.S. Geological Survey Open-File Report 87-359, 33 p.

———1989, Fracture history of the Redwall Limestone, lower Supai Group, and Tertiary units on the Hualapai Indian Reservation, northwestern Arizona—Additional data to Open-File Report 87-359: U.S. Geological Survey Open-File Report 89-463, 42 p.

Rumsey, I.A.P., 1971, Relationship of fractures in unconsolidated superficial deposits to those in the underlying bedrock: Modern Geology, v. 3, p. 25–41.

Sears, J.W., 1973, Structural geology of the Precambrian Grand Canyon Series, Arizona: Laramie, Wyo., University of Wyoming M.S. thesis, 100 p.

Seccombe, J.C., and Decker, A.D., 1986, Geologic and reservoir characteristics of the Red Mountain coalbed methane test site in the Piceance Basin: Chicago, Ill., Gas Research Institute, Topical Report GRI–86/0109, 102 p. Available from National Technical Information Service, Springfield, VA 22161, as Report PB86–217098/XAB.

Shoemaker, E.M., Squires, R.L., and Abrams, M.J., 1978, The Bright Angel and Mesa Butte fault systems of northern Arizona, *in* Smith, R.B., and Eaton, G.P., eds., Cenozoic tectonics and regional geophysics of the western Cordillera: Geological Society of America Memoir 152, p. 341–368.

Stone, D.S., 1969, Wrench faulting and Rocky Mountain tectonics: The Mountain Geologist, v. 6, no. 2, p. 67–79.

Sturgul, J.R., and Irwin, T.D., 1971, Earthquake history of Arizona and New Mexico, 1850–1966: Arizona Geological Society Digest, v. 9, p. 1–22.

Sutphin, H.B., 1986, Occurrence and structural control of collapse features on the southern Marble Plateau, Coconino County, Arizona: Flagstaff, Ariz., Northern Arizona University M.S. thesis, 139 p.

Sutphin, H.B., and Wenrich, K.J., 1983, Structural control of breccia pipes on the southern Marble Plateau, Arizona: U.S. Geological Survey Open-File Report 83-908, 7 p., 2 plates, scale 1:50,000.

———1988, Map showing structural control of breccia pipes on the southern Marble Plateau, north-central Arizona: U.S. Geological Survey Miscellaneous Investigations Series Map I–1778, 2 sheets, scale 1:50,000.

Sutphin, H.B., Wenrich, K.J., and Verbeek, E.R., 1983, Structural control of breccia pipes on the southern Marble Plateau, Arizona: Geological Society of America Abstracts with Programs, v. 115, no. 5, p. 376.

Tisot, P.R., and Murphy, W.I.R., 1962, Physical structure of Green River oil shale from Colorado: U.S. Bureau of Mines Report of Investigations 6184, 24 p.

Tisot, P.R., and Sohns, H.W., 1971, Structural deformation of Green River oil shale as it relates to in-situ retorting: U.S. Bureau of Mines Report of Investigations 7576, 27 p.

Ulrich, G.E., Billingsley, G.H., Hereford, Richard, Wolfe, E.W., Nealey, L.D., and Sutton, R.L., 1984, Map showing geology, structure, and uranium deposits of the Flagstaff 1°×2°

quadrangle, Arizona: U.S. Geological Survey Miscellaneous Investigations Series Map I–1446, 2 sheets, scale 1:250,000.

Van Gosen, B.S., and Wenrich, K.J., 1988, Ground magnetometer surveys over known and suspected breccia pipes on the Coconino Plateau, northwestern Arizona: U.S. Geological Survey Bulletin 1683–C, 31 p.

Verbeek, E.R., and Grout, M.A., 1984a, Fracture studies in Cretaceous and Paleocene strata in and around the Piceance Basin, Colorado—Preliminary results and their bearing on a fracture-controlled natural-gas reservoir at the MWX site: U.S. Geological Survey Open-File Report 84-156, 30 p.

———1984b, Prediction of subsurface fracture patterns from surface studies of joints—An example from the Piceance Creek basin, Colorado, *in* Spencer, C.W., and Keighin, C.W., eds., Geologic studies in support of the U.S. Department of Energy Multiwell Experiment, Garfield County, Colorado: U.S. Geological Survey Open-File Report 84-757, p. 75–86.

———1992, Structural evolution of gilsonite dikes, eastern Uinta Basin, Utah, *in* Fouch, T.D., Nuccio, V.F., and Chidsey, T.C., Jr., eds., Hydrocarbon and mineral resources of the Uinta Basin, Utah and Colorado: Utah Geological Association Guidebook 20, p. 237–255.

———1993, Geometry and structural evolution of gilsonite dikes in the eastern Uinta Basin, Utah: U.S. Geological Survey Bulletin 1787–HH, 42 p.

Verbeek, E.R., Grout, M.A., and Van Gosen, B.S., 1988, Structural evolution of a Grand Canyon breccia pipe—The Ridenour copper-vanadium-uranium mine, Hualapai Indian Reservation, Coconino County, Arizona: U.S. Geological Survey Open-File Report 88-006, 75 p.

Voight, Barry, and St. Pierre, H.P.B., 1974, Stress history and rock stress, *in* Advances in rock mechanics—Proceedings of the Third Congress of the International Society for Rock Mechanics, Denver, Colo., Sept. 1–7, 1974: Washington, National Academy of Sciences, v. 2, pt. A, p. 580–582.

Waechter, N.B., and Johnson, W.E., 1986, Pennsylvanian-Permian paleostructure and stratigraphy as interpreted from seismic data in the Piceance Basin, Colorado, *in* Stone, D.S., ed., New interpretations of northwest Colorado geology: Denver, Colo., Rocky Mountain Association of Geologists, p. 51–64.

Wenrich, K.J., 1985, Mineralization of breccia pipes in northern Arizona: Economic Geology, v. 80, no. 6, p. 1722–1735.

Wenrich, K.J., Billingsley, G.H., and Huntoon, P.W., 1995a, Breccia pipe and geologic map of the northwestern Hualapai Indian Reservation and vicinity, Arizona: U.S. Geological Survey Miscellaneous Investigations Series Map I–2522, 32 p., 2 plates, scale 1:48,000.

———1995b, Breccia pipe and geologic map of the northeastern Hualapai Indian Reservation and vicinity, Arizona: U.S. Geological Survey Miscellaneous Investigations Series Map I–2440, 60 p., 2 plates, scale 1:48,000.

Wenrich, K.J., Chenoweth, W.L., Finch, W.I., and Scarborough, R.B., 1989, Uranium in Arizona, *in* Jenney, J.P., and Reynolds, S.J., eds., Geologic evolution of Arizona: Arizona Geological Society Digest, v. 17, p. 759–794.

Wenrich, K.J., and Sutphin, H.B., 1989, Lithotectonic setting necessary for formation of a uranium-rich, solution-collapse breccia-pipe province, Grand Canyon region, Arizona, *in* Metallogenesis of uranium deposits (Proceedings of a Technical Meeting, March 9–12, 1987, Vienna): International Atomic Energy Agency, p. 307–344.

Williams, P.L., 1964, Geology, structure, and uranium deposits of the Moab Quadrangle, Colorado and Utah: U.S. Geological Survey Miscellaneous Investigations Series Map I–360, scale 1:250,000.

Wong, I.G., and Humphrey, J.R., 1989, Contemporary seismicity, faulting, and the state of stress in the Colorado Plateau: Geological Society of America Bulletin, v. 101, no. 9, p. 1127–1146.

Zoback, M.L., and Zoback, M.D., 1980, State of stress in the conterminous United States: Journal of Geophysical Research, v. 85, p. 6113–6156.

Tectonic and Paleostress Significance of the Regional Joint Network of the Central Paradox Basin, Utah and Colorado

By Marilyn A. Grout[1] *and* Earl R. Verbeek[2]

CONTENTS

FIGURES

TABLES

ABSTRACT

Middle Pennsylvanian through Upper Cretaceous rocks of the central Paradox Basin, in southeastern Utah and southwestern Colorado, are cut by nine regional sets of extension joints. Stratigraphic evidence shows that the nine sets can be grouped into two major systems, an earlier system comprising three sets that evolved in Permian time, and a later system comprising six sets of middle Tertiary and younger age. Three additional sets of an older, Carboniferous system are present in Mississippian and older rocks along the eastern margin of the basin. An additional joint set of middle Tertiary age also is present there.

The regional joint sets of the Permian and Tertiary systems are areally persistent across the central Paradox Basin and show only broad, gradual changes in character from one area to another. Prominent sets in one area thus tend to be prominent sets in others, whether in the tilted and faulted rocks of the Paradox fold and fault belt, in the laccolith-rich eastern part of the basin, or in the broad expanses of flat-lying rock between. The general lack of correlation between joint-set development and major structural features of the Paradox Basin arises from their different age: joint sets of the Permian system predate the major phase(s) of folding and salt movement that gave rise to the prominent salt-cored anticlines of the Paradox fold and fault belt, and the sets of the Tertiary system largely postdate both laccolith intrusion and regional Laramide compression.

Joints of the most prominent and widely distributed Tertiary set in the Paradox Basin strike N. 52°–62° W., subparallel to the trend of the major, salt-cored basin folds. The approximate parallelism has led many workers to assume that the joints are old and affiliated developmentally with the large folds. The joints, however, are vertical regardless of bed inclination on the fold limbs, and thus apparently postdate the folds. Moreover, the joints maintain nearly constant strike when traced laterally along the lengths of individual folds, though the folds are characteristically sinuous. The prolific joints of this set are also present, in abundance, in flat-lying areas far from the basin folds. We thus find little evidence for fold-controlled development of this regional set of joints and suggest instead that the joints are products of a later period of regional crustal extension during which some of the fold crests were offset by approximately fold-parallel normal faults. The post-folding age of these joints was confirmed along the eastern margin of the basin, where they were traced upward into rocks as young as Miocene.

Regional Tertiary joint sets in the Paradox Basin can be correlated to sets of similar orientation, identical sequence of formation, and demonstrated young geologic age in Eocene and Oligocene strata of the Piceance and Uinta Basins to the north. These sets are present as well, along with older sets, in Cretaceous and older rocks along the uplifts that border all three basins. Collectively these joints and associated structures record a prolonged period of counterclockwise stress rotation during mid- to late Tertiary crustal extension. Their presence at many hundreds of localities throughout an area of at least 80,000 km^2 indicates that this event was widespread and affected most or all of the northern Colorado Plateau.

[1]Consulting Geologist, 5407 S.W. Viewpoint Terrace, Portland, OR 97201.

[2]Geologist, Bureau of Land Management, Bldg. 102, Military Circle, P.O. Box 911, Tonopah, NV 89049-0911.

INTRODUCTION

The Paradox Basin occupies the central part of the Colorado Plateau, in southeastern Utah and southwestern Colorado (fig. 1). Like other such basins in the Colorado Plateau/Rocky Mountain region of the western United States, the Paradox Basin is a large structural depression bordered by Tertiary uplifts (Kelley and Clinton, 1960; Davis, 1978). Sedimentary rocks of Pennsylvanian through Cretaceous age are well exposed over vast areas of the basin, making it an ideal region in which to study the geographic and temporal evolution of regional joint sets. The most outstanding structural features of the basin are the northwest-trending evaporite-cored anticlines that collectively define the area known as the Paradox fold and fault belt. These anticlines, which trend about parallel to the Uncompahgre Uplift bordering the basin on the northeast, are unique to the continental Americas and have been the subject of many geologic studies.

In this paper we examine the relation between the evolution of evaporite ("salt")-cored anticlines and jointing of Pennsylvanian through Tertiary strata across the central part of the Paradox Basin. Our initial studies focused on the Lisbon Valley Anticline (Grout and Verbeek, unpub. data, 1990–1994), one of the southernmost folds of the Paradox fold and fault belt (fig. 1). This anticline was chosen for study because strata preserved on its limbs represent a lengthy time interval from Pennsylvanian to Late Cretaceous (fig. 2), and because its faulted crest had not been breached by the Middle Pennsylvanian evaporites. Subsequent studies along the eastern margin of the Paradox Basin further extended the stratigraphic range of our data from Late Mississippian to Tertiary and allowed us to compare joint-network evolution in a region not affected by salt tectonics to that of the Paradox fold and fault belt. Documentation of joint sets in latest Cretaceous through Miocene volcanic and volcaniclastic rocks of the adjacent San Juan Mountains (Lipman, 1989; Cunningham and others, 1994) helped considerably in elucidating the complex Tertiary history of fracture in the region. We then turned to the Dolores Anticline and southern Dolores River area (fig. 1), between Lisbon Valley and the eastern margin of the Paradox Basin, to connect the two areas studied previously. Joint-set correlations throughout the region could then be made on a geographically continuous basis. Once this link was established, we extended our studies northwest of the Lisbon Valley Anticline to the Green River (fig. 1). All told, fracture data are now available from almost 500 localities within a broad swath, 50 km wide by 250 km long, across the entire width of the central Paradox Basin.

A principal conclusion gained from these data is that stratigraphically equivalent rocks in different parts of the basin share many aspects of their joint history. Individual joint sets commonly can be correlated across structural boundaries, and prominent sets in one area tend to be prominent in others as well. One of the most regionally prominent sets strikes parallel to the salt-cored anticlines of the Paradox fold and fault belt, inviting the oft-expressed hypothesis that folding and jointing were genetically linked in this part of the basin. However, as discussed herein, the fold-parallel joints are equally as common and as stratigraphically persistent in nonfolded areas as within them, and there exists little evidence that their formation was structurally controlled. We suggest instead that the joints postdate the folds and formed during the same period of regional crustal extension that gave rise to the crestal normal faults that offset some of the salt-cored anticlines.

TECTONIC OVERVIEW

Initial flowage of the thick accumulations of Middle Pennsylvanian evaporites (fig. 2) that core the major folds of the Paradox fold and fault belt is widely thought to have been triggered in middle to late Desmoinesian (late Middle

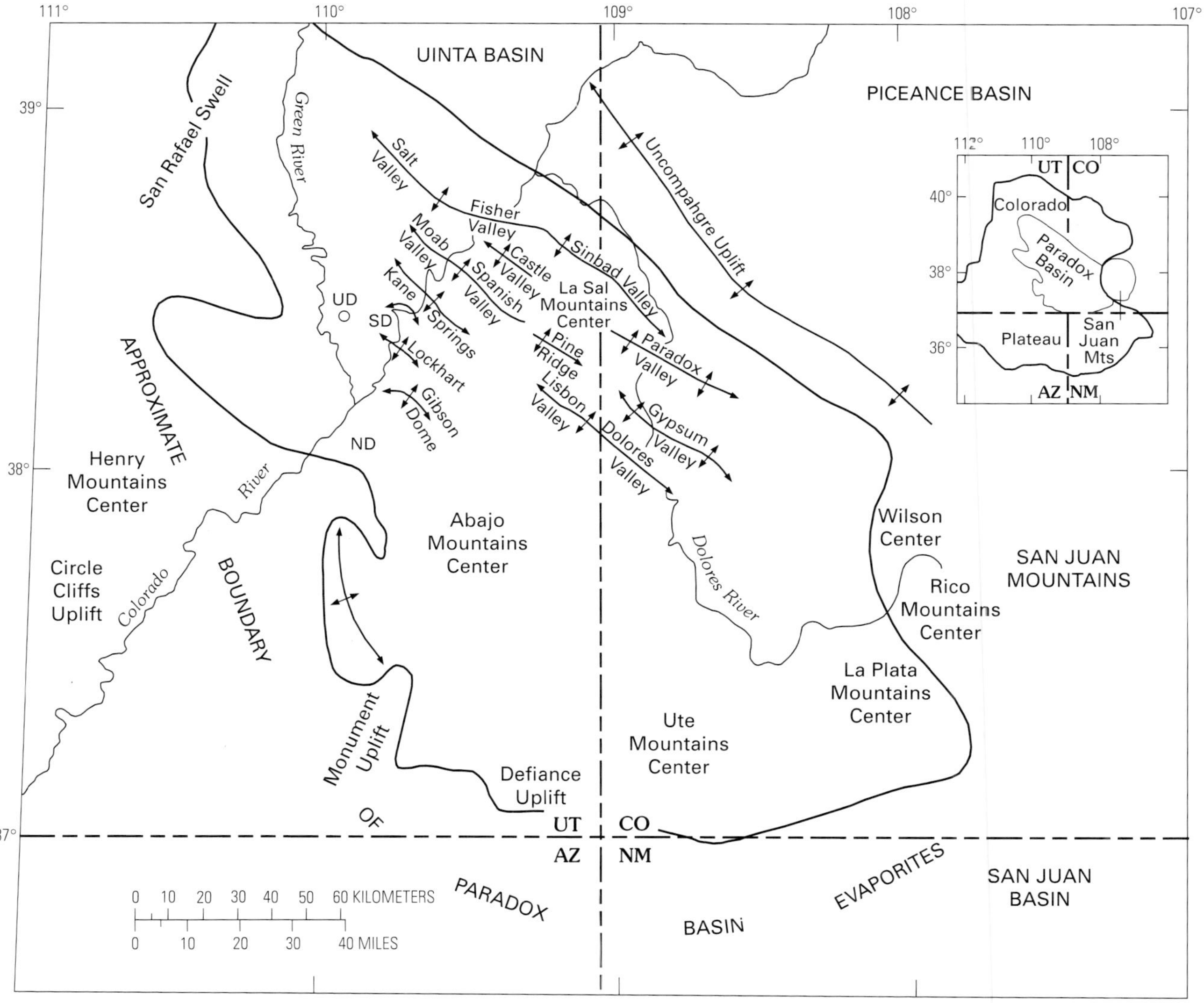

Figure 1. Locations of major salt anticlines, uplifts, igneous laccolithic centers, and other features of interest in the Paradox Basin. Modified from Case and Joesting (1972) with additions from Kelley and Clinton (1960), Haynes and others (1972), Steven and others (1974), and Friedman and Simpson (1980). UD, Upheaval Dome; SD, Shafer Dome; ND, Needles District.

Upper Cretaceous	
Mesaverde Formation	Brown to gray, fine- to medium-grained sandstone, siltstone, and shale; minor carbonaceous shale and coal. Maximum thickness about 100 m
Mancos Shale	Dark-gray to black, soft, fissile marine shale; thin-bedded sandstone near base. Maximum thickness about 1,200 m
Dakota Sandstone	Yellowish-brown and gray quartzitic fluvial sandstone and conglomeratic sandstone in thick beds, with interbedded gray to black carbonaceous nonmarine shale. Maximum thickness about 65 m
Lower Cretaceous	
Burro Canyon Formation	Light-gray and light-brown fluvial quartzitic sandstone and conglomerate interbedded with green and purplish lacustrine siltstone, shale, and mudstone, and thin beds of impure limestone. Maximum thickness about 45 m
Unconformity.	
Upper Jurassic	
Morrison Formation	Variegated shale, massive mudstone, and fine-grained sandstone. Maximum thickness about 240 m
Unconformity.	
Middle Jurassic	
Summerville Formation	Red, gray, green, and brown sandy shale and mudstone; masses of red and white chert near top. Maximum thickness about 45 m
Wanakah Formation	(Lateral equivalent of Summerville Fm.) Greenish-gray limy siltstone and sandstone; light-yellow to white fine-grained sandstone; dark-gray to black bituminous limestone. Maximum thickness about 90 m
Unconformity.	
Entrada Sandstone	Orange, buff, and white fine- to medium-grained, massive and crossbedded eolian sandstone. Maximum thickness about 165 m
Carmel Formation	Red muddy siltstone and shaly sandstone, crossbedded in middle section. Maximum thickness about 40 m
Unconformity.	
Lower Jurassic	
Navajo Sandstone	White, grayish-yellow, gray and pale-orange-pink, fine-grained, crossbedded eolian sandstone. Maximum thickness about 135 m
Kayenta Formation	Red, buff, gray, and lavender irregularly interbedded fluvial shale, siltstone, and fine- to medium-grained sandstone; thin beds of limestone and shale-pellet conglomerate. Maximum thickness about 80 m
Wingate Sandstone	Reddish- to grayish-orange, fine-grained, massive, thick-bedded and prominently crossbedded eolian sandstone. Maximum thickness about 130 m

Figure 2 (above and facing page). Stratigraphic units and brief descriptions of rock types where joints were measured, Paradox Basin, Utah and Colorado. Summarized from Williams (1964), Haynes and others (1972), Steven and others (1974), Tweto and others (1976), Huntoon and others (1982), and O'Sullivan (1991).

Upper Triassic	
Chinle Formation	Red, reddish-brown, and orange-red siltstone interbedded with sandstone, shale, limestone-pebble and shale-pellet conglomerate; gray, brown, and pale-green-gray quartzose sandstone with uranium deposits. Maximum thickness about 200 m
Unconformity.	
Middle(?) and Lower Triassic	
Moenkopi Formation	Brown and reddish-brown shaly siltstone and fine-grained sandstone; purple and reddish-brown arkosic conglomerate. Maximum thickness about 300 m
Unconformity.	
Lower Permian	
Cutler Formation	Red to purple arkose and arkosic fluvial conglomerate; white, gray, and buff quartzose sandstone; gray cherty limestone and dolomite interbedded with sandstone and siltstone. Maximum thickness about 2,400 m
Unconformity.	
Upper and Middle Pennsylvanian	
Honaker Trail Formation	Blue and gray bedded fossiliferous limestone and cherty limestone; gray micaceous sandstone and siltstone; red sandy shale and sandstone; gray arkose and conglomerate. Maximum thickness about 550 m
Middle Pennsylvanian	
Paradox Formation	Salt, anhydrite, and gypsum interbedded with euxinic black shales and limestones. Maximum thickness about 1,500 m
Unconformity.	
Lower Mississippian	
Leadville Limestone	Light- to dark-gray dense to coarsely crystalline limestone and dolomitic limestone; minor intercalated red shale, siltstone, and sandstone; locally cherty; limestone breccia in lower part. Maximum thickness about 55 m
Upper Devonian	
Ouray Limestone	Light-gray, dense to finely crystalline, locally dolomitic limestone. Maximum thickness about 45 m
Elbert Formation	Varicolored calcareous shale, limestone, quartzitic sandstone, and siltstone. Maximum thickness about 30 m

Pennsylvanian) time by rejuvenation of preevaporite basement faults beneath the Paradox Basin. During these movements the evaporites became buried by voluminous arkosic deposits shed from the rising, west-northwest-trending Uncompahgre Highland to the north (Hite, 1961; Peterson, 1989; Huffman and Taylor, 1994). Maximum displacement along the basement faults occurred in Early Permian time (Cater, 1955; Cater and Elston, 1963; Elston and others, 1962). Doelling (1988), however, suggested that the most active salt movements continued until the end of Chinle time (Late Triassic, fig. 2), at least beneath the northwesternmost anticline in the basin (fig. 1). Northeast-trending lineaments that appear in map view to truncate many northwest-trending faults in the basin (as discussed by Friedman and others, 1994) were interpreted by Hite (1975) to reflect intermittent left-lateral movement on underlying basement structures that persisted into Mesozoic time. The rising salt cores affected the thickness and distribution of Mesozoic units until at least pre-Morrison (Late Jurassic) time (Cater, 1970), or even until Mancos (Late Cretaceous) time in the northwesternmost part of the fold and fault belt (Doelling, 1988).

Tectonic activity resumed in latest Cretaceous/Tertiary time when the salt-cored anticlines were rejuvenated, again by movement along basement fault zones. The Uncompahgre Uplift, which borders the Paradox Basin on the north (fig. 1), achieved its final structural configuration at this time. This lengthy feature, which first gained expression in Middle Pennsylvanian time as noted previously, is a nearly rectilinear, west-northwest-trending fold that overlies a blind, basement-cored, basinward-directed overthrust. Seismic-reflection data reveal a vertical component of offset of approximately 6 km across this fault (Frahme and Vaughn, 1983), and a horizontal, mostly left-lateral component of approximately 10 km (Potter and others, 1991).

It has long been thought that the laccolithic centers of the Paradox Basin (fig. 1) were emplaced during this period of crustal compression (see, for example, Peterson, 1989). Some of the K/Ar age determinations indicating a Late Cretaceous age for these centers, however, are now being questioned: new $^{40}Ar/^{39}Ar$ age determinations yield younger, Oligocene and Miocene ages (Nelson and others, 1992; this volume; Mutschler, Larson, and Gaskill, this volume). In light of these new dates, and of new observations on the structural relations between regional joint sets and dikes in the Paradox Basin, Grout and Verbeek (this volume) speculate that the laccoliths formed during a period of middle to late Tertiary crustal extension and are related to the voluminous eruptive deposits along the edges of the Colorado Plateau both to the west (Marysvale volcanic field; Rowley and others, this volume) and to the east (the San Juan Mountains). The structural effects of this extension on the Colorado Plateau, however, were much more subtle than along its margins; its principal expression in the Paradox Basin is a regional set of middle Tertiary extension joints, one of the earliest of the Tertiary sets described in this report.

PREVIOUS FRACTURE STUDIES

Few studies of fractures in the Paradox Basin have been published. Interpretations from one regional and several local studies are summarized herein.

REGIONAL STUDY

The complexity of the fracture network in the Paradox Basin, and the great variation in fracture strike, were early recognized by Kelley and Clinton (1960) during their monumental aerial photographic study of joints, faults, and lineaments on the Colorado Plateau. Kelley and Clinton interpreted the fracture network of the Paradox Basin as the cumulative product of multiple fracture events related chiefly to salt tectonics, to uplift, and to the development of major regional structures. They noted, in particular, the parallelism between the northwest trends of the salt anticlines in the Paradox Basin (fig. 1) and the strikes of numerous joints on the flanks of those folds. Here, then, was suggestive evidence that folding and jointing in at least part of the Paradox Basin were genetically linked. From their map it can be seen, however, that northwest-striking joints are not only missing from many parts of the anticlines, but also that overall they are no more abundant on the folds than away from them. Kelley and Clinton also noted that the strikes of these joints appeared to curve from NNW. through NW. to WNW. along several anticlines, but that the axial traces of the folds remained about N. 45° W. throughout, as, for example, on the Lisbon Valley Anticline. Adding to the evident geometric complexity of the fracture pattern were numerous, visually obvious joints whose strikes and geographic distribution bore no obvious relation to any known structure.

LOCAL STUDIES

The strikes of subsurface fractures in drill core were compared to those of joints in outcrop, and to trends of fracture traces on aerial photographs, by R.J. Warner and T.C. Hansen (Chevron, USA, written commun., 1991) in an unpublished study of the Kane Springs (Cane Creek) Anticline area northwest of, and on trend with, the Lisbon Valley Anticline (fig. 1). Only the longest (≥15 m) and oldest fractures in Permian and Mesozoic rocks were measured during this study, on the assumption that these would best reflect the orientations of regional systematic joint sets. Warner and Hansen concluded that most of the major joints in the study area strike NNW. or NW., regardless of their position relative to the Kane Springs Anticline. Like Kelley and Clinton before them, however, they also recognized the complexity of the fracture network: numerous major joints striking WNW., NNE., and NE. were suggestive of an involved fracture history.

In another recent study of the same area, Morgan and others (1992) showed once again that the dominant regional fracture trend in the Permian and younger surface rocks is NW. In addition they recognized a second regional trend averaging NE., but with considerable azimuthal variation. They concluded that the pattern of surface fractures was unlikely to extend downward through the numerous salt layers to the beds below, and thus that studies of joints in outcrop were of little use in predicting optimal directions for drilling into the "Cane Creek" zone, a stratigraphic interval of current economic interest for petroleum production from the Middle Pennsylvanian Paradox Formation (fig. 2).

The southwest flank of the Salt Valley Anticline (fig. 1) has been the subject of several recent and fascinating studies of fracture evolution on a local scale. Within this area Dyer (1988), Cruikshank and others (1991), Zhao and Johnson (1992), and Cruikshank and Aydin (1995) documented at least five sets of fractures in the Moab Member of the Middle Jurassic Entrada Sandstone. The oldest two sets are not joints but conjugate deformation bands (compressional faults of extremely small displacement; see Aydin, 1978), with average strikes of N. 60° E. and N. 30° E., that resulted from mild northeast-southwest compression normal to the axis of the anticline (Zhao and Johnson, 1992). Younger structures comprise faults, several sets of extension joints, and faulted joints whose sequence of formation and interactions are discussed in detail in the papers cited previously. Among the extension joints are three sets that, from oldest to youngest, strike NNW., NW., and WNW., and which were interpreted by Cruikshank and Aydin (1995) to reflect counterclockwise rotation of the stress field—an effect documented previously on a regional scale in other Tertiary basins farther north (Verbeek and Grout, 1986, this volume; Grout and Verbeek, 1992a; 1992b). Farther south, too, the evidence from hundreds of joint stations in the area discussed in this paper furnishes a strong record of counterclockwise stress rotation during the Tertiary Period.

South of the salt-cored anticline area, McGill and Stromquist (1979) found that incipient graben development in Permian rocks of the Needles district (fig. 1) had dilated two sets of preexisting vertical joints that strike N. 35° E. (oldest) and N. 45° W. (youngest). No evidence was found for a shear origin for either set of joints, or for simultaneous formation suggestive of conjugate sets; further, the joints have the same regular spacing in areas outside the graben fault system as they do within it. McGill and Stromquist concluded (1) that the two sets of joints have an extensional origin, (2) that some of these joints were later reopened and offset to produce the grabens observed, and (3) that this movement probably was due to gliding of the cover rocks on the underlying evaporite layers rather than to upward propagation of basement faults.

FIELD METHODS

Properties of individual joint sets in the Paradox Basin are closely related to lithology, bed thickness, stratal sequence, and previous fracture history in consistent and understandable ways. To document these relations it was first necessary to determine at each outcrop the number of joint sets present and their sequence of formation, thereby reconstructing a local fracture history. Then, for the joints of each set, the following properties were recorded: orientation, size (length, height), spacing, overall shape (planar, subplanar, or nonplanar, with descriptive remarks), surface structures (origin point, plumose structure, twist hackle, arrest lines, slickenside striations, and so on), mineralization and alteration history, relation to other nearby structures such as faults or deformation bands, stratal persistence, and terminating and crosscutting relations with other fractures. Close attention was also paid to how these properties differ between beds of different lithology or thickness in the same outcrop. Though time-consuming, such care is necessary in areas of complex fracture history where joint sets of different age may have nearly identical or overlapping orientations. All of these properties are readily documented in the field and greatly enhance the reliability of correlating sets from one locality to another, and thus of successfully interpreting the regional fracture history.

MAJOR JOINT SYSTEMS AND JOINT-SET NOTATION

The joints of the Paradox Basin initially were studied in two widely separated areas: Lisbon Valley, southeast of Moab, Utah, and the eastern basin margin near Telluride, Colorado. The Middle Pennsylvanian through Cretaceous rocks of the Lisbon Valley area were found to contain nine sets of joints, referred to herein as sets PX_{1-9}. Shortly thereafter an additional four sets were documented in Precambrian through Miocene rocks along the eastern margin of the Paradox Basin. These four sets were labeled P_{1-4}, the lack of the "X" signifying that their relation to the Lisbon Valley sets was then unknown. Subsequent studies over a much wider area clarified the geographic and stratigraphic range of each set and revealed that they can be grouped into three major systems of vastly different age. The oldest of these, a system of three sets (labeled P_{1-3} in table 1), to date has been found only in Mississippian and older rocks along the eastern margin of the basin. Rocks of equivalent age are not exposed farther west; thus it is not known if these early joint sets persist westward beneath the evaporite layers of the basin proper. Joint sets of the next system (sets PX_{1-3}) are of Permian age and predate the major phase of salt movement and anticline growth in the Paradox Basin. The majority of the regional joint sets, however—including sets PX_{4-9} in the basin and set P_4 along its eastern margin—belong to the youngest system and are of middle Tertiary age or younger. For convenience we will refer to these systems as the Carboniferous, Permian, and Tertiary systems, respectively. Their character will be discussed by area, beginning with the Lisbon Valley Anticline in the center of the basin, the area for which the most is known (Grout and Verbeek, unpub. data, 1990–1994).

Table 1. Summary of correlations of regional joint sets in the Paradox Basin.

[Joint sets of Carboniferous age ($P_{1\text{-}3}$) are present in Mississippian and older rocks on the northeast and east margins of the Paradox Basin; those of Permian age ($PX_{1\text{-}3}$) in Permian and older rocks across the basin; and those of Tertiary age ($PX_{4\text{-}9}$) in middle Tertiary and older rocks across the basin. P_4 refers to an additional set on the northeast and east margins but not found elsewhere. Total numbers (n) of outcrops that contain each set are listed in the right-hand column, based on median strikes of joints of sets from 494 outcrops (Grout and Verbeek, unpub. data, 1990–1994). EM, eastern margin of Paradox Basin; LV, Lisbon Valley area; DR, southern Dolores River area; SC, Shafer Dome–Cane Creek Anticline area; LA, Lockhart Basin–Abajo Mountains area; CV, Castle Valley and adjacent Colorado River valley area; UC, Uncompahgre Uplift; PX, Paradox Basin]

	PARADOX BASIN REGIONAL JOINT-SET CORRELATIONS							
Set No.	EM	LV	DR	SC	LA	CV	UC	PX (TOTAL STATIONS)
	Carboniferous sets in Mississippian and older rocks							
P_1	N19W							N19W (*n*=4)
P_2	N64E							N64E (*n*=6)
P_3	N62W						N84W	N68W (*n*=14)
	Permian sets in Permian and older rocks							
PX_1	N18E	N21E		N26E			N11E	N20E (*n*=43)
PX_2	N27W	N29W		N29W			N19W	N28W (*n*=42)
PX_3	N64E	N61E		N64E			N53E	N62E (*n*=34)
	Tertiary sets in Miocene and older rocks							
P_4	N11E						N16E	N11E (*n*=30)
PX_4	N22W	N32W	N30W	N29W	N34W	N33W	N24W	N29W (*n*=151)
PX_5	N53W	N52W	N61W	N57W	N57W	N60W	N62W	N56W (*n*=231)
PX_6	N85W	N84W	N85W	N85W	N82W	N79W	N90W	N85W (*n*=122)
PX_7	N62E	N65E	N66E	N56E	N60E	N56E	N63E	N61E (*n*=189)
PX_8	N25E	N34E	N31E	N26E	N16E	N30E	N13E	N26E (*n*=105)
PX_9	N20W	N38W	N4W	N21W	N27W	N26W	N6W	N18W (*n*=57)

JOINT SETS OF THE CENTRAL PARADOX BASIN AND ADJOINING AREAS

LISBON VALLEY AREA

The Lisbon Valley area contains one of the southernmost of the major evaporite-cored anticlines of the Paradox fold and fault belt (fig. 1). The anticline trends N. 40°–55° W., as do nearly all of the anticlines farther north (Kelley and Clinton, 1960; Cater, 1970; Friedman and Simpson, 1980), and its limbs dip gently, 20° or less (Weir and others, 1961). Its crest is cut by a normal fault zone that trends N. 40°–55° W. and that places Upper Pennsylvanian and Upper Cretaceous rocks in fault contact near the northwestern end of the fold. The crestal fault zone is approximately 1 km across at its widest but rapidly decreases in width toward the anticlinal noses. The zone contains at least seven mappable faults; the average dip of the major strand is 58° to the northeast (Weir and others, 1961). Despite 1,200–1,500 m of displacement across this fault zone (Parker, 1981; Weir and Puffett, 1981), the Middle Pennsylvanian evaporites have not breached the crest of the anticline, in contrast to most of the other evaporite-cored anticlines in the Paradox fold and fault belt.

Two systems of joints have been documented in outcrops of upper Paleozoic and Mesozoic rocks along and near the Lisbon Valley Anticline (Grout and Verbeek, unpub. data, 1990–1994). The older of these is the Permian system, of which all three sets ($PX_{1\text{-}3}$) are present. Median strikes of the restored (to bed-horizontal) attitudes of these sets are, from oldest to youngest, N. 21° E., N. 29° W., and N. 61° E. (table 1). Joints of these sets are present in the upper Paleozoic Honaker Trail and Cutler Formations, but not in overlying strata of the Moenkopi Formation (Triassic) and younger units (fig. 2). The joints of all three sets, regardless of present bed dip on the fold, are almost everywhere nearly perpendicular to bedding. That their attitudes graphically restore to vertical as the anticline is unfolded suggests that the joints predate the major episode(s) of folding, which on stratigraphic grounds postdated the lower Permian beds but predated the uppermost Lower Triassic strata. The present-day attitudes and restricted stratigraphic distribution of the three joint sets thus agree with the tectonic history of the area. Locally, however, rotation of the joints past vertical as the beds are graphically restored to horizontal is suggestive of some early salt movement in the area of the future Lisbon Valley anticline.

The joints of the younger, Tertiary system (the $PX_{4\text{-}9}$ sets, table 1) in the Lisbon Valley Anticline area are present throughout the entire stratigraphic range of strata preserved, from the uppermost Pennsylvanian rocks of the Honaker Trail Formation through the lowermost Upper Cretaceous rocks of the Dakota Sandstone (fig. 2). With the exception of the oldest, PX_4 set, the joints of the Tertiary system are everywhere vertical, regardless of bed dip on the limbs of the fold. In many places they are thus oblique rather than perpendicular to bedding, in contrast to the joints of the older system. Rotation of the PX_4 but not of the PX_5 and later sets brackets the time of the last phase of bed tilting along the anticlinal fold. Further resolution of the age relation between the joints and folds of the Paradox Basin is discussed in later sections.

In addition to these regional sets of joints are local joints near, and subparallel to, some of the crestal normal faults on the Lisbon Valley Anticline. These joints are spatially restricted to within a few meters of each fault and most likely are of similar age to the regional PX_5 set, whose joints likewise strike nearly parallel to the faults (median strike, N. 52° W.). As for folds, the relation of the joints to the regional history will be discussed further in later sections.

The traces of fractures that Kelley and Clinton (1960) compiled from aerial photographs of the Lisbon Valley area are similar in direction to strikes of the $PX_{1\text{-}9}$ sets (table 1) measured for this study. However, a one-to-one correspondence between a given set of fracture traces on the photographs and a joint set documented in the field is not always possible, in part because some sets, though of different age, have similar or overlapping orientations (for example, sets PX_3 and PX_7) and are present in some of the same stratigraphic units. The apparent curving of joint strikes from NNW. to WNW. along the flanks of the anticline, as

Table 2. Median orientations of joint sets in the Lisbon Valley Anticline area.

Median orientations* of joints in each set		Number of localities (n=96)	Percent of localities
Permian joint system			
PX_1	N21E/88NW	9	9.4
PX_2	N29W/88SW	12	12.5
PX_3	N61E/89SE	8	8.3
Tertiary joint system			
PX_4	N32W/90	25	26.0
PX_5	N52W/90	40	41.7
PX_6	N84W/90	20	20.8
PX_7	N65E/90	35	37.6
PX_8	N34E/90	14	14.6
PX_9	N38W/89SW	13	13.5

*N21E/88NW refers to a median strike of N. 21° E. and a dip of 88° to the northwest.

described by Kelley and Clinton (1960), most likely corresponds to some photogeologic combination of the first three sets of the Tertiary system (PX_{4-6}) and one of the Permian system (PX_2).

The most prominent and widely distributed joint set in the Paleozoic rocks of the Lisbon Valley area is the PX_2 set, whose joints have a median strike of N. 29° W. These joints are present in nearly half of the Paleozoic outcrops studied. The most common joints overall, however, are those of the PX_5 set of the Tertiary system (table 2). These joints, with a median strike of N. 52° W., are present in more than half of the outcrops studied, in every rock type of every age. Joints of this set strike subparallel to the axial trace of the Lisbon Valley Anticline, but the implication of a genetic relationship to fold growth is misleading: not only do the joints fail to curve in correspondence to the curved trace of the anticline, but they are fully as abundant many kilometers away from the anticline as they are on it. As we will repeatedly suggest, a genetic connection of jointing to salt-anticline growth in the Paradox Basin does not seem warranted.

EASTERN MARGIN OF THE BASIN

The succession of Precambrian through Miocene rocks (fig. 2) that crops out along the eastern margin of the Paradox Basin (fig. 1) contains joints not only of the two systems known from the Lisbon Valley area, but also joints of the older Carboniferous system. The three sets of the Carboniferous system, P_{1-3} (table 1), are restricted to Lower Mississippian and older rocks; the stratigraphic evidence places their probable age between Late Mississippian and latest Middle Pennsylvanian. Little is known of the properties or tectonic significance of these sets because few outcrops have been studied to date. For the youngest (P_3) set, however, its median strike of N. 62° W. suggests that it may be related to the same period of regional crustal extension that gave rise to the west-northwest-trending Middle Pennsylvanian block faults in the northern Colorado Plateau. Similar block faults are present at depth beneath the Lisbon Valley Anticline (Parker, 1981) and trend N. 60°–65° W. Though mere equivalence in trend hardly constitutes strong evidence of a genetic relation between joints and faults, their probable temporal similarity is likewise suggestive of a common origin.

Among the post-Carboniferous joints, all nine sets known from the Lisbon Valley area are present along the eastern margin of the Paradox Basin as well. Relative-age criteria show that the sets formed in the same sequence between the two areas, greatly strengthening the proposed correlations. The median strikes of all three joint sets of the Permian system (PX_{1-3}) along the eastern margin are within 2°–3° of those for the equivalent sets in the Lisbon Valley area (table 1). Similarly, median strikes for the six sets of the Tertiary system (PX_{4-9}) between the two areas agree within 1°–10° for all but the youngest set. The two areas thus appear to have undergone similar fracture histories. Much of the fracture network is fairly young: all six sets of the Tertiary system have been traced into units as young as Miocene. As discussed elsewhere in this volume (Verbeek and Grout paper), a complex record of Tertiary jointing is a common element of northern Colorado Plateau geology.

Joints of an additional set not known from the Lisbon Valley area (the P_4 set in table 1) are present along the eastern margin of the basin. Relative-age relations with the joints of the other sets establish it as the oldest set of the Tertiary joint system. The P_4 joints are sparse in most formations, and their geographic distribution is poorly known. Joints that appear to be equivalents have been found to date only on the Uncompahgre Uplift (table 1) along the northern basin margin. The P_4 set at present is unknown from the interior of the Paradox Basin.

Kelley and Clinton (1960) noted that joints on the eastern margin of the basin appeared sparse on aerial photographs and generally had different strikes from those in the anticlinal areas farther west, such as on the Lisbon Valley Anticline. The overall fracture density, however, is comparable between the two areas. The apparent shortage of fractures along the eastern basin margin is ascribable to the steep exposures prevalent there, and to a greater average vegetation density, as opposed to the lower-relief, arid land of mesas and common expanses of bare rock farther west. The perceived differences in fracture strike between the two areas probably are ascribable to real differences not in the sets present, but in their relative prominence, and possibly also in part to the difficulty of mapping complex fracture networks from vertical aerial photographs in areas of steep exposure. As previously noted, both areas experienced a comparable fracture history.

SOUTHERN DOLORES RIVER AREA

The drainage area of the southern Dolores River is contiguous with the Lisbon Valley area and the eastern margin of the Paradox Basin (fig. 1). Within this area are several long (>25 km), northwest-trending anticlines and attendant synclines cut by later faults and grabens (Haynes and others, 1972; Williams, 1964). The principal fold, the N. 45° W.-trending, evaporite-cored Dolores Valley Anticline (fig. 1), is truncated on the southeast by a broadly curved zone of N. 30°–80° E.-striking normal faults. To the northwest a short, narrow, shallow graben of N. 70°–75° W. trend cuts acutely across the crest of the anticline.

In the southern Dolores River area, only Triassic and younger rocks are exposed. The joint network consists of six regional sets that appear equivalent in all respects to the PX_{4-9} sets (table 1) of the Tertiary system in adjoining areas. Median strikes of all but the youngest set are within 1°–9° of those for correlative sets in the other two areas discussed previously. The youngest sets show the greatest differences in median strike, a common phenomenon in areas where orientations of young joints are influenced by the presence of several or more older sets. Locally, the Triassic rocks also contain a sparse set of older joints with a median strike of N. 22° E.; these joints may be the link with the middle Tertiary P_4 set of the eastern margin of the basin. Not enough data are as yet available for reliable interpretation.

Local fault-related, reactivated extension joints are prominently exposed near fault zones and the narrow grabens of the southern Dolores River area. All of these joints investigated to date predate the Tertiary regional sets (Grout and Verbeek, unpub. data, 1991–1994). The most prominent examples are in outcrops of the Upper Cretaceous Dakota Sandstone, within the zone of normal faults at the southeast end of the Dolores Anticline. In this area are two well-developed sets that strike N. 40° E. and N. 80° E. Accretionary quartz fibers on the joint surfaces show that the N. 80° E. fractures record minor right-lateral movement, followed by superimposed left-lateral movement. The N. 40° E. fractures record only left-lateral movement. That these movements postdate lithification of the Upper Cretaceous Dakota Sandstone, but predate all six sets of the Tertiary system of joints, suggests that they are associated with Laramide compressive events.

SHAFER DOME–CANE CREEK ANTICLINE AREA

West of the Lisbon Valley Anticline, and on trend with it, are the Cane Creek (also known as Kane Springs) Anticline and Shafer Dome (fig. 1). The Cane Creek Anticline trends N. 40°–60° W., is salt-cored, and has been faulted and breached by the Pennsylvanian evaporites. Shafer Dome, as its name implies, is a nearly equant fold with an ill-defined axial trace mapped as concave either to the north (Huntoon and others, 1982) or to the south (Williams, 1964). Connecting the two structures is the Roberts rift (named by Hite, 1975), a structure of enigmatic origin defined by fracture zones and air-photo lineaments that cut generally N. 30° E. across the area, subparallel to the Colorado River.

Joints in the Middle Pennsylvanian through Middle Jurassic rocks of the Cane Creek–Shafer Dome area correlate well, both in orientation and sequence of formation, with joints of the $PX_{1\text{-}9}$ sets in stratigraphically equivalent rocks in those parts of the basin discussed previously. Median strikes of all but the youngest set are within 10° of those of equivalent sets in all of these areas (table 1). As along the Lisbon Valley Anticline, the dominant sets of the Cane Creek–Shafer Dome area are the PX_4 (N. 29° W.), PX_5 (N. 57° W.), and PX_7 (N. 56° E.) sets.

Although strata in the Cane Creek–Shafer Dome area are faulted, local zones of fault-related joints were noted at few localities. The two most notable exceptions are (1) in Upper Triassic rocks in a faulted area near the southeast end of the Cane Creek Anticline, where local joints strike subparallel to minor normal faults, and (2) southeast of Upheaval Dome, where the joints strike N. 28° E. throughout the local area. These latter joints most likely are products of the same extension that opened the Roberts rift zone.

LOCKHART BASIN–ABAJO MOUNTAINS AREA

The southwesternmost salt-cored structures of the Paradox fold and fault belt are the Lockhart Anticline and Gibson Dome, north of the Abajo Mountains laccolithic complex (fig. 1). The central part of the Lockhart Anticline is spectacularly exposed in the eroded area known as Lockhart Basin. The major folds in this part of the Paradox Basin trend N. 45°–65° W. (Friedman and Simpson, 1980). The strata are cut by minor, N. 30°–50° E.-striking normal faults, some of them forming en echelon grabens, each 1 km or more long. The longest faults form a gently sinuous, N. 40°–80° E.-trending zone 45 km long (Williams, 1964).

Upper Pennsylvanian through Upper Cretaceous rocks in the Lockhart Basin–Abajo Mountains area contain sets of joints of both regional and local extent. The regional joint network consists almost entirely of joints of the Tertiary system; dominant among them are the PX_5 (N. 57° W.) and PX_7 (N. 60° E.) sets. However, all six of the $PX_{4\text{-}9}$ sets are represented, and for the four earliest of them, their median strikes are within 12° of those of the equivalent sets in the other areas studied (table 1). As for the southern Dolores River area, the two younger sets show the greatest strike differences from one area to another.

Some of the oldest Mesozoic rocks exposed in the area contain two sets of joints that predate the regional $PX_{4\text{-}9}$ sets and that apparently are of local extent. Joints of the younger of these sets strike parallel to some of the minor grabens and possibly are related to them, but neither set is well documented. The Lockhart Basin–Abajo Mountains area is the least studied of those so far discussed.

OTHER AREAS

The joint network in the Castle Valley Anticline area, along the Colorado River in the northeastern part of the Paradox fold and fault belt (fig. 1), is dominated by joints of the Tertiary system. Local zones of joints along graben-bounding faults are present also (Grout, unpub. data, 1990), but these make only a minor contribution to the overall fracture pattern. Joint data from this area are sparse and little discussion therefore is warranted, but the fault-associated joints everywhere predate those sets interpreted as regional. Data for the regional sets are included in table 1 for comparison purposes and suggest that the sets probably are equivalents of the PX_{4-9} sets in the rest of the Paradox Basin.

Sparse joint data also are available from the Uncompahgre Uplift (fig. 1), from both the southwest flank bordering the Paradox Basin (Grout, unpub. data, 1994) and the northeast flank bordering the Piceance Basin (Grout, unpub. data, 1981). The fracture history of this uplift, though poorly known, nevertheless shows affinities to that of the Paradox Basin in that the PX_{1-9} sets appear to be present (table 1). Possible correlatives of the P_3 and P_4 sets have also been documented. The Uncompahgre Uplift is the only other area, besides the eastern margin of the Paradox Basin, where the P_4 set is prominent; on the uplift it is best formed in well-cemented, Upper Jurassic and Upper Triassic sandstones.

DISCUSSION

NORTHWEST-TRENDING STRUCTURES IN THE PARADOX BASIN

An overall northwest trend is common to numerous structures in the Paradox Basin. For example, most of the major anticlines and attendant crestal faults in the Paradox fold and fault belt trend N. 40°–60° W. (fig. 1), and median strikes of one of the most prominent joint sets in the same region range from N. 52° W. to N. 62° W. (PX_5, table 1). The joints of a much older set, the P_3 set in Paleozoic rocks along the eastern margin of the basin, have a median strike of N. 68° W. and likely are present at depth beneath parts of the basin as well. Well-log data indicate that large, graben-bounding faults striking N. 60°–65° W. underlie the Lisbon Valley area (Parker, 1981), and fault zones striking northwest are thought to be a major component of the structural framework of the basement rocks beneath the basin (Case and Joesting, 1972). The obvious presence of a northwest trend on satellite images and aerial photographs (Kelley and Clinton, 1960; Friedman and Simpson, 1980; Friedman and others, 1994), on geologic maps of the basin (Williams, 1964), on regional plots of geophysical data (Case and Joesting, 1972), and among structures investigated by generations of geologists in the field (Cater, 1955, 1970; Doelling, 1988; Morgan and others, 1992; Grout and Verbeek, this report) has led many to speculate on possible relations between the diverse features that contribute to this trend. Some of these speculations have been put into print; many have not. We briefly examine some field relations with the intent of providing constraints on possible interpretations.

PALEOZOIC JOINTS AND FAULTS

Prominent joints with a median strike of N. 68° W. form the youngest (P_3) set of the Carboniferous system in Lower Mississippian and older rocks along the eastern margin of the Paradox Basin. Although study of the Carboniferous system of joints in this area is incomplete, the P_3 set was documented at 10 of the 27 outcrops of Mississippian and older rocks studied, and thus is a common one; its probable age is early to mid-Pennsylvanian. Normal faults of similar strike are present in the basin to the west, both on the surface and in the subsurface (Peterson, 1989; Parker, 1981). If both the faults and joints of this orientation prove to be as prominent in the subsurface as along the eastern margin of the basin, a structural link between them in Pennsylvanian time may be substantiated. As previously noted, Pennsylvanian block faults have been documented in other parts of the northern Colorado Plateau and appear to record a widespread episode of crustal extension. Beneath the Piceance Basin, for example, seismic data show that the majority of the fault movement occurred in mid-Pennsylvanian time and that the faults subsequently were buried beneath younger sediments (Waechter and Johnson, 1985, 1986; Grout and others, 1991). It is already suspected that structural troughs bounded by similar block faults beneath the Paradox Basin provided depositional loci for thick sequences of Pennsylvanian evaporites, and that these accumulations later became the sites of the northwest-trending salt anticlines of the Paradox fold and fault belt (Case and Joesting, 1972; Doelling, 1988). Beneath the Lisbon Valley Anticline, however, well-log data indicate that the major subsalt faults trend as much as 25° more westerly than the anticline and are located about 3.8 km southwest of the major fault zone at the surface (Parker, 1981). Thus, a structural link between the northwest-trending Pennsylvanian faults and later structures remains incompletely defined. In any case, it seems highly unlikely that the associated joints bear any genetic relation to later fractures in rocks exposed at the surface, despite the similarity in strike between the P_3 and PX_5 sets. A related report in this volume (Verbeek and Grout) addresses the topic of basement-cover fracture relations in detail for several other areas on the Colorado Plateau.

TERTIARY JOINTS AND FAULTS

N. 52°–62° W.-striking PX_5 joints (table 1) form by far the most common and stratigraphically persistent set in the Paradox Basin. These joints are present in all units and geographic areas studied to date (Grout and Verbeek, unpub. data, 1990–1994) and have been documented at 231 localities. The prominence of this set, and its approximate parallelism to the equally prominent anticlines of the Paradox fold and fault belt, have long fueled speculation that the joints are integrally related to fold generation and most probably formed during stratal stretching along the outer arcs of the anticlinal folds. The following observations, however, negate this conclusion:

1. The joints of the PX_5 set are vertical in both horizontal and inclined beds. On the Lisbon Valley Anticline, for example, vertical PX_5 joints cut obliquely through beds dipping as much as 20° on the flanks of the fold, and they maintain their vertical attitude as bed dips lessen toward the fold crest. The joints of older sets, in contrast, are of variable dip, depending on the degree to which the beds that contain them were tilted during folding.

2. The common range of fold trends in the Paradox fold and fault belt is N. 40°–60° W., but the PX_5 joints are of more constant attitude. In no part of the basin is their median strike more northerly than N. 52° W.

3. Curvature of individual fold axes along their length is not matched by similar curvature in the strike of PX_5 joints.

4. The PX_5 joints are present in all parts of the basin (table 1) and are equally as abundant in areas of unfolded and unfaulted strata as in the Paradox fold and fault belt.

5. Along the eastern margin of the basin, joints of the PX_5 set have been traced upward within the stratigraphic succession into units as young as Miocene. The stratigraphic evidence thus indicates that the joints are far younger than the folds.

From these observations we conclude that the PX_5 joints were superimposed on folds that had formed long before, and that no direct genetic connection exists between them. The tectonic significance of the PX_5 and related joint sets is discussed in the following section.

TECTONIC SIGNIFICANCE OF TERTIARY JOINT SETS IN THE PARADOX BASIN

REGIONAL CORRELATIONS AND YOUNG AGE OF SETS

The evolution of the Tertiary joint system in the Paradox Basin bears intriguing similarities to that of the Piceance and Uinta Basins farther north, in northwestern Colorado and eastern Utah. All three basins and their intervening uplifts contain a complex record of multiple post-Laramide fracture events, during which the regional stress field progressively rotated counterclockwise with time. Specifically, we here suggest that the PX_4, PX_5, and PX_7 sets in the Paradox Basin correlate with the F_1, F_2, and F_3 sets in the Piceance and Uinta Basins as described in papers by Grout and Verbeek (1985, 1992) and Verbeek and Grout (1983, 1984, 1993). The suggested correlations are based not only on similarity in orientation, but more importantly, in all three basins, on *identical sequence of formation* (as documented by abutting relations at many dozens of exposures) and on the *demonstrated young age of the sets.* The F_1 through F_3 sets of the Piceance Basin, for example, are abundantly present in the youngest rocks preserved in that area—the late Eocene beds of the Uinta Formation. The same fracture sets in the neighboring Uinta Basin to the west have been traced upward from the Uinta Formation into younger, Oligocene beds of the Duchesne River Formation (Grout and Verbeek, unpub. data, 1992–1994). The youngest beds in which we have documented them to date have K-Ar ages of 32–34 Ma (Bryant and others, 1989). The present study suggests that the F_1 through F_3 joint sets may be younger still, for their apparent southern correlatives—the PX_4 and younger sets—have been traced eastward across the Paradox Basin into volcanic rocks as young as Miocene along the western edge of the San Juan volcanic field in Colorado. In all three basins, then, there exists widespread evidence of a post-Laramide age for some of the most prominent regional joint sets present. Each of the three joint sets discussed here affected a minimum area of 80,000 km^2, encompassing a very large portion of the northern Colorado Plateau. The outer geographic limits of each set remain undefined.

CENOZOIC STRESS ROTATION

The mid- to late Cenozoic paleostress history of the northern Colorado Plateau, as recorded by the regional fracture network, is one of progressive counterclockwise stress rotation. We first briefly recapitulate the Tertiary structural record in the Piceance Basin, where the F_2 to F_3 transition can be summarized as follows: (a) early, fairly common F_2 joints of N. 50°–60° W. strike; (b) somewhat later, very common F_2 joints of N. 60°–80° W. strike; (c) late, uncommon to rare F_2 joints of N. 80°–90° W. strike; (d) common F_3 joints of N. 80° E. to N. 60° E. strike; and (e) local, left-lateral oblique dilation of F_2 joints, and rarely of F_3 joints, the calcite fibers precipitated within the joints indicating a maximum horizontal compressive stress direction of about N. 50° E. The fracture network thus reveals that the direction of maximum horizontal compressive stress (σ_{hmax}, parallel to joint strike) rotated from northwest through west to east-northeast in mid- to late Cenozoic time. The distribution of these joints within different rock types is revealing as well: the F_2 joints of most northerly strike, for example, are most common in brittle beds such as well-cemented siltstones and marlstones, whereas those of most westerly strike are restricted to exceptionally nonbrittle beds such as high-grade oil shales and weakly cemented sandstones. The most continuous records of stress rotation are preserved in the Eocene oil shales of the Green River Formation, where median strikes of F_2 joints within different beds in the same outcrop differ as a function of organic content of those beds: as organic content increases, F_2 joint strikes swing more westerly (Verbeek and Grout, this volume). On the likely assumption that the most brittle (organic-poor) oil shales fractured first and increasingly less brittle, organic-rich beds progressively later, we interpret these relations as field evidence for counterclockwise stress rotation during the F_2

period of fracture. The amount of rotation, from early F_2 time to the period of joint reactivation following F_3 time, is on the order of 70°–80°. There is some fragmentary evidence as well of progressive stress rotation from the F_1 to F_2 periods of fracture.[3]

The record of stress rotation in the Paradox Basin is not as continuous as that farther north, but the progression from the PX_5 (N. 56° W.) through PX_6 (N. 85° W.) to PX_7 (N. 61° E.) joint sets points to a counterclockwise rotation in the direction of maximum horizontal compressive stress comparable to that indicated by the F_2 to F_3 transition. In the Paradox Basin, however, stresses during the period of time when σ_{hmax} was oriented about east-west were of sufficient magnitude to result in the formation of a discrete joint set represented by abundant joints, whereas farther north this same period is represented only by late, relatively sparse F_2 joints. Such broad, regional changes in the prominence of joint sets are common on the northern Colorado Plateau among all of the sets here discussed.

The notion of time-progressive jointing during a protracted period of stress rotation raises an interesting problem in the practical definition of a joint set. In the Piceance Basin, for example, it is difficult at some outcrops to decide whether a given joint set should be labeled "late F_2" or "early F_3." Similarly, in the Paradox Basin, the distinction between a "late PX_5" or an "early PX_6" set occasionally is difficult to make in the field, and is in fact somewhat arbitrary. Nevertheless, histograms of joint strike in the Paradox Basin show maxima and minima, the maxima corresponding to the joint sets listed in table 1. Individual joint sets in this view do not correspond to the simple conceptual model of discrete fracture "events" separated by long periods of tectonic quiescence. Rather, the mid- to late Cenozoic record of the northern Colorado Plateau is one of progressive and perhaps continuous fracture over a long period of time, each joint set representing some indefinite period when a significant percentage of the strata were strained to the point of fracture. Other beds in the same areas, however, commonly recorded slightly different parts of the fracture history, depending on their mechanical properties. The resultant fracture network, though complex, is a natural and expected product of the noncoaxial tectonic extension of a mechanically diverse sequence of rocks.

[3]Much the same record of stress rotation is preserved within correlative beds of the Uinta Basin farther west. There, however, the early stages of the F_2 period are made more clear by the presence, in the Eocene rocks, of large, abundant, northwest-trending gilsonite (hydrocarbon) dikes, which formed by hydrofracture as pore-fluid pressures built up during maturation of organic matter (Verbeek and Grout, 1992, 1993; Monson and Parnell, 1992). The gilsonite dikes predate the regional F_2 joint set and have more northerly strikes, the angular deviation between them commonly ranging from 10° to 20°, and locally more (Verbeek and Grout, 1992, 1993).

CONCLUSIONS

Thirteen regional sets of joints have been identified in and near the Paradox Basin. Of these, three sets are of Carboniferous (Late Mississippian to Pennsylvanian) age and are restricted in their surface expression to areas where Mississippian and older rocks crop out along the eastern margin of the basin. The westward, subsurface extent of these sets beneath the basin is unknown. An additional three sets are of Permian age and form an important part of the fracture network wherever Lower Permian and older rocks are exposed within the basin. These sets predate the major phase of salt movement and salt-anticline formation in the Paradox fold and fault belt. To date they have not been found above the unconformity that separates the Permian from the Triassic rocks.

The remaining seven sets are of Tertiary age. The oldest of these is known only from the eastern margin of the basin adjacent to the San Juan Mountains in Colorado, and from the Uncompahgre Uplift bordering the basin on the northeast. Its tectonic significance is uncertain. The next oldest set—the earliest Tertiary set of regional significance— appears to be related in time to the mid-Tertiary emplacement of laccolith complexes in the eastern part of the Paradox Basin (Grout and Verbeek, this volume). This and younger sets, all but the last of which are of great prominence, are the major elements of the regional fracture network in Early Triassic through Late Cretaceous rocks in the entire region studied, from the Green River in Utah to the San Juan Mountains in Colorado. Apparent correlatives of these sets have been documented at more than 1,000 localities farther north, in the Piceance and Uinta Basins and their bordering uplifts. Their geologically young age is apparent in many places: they are present in late Eocene beds in the Piceance Basin, in Oligocene beds (32–34 Ma) in the Uinta Basin, and in Oligocene and Miocene rocks in the western part of the San Juan volcanic field bordering the Paradox Basin on the east. Several of these sets provide a structural record of a protracted period of middle to late Tertiary crustal extension during which the regional stress field rotated progressively counterclockwise. This period of noncoaxial crustal extension affected a broad area encompassing at least the northern half of the Colorado Plateau.

ACKNOWLEDGMENTS

This work was supported by the U.S. Geological Survey's Evolution of Sedimentary Basins and National Geologic Mapping Programs. Thomas McCamant, D.L. Brevik, C.T. Throckmorton, and R.V. Grout assisted in the field, the latter as a volunteer. This manuscript was greatly improved by reviews from M.R. Hudson and B.A. Skipp.

REFERENCES CITED

Aydin, Atilla, 1978, Small faults formed as deformation bands in sandstone, *in* Byerlee, J.D., and Wyss, M., eds., Rock friction and earthquake prediction: Pure and Applied Geophysics, v. 116, no. 4–5, p. 913–930.

Carter, W.D., and Gualtieri, J.L., 1965, Geology and uranium-vanadium deposits of the La Sal quadrangle, San Juan County, Utah, and Montrose County, Colorado: U.S. Geological Survey Professional Paper 508, 82 p.

Case, J.E., and Joesting, H.R., 1972, Regional geophysical investigations in the central Colorado Plateau: U.S. Geological Survey Professional Paper 736, 31 p.

Cater, F.W., Jr., 1955, The salt anticlines of southwestern Colorado and southeastern Utah, *in* Geology of parts of the Paradox, Black Mesa, and San Juan Basins: Four Corners Geological Society Field Conference, 1st [Guidebook], p. 125–131.

———1970, Geology of the salt anticline region in southwestern Colorado: U.S. Geological Survey Professional Paper 637, 80 p.

Cater, F.W., Jr., and Elston, D.P., 1963, Structural development of salt anticlines of Colorado and Utah, *in* Backbone of the Americas: American Association of Petroleum Geologists Memoir 2, p. 152–159.

Cruikshank, K.M., and Aydin, Attila, 1995, Unweaving the joints in Entrada Sandstone, Arches National Park, U.S.A.: Journal of Structural Geology, v. 17, no. 3, p. 409–422.

Cruikshank, K.M., Zhao, Guozhu, and Johnson, A.M., 1991, Analysis of minor fractures associated with joints and faulted joints: Journal of Structural Geology, v. 13, no. 8, p. 865–886.

Cunningham, C.G., Naeser, C.W., Marvin, R.F., Luedke, R.G., and Wallace, A.R., 1994, Ages of selected intrusive rocks and associated ore deposits in the Colorado mineral belt: U.S. Geological Survey Bulletin 2109, 31 p.

Davis, G.H., 1978, Monocline fold pattern of the Colorado Plateau, *in* Matthews, Vincent, III, ed., Laramide folding associated with basement block faulting in the Western United States: Geological Society of America Memoir 151, p. 215–233.

Doelling, H.H., 1988, Geology of Salt Valley Anticline and Arches National Park, Grand County, Utah, *in* Salt deformation in the Paradox region: Utah Geological and Mineralogical Survey Bulletin 122, p. 1–58.

Dyer, J.R., 1988, Using joint interactions to estimate paleostress ratios: Journal of Structural Geology, v. 10, no. 7, p. 685–699.

Elston, D.P., Shoemaker, E.M., and Landis, E.R., 1962, Uncompahgre Front and salt anticline region of Paradox Basin, Colorado and Utah: American Association of Petroleum Geologists Bulletin, v. 46, p. 1857–1858.

Frahme, C.W., and Vaughn, E.B., 1983, Paleozoic geology and seismic stratigraphy of the northern Uncompahgre Front, Grand County, Utah, *in* Lowell, J.D., and Gries, Robbie, eds., Rocky Mountain foreland basins and uplifts: Rocky Mountain Association of Geologists Field Conference, Steamboat Springs, Colo., 1983 [Guidebook], p. 201–211.

Friedman, J.D., Case, J.E., and Simpson, S.L., 1994, Tectonic trends of the northern part of the Paradox Basin, southeastern Utah and southwestern Colorado, as derived from Landsat multispectral scanner imaging and geophysical and geologic mapping: U.S. Geological Survey Bulletin 2000–C, p. C1-C30.

Friedman, J.D., and Simpson, S.L., 1980, Lineaments and structure of the northern Paradox Basin, Colorado and Utah: U.S. Geological Survey Miscellaneous Field Studies Map MF–1221, scale 1:250,000.

Grout, M.A., and Verbeek, E.R., 1983, Field studies of joints—Insufficiencies and solutions, with examples from the Piceance Creek basin, Colorado, *in* Gary, J.H., ed., 16th Oil Shale Symposium, Proceedings: Golden, Colo., Colorado School of Mines Press, p. 68–80.

———1992a, Fracture history of the Divide Creek and Wolf Creek anticlines and its relation to Laramide basin-margin tectonism, southern Piceance Basin, northwestern Colorado: U.S. Geological Survey Bulletin 1787–Z, p. Z1–Z32.

———1992b, Joint-history and orientation data for Upper Cretaceous sandstones, Rawlins and Rock Springs uplifts, Washakie Basin, southern Wyoming: U.S. Geological Survey Open-File Report 92-388, 30 p.

Haynes, D.D., Vogel, J.D., and Wyant, D.G., 1972, Geology, structure, and uranium deposits of the Cortez quadrangle, Colorado and Utah: U.S. Geological Survey Miscellaneous Geologic Investigations Map I–629, sheet 1, scale 1:250,000.

Hite, R.J., 1961, Potash-bearing evaporite cycles in the salt anticlines of the Paradox Basin, Colorado and Utah, *in* Short papers in the geologic and hydrologic sciences: U.S. Geological Survey Professional Paper 424–D, p. D135–D138.

———1975, An unusual northeast-trending fracture zone and its relation to basement wrench faulting in northern Paradox Basin, *in* Fassett, J.E., ed., Canyonlands country guidebook: Four Corners Geological Society Field Conference, 8th, p. 217–223.

Huffman, A.C., Jr., and Taylor, D.J., 1994, Pennsylvanian thrust faulting along the northeastern margin of the Paradox Basin, Colorado and Utah: Geological Society of America Abstracts with Programs, v. 26, no. 6, p. 19.

Huntoon, P.W., Billingsley, G.H., Jr., and Breed, W.B., 1982, Geologic map of Canyonlands National Park and vicinity, Utah: Moab, Utah, Canyonlands Natural History Association, scale 1:62,500.

Kelley, V.C., and Clinton, N.J., 1960, Fracture systems and tectonic elements of the Colorado Plateau: Albuquerque, N. Mex., University of New Mexico Press, 104 p.

Lipman, P.W., ed., 1989, Oligocene-Miocene San Juan volcanic field, Colorado, IAVCEI May field trip guide, excursion 16B: New Mexico Bureau of Mines and Mineral Resources Memoir 46, p. 303–380.

McGill, G.E., and Stromquist, A.W., 1979, The grabens of Canyonlands National Park, Utah—Geometry, mechanics, and kinematics: Journal of Geophysical Research, v. 84, no. B9, p. 4547–4563.

Morgan, C.D., Allison, M.L., and Yonkee, W.A., 1992, Optimum directions for horizontal drilling in fractured Cane Creek Shale, Paradox Basin, Utah: American Association of Petroleum Geologists Bulletin, v. 76, no. 8, p. 1265.

Nelson, S.T., Davidson, J.P., and Sullivan, K.R., 1992, New age determinations of central Colorado Plateau laccoliths, Utah—Recognizing disturbed K-Ar systematics and re-evaluating tectonomagmatic relationships: Geological Society of America Bulletin, v. 104, no. 12, p. 1547–1560.

O'Sullivan, R.B., 1991, Correlation of Middle Jurassic San Rafael Group and related rocks from Kane Springs, Utah, to Uravan,

Colorado: U.S. Geological Survey Oil and Gas Investigations Chart OC–134.

Parker, J.M., 1981, Lisbon field area, *in* Wiegard, D.L., ed., Geology of the Paradox Basin: Rocky Mountain Association of Geologists, Denver, Colo., p. 89–100.

Peterson, J.A., 1989, Geology and petroleum resources, Paradox Basin province: U.S. Geological Survey Open-File Report 88–450–U, 69 p.

Potter, C.J., Tang, Rex, and Hainsworth, T.J., 1991, Late Paleozoic structure of the southern part of the Uinta Basin, Utah, from seismic reflection data: U.S. Geological Survey Bulletin 1787-V, p. V1–V10.

Steven, T.A., Lipman, P.W., Hail, W.J., Jr., Barker, Fred, and Luedke, R.G., 1974, Geologic map of the Durango quadrangle, southwestern Colorado: U.S. Geological Survey Miscellaneous Investigations Series Map I–764, scale 1:250,000.

Throckmorton, C.K., and Verbeek, E.R., 1995, Joint networks in the Tiva Canyon and Topopah Spring Tuffs of the Paintbrush Group, southwestern Nevada: U.S. Geological Survey Open-File Report 95-2, 182 p.

Tweto, Ogden, Steven, T.A., Hail, W.J., Jr., and Moench, R.H., 1976, Preliminary geologic map of the Montrose 1°×2° quadrangle, southwestern Colorado: U.S. Geological Survey Miscellaneous Field Studies Map MF–761, scale 1:250,000.

Verbeek, E.R., and Grout, M.A., 1983, Fracture history of the northern Piceance Creek basin, northwestern Colorado, *in* Gary, J.H., ed., 16th Oil Shale Symposium, Proceedings: Golden, Colo., Colorado School of Mines Press, p. 26–44.

———1984, Fracture studies in Cretaceous and Paleocene strata in and around the Piceance Basin, Colorado—Preliminary results and their bearing on a fracture-controlled natural-gas reservoir at the MWX site: U.S. Geological Survey Open-File Report 84-156, 30 p.

———1986, Cenozoic stress rotation, northeastern Colorado Plateau, *in* Stone, D.S., ed., New interpretations of northwest Colorado geology: Rocky Mountain Association of Geologists, p. 97.

———1991, Some general properties of joints and joint networks in horizontally layered sedimentary and volcanic rocks—An overview, *in* Gomberg, J.S., ed., U.S. Geological Survey Committee for the Advancement of Science in the Yucca Mountain Project, Symposium on "Fractures, hydrology, and Yucca Mountain"—Abstracts and summary: U.S. Geological Survey Open-File Report 91-125, p. 10–12.

——— 1992, Structural evolution of gilsonite dikes, eastern Uinta Basin, Utah, *in* Fouch, T.D., Nuccio, V.F., and Chidsey, T.C., Jr., eds., Hydrocarbon and mineral resources of the Uinta Basin, Utah and Colorado: Utah Geological Association Guidebook 20, p. 237–255.

———1993, Geometry and structural evolution of gilsonite dikes in the eastern Uinta Basin, Utah: U.S. Geological Survey Bulletin 1787–HH, p. HH1–HH42.

Verbeek, E.R., Grout, M.A., and Van Gosen, B.S., 1988, Structural evolution of a Grand Canyon breccia pipe—The Ridenour copper-vanadium-uranium mine, Hualapai Indian Reservation, Coconino County, Arizona: U.S. Geological Survey Open-File Report 88-006, 75 p.

Weir, G.W., and Puffett, W.P., 1981, Incomplete manuscript on stratigraphy and structural geology and uranium-vanadium and copper deposits of the Lisbon Valley area, Utah-Colorado: U.S. Geological Survey Open-File Report 81-39, 292 p.

Weir, G.W., Puffett, W.P., and Dodson, C.L., 1961, Preliminary geologic map and section of the Mount Peale 4 NW quadrangle, San Juan County, Utah: U.S. Geological Survey Mineral Investigations Field Studies Map MF–151, 1 sheet, scale 1:24,000.

Williams, P.L., 1964, Geology, structure, and uranium deposits of the Moab quadrangle, Colorado and Utah: U.S. Geological Survey Miscellaneous Geologic Investigations Map I–360, 2 sheets, scale 1:250,000.

Zhao, Guozhu, and Johnson, A.M., 1992, Sequence of deformations recorded in joints and faults, Arches National Park, Utah: Journal of Structural Geology, v. 14, no. 2, p. 225–236.

Cenozoic Igneous and Tectonic Setting of the Marysvale Volcanic Field and Its Relation to Other Igneous Centers in Utah and Nevada

By Peter D. Rowley, Charles G. Cunningham, Thomas A. Steven, Harald H. Mehnert, *and* Charles W. Naeser

CONTENTS

FIGURES

ABSTRACT

The Marysvale volcanic field of southwestern Utah, largely in the High Plateaus transition zone of the Colorado Plateau, lies at the east-northeastern end of the Pioche-Marysvale igneous belt. This belt consists of exposures of mostly Cenozoic volcanic rocks and is underlain by a batholith complex of even greater volume. The volume of volcanic rocks in the Marysvale field totals at least 12,000 km^3. The field consists mostly of middle Cenozoic, intermediate-composition, fundamentally calc-alkaline rocks erupted at 34(?)–22 Ma; associated mineral deposits are mostly of chalcophile elements. Stratovolcano deposits, especially volcanic mudflow breccia and lava flows, dominate; ash-flow tuffs derived from several calderas make up less than 10 percent of the volume of the volcanic rocks. Most of the central and northern part of the field consists of the Bullion Canyon Volcanics of relatively crystal-rich dacite and andesite; source cupolas reached shallow levels, and many intrusions have associated mineral deposits. The southern part of the field is dominated by crystal-poor andesite of the Mount Dutton Formation; postulated source cupolas are deep and unexposed, and associated mineral deposits are sparse. Shallow laccoliths that are unrelated to and south of the deep Mount Dutton sources were emplaced into lower Tertiary sedimentary rocks at the same time. These laccoliths are comagmatic with laccoliths and stocks of the "Iron Axis," in the Great Basin southwest of the Marysvale field. Large underlying batholiths fed both the southern Marysvale field and Iron Axis laccoliths.

The calc-alkaline igneous rocks of the Pioche-Marysvale igneous belt and the slightly younger Delamar–Iron Springs igneous belt to the south are part of a generally middle Cenozoic igneous sequence that spans much of the Western United States. The igneous rocks probably originated by oblique convergence during subduction of oceanic lithosphere beneath western North America. The overall area underlain largely by the igneous rocks is anomalously wide when compared with igneous areas in other parts of the Pacific rim, apparently because the subducted slab continued at a shallow depth as far to the east as the longitude of the Rocky Mountains. The Pioche-Marysvale and Delamar–Iron Springs igneous belts formed under extension parallel to the subduction direction and were bounded by transverse structures ("lineaments") of the same trend that separate areas of different amounts and types of extension. The generally east-northeast-trending crustal extension of the same age as calc-alkaline magmatism was especially profound in the Basin and Range province, but extensional faults were dominant only in some areas, whereas elsewhere extension was accomplished by passive emplacement of shallow intrusions.

The middle Cenozoic calc-alkaline rocks at Marysvale are overlain by an upper Cenozoic, fundamentally bimodal (rhyolitic and basaltic) volcanic association of rocks ranging in age from at least 23 Ma to Quaternary and intertongued in part with coeval clastic basin-fill sedimentary rocks. The bimodal association is dominated by high-silica rhyolite ash-flow tuff and volcanic domes, potassium-rich mafic rocks, and basalt cinder cones. Rocks of the bimodal association are much less voluminous (about 5 percent of the total volume of the Marysvale field) than the older calc-alkaline rocks, but they host many metallic mineral deposits, mostly of gold, silver, and lithophile elements. Bimodal rocks are also irregularly distributed elsewhere in the Pioche-Marysvale and Delamar–Iron Springs igneous belts. The late Cenozoic volcanism and extension began after subduction had started to diminish, as transform motion became significant on the San Andreas transform fault zone. Regional oblique extension, referred to as the basin-range episode and involving mostly faults under a normal-fault stress regime, was oriented generally east-west, and it began in the Marysvale area and the southeastern Great Basin at about 10 Ma, as faults formed the present topography. Transverse structures continued to be active in the Great Basin but most were oriented east-west, parallel to the new extension direction. The eastern Snake River Plain is a youthful model for volcanism and extension along probable underlying transverse structures oriented parallel to its current northeastern extension direction.

The overall effect of extension and volcanism in the Great Basin during the middle and late Cenozoic has been one of east-west spreading (widening) of the Great Basin, creating a bilateral symmetry. Any single axis of spreading is unlikely; probably there were many north-northwest- to north-northeast-trending axes perpendicular to the then-active extension direction. Extension related to these axes was expressed by faulting, igneous intrusions, or both. Under each axis, heat flow increased and the brittle-ductile transition zone rose. Extensional stress migrated with time, parallel to the extension direction, toward the cooler, more brittle margins of the Great Basin. The transverse structures, oriented parallel to the extension direction of the time, accommodated differing amounts of extension and magmatism north and south of them; they are fundamental features probably passing down to the brittle-ductile transition.

INTRODUCTION

The Marysvale volcanic field of southwestern Utah is one of the largest Cenozoic volcanic fields in the Western United States; it largely straddles the High Plateaus section of the Colorado Plateaus province, but the western part of the field is in the Great Basin section of the Basin and Range province (fig. 1). The High Plateaus is geologically a transition zone between the Great Basin and the main, much less deformed part of the Colorado Plateau. The volcanic field overlies Paleozoic to lower Tertiary sedimentary rocks mostly east of the northeast-trending Cordilleran hingeline;

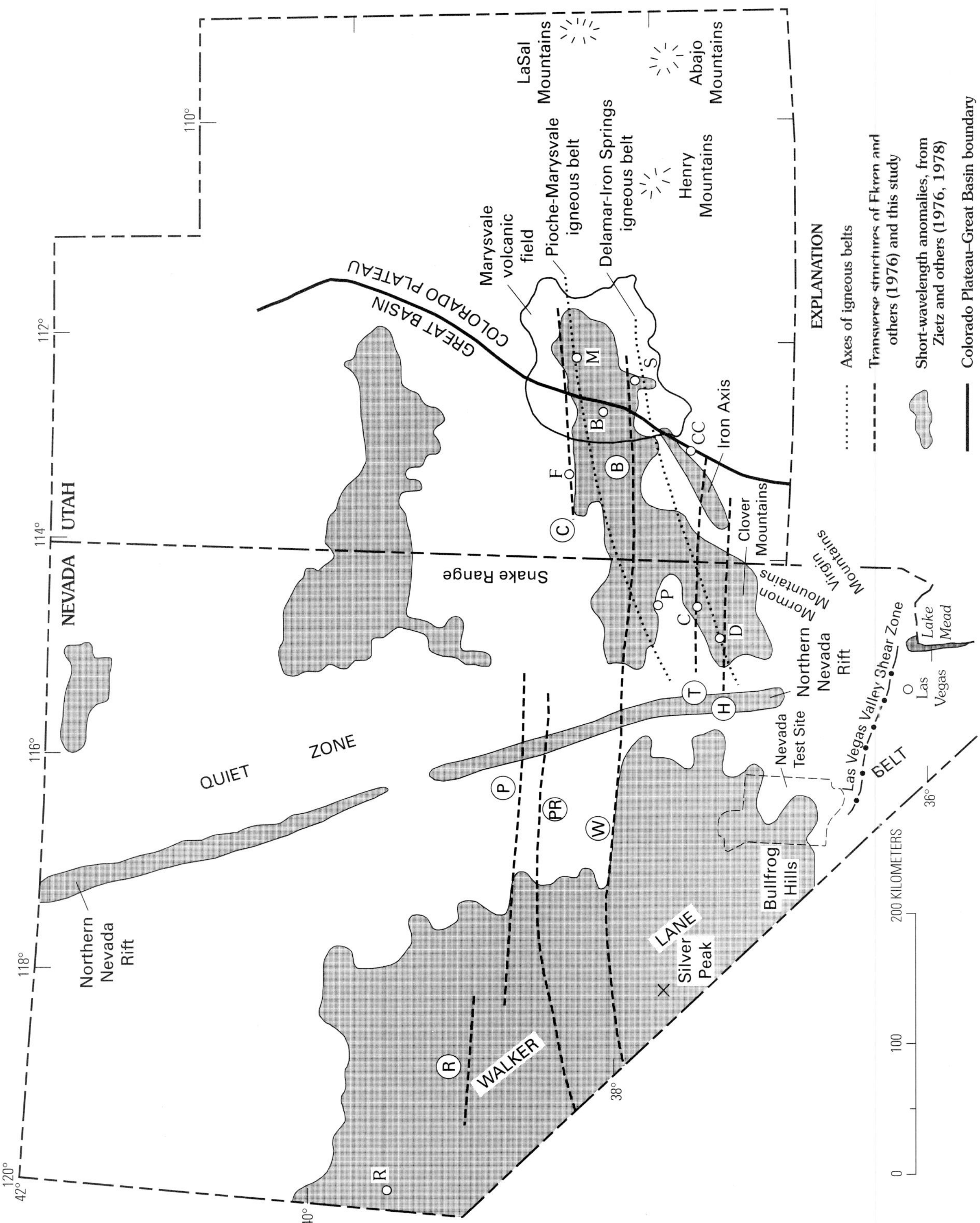

Figure 1. Major igneous and tectonic features of Nevada and Utah. Some additional igneous belts not identified here are shown in Rowley and others (1978, fig. 2). Colorado Plateau–Basin and Range boundary from Fenneman (1931). R, Reno; C, Caliente; P, Pioche; D, Delamar; F, Frisco; M, Marysvale; B, Beaver; S, Spry; CC, Cedar City. Transverse structures: R, Rawhide; P, Prichards Station; PR, Pancake Range; B, Blue Ribbon; W, Warm Springs; C, Cove Fort; T, Timpahute; H, Helene.

Sevier compressional deformation formed southeast-directed thrust sheets that underlie at least the northern part of the field, and apparent Laramide upwarps shed coarse clastic material that is now incorporated in lower Tertiary rocks north of the field (R.E. Anderson and Barnhard, 1992). The volcanic field lies at the eastern end of the east-northeast-trending Pioche-Marysvale igneous belt, the site of extensive, mostly Oligocene and Miocene, volcanism and mineralization above a major batholith complex (fig. 1). Tertiary rocks in the Marysvale field consist of three main sequences: (1) lower Cenozoic (Paleocene to lower Oligocene) fluvial-lacustrine sedimentary rocks (mostly Claron Formation), which overlie in angular unconformity Paleozoic and Mesozoic sedimentary rocks; (2) thick middle Cenozoic (Oligocene and lower Miocene, 34(?) to 22 Ma) intermediate-composition calc-alkaline igneous rocks; and (3) upper Cenozoic (23 Ma to Quaternary), compositionally bimodal volcanic rocks and source intrusions (consisting of basaltic rocks and high-silica rhyolite), which intertongue with and overlie clastic basin-fill sedimentary rocks. Sources of the middle Cenozoic calc-alkaline rocks were mostly stratovolcanoes, shield volcanoes, volcanic domes, and calderas. Some pyroclastic eruptions of the upper Cenozoic rhyolite magma formed calderas; other eruptions formed lava flows and volcanic domes. The upper Cenozoic basaltic lava flows came from fissure eruptions and central vents marked by cinder cones. A total volume of at least 12,000 km^3 of volcanic rocks erupted in the field, mostly between 26 and 23 Ma. The main pulse of lesser-volume bimodal rocks took place at 20–19 Ma (fig. 2), and episodic minor eruptions of rhyolite and/or basalt continued through the remainder of the Cenozoic. Ash-flow tuff makes up less than 10 percent of the volume of the field.

The geologic studies leading to this report began in 1963, when J.J. Anderson and P.D. Rowley, then from the University of Texas and later with the U.S. Geological Survey (USGS), started geologic mapping in the Markagunt and Sevier Plateaus on the southern flank of the Marysvale volcanic field (fig. 3). The USGS began mapping and related studies in the Tushar Mountains and Sevier Valley in the central part of the field in 1975, when T.A. Steven and C.G. Cunningham initiated work on the volcanic rocks and associated mineral deposits. The USGS work was expanded into a broader study of the geology and mineral resources of the whole Richfield 1°×2° quadrangle in 1978. The Marysvale field segment of the quadrangle was studied generally full time by Steven, Cunningham, and Rowley and part time by many others from the USGS and academia. H.T. Morris (USGS), J.J. Anderson (Kent State University and USGS), and M.G. Best (Brigham Young University and USGS) were prominent among the many participants of the Richfield work. K-Ar and fission-track dating by H.H. Mehnert and C.W. Naeser (both USGS) were critical to the study. The overall Richfield investigation resulted in more than 200 open-file and published reports and maps; many of these are cited in Steven and Morris (1987). Most of our conclusions about the Marysvale volcanic field are based on detailed geologic mapping, although field relations that support individual statements are not generally documented here. Our speculative ideas about regional geology that make up the second half of this report are based on our work and on reports by others about nearby areas, and they were written mostly by the senior author; not all coauthors agree with these interpretations. Much of the field data given here come from a series of

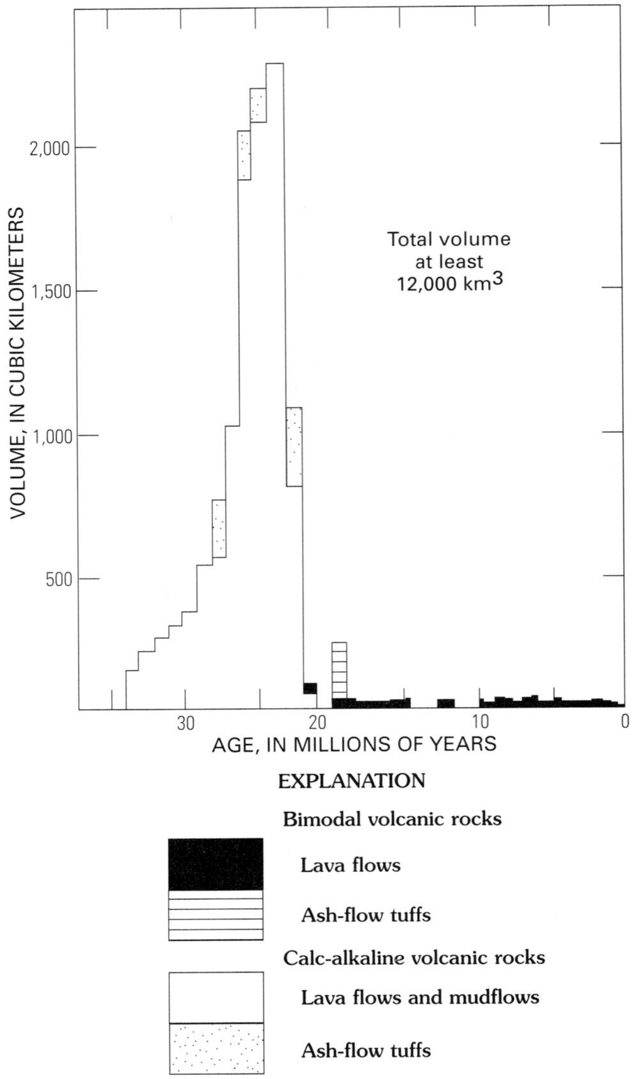

Figure 2. Volumes of volcanic rocks versus age, Marysvale volcanic field, Utah. Each bar represents the estimated volume erupted during a 1-million-year period.

geologic, geophysical, and mineral-resources maps of the central part of the field that started with Cunningham and others (1983) and from a summary map by Steven and others (1990).

This report summarizes field and other data dealing with the timing and character of Marysvale igneous and tectonic activity and places this activity in a larger igneous and tectonic context that includes laccolith clusters farther east in the Colorado Plateau and other igneous centers farther west in the Great Basin. In a companion report (Cunningham and others, this volume), we summarize chronologic patterns of igneous geochemistry and ore deposits. Isotopic ages given in this report without citing references are mostly from K-Ar and fission-track determinations discussed in various of our reports, most recently in Rowley, Mehnert, and others

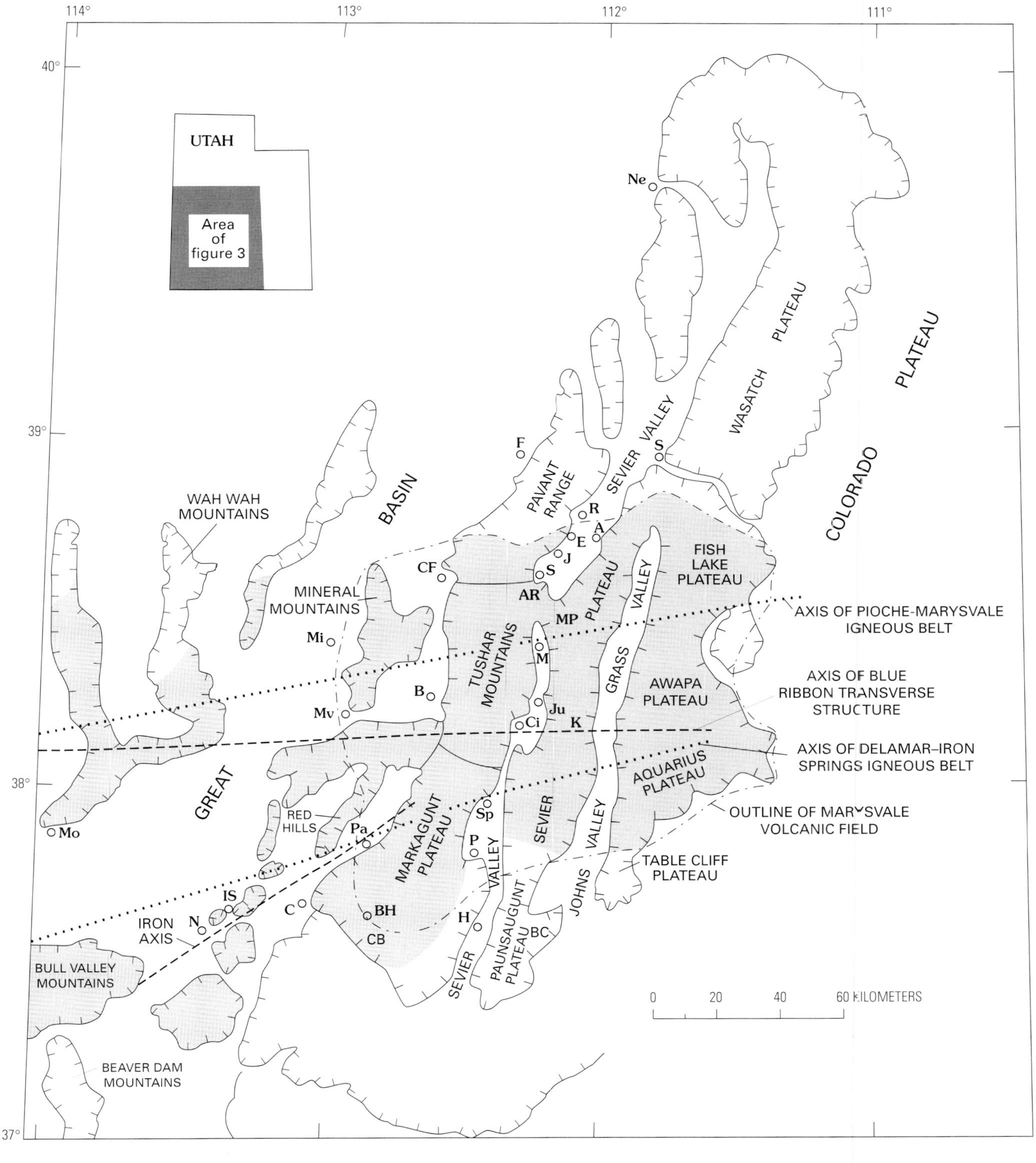

Figure 3. The Marysvale volcanic field and other features of interest in southwestern Utah (after J.J. Anderson and Rowley, 1975, fig. 1, and Steven and others, 1979, fig. 1). Shading shows areas underlain largely by volcanic rocks. The Colorado Plateau–Basin and Range boundary is just west of the towns of Nephi, Fillmore, Cove Fort, Beaver, Parowan, and Cedar City. A, Annabella; B, Beaver; BC, Bryce Canyon National Park; C, Cedar City; CB, Cedar Breaks National Monument; CF, Cove Fort; Ci, Circleville; E, Elsinore, F, Fillmore; H, Hatch; IS, Iron Springs; J, Joseph; Ju, Junction; K, Kingston Canyon; M, Marysvale; Mi, Milford; Mo, Modena; MP, Monroe Peak; Mv, Minersville; N, Newcastle; Ne, Nephi; P, Panguitch; Pa, Parowan; R, Richfield; S, Sevier; Sp, Spry.

(1994) . We are currently compiling data on the igneous isotope geochemistry of the Marysvale volcanic field, and S.R. Mattox and J.A. Walker (Northern Illinois University) are compiling similar data on some of the middle and upper Cenozoic volcanic units of the field (see Mattox, 1992, for instance).

ACKNOWLEDGMENTS

We thank many colleagues for ideas that have become so intimately intertwined with ours that we cannot separate them. Rowley, in particular, appreciates the understanding of structural concepts he gained from patient conversations over many years with the two Andersons, J.J. and R.E. For free exchange of unpublished data that helped us formulate and clarify many of the concepts discussed in this report, we thank J.H. Mackin, Warren Hamilton, H.R. Blank, E.H. McKee, L.W. Snee, K.L. Pierce, D.A. Sawyer, R.B. Scott, R.J. Blakely, M.G. Best, E.B. Ekren, M.A. Kuntz, C.J. Nutt, M.R. Hudson, R.A. Thompson, J.E. Faulds, M.L. Zoback, L.D. Nealey, F.E. Mutschler, Viki Bankey, L.W. Snee, S.A. Minor, D.M. Unruh, D.B. Hacker, A.M. Sarna-Wojcicki, D.S. Barker, D.B. Stoeser, M.C. Reheis, J.K. Otton, M.A. Siders, P.L. Williams, D.M. Miller, C.H. Thorman, W.E. Brooks, J.R. Budahn, F.G. Poole, S.R. Mattox, and I.J. Witkind. We are especially fortunate in being provided unpublished isotopic dates by L.W. Snee, H.H. Mehnert, and A.L. Deino and paleomagnetic data by C.S. Grommé, M.R. Hudson, and J.G. Rosenbaum. We thank R.E. Anderson, M.L. Zoback, and D.B. Stoeser for preprints of their reports in press and for their comments and suggestions on interpretations. We gratefully acknowledge excellent, constructive technical reviews by R.E. Anderson, Warren Hamilton, M.G. Best, and L.D. Nealey, who, although they did not agree with some of our interpretations, gave us much counsel on how to present them. We thank P.L. Martin for a detailed editorial review.

IGNEOUS ROCKS

MIDDLE CENOZOIC CALC-ALKALINE ROCKS

Middle Cenozoic, fundamentally calc-alkaline volcanism centered near Marysvale resulted in huge volumes of dacitic and andesitic lava flows, ash-flow tuff, flow breccia, and volcanic mudflow breccia from clustered stratovolcanoes and calderas and to a lesser extent from shield volcanoes and volcanic domes. The sequence corresponds to voluminous calc-alkaline igneous rocks common throughout the Western United States that probably formed during subduction of the Farallon/Vancouver plates (as described, for instance, by Lipman and others, 1972). Most rocks in the center of the Marysvale volcanic field contain abundant phenocrysts, commonly plagioclase, biotite, and hornblende. These moderately evolved rocks were derived from shallow magma chambers, and moderate erosion has exposed numerous plutons. Locally the roofs of high-level cupolas of the batholith failed, and calderas (Steven and others, 1984) formed as voluminous ash flows were erupted. The most voluminous stratigraphic unit in the central and northern part of the field (Tushar Mountains) is the Bullion Canyon Volcanics (Callaghan, 1939; Steven and others, 1979, 1984), which formed from magmas erupted from stratovolcano and caldera sources. The thickness of the formation is at least 1,500 m, and its original volume was at least 1,700 km^3. Basal parts of the unit underlie a regional stratigraphic marker, the Wah Wah Springs Formation (>29.5 Ma) of the Needles Range Group, an ash-flow unit derived from the Indian Peak caldera complex on the Utah-Nevada State line (Best, Christiansen, and Blank, 1989). These basal, weathered, and locally altered parts of the Bullion Canyon may be as old as 34 Ma (Willis, 1985; Kowallis and Best, 1990). Above the level of the Wah Wah Springs, the Bullion Canyon includes two ash-flow tuff members: the Three Creeks Tuff Member (27.5 Ma), which was derived from the Three Creeks caldera in the southern Pavant Range and had an estimated volume of 200 km^3 (Steven and others, 1984); and the Delano Peak Tuff Member (about 24 Ma), which was derived from the Big John caldera in the central Tushar Mountains and had an initial volume of 100 km^3. Mineralization associated with some of the shallow plutons that produced the Bullion Canyon Volcanics formed replacement alunite and gold deposits in many local districts in the Bullion Canyon terrain.

In contrast to the crystal-rich, moderately evolved rocks that formed the Bullion Canyon Volcanics, the rocks on the southern side of the Marysvale field (Markagunt Plateau and southern Sevier Plateau) are dominated by crystal-poor andesite and subordinate dacite of the Mount Dutton Formation (J.J. Anderson and Rowley, 1975; Fleck and others, 1975; Mattox and Walker, 1989; Mattox, 1992). Mapping shows that most of these generally pyroxene-bearing, less evolved rocks formed clustered stratovolcanoes whose centers occur along an east-trending belt, the Blue Ribbon transverse structure. (See "Transverse Structures" section, p. 181.) The Mount Dutton has a span of ages similar to those of the Bullion Canyon Volcanics, and the two formations intertongue from base to top across the south-central part of the volcanic field. Some Mount Dutton deposits underlie the Wah Wah Springs Formation (>29.5 Ma), but most do not; isotopic ages of those deposits above the Wah Wah Springs range from 26.7 to 21.2 Ma (Fleck and others, 1975). The total volume of the Mount Dutton Formation was at least 5,000 km^3. Plutons and mineral deposits are rarely exposed in Mount Dutton rocks, as magma chambers were deep and few convecting hydrothermal cells were formed. Members of the Mount Dutton Formation include the Beaver Member (25.6 Ma), which consists of volcanic domes with an estimated volume of 20 km^3 in the eastern Black Mountains, and the Kingston Canyon (25.8±0.4 Ma; volume 20 km^3) and Antimony (25.4±0.9 Ma; volume 50 km^3) Tuff Members, which are thin, densely welded, trachytic ash-flow tuffs that intertongue with the Mount Dutton, Bullion Canyon, and other sequences of mudflows and lava flows. The tuff members are synchronous with and petrologically similar to the tuff of Albinus Canyon (25.3±1.3 Ma; volume 60 km^3) in the northern part of the field and probably will eventually be determined to be identical. Probably all were derived from a common as-yet undiscovered source in the southern Pavant Range or adjacent Sevier Valley; Ekren and others (1984) suggested deep sources for petrologically similar tuffs in Idaho.

The Marysvale volcanic field contains many local calc-alkaline volcanic centers; products from these centers were separated during the mapping from the much more voluminous Bullion Canyon Volcanics and Mount Dutton Formation. Among the oldest of these are three informal units in the northern Tushar Mountains and southern Pavant Range that consist of dacitic flows, breccia, and tuff derived from stratovolcanoes and volcanic domes. These units were included in the Bullion Canyon Volcanics by Steven and others (1990) but they were broken out by Cunningham and others (1983) as the volcanic rocks of Dog Valley, the overlying volcanic rocks of Wales Canyon, and an unnamed dome. They underlie the Three Creeks Member and have been deeply eroded; their total volume was at least 300 km^3. The volcanic rocks of Signal Peak overlie the Three Creeks Member mostly in the northern Sevier Plateau. They consist of an andesite to basaltic andesite volcanic plateau and shield volcano with a volume of at least 400 km^3, not including unmapped rocks in areas to the east. The Bullion Canyon Volcanics and the Mount Dutton Formation intertongue eastward with coeval stratovolcano and shield-volcano units, whose extensions east of long 112° W. are as yet unmapped. These units include the volcanic rocks of Little Table, of Willow Spring, and of Langdon Mountain; where we mapped them in the Sevier Plateau, they have a combined volume of 400 km^3. Stratovolcano rocks of the eastern Marysvale field (Williams and Hackman, 1971; Mattox, 1991a) have a volume of about 2,500 km^3 and undoubtedly include deposits of these three units as well as of the volcanic rocks of Signal Peak and the Mount Dutton Formation. In the southern Tushar Mountains, three younger local units overlie the Osiris Tuff (see below): (1) the formation of Lousy Jim, a trachydacite volcanic dome (22 Ma; 25 km^3); (2) the tuff of Lion Flat, a local rhyolite ash-flow tuff with a volume of 8 km^3, which may represent deposits in and near a small concealed caldera or may represent the tuff-ring deposits (C.G. Cunningham, unpub. data, 1992) of the overlying formation of Lousy Jim; and (3) the andesitic lava flows of Kents Lake (12 km^3).

A series of shallow, calc-alkaline laccoliths were emplaced into lower Tertiary sedimentary rocks in the northern Markagunt Plateau, south of the deep-seated Mount Dutton sources and on the southern flank of the Marysvale field, where the volcanic rocks are relatively thin (J.J. Anderson and Rowley, 1975). The largest of these laccoliths is the Spry intrusion (26–25 Ma; fig. 3), which erupted to form a dacitic vent complex containing lava flows and tuffs that intertongue with the lower part of the Mount Dutton Formation (J.J. Anderson and others, 1990a). Breccia and flows from the complex are called the volcanic rocks of Bull Rush Creek, and a regional ash-flow tuff from the complex is called the Buckskin Breccia; their combined volume is about 60 km^3. A prominent short-wavelength gravity low (Blank and Kucks, 1989; Cook and others, 1989, 1990; Saltus and Jachens, 1995; Blank and others, this volume; Viki Bankey, USGS, unpub. data, 1992) and short-wavelength aeromagnetic anomalies (Zietz and others, 1976; Blank and Kucks, 1989; Viki Bankey, USGS, unpub. data, 1992) underlie and extend 50 km south of the vent complex. These anomalies probably mark the upper part of a source batholith that underlies and fed the laccolith. Another laccolith on the southern flank of the Marysvale field is the Iron Peak intrusion (formerly called Iron Point intrusion by J.J. Anderson and Rowley, 1975) of apparently 21–20 Ma, which fed a series of flows and breccias that intertongue with the upper part of the Mount Dutton Formation (Spurney, 1984). This pluton formed at the northeastern end of the "Iron Axis," a string of a dozen laccoliths and other plutons in the Great Basin that formed major iron deposits (Mackin, 1947; Blank and Mackin, 1967; Rowley and Barker, 1978; Van Kooten, 1988; Rowley and others, 1989; Barker, 1991; Blank and others, 1992) in the Iron Springs and Bull Valley mining districts west and southwest of Cedar City (fig. 3). Geophysical data suggest that most if not all of the Iron Axis laccoliths (22 Ma) are interconnected at depth, and the nearby younger (20.5-Ma) Pine Valley laccolith (Cook, 1957; D.B. Hacker, Kent State University, and L.W. Snee, USGS, unpub. data, 1992) is also part of the trend. All are interpreted to be fed by a large underlying composite batholith (Blank and Mackin, 1967; Blank and others, 1992), and this batholith may have extended as far northwest as a pluton (Grant and Proctor, 1988) exposed 30 km northwest of Cedar City (H.R. Blank, unpub. data, 1992). Borrowing an idea of H.R. Blank (oral commun., 1991), which was developed in greater detail by Nickelsen and Merle (1991), Merle and others (1993), J.J. Anderson (1993), and Maldonado (1995), we suggest that the Spry and the other plutons in the northern Markagunt Plateau may be cupolas on a large composite batholith that underlies the northern Markagunt Plateau and adjacent areas and led to a variety of deformational features in roof rocks above the batholith. This batholith would be similar to the one that fed the Iron Axis plutons and may even be connected to it. Rocks of the Mount Dutton Formation as well as the Iron Axis and northern Markagunt

Plateau plutons are part of the Delamar–Iron Springs igneous belt, which is south of the Pioche-Marysvale igneous belt (figs. 1, 3).

The Osiris Tuff (23 Ma) intertongues with the tops of the Bullion Canyon Volcanics and the Mount Dutton Formation. Field relations and petrologic studies show that this distinctive rhyolitic to trachytic tuff was erupted from the Monroe Peak caldera in the central and northern Sevier Plateau and probably had an original volume of at least 250 km^3. Andesite to rhyolite lava flows overlie intracaldera deposits of the Osiris Tuff and are considered to represent the final stages of caldera volcanism; they have isotopic ages of 22–21 Ma and a total volume of about 100 km^3. The Osiris and its associated lava flows represent the last major calc-alkaline volcanic event in the Marysvale field. The magmatic cupola whose top collapsed to form the Monroe Peak caldera resurgently invaded the caldera fill, where it formed intracaldera plutons (23–22 Ma) and it hydrothermally altered the rocks and mineral deposits of the Marysvale district in the western part of the caldera (Steven and others, 1984; Rowley and others, 1986a, b, 1988a, b).

As reported by other participants of this workshop (Sullivan and others, 1991; Sullivan, this volume; Nelson and others, 1992; Nelson, this volume; Ross, this volume), Colorado Plateau laccoliths in southeastern Utah have similar ages not only to the plutons of the Marysvale volcanic field described here, but to volcanic and intrusive rocks in the San Juan volcanic field in southwestern Colorado. Similar ties appear to exist between these igneous centers and centers in the Great Basin extending west to the Reno area. The Colorado Plateau laccoliths also are closely similar in composition to the calc-alkaline rocks of the central Marysvale area, Markagunt Plateau, and Iron Axis (fig. 4). Chemically, the Spry intrusion and the plutons of the Iron Axis are closely affiliated with each other, and they differ somewhat from Marysvale plutons. Plutons from the central Marysvale field and the Henry, Abajo, and La Sal Mountains are progressively more alkalic eastward (fig. 4). The only highly alkaline rocks known in the Marysvale field belong to a small alkali breccia pipe (24 Ma; Rowley, Mehnert, and others, 1994) containing nepheline and corundum 20 km southeast of Marysvale (Agrell and others, 1986).

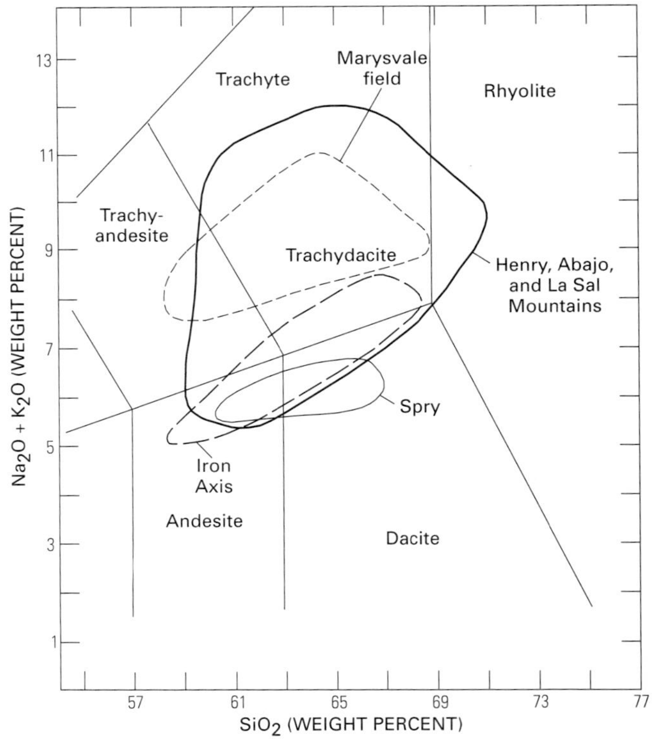

Figure 4. Total alkalies versus silica (Le Bas and others, 1986) of calc-alkaline rocks from the Marysvale volcanic field, Utah, compared with those from the Henry, Abajo, and La Sal Mountains, from the Iron Axis, and from the Spry intrusion. Data from Witkind (1964), C.B. Hunt (1958, 1980), G.L. Hunt (1988), D.S. Barker (University of Texas at Austin, unpub. data, 1985), and other sources.

UPPER CENOZOIC BASALT AND ALKALI RHYOLITE

As in other parts of the Western United States (see, for example, Christiansen and Lipman, 1972; Christiansen and McKee, 1978; McKee and Noble, 1986), the petrologic regime of the Marysvale volcanic field changed dramatically (Cunningham and others, this volume) from calc-alkaline to fundamentally bimodal (basalt and high-silica rhyolite) magmatism, starting no later than 23–22 Ma. The change occurred at different times in different parts of the field. Bimodal volcanic rocks make up about 5 percent of the total volume of the Marysvale field. The change to a bimodal petrologic regime is inferred to broadly coincide with the inception of crustal extension in a brittle crust that was cooler and thicker overall than the crust formed during calc-alkaline magmatism (Lucchitta, 1990; Wernicke, 1992; Lucchitta and Suneson, 1993). As a result, little-contaminated basaltic magmas rose along more deeply penetrating fractures to higher in the crust, leading to a rise in isotherms and the development of alkali-rhyolite eutectic partial melts (C.G. Cunningham, unpub. data, 1988). An alternative model, which we do not favor, suggests that an increase in the emplacement of basalts formed some rhyolite calderas predominantly by fractional crystallization (Perry and others, 1993). Few faults that record extension beginning at 23–22 Ma at Marysvale have been identified, whereas those faults that record the main phase of deformation (basin-range extension), starting about 9 Ma in this area, are numerous and obvious.

Generally the first product of bimodal magmatism consists of potassium-rich mafic rocks from local centers, including the mafic lava flows of Birch Creek Mountain

(22.5 Ma; 40 km^3) and the mafic lava flows of Circleville Mountain (23 Ma; 25 km^3) in the southern Tushar Mountains. These rocks were called by several names, including "older basalt flows" (J.J. Anderson and Rowley, 1975), until we settled on "potassium-rich mafic volcanic rocks," following Best and others (1980); they are shown as such by Steven and others (1990). Best and others (1980), Mattox and Walker (1989, 1990), Walker and Mattox (1989), and Mattox (1991b, 1992) discussed their chemical composition. Potassium-rich mafic volcanic rocks with a volume of about 10 km^3 and similar ages (Best and others, 1980) also occur in the eastern Marysvale field (Williams and Hackman, 1965; Mattox, 1991a, 1992); some of these rocks may be as old as 26 Ma (Mattox, 1991a). Lava flows erupted from other centers (22–21 Ma; Best and others, 1980), on the southern and northern edges of the field, have a total volume of about 10 km^3. Mattox and Walker (1989) ascribed all this potassic mafic volcanism to the final demise of the subduction zone. Mattox (1992), however, noted a similarity in chemical and isotopic composition between the potassium-rich mafic volcanic rocks and the Mount Dutton Formation, which suggests that both are products of calc-alkaline magmatism. The upward movement of potassium-rich mafic magmas at depth may have supplied the heat to produce partial crustal melts erupted as early bimodal rhyolites.

The largest accumulation of volcanic rocks of the bimodal assemblage in the Marysvale field consists of rhyolite of the Mount Belknap Volcanics (Callaghan, 1939; Cunningham and Steven, 1979a), which was erupted from two concurrently active source areas in the central Tushar Mountains and Sevier Valley from 22 to 14 Ma. The westernmost of these two source areas subsided to form the large Mount Belknap caldera when its main ash-flow tuff, the Joe Lott Tuff Member (19 Ma), was erupted (Steven and others, 1984; Budding and others, 1987). Including associated intracaldera rhyolite flows and plugs, the original volume of rocks of the western source area was about 300 km^3. Heterogeneous silicic lava flows, volcanic domes, and tuff were erupted from the eastern source area. The small Red Hills caldera, in Sevier Valley, subsided with eruption of the Red Hills Tuff Member at 19 Ma. The volume of rhyolites of the eastern source area is about 25 km^3. Silicic intrusive rocks (rhyolite dikes and plugs and granitic plutons) were emplaced into the two source areas. The main mineral deposits from Mount Belknap magmatism are of uranium and molybdenum in the Central Mining Area of the Marysvale district (Cunningham and Steven, 1979b; Cunningham and others, 1982). In the eastern Tushar Mountains 10 km south-southwest of Marysvale, an intrusive cupola domed rocks of the Bullion Canyon Volcanics (including the Three Creeks Member) at Alunite Ridge (Cunningham and Steven, 1979c) and formed a halo of gold-bearing veins, a base- and precious-metal manto (Beaty and others, 1986), and, above that, coarse-grained vein alunite dated at 14 Ma (Cunningham and others, 1984).

The Mineral Mountains, at the western edge of the Marysvale field, are underlain by a 25- to 9-Ma composite batholith, the largest exposed in Utah (Nielson and others, 1986; Steven and others, 1990; Coleman and Walker, 1992). It is probably representative of the intrusive rocks that underlie the Pioche-Marysvale igneous belt and is exposed over a large area here only because the Mineral Mountains underwent huge amounts of uplift when the batholith was emplaced as part of a probable metamorphic core complex (Price and Bartley, 1992). Minor early phases (about 25 Ma) represent calc-alkaline intrusions, but later intrusions (22–11 Ma) are more silicic and alkaline and appear to be part of the bimodal association; still later granites (9 Ma) are clearly bimodal.

Beginning at 9 Ma, when the crust began to extend rapidly and a surge of basin-range faulting took place, small rhyolite and basalt eruptions occurred at many scattered locations around the edges of the Marysvale field. The largest of the rhyolite bodies is in the Kingston Canyon area of the southern Sevier Plateau, where rhyolite flows and a volcanic dome of 8–5 Ma make up a total volume of about 50 km^3 (Rowley, 1968; Rowley and others, 1981). Small rhyolite domes (9–8 Ma) that total about 25 km^3 are scattered in the Basin and Range province on the western side of the Marysvale field (J.J. Anderson and others, 1990b). The rhyolite of Gillies Hill (9–8 Ma; 20 km^3) of Evans and Steven (1982) was emplaced along north-striking basin-range faults on the northwestern side of the Beaver Basin, just east of the Mineral Mountains; subsurface plutons of the same age and same presumed lithology have associated mineralized rocks at the Sheeprock mine along the eastern side of the basin (Cunningham and others, 1984). The Sheeprock plutons, which are the intrusive equivalent of the rhyolite of Gillies Hill, may correlate with the youngest plutons of the Mineral Mountains batholith to the west (Coleman and Walker, 1992). Capping this batholith, on the crest and western flank of the Mineral Mountains, is the rhyolite of the Mineral Mountains, a series of Pleistocene (0.8–0.5 Ma) rhyolite domes, lava flows, and tuff (Lipman and others, 1978).

Basalt fields as young as Quaternary (see, for instance, Best and others, 1980, or Mattox, 1992) are scattered through and around the Marysvale field. Sites of eruption do not coincide with older vent areas of calc-alkaline rocks or of the oldest (23–16 Ma) high-silica rhyolite flows. Some basalt vents do coincide, however, with post-9-Ma high-silica rhyolite sources, suggesting that basalt emplacement supplied the heat for those rhyolite partial melts. We calculate the volume of basalt rocks to be at least 100 km^3 in the field and more than this beyond the field. Basalts in the field in general represent asthenospheric melts that worked their way to the surface with relatively little contamination (Fitton and others, 1991) along northerly striking fractures created during basin-range extension. Although erosion has removed or obscured many vents, the presence of basalt flows as louderbacks on the major tilt blocks suggests that many vents were on the edge of the upthrown (footwall) fault blocks in the manner observed in the field by R.E. Anderson (1988) and explained theoretically by Ellis and King (1991) as due to dilation in footwall blocks at depth during normal faulting.

REGIONAL VERSUS LOCAL EXTENSIONAL DEFORMATION

Extensional deformation in the Marysvale volcanic field took place during two different episodes that correspond roughly to the magmatic episodes. The concept of the two extensional episodes is hardly new and now is accepted by most workers (for example, Zoback and others, 1981), although many (for example, Wernicke, 1981, 1985; Von Tish and others, 1985) consider that extension in the Basin and Range province has taken place during a single long event. The evidence for two episodes in the Marysvale area is sketchy, so in the sections below, we bring in map evidence from other parts of Utah and Nevada where we have worked. From there, we expand our view of the Marysvale area and offer ideas on the Cenozoic geologic evolution of much of the Great Basin and Colorado Plateau.

MIDDLE CENOZOIC FAULTS

Local extensional deformation took place during middle Cenozoic calc-alkaline magmatism in the Marysvale volcanic field, as elsewhere in the transition zone and the Great Basin. The orientation of such pre-basin-range, syn-calc-alkaline faults indicates east-west extension at Marysvale, but the faults are sparse and widely scattered and are outside the main intrusive centers. More of these middle Cenozoic faults may exist, but either (1) they cannot be distinguished in age from the dominant late Cenozoic (basin-range) faults in the area or (2) field evidence for constraining the age of the faults of this episode was covered by volcanic rocks, obscured by the poorly known stratigraphy of these thick volcanic piles, or obscured by intrusive structures.

West of the town of Antimony in the southern Sevier Plateau, an angular unconformity due to a northerly striking fault zone separates units having 30° difference in dip. The age of movement is constrained by the Kingston Canyon Tuff Member (26 Ma) below and the Osiris Tuff (23 Ma) above (Rowley, 1968, p. 144, 167). Several west-northwest-striking horsts and grabens of 26–25 Ma were mapped in the northern Markagunt Plateau of the southern Marysvale field by J.J. Anderson (1971, 1985, 1988). Their unusual strike suggests that they represent deformation along the broad west-striking Blue Ribbon transverse structure (see "Transverse Structures" section, p. 181), which underlies the area of the horsts and grabens and was active during and after this time. Other pre-basin-range faults include local "mosaic faults" of numerous strikes that represent shattering of roof rocks above many calc-alkaline intrusions on the batholith complex of the Pioche-Marysvale igneous belt (Steven, 1988).

The northern Markagunt Plateau and adjacent areas also contain the following, perhaps genetically related, pre-basin-range structures, which probably resulted at least partly from the emplacement of a batholith complex under the eastern Delamar–Iron Springs igneous belt: (1) the Markagunt megabreccia, a major pre-24-Ma gravity-slide deposit (Sable and Anderson, 1985; J.J. Anderson, 1993; L.W. Snee, written commun., 1994); (2) the low-angle, 22.5–20 Ma, extensional Red Hills shear zone (Maldonado and others, 1989, 1990, 1992; Maldonado, 1995); and (3) 30- to 20-Ma thin-skinned thrust faults (Davis and Krantz, 1986; Lundin and Davis, 1987; Lundin, 1989; Bowers, 1990; Nickelsen and Merle, 1991; Merle and others, 1993). These diverse structures may represent gravity-driven volcanic spreading southward off the Marysvale volcanic field (Davis and Rowley, 1993).

The near absence of middle Cenozoic extensional faults in the Marysvale volcanic field, despite the abundance of faults of this age farther west in the Great Basin, was a significant conundrum for us for a while. Gans and others (1989, p. 48) best defined the problem when they asked why middle Cenozoic faults are least common in the largest volcanic fields in the West, such as the Marysvale and San Juan fields. We conclude that part of the answer to the puzzle derives from calculations and interpretations by Lachenbruch and Sass (1978) showing that emplacement of basaltic dikes can accommodate extension to the exclusion of normal faulting. Bacon (1982) applied these ideas to the Pleistocene Coso volcanic field of California, where he found a constant rate of emplacement of basalt and rhyolite feeder dikes, which he attributed to relief of crustal least principal stress normal to these dikes. Bursik and Sieh (1989) extended these concepts to the Long Valley–Mono Basin volcanic field in California. There, they concluded, Quaternary oblique faults of the same age as the volcanic field die out along strike into eruptive centers because dikes in the centers take up all extensional strain. (See also Moos and Zoback, 1993.) Thompson and others (1990), Parsons and Thompson (1991), and Parsons and others (1992) expanded the calculations into a more elaborate hypothesis that states that emplacement of basaltic dikes will increase horizontal stress (relieve least principal stress) so that normal faults will not form; this effect may explain the lack of seismicity in the eastern Snake River Plain of Idaho (Thompson and others, 1990). Pierce and others (1991, fig. C7), Pierce and Morgan (1992, plate 1, p. 41–42), and Pierce (oral commun., 1994) explained faults that die out into the Pleistocene Yellowstone caldera of Wyoming by the same mechanisms. In a summary of the calderas (15–9 Ma) of the Nevada Test Site, Sawyer and others (1994, p. 1316) noted that the areas of greatest volume of intracaldera intrusions contain the fewest faults;

they allowed that this could be explained by emplacement of intracaldera intrusions relieving the least principal stress but thought it more likely was due to the cooled and solidified intrusions and their underlying batholith creating a massive buttress that deflected faults around them. In contrast, we here expand on the ideas cited above to suggest that middle Cenozoic, syn-calc-alkaline faults are sparse in the Marysvale field and many other areas because most east-west least principal stress was taken up by passive intrusion of *shallow* batholiths, stocks, and dikes of all calc-alkaline compositions, rather than by faults. In other words, relief of least principal stress in the brittle upper crust can be, depending on the local stress field, by faults or by emplacement of hypabyssal intrusions, or by both. Most dikes and perhaps some shallow intrusions would be expected to strike perpendicular to the extension direction, assuming the general case that they filled joints that propagated during intrusion (Delaney and others, 1986; Grout, this volume). In the Marysvale field, however, dikes are uncommon, and we have noticed no preferred northerly trend for dikes or for the long directions of intrusions.

Faults of all types also are uncommon in the Iron Springs mining district and other parts of the Iron Axis, which are underlain by a large terrain of shallow, calc-alkaline intrusive rocks; we attribute the lack of syn-calc-alkaline (middle Cenozoic) faults also to stress relief by the shallow intrusions, although a general lack also of many younger basin-range faults is better explained by the buttress effect of Sawyer and others (1994; see also Blank, 1994). After our ideas were formulated, J.E. Faulds (University of Iowa, oral commun., 1995) alerted us to similar ideas by him and his colleagues (Faulds, Olson, and Litterell, 1994) in which they refer to "magmatic extension" as an alternative to "mechanical extension" (that is, faulting).

Deeper intrusions are an entirely different matter, but still important during extension. Thompson and Burke (1974), Eaton (1980), Gans (1987), McCarthy and Parsons (1994), and others recognized that emplacement and underplating by deeper intrusions and lateral flow of lower crustal material are the main reasons why rapidly extending crust in areas such as the Great Basin is not also drastically thinned. Catchings and Mooney (1993) cited seismic data that suggest that deep and shallow intrusions occur where intensely faulted areas occur. When extending, brittle upper crust may be "decoupled from a more uniformly deforming (ductile) middle and lower crust" (Gans, 1987, p. 1), as also suggested by Eaton (1980, 1982) and Hamilton (1988a). The laccolith fields of the Colorado Plateau and, of course, the plateau itself, also are little broken by faults. One of us (T.A. Steven) suggests here that the Plateau "floated on" a hot, plastic, periodically and locally partially melted substratum that took up least principal stress in the area of the laccoliths, and Blank and others (this volume) propose that large masses of magma underlay the Colorado Plateau during the early and middle Miocene.

Although syn-calc-alkaline extensional faults in the Marysvale field are sparse, faults of this age are abundant in an area farther west where Rowley is mapping in the western (Nevada) part of the Delamar–Iron Springs igneous belt. This area, near Caliente, Nev. (fig. 1), was deformed by two episodes of east-west extension, preceded by a deformational episode whose age and explanation are not understood. This oldest deformation resulted in the mostly bedding-parallel Stampede fault (Axen and others, 1988), in which the transport direction of the upper plate may have been eastward. All that can be told about the age of deformation is that it preceded the oldest volcanic rocks (31 Ma; Rowley, Shroba, and others, 1994) in the area.

The Stampede was considered to be a detachment fault (Axen and others, 1988; Bartley and others, 1988; Taylor and others, 1989; Rowley, Shroba, and others, 1994), which Taylor and Bartley (1992) and Axen and others (1993) correlated with the Snake Range décollement (Miller and others, 1983; Gans and others, 1985) to the north (fig. 1), and thus they considered that it represents a significant early Tertiary episode of extension. However, these interpretations are questionable, for the Stampede also resembles bedding-parallel attenuation faults found in association with compressional structures that were mapped by Miller (1991), Nutt and Thorman (1992, 1993, 1994), and Nutt and others (1992, 1994) in northwestern Utah and northeastern Nevada. These authors showed the structures to be pre-lower(?) Eocene and suggested that they are of Sevier age. New data suggest that the structures predate a Cretaceous intrusion and perhaps a Jurassic intrusion (C.J. Nutt and C.H. Thorman, oral commun., 1995). The faults involved in these structures in the northeastern Great Basin are similar to mapped younger-on-older, bedding-parallel faults considered by Allmendinger and Jordan (1984) to be pre-Late Jurassic and by Miller (1991) to be Mesozoic. Similar faults associated with general compressional (Sevier) thrusts in the Raft River area of northwestern Utah and southern Idaho have been isotopically dated at Late Cretaceous (Wells and others, 1990). "Extensional faults" mapped in the Devonian Guilmette Formation south of the western Caliente caldera complex by Page and Scott (1991) were interpreted by Axen and others (1993) to be part of their early Tertiary extensional event, but these and other such faults in the area are now considered to be of Sevier age (Swadley and others, 1994; Page, 1995). This deformational episode will not be considered further because of uncertainties in its age and interpretation.

The faults of the two clear Cenozoic extensional episodes studied during mapping in progress by Rowley and R.E. Anderson are well exposed in the western Caliente caldera complex, Nevada-Utah (at, south, and east of the town of Caliente, fig. 1) and adjacent areas. The older episode, representing syn-calc-alkaline, pre-basin-range faults, is the main deformational event in the area and resulted in high-angle, mostly north-northwest- and north-northeast-striking oblique-slip, strike-slip, and normal-slip faults and in north-striking detachment faults (Axen and others, 1988; Burke, 1991; Rowley, Snee, and others, 1992). The ages of these older faults are well constrained by dated dikes (20–16 Ma; Rowley, Snee, and others, 1992; L.W. Snee, written commun., 1993) intruded into the fault zones, and field evidence suggests that the episode probably began close to 25 Ma; faulting continued until about 12 Ma. This episode coincided with magmatism in the Caliente caldera complex (23–13 Ma), a large (80 by 35 km) complex of inset calderas that erupted low-silica rhyolite (calc-alkaline) tuffs until about 18 Ma, then high-silica rhyolite (bimodal) tuffs after about 15.5 Ma (Rowley, Snee, and others, 1992; Rowley and others, 1995).

Middle Cenozoic strain in the Caliente area was complicated and heterogeneous, and it was similar to that described and interpreted by R.E. Anderson (1987), Siders and Shubat (1986), R.E. Anderson and Bohannon (1993), R.E. Anderson and Barnhard (1993a, b), R.E. Anderson and others (1994), and Scott, Grommé, and others (1995) east and south of the Caliente caldera complex (see next paragraph). On the basis of mapped structures and of fault kinematic data analyzed according to the methods of Angelier and others (1985) and Petit (1987), we follow the reasoning, though not necessarily the orientation of the extension direction, of Wright (1976; his "field II" faults), R.E. Anderson (1973, 1984, 1986, 1987, 1989, 1990), Angelier and others (1985), Zoback (1989), R.E. Anderson and Barnhard (1993a, b), and R.E. Anderson and others (1994). The middle Cenozoic extension direction (that is σ_3, the least-principal-stress axis) in the Caliente area is interpreted to be horizontal and east-west on the basis of a conjugate set of predominant north-northwest oblique (right-lateral, normal) and subordinate north-northeast oblique (left lateral, normal) faults (Rowley, unpub. data). This orientation differs from the east-northeast extension direction suggested by R.E. Anderson and Ekren (1977), Zoback and others (1981), and Michel-Noël and others (1990), all of whom interpreted the north-northwest-striking faults to be normal faults. The σ_1 (greatest-principal-stress axis) was locally or periodically vertical, in order to explain north-striking normal faults, but more generally it was oriented horizontally and trended generally northward (that is, it reflects north-south shortening); furthermore, σ_1 and σ_2 (intermediate-principal-stress axis) may have been similar to each other in magnitude and they may have interchanged at times (Wright, 1976; Zoback, 1989). When σ_1 and σ_2 were equal, north-striking dikes and intrusions would be most likely to form; repeated injections of magma as extension continued would have resulted in batholith complexes (igneous belts) elongated parallel to σ_3. We agree with Angelier and others (1985) that normal and strike-slip faults in the same area "represent stress oscillations in time and space rather than discrete stress reorganizations." The faults of this main episode occur both inside and outside the Caliente caldera complex and so are not a consequence of magmatism. The Caliente caldera complex differs from the northern and central Marysvale volcanic field in that intrusions are not commonly exposed at Caliente and thus are not shallow. The strike-slip faults in the Caliente area do not parallel the extensional kinematic axis (the direction of extension), so we do not interpret them as tear faults in the upper plates of detachment faults, as did Michel-Noël and others (1990).

There is a considerable literature documenting extension on pre-basin-range faults throughout other parts of the Great Basin. These are all called middle Cenozoic faults here, although in the northern Great Basin, they are as old as Eocene, as are the Tertiary igneous rocks there. One of the first to report these faults was Ekren and others (1968) from the Nevada Test Site (fig. 1); their conclusions were further documented by Ekren and others (1971), R.E. Anderson and Ekren (1977), and R.E. Anderson (1978). In fact, extension is a characteristic feature of subduction-related arc magmatism (Hamilton, 1988c, 1989, 1995). Like Elston (1984), we consider that in some parts of the Basin and Range province, including Caliente, middle Cenozoic extension was greater than late Cenozoic extension. Basins containing clastic sedimentary fill, smaller than those of the later basin-range episode, have been identified in some areas (for example, Bohannon, 1984; Seedorf, 1991; Christiansen and Yeats, 1992; Axen and others, 1993). Miller (1991) and Liberty and others (1994) interpreted seismic data in the central and northern Great Basin to indicate that there, as in the Marysvale area (see section on "Late Cenozoic Faults and Basin-Fill Deposits") and some other parts of the central Great Basin that they cited, broad sag basins preceded the narrow fault-bounded basins of the basin-range episode. Other basins of this age may be unrecognized or buried under younger basin-fill sediments deposited in the same place. Metamorphic core complexes and both related and unrelated low-angle detachment faults (for example, Proffett, 1977; Crittenden and others, 1980; Wernicke, 1981, 1985, 1992; Wernicke and others, 1985, 1988; Allmendinger and others, 1983; Hamilton, 1987, 1988a, b; Davis and Lister, 1988; Axen, 1991; Wernicke and Axen, 1988), developed in many places in the Basin and Range province during calc-alkaline magmatism (Lipman, 1992; Lister and Baldwin, 1993). The Basin and Range province now is characterized by high heat flow due to extension (Lachenbruch and Sass, 1978; Lachenbruch and others, 1994), and heat flow was probably greater during the middle Cenozoic (Lachenbruch and others, 1994), with a resulting high level for the brittle-ductile transition, thermal softening of brittle crust, and the development of voluminous plutons and metamorphic core complexes (Hamilton, 1988a; Armstrong and Ward, 1991). R.E. Anderson (1990) and R.E. Anderson and others (1994) theorized a method of middle to late Cenozoic "tectonic escape" and structural rafting of the brittle upper crust on a laterally (westward to southwestward) flowing mass of mid-crustal material in order to explain a complex array of faults characterized by north-south shortening and east-west extension and of synextensional intrusions in the Lake Mead–Las Vegas area (fig. 1). This idea is attractive for the extreme southern Great Basin and northern part of the southern Basin and Range province, but for other parts of the Great Basin during the middle Cenozoic, any such tectonic escape must be superimposed on a more regional pattern of subduction and extension to explain regional structural trends of which the extreme southern Great Basin is a part.

Zoback and others (1981) compiled data on faults of 20–10 Ma in the Basin and Range province and concluded that the overall orientation of σ_3 then was east-northeast, parallel to the subduction direction (Hamilton, 1989). Henry and Aranda-Gomez (1992) noted the same orientation for middle Cenozoic faults in the Basin and Range province of Mexico. This orientation appears to be valid as an average, but there are some exceptions. One is the east-west extension direction in the Caliente area. The Marysvale field also may be an exception, although the data are too sparse to constrain the extension direction to better than a general east-west trend.

In the Great Basin and other areas of major extension, a direct, one-to-one causal correlation between extensional faulting and older and synchronous calc-alkaline magmatism has been proposed (Gans and others, 1989), by which magmatism thermally weakens the crust and leads to faulting. Nonetheless, the hypothesis is difficult to prove (Best and Christiansen, 1991; Wernicke, 1992; Axen and others, 1993): 35–20 Ma ash-flow tuffs were spread over large parts of the central Great Basin but were accompanied by little evidence of regional extension or topographic relief, such as angular unconformities or clastic sedimentary rocks between them (McKee and others, 1970; Best and Christiansen, 1991; McKee and Noble, 1986; E.H. McKee, oral commun., 1992). However, local angular unconformities are commonly obscure and difficult to find, and local clastic sedimentary units are easily removed by later erosion, so this negative evidence should be used with caution. Shallow intrusions are closely associated in time and place with significant extension in some nearby areas, as in the Caliente caldera complex (Rowley and R.E. Anderson, mapping in progress), Kane Springs Wash caldera (Scott, Grommé, and others, 1995) just to the south, and Wilson Ridge pluton in the Lake Mead area (R.E. Anderson and Barnhard, 1991; Barnhard and R.E. Anderson, 1991; R.E. Anderson, 1993; R.E. Anderson and others, 1994). A close association has been documented between shallow intrusions and extensional faulting in other areas in the West, including those described by Tobisch and others (1986), Hutton (1988), Lipman (1988), Glazner and Ussler (1989), Armstrong (1990), Ferguson (1990), Haxel and others (1990), Armstrong and Ward (1991), Hardyman and Oldow (1991), Meyer and Foland (1991), Faulds (1993), Lister and Baldwin (1993), Sawyer and others (1994), and Minor (1995). Thus we conclude that extensional faulting in the southern Great Basin and Marysvale field took place throughout the episode of calc-alkaline magmatism, but either could dominate in any one place, depending on whether faults or passive intrusions took up the stress. Where they both occur in a local area, they will be the same age. Also, basement structures probably deflected stress trajectories and encouraged heterogeneous deformation. In other words, magmatism is not due to faulting nor vice versa; both result from extension in the brittle upper crust and they may or may not occur in the same place. As subduction proceeded, the extension of crust beneath the Basin and Range province resulted in different combinations of normal, oblique, strike-slip, and even reverse faults of different dip angles (all depending on local stress conditions) and in the generation and shallow emplacement of magma.

LATE CENOZOIC FAULTS AND BASIN-FILL DEPOSITS

In contrast to the poorly exposed middle Cenozoic faults in the Marysvale volcanic field, the evidence for late Cenozoic faults is strong. Faults of this episode largely postdate 10 Ma and produced the present topography of fault-block plateaus, ranges, and basins (Stewart, 1971). They are called basin-range faults following the definition of Gilbert (1928) and the spelling of Mackin (1960). The best evidence for the age of the main phase of basin-range faulting in the Marysvale area comes from the rhyolite in and north of Kingston Canyon, an antecedent canyon that cuts west through the southern Sevier Plateau (Rowley and others, 1981). Here rhyolite of about 8 Ma that caps the plateau predated the main part of basin-range extension. Canyon cutting took place during and after uplift of the plateau fault block at least 2,000 m along the Sevier fault zone. A rhyolite volcanic dome was emplaced in the bottom of the canyon at 5 Ma, after main-phase faulting. Abundant smaller basin-range faults in the area, however, are as young as Quaternary.

Grabens formed closed basins during basin-range extension in the eastern Great Basin and the transition zone, and they were filled with poorly to moderately consolidated, mostly clastic sedimentary rocks. Most basins in the Great Basin started to form at about 10 Ma (Zoback and others, 1981; R.E. Anderson and others, 1983; R.E. Anderson, 1989). The Beaver Basin near the western edge of the Marysvale field is typical. The upper part of its fill contains soil horizons, air-fall tuff, basalt flows, and fossils that constrain its age between Pliocene and Quaternary (R.E. Anderson and others, 1978; Machette and others, 1984; Machette, 1985; J.J. Anderson and others, 1990b). The beds in the lower part of the basin are not exposed, but the age of 9 Ma for the rhyolite of Gillies Hill, which is emplaced along boundary faults on both sides of the basin (Evans and Steven, 1982), and the age of 10 Ma for basalts adjacent to the basin in the Cedar City area (R.E. Anderson and

Mehnert, 1979) suggest that the basin began to form then. In most of the High Plateaus, where drainage is integrated, basin-fill deposits of the Sevier River Formation (Callaghan, 1938; J.J. Anderson and Rowley, 1975; J.J. Anderson, 1987; Rowley and others, 1988a, b) contain tuff beds and basalt flows ranging from 14.2 to 7.1 Ma (Steven and others, 1979; Best and others, 1980). The Sevier River Formation predates basin-range faults. Rowley and others (1979, p. 16; 1981, p. 600) suggested that lower parts of the formation were deposited in basins formed by broad warping because we were unable to find basin-bounding faults of that age. Overlying basin-fill deposits that formed synchronously with basin-range faults are probably Pliocene and Quaternary.

Most basin-range faults in the Marysvale volcanic field are high-angle normal faults that strike from north-northwest to north-northeast. Many of these faults are subvertical at the surface and may represent movement on older tensional fractures (E.M. Anderson, 1951) that formed perpendicular to a horizontal σ_3 when the amounts of stress in the other two axes (σ_1 and σ_2) were equal (J.J. Anderson and Rowley, unpub. data, 1992). Alternatively, the faults are subvertical where seen because they are close to their original surface of erosion, where σ_1 equals σ_2; only at greater depth did lithostatic pressure increase to produce a greater stress (vertical σ_1), and there the faults dip at closer to their ideal 60° (R.B. Scott, oral commun., 1994). Tensional fractures related to the faults probably supplied subvertical feeders for the many basalt flows in the eastern Great Basin and the High Plateaus. Some faults may represent movement on older subvertical torsional fractures formed during twisting due to uneven amounts of vertical uplift on various parts of some fault blocks (Rowley, 1968). In most places in the area, the trend of σ_3 apparently was generally eastward, and σ_1 was vertical in late Miocene to Pliocene time. Strain, however, was heterogeneous and complex, and it has not been studied in detail. R.E. Anderson (1986) and R.E. Anderson and Barnhard (1992) ascribed a north-northeast-striking zone of left slip in the southern Pavant Range and Sevier Valley to east-west σ_3 and variation between north-trending σ_1 and σ_2, similar to relationships mentioned above for the Caliente area and the Nevada-Utah-Arizona border area. Similar complications are indicated by left slip along the north-northeast-striking Hurricane fault zone near Cedar City (R.E. Anderson and Mehnert, 1979) and the north-northeast-striking Paunsaugunt fault zone in Johns Valley (R.E. Anderson and Barnhard, 1993b).

On the basis of mapping by Rowley in the Caliente area, the third and youngest episode of extension, that of basin-range normal faulting, formed north-trending basins and ranges there (Rowley, Snee, and others, 1992). The explanation there seems relatively simple: namely, east-trending σ_3 and vertical σ_1. In the northeastern Great Basin, the extension direction is east-northeast and east (Pierce and Morgan, 1992, fig. 22). Overall in the Great Basin, Zoback and others (1981) interpreted a general west-northwest direction of σ_3, partly because many basin ranges trend north-northeast. More ranges trend north, however, and partly for this reason, an east-west extension direction is more common. This direction represents a clockwise change with time of about 30°–45° from the σ_3 direction for pre-basin-range faults. The change took place when extension continued but resulted from an entirely different tectonic picture, that of oblique extension (Hamilton and Myers, 1966) of the Great Basin resulting from right slip along the San Andreas transform and along parallel fault zones of significant right-slip transform motion as far east as the Walker Lane belt. This idea is not much different from the concepts of Hamilton and Myers (1966), Atwater (1970, fig. 14), Christiansen and McKee (1978), and McKee and Noble (1986), which in turn relate to suggestions by Wise (1963) that the Western United States is a megashear. The Great Basin became a rift, with raised shoulders—the Sierra Nevada and Wasatch Front—on its western and eastern sides, respectively (Eaton, 1982). Such shoulders are typical of rift margins and may be related to isostatic responses in the lithosphere due to lithosphere changes (Schmidt and Rowley, 1986) or to unloading of footwalls by normal faults (May and others, 1994).

IGNEOUS BELTS AND MINERAL BELTS

Magmatism and mineralization in the Marysvale volcanic field were partly controlled by long-lived regional igneous and tectonic features. The main regional feature is the Pioche-Marysvale igneous belt, which was originally called the Pioche mineral belt by Shawe and Stewart (1976). They also noted a parallel subbelt to the south, which they called the Delamar–Iron Springs mineral belt. Most igneous rocks of both belts are middle Cenozoic calc-alkaline rocks, but some are upper Cenozoic high-silica rhyolite. We have modified the names to call them igneous belts because virtually all the mineral deposits that Shawe and Stewart plotted were localized by intrusive rocks, which we regard as cupolas on batholiths that largely underlie the belts (Steven and others, 1984). The intrusive rocks probably have a volume considerably greater than that of the overlying volcanic rocks. A north-south partial cross section across the Pioche-Marysvale batholith complex is exposed in the Mineral Mountains, where basin-range uplift as a core complex (Price and Bartley, 1992) led to exposures of 25- to 9-Ma plutons (Nielson and others, 1986; Coleman and Walker, 1992). Magmatism in the Pioche-Marysvale belt generally migrated eastward over time (Steven and others, 1984). The plutons of the Pioche-Marysvale and Delamar–Iron Springs igneous belts are clearly delineated by short-wavelength aeromagnetic anomalies, and thus we have modified Shawe and Stewart's shape of the two belts (fig. 1), but not their overall trend, to reflect these aeromagnetic data (Zietz and others, 1976, 1978; Mabey and others, 1978; Hildenbrand and Kucks, 1988a, b; Blank and Kucks, 1989).

The Pioche-Marysvale and slightly younger Delamar–Iron Springs igneous belts (fig. 1) are responsible for only some of the short-wavelength aeromagnetic anomalies in the Great Basin that are interpreted to indicate the presence of igneous rocks. Maps of mineral belts by Shawe and Stewart (1976) and Bagby (1989) document other east- and east-northeast-trending belts in the eastern half of the Great Basin, as well as complex, dismembered, northwest- to west-northwest-trending belts in the Walker Lane belt. These alignments of mining districts, which reflect igneous centers (most of them middle Cenozoic), are also shown by aeromagnetic anomalies (Stewart and others, 1977; Zietz and others, 1976, 1978; Hildenbrand and Kucks, 1988a, b; Blank and Kucks, 1989; Viki Bankey, unpub. data, 1992). The two halves of the Great Basin are separated by a north-trending belt called the "quiet zone" by Stewart and others (1977), which is characterized aeromagnetically by rocks of low magnetic intensity and relief (fig. 1). The shape of the aeromagnetic anomalies indicates a bilateral symmetry to the Great Basin about the quiet zone, and this symmetry is even more obvious in the regional (long wavelength) gravity field, the topography, and other parameters, but not in the exposed geology (Eaton and others, 1978). The mostly east-northeast-trending belts in the eastern Great Basin parallel the middle Cenozoic extension direction of Zoback and others (1981), as do eastern parts of some belts just west of the quiet zone (fig. 5 of Stewart and others, 1977). Thus all belts probably are remnants of middle Cenozoic magmatism.

The Pioche-Marysvale and Delamar–Iron Springs igneous belts are integral parts of a great east-trending arcuate swath of middle Cenozoic igneous rocks, which has been described by Stewart and others (1977) and Best, Christiansen, and others (1989) to extend from west of Reno to Marysvale. It contains the most significant Oligocene and early Miocene igneous centers in the Great Basin. Best (1988) and Sullivan and others (1991) noted that the Colorado Plateau laccoliths and the San Juan volcanic field seem to be an eastward extension of the swath, and they referred to the overall belt as the Reno–San Juan magmatic zone. This zone reflects not only intensity of volcanic activity, but also time; it is only the 31- to 20-Ma part of the overall Great Basin calc-alkaline volcanic province, and represents the "ignimbrite flareup" of peak volcanism in the Great Basin (Best and Christiansen, 1991). Parallel igneous belts in the Great Basin are progressively older (as old as Eocene) to the north and northeast and younger to the south and southwest (Stewart and others, 1977; Cross and Pilger, 1978; Christiansen and Yeats, 1992). This pattern of migration of magmatism is but one of several in the west (Cross and Pilger, 1978; fig. 4; R.E. Anderson, 1989, fig. 6; Best, Christiansen, and others, 1989, fig. 2), some of which diverge from others at high angles. When taken together, these complicated patterns are confusing. The southward and southeastward migration of igneous belts or zones in the Great Basin, which is at an angle to an expected eastward to northeastward subduction trend, has been explained as a progressive steepening or lateral foundering (first in the north, then progressing southward) of the northern part of a subducted slab of oceanic lithosphere (Best and Christiansen, 1991, fig. 2); such foundering allowed a southward-moving wedge of asthenosphere (Hamilton, 1995, figs. 5–6) to move upward and to generate mafic magmas, apparently by supplying heat to melt the dewatering slab of crustal material (Lipman, 1980, fig. 14b, c; Best and Christiansen, 1991, fig. 2). This explanation is one of several (Cross and Pilger, 1978; Stewart, 1983; Henderson and others, 1984; Gans and others, 1989) that have been proposed.

The igneous belts in the Great Basin, which are younger southward, are accompanied by mostly northerly striking normal faults of the same age as the magmatism. This idea, stated by Gans and others (1989), Dilles and Gans (1995), and Scott, Unruh, and others (1995), can be further demonstrated by mapping in progress by R.B. Scott, R.E. Anderson, and Rowley in southeastern Nevada. The middle Cenozoic calc-alkaline igneous belts in the Great Basin have generally been explained by a petrogenetic model involving a subduction zone dipping east-northeast beneath the Western United States (Stewart, 1983). Although this subduction model is not universally accepted, it finds support in the igneous geochemistry, which indicates subduction signatures (Gill, 1981; Fitton and others, 1991). Furthermore, when we look at the larger picture, we note that the volcanic rocks of the Mogollon-Datil field in Arizona and New Mexico (McIntosh and others, 1992) and the Sierra Madre Occidental field in Mexico (Ward, 1991) have ages and igneous compositions similar to those of the Reno–San Juan magmatic zone (Armstrong and Ward, 1991). Thus the various trends of southward younging calc-alkaline activity in the Basin and Range province are smaller patterns within the more regional pattern of volcanism in North America and elsewhere along the Pacific rim. These smaller patterns seem to be most abundant where this global belt is anomalously wide, and their explanation is tied probably to the combination of processes responsible for the large width of the calc-alkaline province in the West. Combining the geochemical signatures linking these rocks with subduction and the overall tectonic evidence, we favor an origin for all these igneous rocks related ultimately to a subduction zone dipping eastward to northeastward beneath the continent. The flat subducted slab that existed during Sevier and Laramide deformation and during the early Tertiary apparently slowed, shortened, and perhaps steepened in the middle Cenozoic as the East Pacific Rise approached the trench, and thus the oceanic lithosphere became younger and less likely to form coherent, long-lasting slabs when subducted (Severinghaus and Atwater, 1990).

TRANSVERSE STRUCTURES

In addition to igneous belts, important tectonic features in the Great Basin that appear to be locally associated with volcanism are long-lived, east-northeast- to east-southeast-striking transverse structures (fig. 1). These have been called "lineaments" by many of us for years but, because many of these structures have now been mapped and therefore evidence for some is finally solid (as opposed to those types of "lineaments" observed only on aerial photos or by remote sensing), the nongenetic term "transverse structure," from Duebendorfer and Black (1992) and some others, is here adopted. Many transverse structures have been recognized for years by mining and petroleum geologists (such as Fuller, 1964; Hilpert and Roberts, 1964; Roberts, 1964, 1966), and citations to many others were given by Eaton and others (1978, fig. 3–11B), Mabey and others (1978), Rowley and others (1978), Stewart (1983), and Duebendorfer and Black (1992). Use of a nongenetic term is required because we now know that the structures have different types, scales, and origins, but all are similar in that they strike transverse to most Great Basin topographic and geologic features and, importantly, in that they formed parallel to the extension direction of the time. They can be thought of as "continental transform" structures because, on the one hand, they are like transform faults in the ocean basins in that they may form as the result of different amounts and directions of magmatic spreading north and south of them. In the Great Basin, on the other hand, only part of this spreading is magmatic, and most is due to faulting. (R.N. Anderson and Noltimier, 1973, showed that some spreading along ocean-basin spreading centers also is by faults.) Transverse structures terminate abruptly along strike as their displacement is taken up by other structures or when spreading or extension north and south of them is the same. In both transform faults (Wilson, 1965) and transverse structures, actual slip may be opposite to the apparent displacement.

The transverse structures were best documented by Ekren and others (1976), who showed that they include some broad and some concentrated zones of recurrent fault offset and local folding. Most of the transverse structures began in the Oligocene, though some probably were late Mesozoic, and most are so young that basin ranges are offset or terminated along them. In other places along strike, the structures mark terminations or interruptions of topographic features and of magnetic or other geophysical anomalies. Tertiary volcanic units terminate or thin across the transverse structures in some places, generally reflecting pre-depositional topographic highs formed by them. The structures have localized plutons and volcanic centers over a significant period, which implies that they have deep crustal control (Ekren and others, 1976). We would not, however, call them rifts, as Bartley and others (1992) did, nor think of them as spreading centers, as Bartley (1989) did, because they formed parallel to the extension direction.

Offset along the transverse structures is difficult to determine, probably because some are large, long-lived shear zones that are so badly deformed and hydrothermally altered that kinematic indicators were destroyed. Some transverse structures, however, seem to have undergone oblique slip or strike slip, although the sense of strike slip may reverse itself along strike. Strike-slip motion, where present, does not seem to be great in many transverse structures; those structures that cross the northern Nevada rift (see section on "Spreading") show little strike-slip offset of the rift. Some transverse structures are well-known zones that transfer different amounts and types of extension on either side of them, such as the east-striking, left-lateral Garlock fault of California (Davis and Burchfiel, 1973), the west-northwest-striking right-lateral Las Vegas Valley shear zone (fig. 1; Fleck, 1970; R.E. Anderson and others, 1972; Liggett and Childs, 1977), or the related northeast-striking, left-lateral Lake Mead fault system (Duebendorfer and Simpson, 1994; R.E. Anderson and others, 1994) just to the south. Other transverse structures are broad zones of distributed shear, and these may include individual faults that are locally concealed beneath the volcanic rocks that erupted along them (Rowley and others, 1978; Hudson and others, 1993, 1995; Hudson and Rosenbaum, 1994). The term *accommodation zone*, which has been around a long time (for example, Davis and Burchfiel, 1973) and is now applied by Faulds and others (1990) to a tilt-block domain boundary, is an appropriate genetic name for the type of deformation seen in some transverse structures. Perhaps the best example of an accommodation zone is the 75-km-long Black Mountains–Highland Spring Range accommodation zone of Arizona and Nevada, which separates rocks of opposite structural polarity (domino-like west-dipping faults with east-dipping beds north of the zone, and east-dipping faults with west-dipping beds south of the zone). Faulds and others (1990) originally explained this zone as an expression of torsion between two oppositely dipping detachment faults. In a drastic modification of his model, Faulds (1992, 1994) showed that detachment faults and strike-slip faults are not required under the accommodation zone. His zone thus may actually represent *less* deformed rock than the terrains north and south of it, inasmuch as north-striking faults to the north and south die out, and beds flatten, into the zone. In another modification of his model, Faulds, Olson, and Littrell (1994) used "rupture barrier" as a synonym for accommodation zone and suggested that the control on a barrier's location is a belt of igneous rocks. Not all accommodation zones are transverse structures because some are not transverse to regional structures or parallel to the regional extension direction (Faulds, Olson, and Littrell, 1994), and others may be relatively minor structures. Most transverse structures represent different amounts of spreading (which we define to consist of both faulting and magmatism) north and south of them either by shearing or by scissor-like torsion.

Davis and Burchfiel (1973) and Duebendorfer and Black (1992) explained how transverse structures can terminate along strike or exhibit opposite directions of strike slip along strike. Mapping by Bartley (1989, 1990), Bartley and others (1992, 1994), Overtoom and others (1993), Overtoom and Bartley (1994), and Brown and Bartley (1994) found little evidence of significant strike slip along easterly striking faults within the Blue Ribbon transverse structure, and thus they interpreted them to be normal faults oriented perpendicular to the extension direction. We instead interpret them to be examples of transverse structures along which accommodation was done with little if any strike-slip motion and perhaps even with little shearing, as explained in the models of Davis and Burchfiel (1973), Duebendorfer and Black (1992), and Faulds (1992, 1994). Stewart (1980) and Stewart and Roldan-Quintana (1994) plotted tilt-block domains in the Basin and Range province and showed that transverse structures bound many regions of opposite structural polarity north and south of them. Hurtubise (1994) documented opposite polarity north and south of the Blue Ribbon transverse structure. Detailed mapping by Stoeser (1993) and C.J. Nutt (oral commun., 1994) found that some transverse structures accommodate different amounts and types of low-angle attenuation and detachment faulting north and south of them.

Eaton and others (1978) and Eaton (1979a) suggested that transverse structures are transform faults oriented perpendicular to a possible north-northwest-striking spreading ridge that was coaxial with or near the quiet zone and around which the bilateral symmetry developed in the Great Basin (see next section); they concluded that the transverse structures are analogous to transforms perpendicular to spreading ridges in ocean basins. We find that, during the middle Tertiary, transverse structures formed parallel to the igneous belts and probably localized the igneous belts, but the transverse structures now strike east and cut the igneous belts at angles of as much as 30° because of the clockwise change in σ_3 noted by Zoback and others (1981). Because they formed parallel to the subduction direction, perhaps transverse structures formed above, and were localized by, oceanic fracture zones (that is, fossil transform faults; Menard and Chase, 1970) that formed in the Farallon/Vancouver plates and then were subducted under the Western United States. Severinghaus and Atwater (1990, figs. 8–13) suggested that such fracture zones controlled the geometry of subducting slabs and that by about 10 Ma, the eastward projection of these fracture zones under the Western United States had changed from east-northeast to east, like the overlying transverse structures.

Whatever the origin of the transverse structures, by the late Cenozoic they were oriented east, generally parallel to the late Cenozoic extension direction, and most functioned to accommodate adjacent areas that had different amounts of extension. In other words, they separated domains of different geologic tilting, extension, magmatism, and history. In the same way that the bases of strike-slip faults require some kind of subhorizontal plane of uncoupling (Stone, 1986) to allow their movement, so also do many low-angle faults in the crust require some kind of bounding subvertical fault zone to allow their movement. Transverse structures are to a degree analogous to tear faults (parallel to the transport direction) in hanging walls of low-angle faults (see examples in R.E. Anderson, 1971), but most are major features, so the underlying subhorizontal zone of detachment into which they feed is probably the brittle-ductile transition zone in the crust. Detachment, attenuation, and high-angle faults are lesser features bounded by transverse structures.

One transverse structure in the Marysvale volcanic field is the Blue Ribbon transverse structure (formerly called the Blue Ribbon lineament). Rowley and others (1978) concluded that this structure is about 25 km wide (north-south) and 280 km long (east-west) in Utah-Nevada and connects across the quiet zone with the Warm Springs "lineament" in Nevada (Ekren and others, 1976), for a combined length of about 600 km. Hurtubise (1989, 1994) provided valuable data about the feature by mapping it across the quiet zone east of where Ekren and others (1976) showed it, and he called it the Silver King lineament. We retain the name Blue Ribbon transverse structure for the part of the overall zone that is east of Ekren's Warm Springs transverse structure. The feature extends eastward through the southern side of the Marysvale volcanic field (figs. 1, 3), where Rowley, Mehnert, and others (1994) interpreted it to control features as different in age as the string of stratovolcano vents (32(?)–21 Ma) for the Mount Dutton Formation and several rhyolite centers (9–5 Ma). Another transverse structure in the Marysvale field is the east-striking Cove Fort transverse structure, which is at least 150 km long. It contains the east-striking Clear Creek downwarp between the Tushar Mountains and Pavant Range. It also includes the east-striking Cove Creek fault of Steven and Morris (1983), which is marked by hot springs (Ross and Moore, 1994); an extensive zone of faults, fractures, and hydrothermally altered rocks; and down-to-the-south, pre-Osiris Tuff offset. The transverse structure defines the northern side of the igneous centers of the Pioche-Marysvale igneous belt, and, like most transverse structures, it has strong geophysical expression on both regional (Zietz and others, 1976; Blank and Kucks, 1989; Saltus and Jachens, 1995; Viki Bankey, USGS, unpub. data, 1992) and local (Campbell and others, 1984; Cook and others, 1984) scales. Additional evidence that the Cove Fort transverse structure has first-order structural significance was the recognition by T.A. Steven (written commun., 1985) of what he called the "Sevier River oval," just north of the transverse structure. He defined this as a north-northeast-striking oval-shaped domal uplift about 115 km long, whose eastern side is the central and southern Canyon Range, Valley Mountains, and Pavant Range; whose center is the southern Sevier Desert and Black Rock Desert; and whose western side includes the Cricket Mountains (fig. 3). The Sevier River oval has strong geophysical expression (Thompson and Zoback, 1979; Saltus and Jachens, 1995). Steven interpreted it to be a late Miocene and younger, incipient (little eroded) metamorphic core complex. He interpreted the west-dipping Sevier Desert detachment of McDonald (1976), which has been imaged by seismic reflection data west of there (Allmendinger and others, 1983; Von Tish and others, 1985), to be genetically related to the core complex. Otton (1995) may have found the "breakaway" scarp of this detachment on the western side of the Canyon Range, and he has dated large rock avalanches shed from the scarp at 13–12 Ma.

In previous discussions of Marysvale geology, we have not stressed the significance of transverse structures. However, recent mapping of the Caliente caldera complex near the western edge of the Delamar–Iron Springs igneous belt by Rowley has emphasized their significance (Best and others, 1993). Thus, for example, the east-striking Timpahute "lineament" of Ekren and others (1976), which they recognized to be at least 140 km long, extends another 30 km east to the Utah border and defines the northern side of the Caliente caldera complex, the Chief mining district, plutons, and other features (Ekren and others, 1977; Rowley, mapping in progress).

Similarly, the newly recognized east-striking Helene transverse structure (Rowley, unpub. mapping, 1992) defines the southern side of the Caliente caldera complex, the major Delamar gold mining district, two smaller gold districts, many east-striking faults, east-striking rhyolite dikes, rhyolite domes, plutons that fed the dikes and domes and localized the gold, a diatreme, and other features for at least 40 km (fig. 1). In fact, the Caliente caldera complex can be thought of as an east-trending volcano-tectonic trough, in the sense of those described by Burke and McKee (1979), bounded by the transverse structures. The Caliente caldera complex thus can be considered to be a smaller version of an igneous belt: although it has been recognized for years that most simple calderas in the Great Basin are elongate east-west (Best, Christiansen, and others, 1989) due to later extension, the Caliente caldera complex is much more asymmetrical than most (80 km east-west versus 35 km north-south). This is because recurring, simultaneous faulting and magmatism extended the complex parallel to the direction of middle Cenozoic extension. In partly the same way, igneous belts may grow or "spread" parallel to the extension direction by combined north-striking dikes, shallow intrusions, and northwest- to northeast-striking faults in zones bounded by transverse structures.

PLATE-TECTONIC HISTORY

Locally significant Late Cretaceous and early Cenozoic calc-alkaline magmatism (Lipman, 1992) and Sevier and Laramide deformation are interpreted to be the results of rapid subduction along a shallow east- to northeast-dipping slab of oceanic lithosphere that reached as far east as Colorado (see, for example, Atwater, 1989; Severinghaus and Atwater, 1990; Helmstaedt and Schulze, 1991). The stress regime at this time consisted of generally east-west compression, yet transverse structures probably started forming by Laramide and Sevier time.

Starting in the middle Cenozoic (Eocene to Miocene), voluminous calc-alkaline magmatism became widespread in the Western United States as far east as Colorado (Lipman and others, 1972), but the origin of the magmas and their plate-tectonic setting are much less clear than is the plate-tectonic setting of the West during the Late Cretaceous and early Cenozoic. Sevier- and Laramide-age east-west compression in the Great Basin had changed to east-northeast extension during middle Cenozoic calc-alkaline magmatism. The subduction model requires that the shallow-dipping, obliquely subducting slab continued to reach as far east as Colorado (Atwater, 1989; Severinghaus and Atwater, 1990), but some workers rejected the model and, following Coney and Reynolds (1977), concluded that the slab steepened through time and did not continue far inland. We endorse a subduction model (Atwater, 1989) in which the slab was far inland in the early and middle Cenozoic, then shortened and perhaps steepened throughout the middle Cenozoic. Regardless of the model, thermal softening of the lithosphere may have allowed diapir-like masses of asthenospheric mantle (Eaton, 1979a) to follow various upward paths, influenced perhaps in part by structures in the lithosphere (Lipman, 1980, 1992), by warps in the slab (Best and Christiansen, 1991), or by foundered pieces of the subducted slab. Warps and foundered pieces of the slab, in turn, may have been localized by the dominant structures of the slab, namely oceanic fracture zones (Menard and Chase, 1970) and transform faults. Severinghaus and Atwater (1990) projected these structures east-northeast from the Pacific plate under the North American plate, and some evidence for this interpretation comes from teleseismic *P*-wave images that reveal a northeast grain to the upper mantle under the Great Basin (Humphreys and Duecker, 1994, fig. 7). The transverse structures and the bilateral symmetry of the Great Basin both developed largely during the middle Cenozoic, although they probably began to form earlier. Nd isotope data indicate that voluminous middle Cenozoic calc-alkaline magmas throughout the Western United States and Mexico contain a major, commonly dominant, mantle component; this component most likely was derived by fractional crystallization of mantle-derived basaltic magmas and assimilated crustal rocks melted by the heat of the basaltic magmas (Smith, 1979; Gans and others, 1989; Fitton and others, 1991; Johnson, 1991; Coleman and Walker, 1992; Lipman, 1992; Wiebe, 1993). After about 28 Ma, the subduction system began to be replaced by the San Andreas transform fault system, and subduction rates slowed as the Farallon/Vancouver plates were progressively consumed (Atwater, 1970; Christiansen and Lipman, 1972; Best and Christiansen, 1991).

Beginning no later than 23 Ma, compositionally bimodal magmas (Christiansen and Lipman, 1972) that lack geochemical subduction signatures were emplaced in the Marysvale area. This change in the Basin and Range province broadly coincided with progressing diminution of the Farallon/Vancouver plates west of the province as the San Andreas transform fault expanded northward and southward along the western edge of the continent; transform motion was accommodated by other faults parallel to the San Andreas fault, extending east to include the Walker Lane belt (Atwater, 1970; Eaton, 1980), which defines the western side of the Great Basin. Subduction progressively ceased northward in the Western United States except under the Cascade Range as the transform lengthened (Atwater, 1970; Christiansen and Lipman, 1972; Christiansen and McKee, 1978; Dickinson and Snyder, 1979). The Great Basin in the late Cenozoic underwent oblique extension as a broad megashear (Atwater, 1970, fig. 14), resulting in bimodal magmatism that began locally before 20 Ma and in basin-range deformation that culminated sometime after 9 Ma in the Marysvale area and most other areas; the Great Basin became a very broad rift. Extension was oriented east to east-southeast on average, though in some places, it was in a somewhat different direction. Some faulting continued along the previously formed transverse structures (Ekren and others, 1976), which now took on a generally eastward strike, synchronous with north- to north-northeast-striking normal faults that were concentrated along and near the older northern Nevada rift (Wallace, 1984; Blakely, 1988; Blakely and Jachens, 1991; Catchings, 1992) and were broadly distributed elsewhere through the Basin and Range province.

SPREADING

We propose a variant of the subduction model for the middle Cenozoic rocks of the Great Basin that draws on the ideas of Scholz and others (1971), Proffett (1977), Eaton and others (1978), Eaton (1979a, 1982, 1984), and Mabey and others (1978). These geologists and geophysicists proposed that the calc-alkaline igneous rocks in the Great Basin represent magmatic spreading about a single axis, during subduction of the Farallon/Vancouver plates, leading to the bilateral symmetry in the Great Basin. Upwelling asthenosphere underlying the symmetry axis was implied (Scholz and others, 1971). In this model, the Great Basin was underlain by a north-northwest-trending spreading axis, oriented perpendicular to the subduction direction, and hydrothermal activity along the axis altered the ferromagnesian minerals in the rocks and reduced their magnetic susceptibility, thereby forming the magnetically quiet zone. Supposedly, as the elongate asthenosphere diapir hit the base of the crust, it spread out horizontally in both directions, and outward-directed tractional stresses on the crust led to extensional faulting and bilateral symmetry. The east-striking transverse structures in this model represent transform faults that strike perpendicular to the spreading axis (Eaton and others, 1978); the intrusive bodies of the igneous belts were emplaced along these faults.

A north-northwest-trending aeromagnetic high just west of the magnetic quiet zone was noted by Mabey and others (1978), who suggested that it represents a ridgelike product of arc spreading. Zoback and Thompson (1978) called the aeromagnetic high the northern Nevada rift (fig. 1) and noted that (1) it is filled with 17- to 14-Ma basaltic and rhyolite dikes, and (2) its trend, like that of normal faults and dikes in tension fractures, is perpendicular to the middle Cenozoic direction of extension. McKee and Noble (1986), Blakely (1988), Blakely and Jachens (1991), and Zoback and others (1994) described the rift in greater detail and extended it to southern Nevada, near the geographic axis of the middle Cenozoic Great Basin. R.E. Anderson and others (1994) extended it into the Lake Mead area of southern Nevada and adjacent Arizona and concluded that here the rift had at least 15 km of extension. It may pass northward into Oregon; Stewart and others (1975) connected it with the west-northwest-striking Brothers fault zone of Oregon and called it the Oregon-Nevada lineament, whereas Pierce and Morgan (1992) called it the Nevada-Oregon rift zone. But because it is best exposed in northern Nevada and its continuation into Oregon is in doubt (Mabey and others, 1978), we will resist the temptation to rename it. The northern Nevada rift is cogenetic with the McDermitt caldera, at its northern end, and the Columbia River Basalt Group, north of it (Zoback and others, 1994). Christiansen and Yeats (1992) suggested that the north-northwest-striking western Snake River Plain graben was formerly the northern part of the northern Nevada rift but that the graben was offset 100 km by an east-northeast right-lateral fault or transform fault that underlies the eastern Snake River Plain. The northern Nevada rift and its parallel features are similar in size to the Rio Grande Rift, but are less eroded. On the basis of paleomagnetic data, Li and others (1990) suggested counterclockwise rotation of the rift, but this interpretation was rebutted by Zoback and others (1994).

A setback to the spreading model of Scholz and others (1971) came from Blakely (1988), who concluded from analysis of geophysical data that the low magnetic susceptibility of the rocks in the quiet zone is not due to former high heat flow but may be due instead to a low proportion of magnetite relative to ilmenite in igneous rocks of the quiet zone. Certainly an interpretation for spreading of a type similar to sea-floor spreading, involving convective upwelling of magma along a single mid-oceanic type ridge spreading center, remains speculative and unlikely (R.J. Blakely and E.H. McKee, oral commun., 1992).

In our variant of the model of Scholz and others, we follow Hamilton (1988a, 1989) and perhaps others and use "spreading" in the sense of a widening of the Great Basin by a combination of pure-shear extension and voluminous magmatism about many axes perpendicular to the extension direction during both the middle and late Cenozoic. The axes include, in addition to the northern Nevada rift, other parallel magnetic anomalies and zones of seismic activity and heat flow that have been described by Wallace (1984), Blakely (1988), McKee and Blakely (1990), Blakely and Jachens (1991), and Catchings (1992) in northern and central Nevada and interpreted by them to indicate zones of concentrated extension. We suggest that, whether by faults or by intrusions, heat flow increased (Lachenbruch and Sass, 1978) under these axes and the brittle-ductile transition zone (below the seismogenic crust; Sibson, 1989) rose and arched under them to form blisterlike forms (metamorphic core complexes) or ridgelike forms. Perhaps with continued extension under these axes, the brittle-ductile transition zone rose higher and broadened perpendicularly outward, providing a driving force for extension and magmatism outward in both directions. Either the crust at and above the raised brittle-ductile transition zone softened, leading to a reduction in shear strength, or stresses dropped after extension along the axis (see Mandl, 1987) and thus extension ceased. East-west least principal stress continued in the Great Basin, however, and thus extension migrated generally toward the cooler, more brittle crust at the margins of the Great Basin. Hamilton (1985, 1988c) pointed out that spreading in North Island, New Zealand (Stern, 1985), may be similar to that in the Great Basin, and that, according to Hamilton, spreading in both areas may actually be due to extension outpacing magmatism. The result in New Zealand is crustal thinning and subsidence, the opposite to the development of most continental magmatic arcs.

Evidence exists for Cenozoic magmatic spreading that generally migrated progressively from northerly striking centers near the central Great Basin outward to its western and eastern boundaries (the motion is relative to the centers, if we hold the North American plate motionless with respect to the Pacific plate) (Eaton, 1980). Early compilation of isotopic dates (Armstrong and others, 1969; McKee, 1971) noted such an outward migration of magmatism and faulting. Later compilations (Armstrong and Ward, 1991) showed that this picture is not that simple, because spreading as it is known in the ocean basins probably does not occur on continents, because a single center at any one time is unlikely, and because younger centers may have been superimposed on or near older ones. The best evidence of progressive migrations of eruptive centers is from the northwestern Great Basin, where MacLeod and others (1976) noted rhyolite domes and centers just southwest of the Brothers fault zone that become progressively younger west-northwest, from 10 Ma in southeastern Oregon to Quaternary in central Oregon. Good data from the eastern Snake River Plain just north of the northeastern Great Basin, summarized by Pierce and Morgan (1992), shows that rhyolite calderas and other eruptive centers there, in contrast, migrated east-northeast from 16 Ma at the Nevada-Oregon-Idaho border to Quaternary at Yellowstone National Park. Other migrations have been documented in the Great Basin, although some of these are subtle, as in northeastern Utah (Miller, 1991) and in the Tintic–Deep Creek igneous belt of west-central Utah (Stoeser, 1993). Farther south, Steven and others (1984) noted that at least the eastern two-thirds of the middle Cenozoic part of the Pioche-Marysvale igneous belt is generally younger (35 to 23 Ma) eastward. On the southwestern side of the Great Basin, Silberman and others (1975) observed andesites that are older (27 Ma) in the east and younger (25 to 20 and 8 to 7 Ma) in the west. Dickinson and Snyder (1979, fig. 8) also recognized a progressive north-to-northwest migration of igneous centers with time (16 to 5 Ma). More details on these northwest migrations of rhyolite centers were noted by Luedke and Smith (1981), Sawyer and others (1994), and A.M. Sarna-Wojcicki (written commun., 1994), starting with eruptions from the 15- to 11-Ma calderas of the Nevada Test Site (Sawyer and others, 1994), the 9-Ma Black Mountain caldera (Sawyer and others, 1994) 5 km northwest of the Test Site, the 7.5-Ma Stonewall Mountain volcanic center (Weiss and others, 1993; Sawyer and others, 1994) 40 km farther northwest, the 7- to 3-Ma Mount Jackson dome field (Weiss and others, 1993) 20 km farther west, the 6-Ma Silver Peak volcanic center (Robinson and others, 1968; Sawyer and others, 1994) 40 km farther northwest (fig. 1), and ending with the Pliocene and Quaternary Long Valley–Mono Basin volcanic center (Bailey, 1989; Bursik and Sieh, 1989) 80 km farther west-northwest, in California. Thompson and others (in press) documented a westward and northwestward migration of 8- to 3-Ma volcanic rocks and post-6.5-Ma faults in the Death Valley area. Except for young basalts and faults that form an axial fracture in the central Great Basin (Smith and Luedke, 1984; Wallace, 1984; Blakely, 1988; Fitton and others, 1991), it has long been known that the youngest basalts in the Great Basin occur on its eastern and western boundaries (Christiansen and McKee, 1978; Eaton and others, 1978; Eaton, 1979b; Smith and Luedke, 1984; Fitton and others, 1991). Within these Quaternary basalt fields at the eastern boundary, Luedke and Smith (1978) and Nealey and others (1994) noted an eastward migration with time, as has been observed at the eastern edge of the southern Basin and Range province in northern Arizona (Best and Brimhall, 1974; Tanaka and others, 1986; Condit and others, 1989).

Spreading due to faulting probably also migrated outward from centers in the central Great Basin, and this scenario is certainly consistent with the current high seismicity along the basin's western and eastern boundaries (Christiansen and McKee, 1978; Eaton, 1979b). Evidence within the basin, however, is poor, with one main exception: Hamilton (1988b) showed a westward younging of termination of major extension, from middle Miocene in the Nevada Test Site, to late Miocene in the

Bullfrog Hills (fig. 1) and in the Funeral Mountains of California, to still active in the Death Valley of California. Also, on the eastern side of the southern Basin and Range province in northern Arizona, Faulds, Gans, and Smith (1994) noted progressively younger faulting eastward (16 to 10 Ma).

Formerly, various names were applied to the type of middle Cenozoic calc-alkaline magmatic arc in the Western United States that contains components of spreading or extension. The terms "ensialic back arc" or "marginal basin" were most commonly used, but these seem inappropriate because most back arcs (as described by Saunders and Tarney, 1984) are underlain by oceanic lithosphere and contain thick marine sedimentary rocks and local basalt rather than subaerial calc-alkaline volcanic rocks (see, for example, Rowley and others, 1991; Rowley, Kellogg, and others, 1992). Furthermore, when applied to the Western United States, such terms leave unconsidered the location of the concurrent magmatic arc west of any back arc. Thus Elston (1984) instead proposed "extensional orogen," but we prefer the name "magmatic arc." Neglecting such semantic matters, however, the point to be made here is that middle Miocene calc-alkaline magmatism and late Cenozoic bimodal magmatism in the Great Basin are associated with significant extension, possibly along many north-northwest- to north-northeast-trending, perhaps diffuse, central axes, and thus a spreading analogy can be made. Thus the Great Basin extended much more than adjacent areas on the Pacific rim.

Spreading was partly accommodated by transverse structures that formed parallel to the spreading direction, that occur throughout the Great Basin, and that partly bound the basin on the north (Christiansen and McKee, 1978; McKee and Noble, 1986, fig. 6) and the south (Davis and Burchfiel, 1973). Accommodation by transverse structures allowed igneous belts to grow parallel to the spreading direction along centers oriented perpendicular to the spreading direction, whereas areas north and south of the igneous belts spread by faulting.

A model for the development of igneous belts is provided by the eastern Snake River Plain. This broad lava plain is generally shown on physiographic maps to bound the Great Basin on the northeast, but geologically it is an uneroded version of the igneous belts within the Great Basin, and to its north, basin-range topography continues (Eaton and others, 1978; Kuntz and others, 1992). The plain is oriented east-northeast to northeast, parallel to the extension direction in the area (Kuntz, 1992; Kuntz and others, 1992). Southward from the eastern Snake River Plain, this direction gradually swings about 40° into the present east-west extension direction of the northeastern Great Basin (Zoback and others, 1981; Zoback, 1989; Pierce and Morgan, 1992, fig. 22). Most rocks at the surface of the plain are Quaternary basalt lava flows that erupted along north-northwest-striking linear vents; basin-range faults north of the plain strike in the same direction, whereas those south of the plain strike about north, all being perpendicular to the extension direction (Kuntz and others, 1992, fig. 11). Calderas or other vents generally below the surface of the plain erupted mostly rhyolite ash-flow tuffs and record the progressive east-northeast to northeast migration of eruptions, from 16 Ma to Quaternary (Pierce and Morgan, 1992). In the same way that we would not give the name "rift" to transverse structures, we agree with Kuntz and others (1992) and Pierce and Morgan (1992) that Hamilton's (1989) use of "rift" for the eastern Snake River Plain is inappropriate; the plain was oriented parallel, not perpendicular, to the extension direction when it formed. We speculate instead that the eastern Snake River Plain is bound on its north and south sides by buried east-northeast- to northeast-striking transverse structures. No faults of the same strike are exposed, although they have been modeled by Sparlin and others (1982). The transverse structures that we propose may not be expressed primarily by faults. These zones allowed east-northeast spreading to be taken up primarily by magmatism within the plain, and primarily by normal faulting to the north and south (Kuntz and others, 1992). The plain is thus an igneous belt formed by basalt feeder dikes oriented perpendicular to the spreading direction and by shallow rhyolitic intrusions whose roofs collapsed to form calderas. The eastward progression in rhyolite eruptions here in the (geologically defined) eastern part of the Great Basin is contrasted with the westward progression in rhyolite eruptions in the western part of the Great Basin south of the Brothers fault zone. The Brothers fault zone is a transverse structure that includes right slip and defines part of the northwestern edge of the Great Basin (Lawrence, 1976; McKee and Noble, 1986).

The two transverse structures that we suggest underlie the eastern Snake River Plain continue northeast of Yellowstone into Montana (Eaton and others, 1975; Mabey and others, 1978), parallel to Precambrian structures (Erslev and Sutter, 1990; Pierce and Morgan, 1992, p. 32). One of these structures, the northeast-striking Proterozoic Madison mylonite zone of Erslev and Sutter (1990) in the southern Madison Range, may have been followed by our suggested transverse structure that bounds the northwest side of the plain; farther northeast, the transverse structure may even control the northeast-striking Holocene Emigrant fault of Pierce and Morgan (1992, plate 1) south of Bozeman. Our speculations thus are similar to those of Eaton and others (1975), Mabey and others (1978), and Christiansen and McKee (1978), who suggested that Precambrian crustal flaws controlled the location of the eastern Snake River Plain. We would emphasize, however, that parallelism to the extension direction is the dominant control of the magmatism in the plain. The two transverse structures also continue southwest into Nevada, and the overall 1,000-km-long belt in Montana, Idaho, and Nevada was called the Humboldt zone by Mabey and others (1978) and Rowan and Wetlaufer (1981). The zone is characterized by high heat flow (Lachenbruch and Sass, 1978; Blakely, 1988) of the Battle Mountain high. Pierce and Morgan (1992, p. 32), however, doubted that the two "lineaments" of Eaton and others (1975) and Mabey and others (1978) were the dominant control of the eastern Snake River Plain. "Why," they asked, "would two such lineaments be exploited simultaneously rather than the failing of one ***?" This remains a good question.

Pierce and Morgan (1992), Zoback and others (1994), and others ascribed the northeast migration of igneous activity in the eastern Snake River Plain and Yellowstone National Park area to the passage of the continental plate over a "hot spot" from an underlying mantle plume. Their evidence is compelling (see also Pierce and others, 1992), but the theory does not explain the synchronous but symmetrically opposite migration along and near the Brothers fault zone (Lipman, 1992, p. 491), as well as other igneous belts in the Great Basin.

Was subduction a prerequisite for volcanism in the Great Basin and nearby areas? Mutschler and others (1987, 1991, this volume) answered "no" and proposed instead that volcanism is due to subcrustal lithospheric thinning and mantle upwelling above passive hot spots. Lipman (1992) and we disagree with their model (see section on Igneous Belts and Mineral Belts, p. 180). Mutschler and his colleagues identified several passive hot spots in the Western United States, including the possible Great Basin spreading center discussed above, which they called the Great Basin regional gravity low. Blank and others (this volume) propose another to underlie the entire Colorado Plateau. Asthenospheric upwarps have been suggested by others, including Eaton (1987), who proposed a large bulge that uplifted the southern Rocky Mountains and formed a crestal graben, the Rio Grande Rift. White and others (1987) traced the development of ocean basins that started to form by upwelling of asthenosphere along continental rifts; decompression of the upwelled material theoretically produced magmas by partial melting. They suggested that the Great Basin is an example of a continental rift that did not stretch sufficiently to form a new ocean basin. According to White and others (1987), if stretching stops before the lithosphere has been reduced to less than half its original thickness, the thinned region subsides to form a sedimentary basin like the North Sea, whereas if stretching continues and the continental lithosphere breaks, new oceanic lithosphere is formed by sea-floor spreading, bordered on either side by rifted continental margins, as in the Red Sea or the Atlantic Ocean. Rosendahl (1987, p. 461) coined the word "pretransform" to describe transverse structures in the East African rift that parallel the extension direction and in the future will become true transforms in oceanic crust when rifting advances to the stage that the area becomes an ocean basin.

CONCLUSIONS

The large Marysvale and San Juan volcanic fields occur on the edges of the Colorado Plateau, where voluminous hot mantle-derived magma formed shallow crustal calc-alkaline magma chambers that congealed to form composite batholiths. In contrast, the laccolith clusters in the rigid Colorado Plateau between these volcanic fields are minor, widely scattered igneous features that had few effusive products. According to Gilbert (1877) and Corry (1988), laccoliths form, whether in the Colorado Plateau or elsewhere, when magma rises to a level in the crust where the rocks have about the same specific gravity as the melt. At about that level, relatively close to the surface (about 1–4 km or less in southeastern Utah, the northern Markagunt Plateau, and the Iron Axis), the magma may spread out along low-dipping incompetent sedimentary beds. In this near-surface environment alone, σ_3 becomes vertical and σ_1 equals σ_2, so magma spreads out laterally and lifts thin roof rocks. Parsons and others (1992), however, argued that rheological boundaries between horizontal packages of host rocks are sites for emplacement of horizontal sills or laccoliths only after the emplacement of vertical dikes (perpendicular to horizontal extension) has relieved tension and caused the least principal stress above this boundary to become vertical, whereas the vertical stress axis below the boundary remains as σ_2. Regardless of the dominating mechanisms, most of the laccoliths of the Iron Axis spread out in gypsiferous shale in the lower part of the Carmel Formation and above the massive Navajo Sandstone, the laccoliths on the southern flank of the Marysvale field spread out in shale in the lower part of the Claron Formation, and laccoliths of the Colorado Plateau and locally in the San Juan area spread out in the Mancos Shale and underlying units. Laccoliths differ from unfloored stocks in having much smaller volumes of magma available for eruption, so attendant volcanic units tend to be smaller, although the products erupted from both sources are similar. Such volcanic rocks of limited extent were erupted from the Spry and Iron Peak laccoliths along the southern side of the Marysvale field (J.J. Anderson and Rowley, 1975; J.J. Anderson and others, 1990a), at some laccoliths of the Iron Axis farther south (Blank and others, 1992; D.B. Hacker, Kent State University, written commun., 1992; Rowley, unpub. mapping, 1992), at laccoliths in West Texas (Henry and Price, 1988; Henry and others, 1991), and in some other areas (Corry, 1988). Laccoliths form some significant ore bodies, as at Iron Springs (Mackin, 1947), but most of their hydrothermal-type deposits are small.

Generally, middle Cenozoic calc-alkaline volcanic rocks similar to those of the Marysvale field are voluminous elsewhere in the Western United States (Johnson, 1991), and their eruptive centers are underlain by even larger batholith complexes. The great breadth of these similar types of igneous rocks is best explained as a result of magmatism above a subduction zone, as elsewhere around the Pacific margin and other parts of the world. The generation of calc-alkaline rocks appears to be driven by "large fluxes of mantle-derived basaltic magmas" (Smith, 1979; Johnson, 1991). The intrusive rocks and related hydrothermal mineral deposits constitute a treasure house of valuable resources. During the middle Cenozoic, extension oriented generally east-northeast was expressed by local high-angle normal faults, many of which flatten at depth, by metamorphic core complexes, and by shallow intrusions. In many places, depending on the stress field, oblique- and strike-slip faults were also produced. Some of the most productive mineral belts in the Great Basin, mostly of gold, silver, and chalcophile elements, resulted where permeability due to faulting and brecciation coincided in space and time with magmatism and resulting hydrothermal convection cells (John and others, 1989, 1991; Hardyman and Oldow, 1991; Henley and Adams, 1992; Rowley, Snee, and others, 1992; Willis and Tosdal, 1992).

Upper Cenozoic bimodal igneous rocks are less voluminous than calc-alkaline rocks, but they also generated many deposits of base and precious metals, as well as lithophile elements. Bimodal volcanism began in the Marysvale area by about 23 Ma and in the Caliente area no later than 15.5 Ma. Bimodal magmatism in the Basin and Range province correlates with transform movement on the San Andreas fault zone and the end of subduction. Most rhyolites probably resulted from partial melting of local areas of crust. Extension in the Great Basin continued, although now oriented east-west or perhaps east-southeast, yet the main late Cenozoic deformational episode that established the present topography, that of basin-range faulting, did not begin until later (after 10 Ma in the Marysvale area and the southeastern Great Basin). Most basin-range faults are relatively high-angle normal faults where exposed, probably decreasing to lower angles with depth, but strike-slip faults occur locally.

The result of volcanism and major extension in the Great Basin during the late Cenozoic and probably the middle Cenozoic can be considered a type of east-west "spreading" or widening of continental crust, creating bilateral symmetry. Numerous northerly trending axes involving different combinations of spreading by faulting and hypabyssal intrusion appear to occur in the central Great Basin but may also be present throughout the subprovince. Long-lasting, recurring combinations of faults and shallow intrusions, as in the northern Nevada rift, Wilson Ridge pluton, and Caliente caldera complex represent examples of local extensional spreading by this twofold combination. Transverse structures oriented east-northeast in the middle Cenozoic, parallel to the subduction direction, represent zones separating different amounts of extension and calc-alkaline magmatism north and south of them. They controlled the shape of the east-northeast-trending igneous belts within the Great Basin. The igneous belts became younger to the south-southeast, but each belt grew eastward or westward (generally westward in the western Great Basin and eastward in the eastern Great Basin) away from a central axis. In the late Cenozoic, during oblique extension that culminated in basin-range deformation, accommodation offset continued along some of the transverse structures, bringing them to their present east-west trend, and some bimodal volcanism also took place along them. The Great Basin continues to spread about north-northeast- or north-trending axes, and most recent basalts and faults occur at the basin's axis and eastern and western margins.

REFERENCES CITED

Agrell, S.O., Charnley, N.R., and Rowley, P.D., 1986, The occurrence of hibonite, perovskite, zirconolite, pseudobrookites and other minerals in a metamorphosed hydrothermal system at Pine Canyon, Piute County, Utah, U.S.A. [abs.]: Mineralogical Society (London) Bulletin 72, p. 4.

Allmendinger, R.W., and Jordan, T.E., 1984, Mesozoic structure of the Newfoundland Mountains, Utah—Horizontal shortening and subsequent extension in the hinterland of the Sevier orogenic belt: Geological Society of America Bulletin, v. 95, p. 1280–1292.

Allmendinger, R.W., Sharp, J.W., Von Tish, Douglas, Serpa, Laura, Brown, Larry, Kaufman, Sidney, Oliver, Jack, and Smith, R.B., 1983, Cenozoic and Mesozoic structure of the eastern Basin and Range province, Utah, from COCORP seismic-reflection data: Geology, v. 11, p. 532–536.

Anderson, E.M., 1951, The dynamics of faulting and dyke formation with applications to Britain: London, Oliver and Boyd, 206 p.

Anderson, J.J., 1971, Geology of the southwestern High Plateaus of Utah—Bear Valley Formation, an Oligocene-Miocene volcanic arenite: Geological Society of America Bulletin, v. 82, p. 1179–1205.

———1985, Mid-Tertiary block faulting along west and northwest trends, southern High Plateaus, Utah: Geological Society of America Abstracts with Programs, v. 17, no. 7, p. 513.

———1987, Late Cenozoic drainage history of the northern Markagunt Plateau, Utah, *in* Kopp, R.S., and Cohenour, R.E., eds., Cenozoic geology of western Utah—Sites for precious metal and hydrocarbon accumulations: Utah Geological Association Publication 16, p. 271–278.

———1988, Pre-basin-range block faulting along west and northwest trends, southeastern Great Basin and southern High Plateaus: Geological Society of America Abstracts with Programs, v. 20, no. 3, p. 139.

———1993, The Markagunt megabreccia—Large Miocene gravity slides mantling the northern Markagunt Plateau, southwestern Utah: Utah Geological Survey Miscellaneous Publication 93-2, 37 p.

Anderson, J.J., and Rowley, P.D., 1975, Cenozoic stratigraphy of the southwestern High Plateaus of Utah, *in* Anderson, J.J., Rowley, P.D., Fleck, R.J., and Nairn, A.E.M., Cenozoic geology of southwestern High Plateaus of Utah: Geological Society of America Special Paper 160, p. 1–52.

Anderson, J.J., Rowley, P.D., Blackman, J.T., Mehnert, H.H., and Grant, T.C., 1990a, Geologic map of the Circleville Canyon area, southern Tushar Mountains and northern Markagunt Plateau, Beaver, Garfield, Iron, and Piute Counties, Utah: U.S. Geological Survey Miscellaneous Investigations Series Map I–2000, scale 1:50,000.

Anderson, J.J., Rowley, P.D., Machette, M.N., Decatur, S.H., and Mehnert, H.H., 1990b, Geologic map of the Nevershine Hollow area, eastern Black Mountains, southern Tushar Mountains, and northern Markagunt Plateau, Beaver and Iron Counties, Utah: U.S. Geological Survey Miscellaneous Investigations Series Map I–1999, scale 1:50,000.

Anderson, R.E., 1971, Thin skin distension in Tertiary rocks of southeastern Nevada: Geological Society of America Bulletin, v. 82, p. 43–58.

———1973, Large-magnitude late Tertiary strike-slip faulting north of Lake Mead, Nevada: U.S. Geological Survey Professional Paper 794, 18 p.

———1978, Chemistry of Tertiary volcanic rocks in the Eldorado Mountains, Clark County, Nevada, and comparisons with rocks from some nearby areas: U.S. Geological Survey Journal of Research, v. 6, p. 409–424.

———1984, Strike-slip faults associated with extension in and adjacent to the Great Basin: Geological Society of America Abstracts with Programs, v. 16, no. 6, p. 429.

———1986, Coeval mixed-mode dip-slip and strike-slip faulting in and adjacent to the Basin and Range, Utah-Nevada: Geological Society of America Abstracts with Programs, v. 18, no. 5, p. 338.

———1987, Neogene geologic history of the Nevada-Utah border area at and near latitude 37°30′N.: Geological Society of America Abstracts with Programs, v. 19, no. 7, p. 572.

———1988, Hazard implications of joint-controlled basaltic volcanism in southwestern Utah: Geological Society of America Abstracts with Programs, v. 20, no. 7, p. A115.

———1989, Tectonic evolution of the Intermontane System, Basin and Range, Colorado Plateau, and High Lava Plains, chap. 10 *of* Pakiser, L.C., and Mooney, W.D., eds., Geophysical framework of the continental United States: Geological Society of America Memoir 172, p. 163–176.

———1990, West-directed tectonic escape between the northern and southern sectors of the Basin and Range province: Geological Society of America Abstracts with Programs, v. 22, no. 3, p. 2.

———1993, Tectonic significance of a Miocene dike swarm and its post-emplacement vertical and meridional collapse, Lake Mead area, Nevada, Arizona: Geological Society of America Abstracts with Programs, v. 25, no. 5, p. 3.

Anderson, R.E., and Barnhard, T.P., 1991, Relationship between Miocene plutonism, uplift, and extension, Lake Mead area, northernmost Arizona and adjacent Nevada: Geological Society of America Abstracts with Programs, v. 23, no. 5, p. 245.

———1992, Neotectonic framework of the central Sevier Valley area, Utah, and its relationship to seismicity, *in* Gori, P.L., and Hays, W.W., eds., Assessment of regional earthquake hazards and risk along the Wasatch Front, Utah: U.S. Geological Survey Professional Paper 1500–F, 47 p.

———1993a, Heterogeneous Neogene strain and its bearing on horizontal extension and horizontal and vertical contraction at the margin of the extensional orogen, Mormon Mountains area, Nevada and Utah: U.S. Geological Survey Bulletin 2011, 43 p.

———1993b, Aspects of three-dimensional strain at the margin of the extensional orogen, Virgin River depression area, Nevada, Utah, and Arizona: Geological Society of America Bulletin, v. 105, p. 1019–1052.

Anderson, R.E., Barnhard, T.P., and Snee, L.W., 1994, Roles of plutonism, mid-crustal flow, tectonic rafting, and horizontal collapse in shaping the Miocene strain field of the Lake Mead area, Nevada and Arizona: Tectonics, v. 13, p. 1381–1410.

Anderson, R.E., and Bohannon, R.G., 1993, Three-dimensional aspects of the Neogene strain field, Nevada-Utah-Arizona tricorner area, *in* Lahren, M.M., Trexler, J.H., Jr., and Spinosa, Claude, eds., Crustal evolution of the Great Basin and the Sierra Nevada; Geological Society of America Cordilleran/Rocky Mountain Sections Joint Meeting, Reno, Nev., 1993, Field Trip Guidebook: Reno, Nev., University of Nevada, Mackay School of Mines, p. 167–196.

Anderson, R.E., Bucknam, R.C., and Hamblin, Kenneth, 1978, Road log to the Quaternary tectonics of the Intermountain seismic belt between Provo and Cedar City, Utah: Geological Society of America Rocky Mountain Section Annual Meeting, Field Trip 8, 50 p.

Anderson, R.E., and Ekren, E.B., 1977, Comment *on* Late Cenozoic fault patterns and stress fields in the Great Basin and westward displacement of the Sierra Nevada block (Wright, 1976): Geology, v. 5, p. 388–389.

Anderson, R.E., Longwell, C.R., Armstrong, R.L., and Marvin, R.F., 1972, Significance of K-Ar ages of Tertiary rocks from the Lake Mead region, Nevada-Arizona: Geological Society of America Bulletin, v. 83, p. 273–288.

Anderson, R.E., and Mehnert, H.H., 1979, Reinterpretation of the history of the Hurricane Fault in Utah, *in* Newman, G.W., and Goode, H.D., eds., Basin and Range Symposium and Great Basin Field Conference: Denver, Colo., Rocky Mountain Association of Geologists, p. 145–165.

Anderson, R.E., Zoback, M.L., and Thompson, G.A., 1983, Implications of selected subsurface data on the structural form and evolution of some basins in the northern Basin and Range province, Nevada and Utah: Geological Society of America Bulletin, v. 94, p. 1055–1072.

Anderson, R.N., and Noltimier, H.C., 1973, A model for the horst and graben structure of midocean ridge crests based upon spreading velocity and basalt delivery to the oceanic crust: Geophysics Journal of the Royal Astronomical Society, v. 34, p. 137–147.

Angelier, Jacques, Coletta, Bernard, and Anderson, R.E., 1985, Neogene paleostress changes in the Basin and Range—A case study at Hoover Dam, Nevada-Arizona: Geological Society of America Bulletin, v. 96, p. 347–361.

Armstrong, R.L., 1990, Cenozoic magmatism in the North America Cordillera and the origin of metamorphic core complexes: Geological Society of America Abstracts with Programs, v. 22, no. 3, p. 4.

Armstrong, R.L., Ekren, E.B., McKee, E.H., and Noble, E.C., 1969, Space-time relations of Cenozoic silicic volcanism in the Great Basin of the Western United States: American Journal of Science, v. 267, p. 478–490.

Armstrong, R.L., and Ward, Peter, 1991, Evolving geographic patterns of Cenozoic magmatism in the North American cordillera—The temporal and spatial association of magmatism and metamorphic core complexes: Journal of Geophysical Research, v. 96, no. B8, p. 13201–13224.

Atwater, Tanya, 1970, Implications of plate tectonics for the Cenozoic tectonic evolution of western North America: Geological Society of America Bulletin, v. 81, p. 3513–3536.

———1989, Plate tectonic history of the northeast Pacific and western North America, *in* Winterer, E.L., Hussong, D.M., and Decker, R.W., eds., The eastern Pacific Ocean and Hawaii: Geological Society of America, The geology of North America, v. N, p. 21–72.

Axen, G.J., 1991, Tertiary extension, magmatism, and thrust reactivation in the southern Great Basin, and a mechanical model for detachment faulting: Cambridge, Mass., Harvard University Ph. D. dissertation, 235 p.

Axen, G.J., Lewis, P.R., Burke, K.J., Sleeper, K.G., and Fletcher, J.M., 1988, Tertiary extension in the Pioche area, Lincoln County, Nevada, *in* Weide, D.L., and Faber, M.L., eds., This extended land—Geological journeys in the southern Basin and Range: Geological Society of America, Cordilleran Section Meeting, Las Vegas, Nev., 1988, Field Trip Guidebook, p. 3–5.

Axen, G.J., Taylor, W.J., and Bartley, J.M., 1993, Space-time patterns and tectonic controls of Tertiary extension and magmatism in the Great Basin of the Western United States: Geological Society of America Bulletin, v. 105, p. 56–76.

Bacon, C.R., 1982, Time-predictable bimodal volcanism in the Coso Range, California: Geology, v. 10, p. 65–69.

Bagby, W.C., 1989, Patterns of gold mineralization in Nevada and Utah, *in* Shawe, D.R., and Ashley, R.P., eds., United States gold terranes—Part I, Chapter B: U.S. Geological Survey Bulletin 1857, p. B11–B21.

Bailey, R.A., 1989, Geologic map of the Long Valley caldera, Mono–Inyo Craters volcanic chain and vicinity, eastern California: U.S. Geological Survey Miscellaneous Investigations Series Map I–1933.

Barker, D.S., 1991, Quartz monzonite and associated iron deposits, Iron Springs, Utah: Geological Society of America Abstracts with Programs, v. 23, no. 5, p. A388.

Barnhard, T.P., and Anderson, R.E., 1991, Tectonic significance of dike orientations in the Lake Mead area of Nevada and Arizona: Geological Society of America Abstracts with Programs, v. 23, no. 5, p. 233.

Bartley, J.M., 1989, Changing Tertiary extension directions in the Dry Lake Valley area, Nevada, and a possible dynamic model, *in* Garside, L.J., and Shaddrick, D.R., eds., Compressional and extensional structural styles in the northern Basin and Range province—Seminar Proceedings: Reno, Nev., Nevada Petroleum Society and Geological Society of Nevada, p. 35–39.

———1990, Hypothetical mid-Tertiary synvolcanic stress rotation in the southeastern Great Basin: Geological Society of America Abstracts with Programs, v. 22, no. 3, p. 6.

Bartley, J.M., Axen, G.J., Taylor, W.J., and Fryxell, J.E., 1988, Cenozoic tectonics of a transect through eastern Nevada near 38° N. latitude, *in* Weide, D.L., and Faber, M.L., eds., This extended land—Geological journeys in the southern Basin and Range; Geological Society of America Field Trip Guidebook, Cordilleran Section Meeting, Las Vegas, Nev., 1988: Las Vegas, Nev., University of Nevada, Department of Geosciences, p. 1–20.

Bartley, J.M., Friedrich, A.M., Walker, J.D., Coleman, D.S., and Price, D.E., 1994, Post-Sevier belt normal faulting in southwestern Utah: Geological Society of America Abstracts with Programs, v. 26, no. 2, p. 37.

Bartley, J.M., Taylor, W.J., and Lux, D.R., 1992, Blue Ribbon volcanic rift in southeastern Nevada and its effects on Basin and Range fault segmentation: Geological Society of America Abstracts with Programs, v. 24, no. 6, p. 2.

Beaty, D.W., Cunningham, C.G., Rye, R.O., Steven, T.A., and Gonzalez-Urien, Eliseo, 1986, Geology and geochemistry of the Deer Trail Pb-Zn-Ag-Cu manto deposits, Marysvale district, west-central Utah: Economic Geology, v. 81, p. 1932–1952.

Best, M.G., 1988, Early Miocene change in direction of least principal stress, Southwestern United States—Conflicting inferences from dikes and metamorphic core-detachment fault terranes: Tectonics, v. 7, no. 2, p. 249–259.

Best, M.G., and Brimhall, W.H., 1974, Late Cenozoic alkalic basaltic magmas in the western Colorado Plateaus and the Basin and Range transition zone, U.S.A., and their bearing on mantle dynamics: Geological Society of America Bulletin, v. 85, p. 1677–1690.

Best, M.G., and Christiansen, E.H., 1991, Limited extension during peak Tertiary volcanism, Great Basin of Nevada and Utah: Journal of Geophysical Research, v. 96, no. B8, p. 13509–13528.

Best, M.G., Christiansen, E.H., and Blank, R.H., Jr., 1989, Oligocene caldera complex and calc-alkaline tuffs and lavas of the Indian Peak volcanic field, Nevada and Utah: Geological Society of America Bulletin, v. 101, p. 1076–1090.

Best, M.G., Christiansen, E.H., Deino, A.L., Grommé, C.S., McKee, E.H., and Noble, D.C., 1989, Eocene through Miocene volcanism in the Great Basin of the Western United States: New Mexico Bureau of Mines Memoir 47, p. 91–133.

Best, M.G., McKee, E.H., and Damon, P.E., 1980, Space-time-composition patterns of late Cenozoic mafic volcanism, southwestern Utah and adjoining areas: American Journal of Science, v. 280, p. 1035–1050.

Best, M.G., Scott, R.B., Rowley, P.D., Swadley, W C, Anderson, R.E., Grommé, C.S., Harding, A.E., Deino, A.L., Christiansen, E.H., Tingey, D.G., and Sullivan, K.R., 1993, Oligocene-Miocene caldera complexes, ash-flow sheets, and tectonism in the central and southeastern Great Basin, *in* Lahren, M.M., Trexler, J.H., Jr., and Spinosa, Claude, Crustal evolution of the Great Basin and the Sierra Nevada; Geological Society of America Cordilleran/Rocky Mountain Sections Joint Meeting, Reno, Nev., 1993, Field Trip Guidebook: Reno, Nev., University of Nevada, Mackay School of Mines, p. 285–311.

Blakely, R.J., 1988, Curie temperature isotherm analysis and tectonic implications of aeromagnetic data from Nevada: Journal of Geophysical Research, v. 93, no. B10, p. 11817–11832.

Blakely, R.J., and Jachens, R.C., 1991, Regional study of mineral resources in Nevada—Insights from three-dimensional analysis of gravity and magnetic anomalies: Geological Society of America Bulletin, v. 103, p. 795–803.

Blank, H.R., 1994, Rafting and magmatism in E. Central Nevada—Insights from gravity and magnetic data: Geological Society of America Abstracts with Programs, v. 26, no. 2, p. 39.

Blank, H.R., and Kucks, R.P., 1989, Preliminary aeromagnetic, gravity, and generalized geologic maps of the USGS Basin and Range–Colorado Plateau transition zone study area in southwestern Utah, southeastern Nevada, and northwestern Arizona (the "BARCO" project): U.S. Geological Survey Open-File Report 89–432, 16 p.

Blank, H.R., Jr., and Mackin, J.H., 1967, Geologic interpretation of an aeromagnetic survey of the Iron Springs district, Utah: U.S. Geological Survey Professional Paper 516–B, 14 p.

Blank, H.R., Rowley, P.D., and Hacker, D.B., 1992, Miocene monzonitic intrusions and associated megabreccias of the Iron Axis region, southwestern Utah, *in* Wilson, J.R., ed., Field guide to geologic excursions in Utah and adjacent areas of Nevada, Idaho, and Wyoming; Geological Society of America, Rocky Mountain Section Meeting, Ogden, Utah, 1992: Utah Geological Survey Miscellaneous Publication 92–3, p. 399–420.

Bohannon, R.G., 1984, Nonmarine sedimentary rocks of Tertiary age in the Lake Mead region, southeastern Nevada and northwestern Arizona: U.S. Geological Survey Professional Paper 1259, 69 p.

Bowers, W.E., 1990, Geologic map of Bryce Canyon National Park and vicinity, southwestern Utah: U.S. Geological Survey Miscellaneous Investigations Series Map I–2108, scale 1:24,000.

Brown, C.L., and Bartley, J.M., 1994, Multiple-phase extension in the Sheeprock and Simpson Mountains of west-central Utah: Geological Society of America Abstracts with Programs, v. 26, no. 2, p. 41.

Budding, K.E., Cunningham, C.G., Zielinski, R.A., Steven, T.A., and Stern, C.R., 1987, Petrology and chemistry of the Joe Lott Tuff Member of the Mount Belknap Volcanics, Marysvale volcanic field, west-central Utah: U.S. Geological Survey Professional Paper 1354, 47 p.

Burke, D.B., and McKee, E.H., 1979, Mid-Cenozoic volcano-tectonic troughs in central Nevada: Geological Society of America Bulletin, v. 90, p. 181–184.

Burke, K.J., 1991, Tertiary extensional tectonism in the northern Chief and southernmost Highland Ranges, Lincoln County, Nevada: Flagstaff, Ariz., Northern Arizona University M.S. thesis, 91 p.

Bursik, Marcus, and Sieh, Kerry, 1989, Range front faulting and volcanism in the Mono Basin, eastern California: Journal of Geophysical Research, v. 94, no. B11, p. 15587–15609.

Callaghan, Eugene, 1938, Preliminary report on the alunite deposits of the Marysvale region, Utah: U.S. Geological Survey Bulletin 886–D, p. 91–134.

———1939, Volcanic sequence in the Marysvale region in south-west-central Utah: American Geophysical Union Transactions, 20th Annual Meeting, Washington, D.C., pt. 3, p. 438–452.

Campbell, D.L., Cunningham, C.G., Steven, T.A., Rowley, P.D., Glassgold, L.B., and Anderson J.J., 1984, Aeromagnetic map on a geologic base map of the Tushar Mountains and adjoining areas, Marysvale volcanic field, Utah: U.S. Geological Survey Miscellaneous Investigations Series Map I–1430–D, scale 1:50,000.

Catchings, R.D., 1992, A relation among geology, tectonics, and velocity structure, western to central Nevada Basin and Range: Geological Society of America Bulletin, v. 104, p. 1178–1192.

Catchings, R.D., and Mooney, W.D., 1993, Geophysical indicators of magmatism and tectonism in the Basin and Range: Geological Society of America Abstracts with Programs, v. 25, no. 5, p. 19.

Christiansen, R.L., and Lipman, P.W., 1972, Cenozoic volcanism and plate-tectonic evolution of the Western United States—II, Late Cenozoic: Philosophical Transactions of Royal Society of London, v. A271, p. 249–284.

Christiansen, R.L., and McKee, E.H., 1978, Late Cenozoic volcanic and tectonic evolution of the Great Basin and Columbia intermontane regions, *in* Smith, R.B., and Eaton, G.P., eds., Cenozoic tectonics and regional geophysics of the western Cordillera: Geological Society of America Memoir 152, p. 283–311.

Christiansen, R.L., and Yeats, R.S., 1992, Post-Laramide geology of the U.S. Cordilleran region, *in* Burchfiel, B.C., Lipman, P.W., and Zoback, M.L., eds., The Cordilleran orogen—Conterminous U.S.: Geological Society of America, The Geology of North America, v. G–3, p. 261–406.

Coleman, D.S., and Walker, J.D., 1992, Evidence for the generation of juvenile granitic crust during continental extension, Mineral Mountains batholith, Utah: Journal of Geophysical Research, v. 97, no. B7, p. 11011–11024.

Condit, C.D., Crumpler, L.S., Aubele, J.C., and Elston, W.E., 1989, Patterns of volcanism along the southern margin of the Colorado Plateau—The Springerville field: Journal of Geophysical Research, v. 94, no. B6, p. 7975–7986.

Coney, P.J., and Reynolds, S.J., 1977, Cordilleran Benioff zones: Nature, v. 270, p. 403–406.

Cook, E.F., 1957, Geology of the Pine Valley Mountains, Utah: Utah Geological and Mineralogical Survey Bulletin 58, 111 p.

Cook, K.L., Bankey, Viki, and Mabey, D.R., 1990, Patterns of geologic structure determined from new gravity map of Utah: Society of Exploration Geophysicists, 60th Annual International Meeting, San Francisco, Sept. 23–27, 1990, Abstracts, p. 670–672.

Cook, K.L., Bankey, Viki, Mabey, D.R., and DePangher, Michael, 1989, Complete Bouguer gravity anomaly map of Utah: Utah Geological and Mineral Survey Map 122, scale 1:500,000.

Cook, K.L., Halliday, M.E., Cunningham, C.G., Steven, T.A., Rowley, P.D., Glassgold, L.B., Anderson, J.J., and Coles, L.L., 1984, Complete Bouguer gravity anomaly map on a geologic base map of the Tushar Mountains and adjoining areas, Marysvale volcanic field, Utah: U.S. Geological Survey Miscellaneous Investigations Series Map I–1430–C, scale 1:50,000.

Corry, C.E., 1988, Laccoliths—Mechanics of emplacement and growth: Geological Society of America Special Paper 220, 110 p.

Crittenden, M.D., Jr., Coney, P.J., and Davis, G.H., eds., 1980, Cordilleran metamorphic core complexes: Geological Society of America Memoir 153, 490 p.

Cross, T.A., and Pilger, R.H., Jr., 1978, Constraints on absolute motion and plate interaction inferred from Cenozoic igneous activity in the Western United States: American Journal of Science, v. 278, p. 865–902.

Cunningham, C.G., Ludwig, K.R, Naeser, C.W., Weiland, E.K., Mehnert, H.H., Steven, T.A., and Rasmussen, J.D., 1982,

Geochronology of hydrothermal uranium deposits and associated igneous rocks in the eastern source area of the Mount Belknap Volcanics, Marysvale, Utah: Economic Geology, v. 77, no. 2, p. 453–463.

Cunningham, C.G., Rye, R.O., Steven, T.A., and Mehnert, H.H., 1984, Origins and exploration significance of replacement and vein-type alunite deposits in the Marysvale volcanic field, west-central Utah: Economic Geology, v. 79, p. 50–71.

Cunningham, C.G., and Steven, T.A., 1979a, Mount Belknap and Red Hills calderas and associated rocks, Marysvale volcanic field, west-central Utah: U.S. Geological Survey Bulletin 1468, 34 p.

———1979b, Uranium in the central mining area, Marysvale district, west-central Utah: U.S. Geological Survey Miscellaneous Investigations Series Map I–1177, scale 1:24,000.

———1979c, Geologic map of the Deer Trail Mountain–Alunite Ridge mining area, west-central Utah: U.S. Geological Survey Miscellaneous Investigations Series Map I–1230, scale 1:24,000.

Cunningham, C.G., Steven, T.A., Rowley, P.D., Glassgold, L.B., and Anderson, J.J., 1983, Geologic map of the Tushar Mountains and adjoining areas, Marysvale volcanic field, Utah: U.S. Geological Survey Miscellaneous Investigations Series Map I–1430–A, scale 1:50,000.

Davis, G.A., and Burchfiel, B.C., 1973, Garlock fault—An intracontinental transform structure, southern California: Geological Society of America Bulletin, v. 84, p. 1407–1422.

Davis, G.A., and Lister, G.S., 1988, Detachment faulting in continental extension—Perspectives from the southwestern U.S. Cordillera: Geological Society of America Special Paper 218, p. 133–159.

Davis, G.H., and Krantz, R.W., 1986, "Post-Laramide" thrust faults in the Claron Formation, Bryce Canyon National Park, Utah: Geological Society of America Abstracts with Programs, v. 18, no. 5, p. 98.

Davis, G.H., and Rowley, P.D., 1993, Miocene thrusting, gravity sliding, and near-surface batholithic emplacement, Marysvale volcanic field, southwestern Utah [abs.]: Eos, v. 74, no. 43, p. 647.

Delaney, P.T., Pollard, D.D., Ziony, J.I., and McKee, E.H., 1986, Field relations between dikes and joints—Emplacement processes and paleostress analysis: Journal of Geophysical Research, v. 91, no. B5, p. 4920–4938.

Dickinson, W.R., and Snyder, W.S., 1979, Geometry of subducted slabs related to San Andreas transform: Journal of Geology, v. 87, p. 609–627.

Dilles, J.H., and Gans, P.B., 1995, The chronology of Cenozoic volcanism and deformation in the Yerington area, western Basin and Range and Walker Lake: Geological Society of America Bulletin, v. 107, p. 474–486.

Duebendorfer, E.M., and Black, R.A., 1992, Kinematic role of transverse structures in continental extension—An example from the Las Vegas Valley shear zone, Nevada: Geology, v. 20, p. 1107–1110.

Duebendorfer, E.M., and Simpson, D.A., 1994, Kinematics and timing of Tertiary extension in the western Lake Mead region, Nevada: Geological Society of America Bulletin, v. 106, p. 1057–1073.

Eaton, G.P., 1979a, A plate-tectonic model for late Cenozoic crustal spreading in the Western United States, *in* Riecker, R.E., ed., Rio Grande rift—Tectonics and magmatism: Washington, D.C., American Geophysical Union, p. 7–32.

———1979b, Regional geophysics, Cenozoic tectonics, and geologic resources of the Basin and Range province and adjoining regions, *in* Newman, G.W., and Goode, H.D., eds., Basin and Range Symposium and Great Basin Field Conference: Denver, Colo., Rocky Mountain Association of Geologists, p. 11–39.

———1980, Geophysical and geological characteristics of the crust of the Basin and Range province *in* Continental tectonics, studies in geophysics: Washington, D.C., National Academy of Sciences, p. 96–113.

———1982, The Basin and Range province—Origin and tectonic significance: Annual Reviews of Earth and Planetary Sciences, v. 10, p. 409–440.

———1984, The Miocene Great Basin of western North America as an extending back-arc region: Tectonophysics, v. 102, p. 275–295.

———1987, Topography and origin of the southern Rocky Mountains and Alvarado ridge, *in* Coward, M.P., Dewey, J.F., and Hancock, P.L., eds., Continental extensional tectonics: Geological Society of London Special Publication 28, p. 355–369.

Eaton, G.P., Christiansen, R.L., Iyer, H.M , Pitt, A.M., Mabey, D.R., Blank, H.R., Jr., Zietz, Isidore, and Gettings, M.E., 1975, Magma beneath Yellowstone National Park: Science, v. 188, p. 787–796.

Eaton, G.P., Wahl, R.R., Prostka, H.J., Mabey, D.R., and Kleinkopf, M.D., 1978, Regional gravity and tectonic patterns—Their relation to late Cenozoic epeirogeny and lateral spreading in the western Cordillera, *in* Smith, R.B., and Eaton, G.P., eds., Cenozoic tectonics and regional geophysics of the western Cordillera: Geological Society of America Memoir 152, p. 51–91.

Ekren, E.B., Anderson, R.E., Rogers, C.L., and Noble, D.C., 1971, Geology of northern Nellis Air Force Base Bombing and Gunnery Range, Nye County, Nevada: U.S. Geological Survey Professional Paper 651, 91 p.

Ekren, E.B., Bucknam, R.C., Carr, W.J., Dixon, G.L., and Quinlivan, W.D., 1976, East-trending structural lineaments in central Nevada: U.S. Geological Survey Professional Paper 986, 16 p.

Ekren, E.B., McIntyre, D.H., and Bennett, E.H., 1984, High-temperature, large-volume, lavalike ash-flow tuffs without calderas in southwestern Idaho: U.S. Geological Survey Professional Paper 1272, 76 p.

Ekren, E.B., Orkild, P.P., Sargent, K.A., and Dixon, G.L., 1977, Geologic map of Tertiary rocks, Lincoln County, Nevada: U.S. Geological Survey Miscellaneous Investigations Series Map I–1041, scale 1:250,000.

Ekren, E.B., Rogers, C.L., Anderson, R.E., and Orkild, P.P., 1968, Age of Basin and Range normal faults in Nevada Test Site and Nellis Air Force Range, Nevada, *in* Eckel, E.B., ed., Nevada Test Site: Geological Society of America Memoir 110, p. 247–250.

Ellis, Michael, and King, Geoffrey, 1991, Structural control of flank volcanism in continental rifts: Science, v. 254, p. 839–842.

Elston, W.E., 1984, Subduction of young oceanic lithosphere and extensional orogeny in southwestern North America during mid-Tertiary time: Tectonics, v. 3, no. 2, p 229–250.

Erslev, E.A., and Sutter, J.F., 1990, Evidence for Proterozoic mylonitization in the northwestern Wyoming province: Geological Society of America Bulletin, v. 102, p. 1681–1694.

Evans, S.H., Jr., and Steven, T.A., 1982, Rhyolites in the Gillies Hill–Woodtick Hill area, Beaver County, Utah: Geological Society of America Bulletin, v. 93, p. 1131–1141.

Faulds, J.E., 1992, From accommodation zones to metamorphic core complexes—Tracking the progressive development of major normal fault systems: Geological Society of America Abstracts with Programs, v. 24, no. 7, p. A158.

———1993, The Mt. Perkins block, northwestern Arizona—An exposed cross section of a synextensional volcano in highly extended terrane: Geological Society of America Abstracts with Programs, v. 25, no. 5, p. 36.

———1994, New insights on the geometry and kinematics of the Black Mountains–Highland Spring Range accommodation zone (BHZ), Arizona and Nevada: Geological Society of America Abstracts with Programs, v. 26, no. 2, p. 51.

Faulds, J.E., Gans, P.B., and Smith, E.I., 1994, Spatial and temporal patterns of extension in the northern Colorado River extensional corridor, northwestern Arizona and southern Nevada: Geological Society of America Abstracts with Programs, v. 26, no. 2, p. 51.

Faulds, J.E., Geissman, J.W., and Mawer, C.K., 1990, Structural development of a major extensional accommodation zone in the Basin and Range province, northeastern Arizona and southern Nevada—Implications for kinematic models of continental extension, *in* Wernicke, Brian, ed., Basin and Range extensional tectonics near the latitude of Las Vegas, Nevada: Geological Society of America Memoir 176, p. 37–76.

Faulds, J.E., Olson, E.L., and Littrell, R.L., 1994, Magmatic origin of rupture barriers and accommodation zones in extensional orogens—Analogies between continental and oceanic rifts [abs.]: Eos, v. 75, no. 44, p. 678.

Fenneman, N.M., 1931, Physiography of Western United States: New York, McGraw-Hill Book Company, Inc., 534 p.

Ferguson, C.A., 1990, Localized high extension related to emplacement of a shallow synvolcanic pluton, Rio Grande Rift, New Mexico: Geological Society of America Abstracts with Programs, v. 22, no. 3, p. 22.

Fitton, J.G., James, Dodie, and Leeman, W.P., 1991, Basic magmatism associated with late Cenozoic extension in the Western United States—Compositional variations in space and time: Journal of Geophysical Research, v. 96, no. B8, p. 13693–13711.

Fleck, R.J., 1970, Age and possible origin of the Las Vegas Valley shear zone, Clark and Nye Counties, Nevada: Geological Society of America Abstracts with Programs, v. 2, no. 5, p. 333.

Fleck, R.J., Anderson, J.J., and Rowley, P.D., 1975, Chronology of mid-Tertiary volcanism in High Plateaus region of Utah, *in* Anderson, J.J., Rowley, P.D., Fleck, R.J., and Nairn, A.E.M., Cenozoic geology of southwestern High Plateaus of Utah: Geological Society of America Special Paper 160, p. 53–62.

Fuller, M.D., 1964, Expression of E-W fractures in magnetic surveys in parts of the U.S.A.: Geophysics, v. 29, p. 602–622.

Gans, P.B., 1987, An open-system, two-layer crustal stretching model for the eastern Great Basin: Tectonics, v. 6, p. 1–12.

Gans, P.B., Mahood, G.A., and Schermer, Elizabeth, 1989, Synextensional magmatism in the Basin and Range province—A case study from the eastern Great Basin: Geological Society of America Special Paper 233, 53 p.

Gans, P.B., Miller, E.L., McCarthy, J., and Ouldcott, M.L., 1985, Tertiary extensional faulting and evolving ductile-brittle transition zones in the northern Snake Range and vicinity—New insights from seismic data: Geology, v. 13, p. 189–193.

Gilbert, G.K., 1877, Report on the geology of the Henry Mountains: U.S. Geographical and Geological Survey of the Rocky Mountains Region (Powell), 160 p.

———1928, Basin Range faulting along the Oquirrh Range, Utah: Geological Society of America Bulletin, v. 39, p. 1103–1130.

Gill, J.B., 1981, Orogenic andesites and plate tectonics: Berlin, Springer-Verlag, 390 p.

Glazner, A.F., and Ussler, William, III, 1989, Crustal extension, crustal density, and the evolution of Cenozoic magmatism in the Basin and Range of the Western United States: Journal of Geophysical Research, v. 94, no. B6, p. 7952–7960.

Grant, S.K., and Proctor, P.D., 1988, Geologic map of the Antelope Peak quadrangle, Iron County, Utah: Utah Geological and Mineral Survey Open-File Report 130, 20 p., scale 1:20,000.

Hamilton, Warren, 1985, Subduction, magmatic arcs, and foreland deformation, *in* Howell, D.G., ed., Tectonostratigraphic terranes of the circum-Pacific region: Houston, Texas, Circum-Pacific Council for Energy and Mineral Resources, Earth Science Series, no. 1, p. 259–262.

———1987, Crustal extension in the Basin and Range province, Southwestern United States, *in* Coward, M.P., Dewey, J.F., and Hancock, P.L., eds., Continental extensional tectonics: Geological Society Special Publication 28, p. 155–176.

———1988a, Tectonic setting and variations with depth of some Cretaceous and Cenozoic structural and magmatic systems of the Western United States, *in* Ernst, W.G., ed., Metamorphism and crustal evolution of the Western United States: Englewood Cliffs, N.J., Prentice Hall, p. 1–40.

———1988b, Detachment faulting in the Death Valley region, California and Nevada, *in* Carr, M.D., and Yount, J.C., eds., Geologic and hydrologic investigations of a potential nuclear waste disposal site at Yucca Mountain, southern Nevada: U.S. Geological Survey Bulletin 1790, p. 51–86.

———1988c, Plate tectonics and island arcs: Geological Society of America Bulletin, v. 100, p. 1503–1527.

———1989, Crustal geologic processes of the United States, *in* Pakiser, L.C., and Mooney, W.D., eds., Geophysical framework of the continental United States: Geological Society of America Memoir 172, p. 743–781.

———1995, Subduction systems and magmatism, *in* Smellie, J.L., ed., Volcanism associated with extension at consuming plate margins: Geological Society Special Publication 81, p. 3–28.

Hamilton, Warren, and Myers, W.B., 1966, Cenozoic tectonics of the Western United States: Reviews of Geophysics, v. 4, p. 509–549.

Hardyman, R.F., and Oldow, J.S., 1991, Tertiary tectonic framework and Cenozoic history of the central Walker Lane, Nevada, *in* Raines, G.L., Lisle, R.E., Schafer, R.W., and Wilkinson, W.H., eds., Geology and ore deposits of the Great Basin: Geological Society of Nevada Symposium Proceedings, v. 1, p. 279–301.

Haxel, G.B., Simmons, A.M., and McCarthy, J., 1990, Synextensional dioritic magmatism within and beneath metamorphic core complexes, southern Arizona and southeastern

California—Integration of structural, petrologic, and seismic refraction data: Geological Society of America Abstracts with Programs, v. 22, no. 3, p. 28–29.

Helmstaedt, H.H., and Schulze, D.J., 1991, Early to mid-Tertiary inverted metamorphic gradient under the Colorado Plateau—Evidence from eclogite xenoliths in ultramafic microbreccias, Navajo volcanic field: Journal of Geophysical Research, v. 96, no. B8, p. 13225–13235.

Henderson, L.J., Gordon, R.G., and Engebretson, D.C., 1984, Mesozoic aseismic ridges on the Farallon plate and southward migration of shallow subduction during the Laramide orogeny: Tectonics, v. 3, p. 121–132.

Henley, R.W., and Adams, D.P.M., 1992, Strike-slip fault reactivation as a control on epithermal vein-style gold mineralization: Geology, v. 20, p. 443–446.

Henry, C.D., and Aranda-Gomez, J.J., 1992, The real southern Basin and Range—Mid- to late Cenozoic extension in Mexico: Geology, v. 20, p. 701–704.

Henry, C.D., Muehlberger, W.R., and Price, J.G., 1991, Igneous and structural evolution of the Solitario laccocaldera, Trans-Pecos Texas: Geological Society of America Abstracts with Programs, v. 23, no. 5, p. A451.

Henry, C.D., and Price, J.G., 1988, Laccocaldera—A new caldera type from Trans-Pecos Texas: Geological Society of America Abstracts with Programs, v. 20, no. 7, p. 113.

Hildenbrand, T.G., and Kucks, R.P., 1988a, Total intensity magnetic anomaly map of Nevada: Nevada Bureau of Mines and Geology Map 93A, scale 1:750,000.

———1988b, Filtered magnetic anomaly maps of Nevada: Nevada Bureau of Mines and Geology Map 93B, scale 1:1,000,000.

Hilpert, L.S., and Roberts, R.J., 1964, Geology—Economic geology, *in* Mineral and water resources of Utah: U.S. Congress, 88th, 2d session, Committee Print, p. 28–38.

Hudson, M.R., and Rosenbaum, J.G., 1994, Broad transfer zones—Examples of vertical-axis rotation and discontinuous faulting in the Great Basin: Geological Society of America Abstracts with Programs, v. 26, no. 3, p. 61.

Hudson, M.R., Rosenbaum, J.G., Scott, R.B., Rowley, P.D., and Grommé, C.S., 1993, Paleomagnetic evidence for counterclockwise rotation in an extensional transfer zone, southeastern Great Basin [abs.]: Eos, v. 74, no. 43, p. 207–208.

———1995, The western Delamar–Iron Springs belt—A broad zone of sinistral shear in the Basin and Range province, U.S.A. [abs.]: International Union of Geodesy and Geophysics, 21st Assembly, Boulder, Colo., week B, p. B185.

Humphreys, E.D., and Dueker, K.G., 1994, Physical state of the Western U.S. upper mantle: Journal of Geophysical Research, v. 99, no. B5, p. 9635–9650.

Hunt, C.B., 1958, Structural and igneous geology of the La Sal Mountains, Utah: U.S. Geological Survey Professional Paper 294–I, p. 305–364.

———1980, Structural and igneous geology of the Henry Mountains, Utah, *in* Picard, M.D., ed., Henry Mountains Symposium: Utah Geological Association Publication 8, p. 25–106.

Hunt, G.L., 1988, Petrology of the Mt. Pennell central stock, Henry Mountains, Utah: Brigham Young University Geology Studies, v. 35, p. 81–100.

Hurtubise, D.O., 1989, Stratigraphy and structure of the Seaman Range and Fox Mountain area, Lincoln and Nye Counties, Nevada, with an emphasis on the Devonian System: Golden, Colo., Colorado School of Mines Ph. D. dissertation, 443 p.

———1994, Silver King lineament, the missing link of a 500 km east-trending structure in the southern Great Basin, *in* Dobbs, S.W., and Taylor, W.J., eds., Structural and stratigraphic investigations and petroleum potential of Nevada, with special emphasis south of the Railroad Valley producing trend: Nevada Petroleum Society, Conference Volume 2, p. 127–139.

Hutton, D.H.W., 1988, Granite emplacement mechanisms and tectonic controls—Inferences from deformational studies: Transactions of the Royal Society of Edinburgh, Earth Sciences, v. 79, p. 245–255.

John, D.A., Nash, J.T., Clark, C.W., and Wulftange, N.H., 1991, Geology, hydrothermal alteration, and mineralization at the Paradise Peak gold-silver-mercury deposit, Nye County, Nevada, *in* Raines, G.L., Lisle, R.E., Schafer, R.W., and Wilkinson, W.H., eds., Geology and ore deposits of the Great Basin: Geological Society of Nevada Symposium Proceedings, v. 1, p. 1020–1050.

John, D.A., Thomason, R.E., and McKee, E.H., 1989, Geology and K-Ar geochronology of the Paradise Peak mine and the relationship of pre-basin and range extension to early Miocene precious metal mineralization in west-central Nevada: Economic Geology, v. 84, p. 631–649.

Johnson, C.M., 1991, Large-scale crust formation and lithosphere modification beneath middle to late Cenozoic calderas and volcanic fields, western North America: Journal of Geophysical Research, v. 96, no. B8, p. 13485–13507.

Kowallis, B.J., and Best, M.G., 1990, Fission track ages from volcanic rocks in southwestern Utah and southeastern Nevada: Isochron/West, no. 55, p. 24–27.

Kuntz, M.A., 1992, A model-based perspective of basaltic volcanism, eastern Snake River Plain, Idaho, *in* Link, P.K., Kuntz, M.A., and Platt, L.B., eds., Regional geology of eastern Idaho and western Wyoming: Geological Society of America Memoir 179, p. 289–304.

Kuntz, M.A., Covington, H.R., and Schorr, L.J., 1992, An overview of basaltic volcanism of the eastern Snake River Plain, Idaho, *in* Link, P.K., Kuntz, M.A., and Platt, L.B., eds., Regional geology of eastern Idaho and western Wyoming: Geological Society of America Memoir 179, p. 227–267.

Lachenbruch, A.H., and Sass, J.H., 1978, Models of an extending lithosphere and heat flow in the Basin and Range province, *in* Smith, R.B., and Eaton, G.P., eds., Cenozoic tectonics and regional geophysics of the western Cordillera: Geological Society of America Memoir 152, p. 209–250.

Lachenbruch, A.H., Sass, J.H., and Morgan, Paul, 1994, Thermal regime of the southern Basin and Range province—2, Implications of heat flow for regional extension and metamorphic core complexes: Journal of Geophysical Research, v. 99, no. B11, p. 22121–22133.

Lawrence, R.D., 1976, Strike-slip faulting terminates the Basin and Range province in Oregon: Geological Society of America Bulletin, v. 87, p. 846–850.

Le Bas, M.J., Le Maitre, R.W., Streckeisen, A., and Zanettin, B., 1986, A chemical classification of volcanic rocks based on the total alkali-silica diagram: Journal of Petrology, v. 27, pt. 3, p. 745–750.

Li, Yianping, Geissman, J.W., Nur, Amos, Ron, Hagai, and Huang, Qing, 1990, Paleomagnetic evidence for counterclockwise block rotation in the north Nevada rift region: Geology, v. 18, p. 79–82.

Liberty, L.M., Heller, P.L., and Smithson, S.B., 1994, Seismic reflection evidence for two-phase development of Tertiary basins from east-central Nevada: Geological Society of America Bulletin, v. 106, p. 1621–1633.

Liggett, M.A., and Childs, J.F., 1977, An application of satellite imagery to mineral exploration, *in* Woll, P.W., and Fischer, W.A., eds., Proceedings of the first annual William T. Pecora Memorial Symposium: U.S. Geological Survey Professional Paper 1015, p. 253–270.

Lipman, P.W., 1980, Cenozoic volcanism in the Western United States—Implications for continental tectonics, *in* Studies in geophysics—Continental tectonics: Washington, D.C., National Academy of Sciences, p. 161–174.

———1988, Evolution of silicic magma in the upper crust—The mid-Tertiary Latir volcanic field and its cogenetic granitic batholith, northern New Mexico, U.S.A.: Transactions of the Royal Society of Edinburgh, v. 79, p. 265–288.

———1992, Magmatism in the Cordilleran United States—Progress and problems, *in* Burchfiel, B.C., Lipman, P.W., and Zoback, M.L., eds., The Cordilleran orogen—Conterminous U.S.: Geological Society of America, The Geology of North America, v. G–3, p. 481–514.

Lipman, P.W., Prostka, H.J., and Christiansen, R.L., 1972, Cenozoic volcanism and plate-tectonic evolution of the Western United States—I, Early and middle Cenozoic: Philosophical Transactions of Royal Society of London, ser. A, v. 271, p. 217–248.

Lipman, P.W., Rowley, P.D., Mehnert, H.H., Evans, S.H., Jr., Nash, W.P., and Brown, F.H., 1978, Pleistocene rhyolite of the Mineral Mountains, Utah—Geothermal and archeological significance, *with sections on* Fission-track dating, by G.A. Izett and C.W. Naeser, *and on* Obsidian-hydration dating, by Irving Friedman: U.S. Geological Survey Journal of Research, v. 6, no. 1, p. 133–147.

Lister, G.S., and Baldwin, S.L., 1993, Plutonism and the origin of metamorphic core complexes: Geology, v. 21, p. 607–610.

Lucchitta, Ivo, 1990, Role of heat and detachment in continental extension as viewed from the eastern Basin and Range province in Arizona: Tectonophysics, v. 174, p. 77–114.

Lucchitta, Ivo, and Suneson, N.H., 1993, Dips and extension: Geological Society of America Bulletin, v. 105, p. 1346–1356.

Luedke, R.G., and Smith, R.L., 1981, Map showing distribution, composition, and age of late Cenozoic volcanic centers in California and Nevada: U.S. Geological Survey Miscellaneous Investigations Series Map I–1091–C, scale 1:1,000,000.

Lundin, E.R., 1989, Thrusting of the Claron Formation, Bryce Canyon region, Utah: Geological Society of America Bulletin, v. 101, p. 1038–1050.

Lundin, E.R., and Davis, G.H., 1987, Southeast vergent thrust faulting and folding of the Eocene(?) Claron Formation, Bryce Canyon National Park, Utah: Geological Society of America Abstracts with Programs, v. 19, no. 5, p. 317.

Mabey, D.R., Zietz, Isidore, Eaton, G.P., and Kleinkopf, M.D., 1978, Regional magnetic patterns in part of the Cordillera in the Western United States, *in* Smith, R.B., and Eaton, G.P., Cenozoic tectonics and regional geophysics of the western Cordillera: Geological Society of America Memoir 152, p. 93–106.

Machette, M.N., 1985, Late Cenozoic geology of the Beaver Basin, southwestern Utah: Brigham Young University Studies in Geology, v. 32, pt. 1, p. 19–37.

Machette, M.N., Steven, T.A., Cunningham, C.G., and Anderson, J.J., 1984, Geologic map of the Beaver quadrangle, Beaver and Piute Counties, Utah: U.S. Geological Survey Miscellaneous Investigations Series Map I–1520, scale 1:50,000.

Mackin, J.H., 1947, Some structural features of the intrusions in the Iron Springs district: Utah Geological Society Guidebook 2, 62 p.

———1960, Structural significance of Tertiary volcanic rocks in southwestern Utah: American Journal of Science, v. 258, p. 81–131.

MacLeod, N.S., Walker, G.W., and McKee, E.H., 1976, Geothermal significance of eastward increase in age of upper Cenozoic rhyolitic domes in southeastern Oregon, *in* Second United Nations Symposium on the Development and Use of Geothermal Resources: Washington, D.C., U.S. Government Printing Office, v. 1, p. 465–474.

Maldonado, Florian, 1995, Decoupling of mid-Tertiary rocks, Red Hills–western Markagunt Plateau, southwestern Utah, *in* Scott, R.B., and Swadley, W C, eds., Geologic studies in the Basin and Range–Colorado Plateau transition in southeastern Nevada, southwestern Utah, and northwestern Arizona, 1992: U.S. Geological Survey Bulletin 2056–I, p. 235–254.

Maldonado, Florian, Sable, E.G., and Anderson, J.J., 1989, Evidence for shallow detachment faulting of mid-Tertiary rocks, Red Hills (Basin and Range), with implications for a more extensive detachment zone in the adjacent Markagunt Plateau (Colorado Plateau), southwest Utah [abs.]: Eos, v. 70, no. 43, p. 1336.

———1990, Shallow detachment of mid-Tertiary rocks, Red Hills (Basin and Range), with implications for a regional detachment zone in the adjacent Markagunt Plateau (Colorado Plateau), southwest Utah: Geological Society of America Abstracts with Programs, v. 22, no. 3, p. 63.

———1992, Evidence for a Tertiary low-angle shear zone, Red Hills, Utah, with implications for a regional shear zone in the adjacent Colorado Plateau, *in* Harty, K.M., ed., Engineering and environmental geology of southwestern Utah: Utah Geological Association Publication 21, p. 315–324.

Mandl, G., 1987, Tectonic deformation by rotating parallel faults—The "bookshelf" mechanism: Tectonophysics, v. 141, p. 177–316.

Mattox, S.R., 1991a, Origin of potassium-rich mafic lava flows, Marysvale volcanic field, Utah: AGU 1991 Fall Meeting, San Francisco, Program and Abstracts (supplement to Eos, v. 72, no. 44), p. 561.

———1991b, Petrology, age, geochemistry, and correlation of the Tertiary volcanic rocks of the Awapa Plateau, Garfield, Piute, and Wayne Counties, Utah: Utah Geological Survey Miscellaneous Publication 91–5, 46 p.

———1992, Geochemistry, origin, and tectonic implications of mid-Tertiary and late Tertiary and Quaternary volcanic rocks, southern Marysvale volcanic field, Utah: DeKalb, Ill., Northern Illinois University Ph. D. dissertation, 502 p.

Mattox, S.R., and Walker, J.A., 1989, Geochemistry and tectonism of lavas erupted during the transition from compression to extension, southern Marysvale complex, southwestern Utah, *in* Continental magmatism; International Association of

Volcanology and Chemistry of the Earth's Interior, General Assembly, Santa Fe, N. Mex., 1989, Abstracts: New Mexico Bureau of Mines and Mineral Resources Report 131, p. 178.

———1990, Late Cenozoic lavas of the Utah transition zone: Geological Society of America Abstracts with Programs, v. 22, no. 3, p. 65.

May, S.J., Kelley, S.A., and Russell, L.R., 1994, Footwall unloading and rift shoulder uplifts in the Albuquerque Basin—Their relation to syn-rift fanglomerates and apatite fission-track ages, *in* Keller, G.R., and Cather, S.M., eds., Basins of the Rio Grande Rift—Structure, stratigraphy, and tectonic setting: Geological Society of America Special Paper 291, p. 125–134.

McCarthy, Jill, and Parsons, Tom, 1994, Insights into the kinematic Cenozoic evolution of the Basin and Range–Colorado Plateau transition from coincident seismic refraction and reflection data: Geological Society of America Bulletin, v. 106, p. 747–759.

McDonald, R.E., 1976, Tertiary tectonics and sedimentary rocks along the transition—Basin and Range province to plateau and thrust belt province, Utah, *in* Hill, J.G., ed., Symposium on geology of the Cordilleran hingeline: Denver, Colo., Rocky Mountain Association of Geologists, p. 281–317.

McIntosh, W.C., Chapin, C.E., Ratté, J.C., and Sutter, J.F., 1992, Time-stratigraphic framework for the Eocene-Oligocene Mogollon-Datil volcanic field, southwest New Mexico: Geological Society of America Bulletin, v. 104, p. 851–871.

McKee, E.H., 1971, Tertiary igneous chronology of the Great Basin of Western United States—Implications for tectonic models: Geological Society of America Bulletin, v. 82, p. 3497–3502.

McKee, E.H., and Blakely, R.J., 1990, Tectonic significance of linear, north-trending anomalies in north-central Nevada, *in* Geology and ore deposits of the Great Basin—Program with abstracts: Reno, Nev., Geological Society of Nevada, p. 49.

McKee, E.H., and Noble, D.C., 1986, Tectonic and magmatic development of the Great Basin of Western United States during late Cenozoic time: Modern Geology, v. 10, p. 39–49.

McKee, E.H., Noble, D.C., and Silverman, M.L., 1970, Middle Miocene hiatus in volcanic activity in the Great Basin area of the Western United States: Earth and Planetary Science Letters, v. 8, no. 2, p. 93–96.

Menard, H.W., and Chase, T.E., 1970, Fracture zones, *in* The sea—Ideas and observations on progress in the study of the seas: New York, Wiley-Interscience, v. 4, pt. I, p. 421–443.

Merle, O.R., Davis, G.H., Nickelsen, R.P., and Gourlay, P.A., 1993, Relation of thin-skinned thrusting of Colorado Plateau strata in southwestern Utah to Cenozoic magmatism: Geological Society of America Bulletin, v. 105, p. 387–398.

Meyer, Jeff, and Foland, K.A., 1991, Magmatic-tectonic interaction during early Rio Grande Rift extension at Questa, New Mexico: Geological Society of America Bulletin, v. 103, p. 993–1006.

Michel-Noël, G., Anderson, R.E., and Angelier, Jacques, 1990, Fault kinematics and estimates of strain partitioning of a Neogene extensional fault system in southeastern Nevada, *in* Wernicke, B.P., ed., Basin and Range extensional tectonics near the latitude of Las Vegas, Nevada: Geological Society of America Memoir 176, p. 155–180.

Miller, D.M., 1991, Mesozoic and Cenozoic tectonic evolution of the northeastern Great Basin, *in* Buffa, R.H., and Coyner, A.R., eds., Geology and ore deposits of the Great Basin: Reno, Nev., Geological Society of Nevada, p. 202–228.

Miller, E.L., Gans, P.B., and Garing, John, 1983, The Snake Range décollement—An exhumed mid-Tertiary ductile-brittle transition: Tectonics, v. 2, no. 3, p. 239–263.

Minor, S.A., 1995, Superposed local and regional paleostresses—Fault-slip analysis of Neogene extensional faulting near coeval caldera complexes, Yucca Flat Nevada: Journal of Geophysical Research, v. 100, no. B6, p. 10507–10528.

Moos, Daniel, and Zoback, M.D., 1993, State of stress in the Long Valley caldera, California: Geology, v. 21, p. 837–840.

Mutschler, F.E., Johnson, D.C., and Mooney, T.C., 1991, A speculative plate tectonic model for the central Montana alkalic province and related gold deposits, *in* Baker, D.W., and Berg, R.B., eds., Guidebook of the central Montana alkalic province—Geology, ore deposits and origin: Montana Bureau of Mines and Geology Special Paper 100, p. 121–124.

Mutschler, F.E., Larson, E.E., and Bruce, R.M., 1987, Laramide and younger magmatism in Colorado—New petrologic and tectonic variations on old themes: Colorado School of Mines Quarterly, v. 82, no. 4, p. 1–47.

Nealey, L.D., Maldonado, Florian, Unruh, D.M., and Budahn, J.R., 1994, Tectonic implications of Quaternary volcanism in the western Markagunt Plateau and Red Hills area, southwestern Utah—Geochemical and geochronological evidence, *in* Blackett, R.E., and Moore, J.N., eds., Cenozoic geology and geothermal systems of southwestern Utah, 1994 Field Symposium: Utah Geological Association Publication 23, p. 117–124.

Nelson, S.T., Davidson, J.P., and Sullivan, K.R., 1992, New age determinations of central Colorado Plateau laccoliths, Utah—Recognizing disturbed K-Ar systematics and re-evaluating tectonomagmatic relationships: Geological Society of America Bulletin, v. 104, p. 1547–1560.

Nickelsen, R.P., and Merle, O., 1991, Structural evolution at the tip line of a mid-Tertiary compressional event in south-western Utah: Geological Society of America Abstracts with Programs, v. 23, no. 1, p. 109.

Nielson, D.L., Evans, S.H., Jr., and Sibbett, B.S., 1986, Magmatic, structural, and hydrothermal evolution of the Mineral Mountains intrusive complex, Utah: Geological Society of America Bulletin, v. 97, p. 765–777.

Nutt, C.J., and Thorman, C.H., 1992, Pre-late Eocene structures and their control of gold ore at the Drum Mine, west-central Utah, *in* Thorman, C.H., ed., Application of structural geology to mineral and energy resources of the Central and Western United States: U.S. Geological Survey Bulletin 2012, p. F1–F7.

———1993, Eocene sedimentary and volcanic rocks and their use in dating Mesozoic and Tertiary structures in the southern Deep Creek Range, Nevada and Utah: Geological Society of America Abstracts with Programs, v. 25, no. 5, p. 128.

———1994, Eocene or older low-angle attenuation faults in the eastern Great Basin—Features of regional extension or contraction?: Geological Society of America Abstracts with Programs, v. 26, no. 2, p. 78.

Nutt, C.J., Thorman, C.H., and Brooks, W.E., 1994, Geologic setting of the southern part of the Goshute Indian Reservation, Nevada and Utah, *in* Thorman, C.H., Nutt, C.J., and Potter, C.J., eds., Dating of pre-Tertiary attenuation structures in upper Paleozoic and Mesozoic rocks and the Eocene history in northeast Nevada and northwest Utah: Reno, Nev., Nevada Petroleum Society, Sixth Annual Field Trip, Guidebook, p. 47–57.

Nutt, C.J., Thorman, C.H., Snee, L.W., and Zimmermann, R.A., 1992, Mesozoic to early Tertiary low-angle younger-over-older faults in the Drum Mountains, west-central Utah: Geological Society of America Abstracts with Programs, v. 24, no. 6, p. 56.

Otton, J.K., 1995, Western frontal fault of the Canyon Range—Is it the breakaway zone of the Sevier Desert detachment?: Geology, v. 23, p. 547–550.

Overtoom, G.J., and Bartley, J.M., 1994, Evidence from the Golden Gate Range for Oligocene synvolcanic stress reorientation along the Blue Ribbon lineament in east-central Nevada: Geological Society of America Abstracts with Programs, v. 26, no. 2, p. 79.

Overtoom, G.J., Bartley, J.M., and Geissman, J.W., 1993, Structural and paleomagnetic constraints on the origin and nature of W–NW striking structures in the Blue Ribbon lineament—Evidence from the northern Golden Gate Range, Nevada [abs.]: Eos, v. 74, no. 43, p. 207.

Page, W.R., 1995, Low-angle normal faults in Devonian rocks of the southern Delamar Mountains, Lincoln County, Nevada, *in* Scott, R.B., and Swadley, W C, eds., Geologic studies in the Basin and Range–Colorado Plateau transition in southeastern Nevada, southwestern Utah, and northwestern Arizona, 1992: U.S. Geological Survey Bulletin 2056–G, p. 203–218.

Page, W.R., and Scott, R.B., 1991, Ramping extensional faults in the Devonian Guilmette Formation, southern Nevada: Geological Society of America Abstracts with Programs, v. 23, no. 4, p. 55.

Parsons, Tom, Sleep, N.H., and Thompson, G.A., 1992, Host rock rheology controls on the emplacement of tabular intrusions—Implications for underplating of extending crust: Tectonics, v. 11, p. 1348–1356.

Parsons, Tom, and Thompson, G.A., 1991, The role of magma overpressure in suppressing earthquakes and topography—Worldwide examples: Science, v. 253, p. 1399–1402.

Perry, F.V., DePaolo, D.J., and Baldridge, W.S., 1993, Neodymium isotopic evidence for decreasing crustal contribution to Cenozoic ignimbrites of the Western United States—Implications for the thermal evolution of the Cordilleran crust: Geological Society of America Bulletin, v. 105, p. 872–882.

Petit, J.P., 1987, Criteria for the sense of movement on fault surfaces in brittle rocks: Journal of Structural Geology, v. 9, p. 597–608.

Pierce, K.L., Adams, K.D., and Sturchio, N.C., 1991, Geologic setting of the Corwin Springs Known Geothermal Resources Area–Mammoth Hot Springs area in and adjacent to Yellowstone National Park, *in* Sorey, M.L., ed., Effects of potential geothermal development in the Corwin Springs Known Geothermal Resources Area, Montana, on the thermal features of Yellowstone National Park: U.S. Geological Survey Water-Resources Investigations Report 91–4052, p. C1–C37.

Pierce, K.L., Milbert, D.G., and Saltus, R.W., 1992, Geoid dome culminates on Yellowstone—Yellowstone hotspot fed by a thermal mantle plume? [abs.]: Eos, v. 73, no. 14, p. 284.

Pierce, K.L., and Morgan, L.A., 1992, The track of the Yellowstone hot spot—Volcanism, faulting, and uplift, *in* Link, P.K., Kuntz, M.A., and Platt, L.B., eds., Regional geology of eastern Idaho and western Wyoming: Geological Society of America Memoir 179, p. 1–53.

Price, D.E., and Bartley, J.M., 1992, Three-dimensional extensional structure of the southern Mineral Mountains, southwestern Utah: Geological Society of America Abstracts with Programs, v. 24, no. 6, p. 58.

Proffett, J.M., Jr., 1977, Cenozoic geology of the Yerington district, Nevada, and implications for the nature and origin of Basin and Range faulting: Geological Society of America Bulletin, v. 88, p. 247–266.

Roberts, R.J., 1964, Economic geology, *in* Mineral and water resources of Nevada: U.S. Congress, 88th, 2d session., Senate Document 87, p. 39–48.

———1966, Metallogenic provinces and mineral belts in Nevada, *in* AIME Pacific Southwest Mineral Industry Conference, Sparks, Nev., 1965, Papers: Nevada Bureau of Mines Report 13, pt. A, p. 47–72.

Robinson, P.T., McKee, E.H., and Moiola, R.J., 1968, Cenozoic volcanism and sedimentation, Silver Peak region, western Nevada and adjacent California, *in* Coats, R.R., Hay, R.L., and Anderson, C.A., Studies in volcanology—A memoir in honor of Howel Williams: Geological Society of America Memoir 116, p. 577–611.

Rosendahl, B.R., 1987, Architecture of continental rifts with special reference to East Africa: Annual Reviews of Earth and Planetary Science, v. 15, p. 445–503.

Ross, H.P., and Moore, J.N., 1994, Geophysical investigations of the Cove Fort–Sulphurdale geothermal system, Utah, *in* Blackett, R.E., and Moore, J.N., eds., Cenozoic geology and geothermal systems of southwestern Utah, 1994 field symposium: Utah Geological Association Publication 23, p. 45–59.

Rowan, L.C., and Wetlaufer, P.H., 1981, Relation between regional lineament systems and structural zones in Nevada: American Association of Petroleum Geologists Bulletin, v. 65, p. 1414–1432.

Rowley, P.D., 1968, Geology of the southern Sevier Plateau, Utah: Austin, Texas, University of Texas Ph. D. dissertation, 340 p.

Rowley, P.D., and Barker, D.S., 1978, Geology of the Iron Springs mining district, Utah, *in* Shawe, D.R., and Rowley, P.D., eds., Guidebook to mineral deposits of southwestern Utah—Field excursion C–2: Utah Geological Association Publication 7, p. 49–58.

Rowley, P.D., Cunningham, C.G., Steven, T.A., Mehnert, H.H., and Naeser, C.W., 1988a, Geologic map of the Marysvale quadrangle, Piute County, Utah: Utah Geological and Mineral Survey Map 105, scale 1:24,000.

———1988b, Geologic map of the Antelope Range quadrangle, Sevier and Piute Counties, Utah: Utah Geological and Mineral Survey Map 106, scale 1:24,000.

Rowley, P.D., Kellogg, K.S., Vennum, W.R., Laudon, T.S., Thomson, J.W., O'Neill, J.M., and Lidke, D.J., 1991, Tectonic setting of the English Coast, eastern Ellsworth Land, Antarctica, *in* Thomson, M.R.A., Crame, J.A., and Thomson, J.W., eds., Geological evolution of Antarctica: Cambridge, Cambridge University Press, p. 467–473.

Rowley, P.D., Kellogg, K.S., Williams, P.L., Willan, C.F.H., and Thomson, J.W., 1992, Geological map, Sheet 6, Southern Palmer Land and eastern Ellsworth Land: Cambridge, British Antarctic Survey BAS 500G series, scale 1:500,000.

Rowley, P.D., Lipman, P.W., Mehnert, H.H., Lindsey, D.A., and Anderson, J.J., 1978, Blue Ribbon lineament, an east-trending structural zone within the Pioche mineral belt of southwestern Utah and eastern Nevada: U.S. Geological Survey Journal of Research, v. 6, no. 2, p. 175–192.

Rowley, P.D., McKee, E.H., and Blank, H.R., Jr., 1989, Miocene gravity slides resulting from emplacement of the Iron Mountain pluton, southern Iron Springs mining district, Iron County, Utah [abs.]: Eos, v. 70, no. 43, p. 1309.

Rowley, P.D., Mehnert, H.H., Naeser, C.W., Snee, L.W., Cunningham, C.G., Steven, T.A., Anderson, J.J., Sable, E.G., and Anderson, R.E., 1994, Isotopic ages and stratigraphy of Cenozoic rocks of the Marysvale volcanic field and adjacent areas, west-central Utah: U.S. Geological Survey Bulletin 2071, 35 p.

Rowley, P.D., Nealey, L.D., Unruh, D.M., Snee, L.W., Mehnert, H.H., Anderson, R.E., and Grommé, C.S., 1995, Stratigraphy of Miocene ash-flow tuffs in and near the Caliente caldera complex, southeastern Nevada and southwestern Utah, *in* Scott, R.B., and Swadley, W C, eds., Geologic studies in the Basin and Range–Colorado Plateau transition in southeastern Nevada, southwestern Utah, and northwestern Arizona, 1992: U.S. Geological Survey Bulletin 2056–B, p. 42–88.

Rowley, P.D., Shroba, R.R., Simonds, F.W., Burke, K.J., Axen, G.J., and Olmore, S.D., 1994, Geologic map of the Chief Mountain quadrangle, Lincoln County, Nevada: U.S. Geological Survey Geologic Quadrangle Map GQ–1731, scale 1:24,000.

Rowley, P.D., Snee, L.W., Mehnert, H.H., Anderson, R.E., Axen, G.J., Burke, K.J., Simonds, F.W., Shroba, R.R., and Olmore, S.D., 1992, Structural setting of the Chief mining district, eastern Chief Range, Lincoln County, Nevada, *in* Thorman, C.H., ed., Application of structural geology to mineral and energy resources of the Central and Western United States: U.S. Geological Survey Bulletin 2012, p. H1–H17.

Rowley, P.D., Steven, T.A., Anderson, J.J., and Cunningham, C.G., 1979, Cenozoic stratigraphic and structural framework of southwestern Utah: U.S. Geological Survey Professional Paper 1149, 22 p.

Rowley, P.D., Steven, T.A., and Mehnert, H.H., 1981, Origin and structural implications of upper Miocene rhyolites in Kingston Canyon, Piute County, Utah: Geological Society of America Bulletin, pt. I, v. 92, p. 590–602.

Rowley, P.D., Williams, P.L., and Kaplan, A.M., 1986a, Geologic map of the Greenwich quadrangle, Piute County, Utah: U.S. Geological Survey Geologic Quadrangle Map GQ–1589, scale 1:24,000.

———1986b, Geologic map of the Koosharem quadrangle, Sevier and Piute Counties, Utah: U.S. Geological Survey Geologic Quadrangle Map GQ–1590, scale 1:24,000.

Sable, E.G., and Anderson, J.J., 1985, Tertiary tectonic slide megabreccia, Markagunt Plateau, southwestern Utah: Geological Society of America Abstracts with Programs, v. 17, no. 4, p. 263.

Saltus, R.W., and Jachens, R.C., 1995, Gravity and basin-depth maps of the Basin and Range province, Western United States: U.S. Geological Survey Geophysical Investigations Map GP–1012, scale 1:2,500,000.

Saunders, A.D., and Tarney, J., 1984, Geochemical characteristics of basaltic volcanism within back-arc basins, *in* Kokebar, B.P., and Howells, M.F., eds., Marginal basin geology—Volcanic and associated sedimentary and tectonic processes in modern and ancient marginal basins: Geological Society Special Publication 16, p. 59–76.

Sawyer, D.A., Fleck, R.J., Lanphere, M.A., Warren, R.G., Broxton, D.E., and Hudson, M.R., 1994, Episodic caldera volcanism in the Miocene southwestern Nevada volcanic field—Revised stratigraphic framework, $^{40}Ar/^{39}Ar$ geochronology, and implications for magmatism and extension: Geological Society of America Bulletin, v. 106, p. 1304–1318.

Schmidt, D.L., and Rowley, P.D., 1986, Continental rifting and transform faulting along the Jurassic Transantarctic Rift, Antarctica: Tectonics, v. 5, p. 279–291.

Scholz, C.H., Barazangi, Muawia, and Sbar, M.L., 1971, Late Cenozoic evolution of the Great Basin, Western United States, as an ensialic interarc basin: Geological Society of America Bulletin, v. 82, p. 2979–2990.

Scott, R.B., Grommé, C.S., Best, M.G., Rosenbaum, J.G., and Hudson, M.R., 1995, Stratigraphic relationships of Tertiary volcanic rocks in central Lincoln County, southeastern Nevada, *in* Scott, R.B., and Swadley, W C, eds., Geologic studies in the Basin and Range–Colorado Plateau transition in southeastern Nevada, southwestern Utah, and northwestern Arizona, 1992: U.S. Geological Survey Bulletin 2056–A, p. 5–41.

Scott, R.B., Unruh, D.M., Snee, L.W., Harding, A.E., Nealey, L.D., Blank, H.R., Jr., Budahn, J.R., and Mehnert, H.H., 1995, Relation of peralkaline magmatism to heterogeneous extension during the middle Miocene, southeastern Nevada: Journal of Geophysical Research, v. 100, no. B6, p. 10381–10401.

Seedorff, E., 1991, Magmatism, extension, and ore deposits of Eocene to Holocene age in the Great Basin—Mutual effects and preliminary proposed genetic relationships, *in* Raines, G.L., Lisle, R.E., Schaefer, R.W., and Wilkinson, W.H., eds., Geology and ore deposits of the Great Basin: Reno, Nevada, Geological Society of Nevada Symposium, v. 1, p. 133–178.

Severinghaus, Jeff, and Atwater, Tanya, 1990, Cenozoic geometry and thermal state of the subducting slabs beneath western North America, chap. 1 *of* Wernicke, B.P., ed., Basin and Range extensional tectonics near the latitude of Las Vegas, Nevada: Geological Society of America Memoir 176, p. 1–22.

Shawe, D.R., and Stewart, J.H., 1976, Ore deposits as related to tectonics and magmatism, Nevada and Utah: Transactions of American Institute of Mining, Metallic, and Petroleum Engineers, v. 260, p. 225–232.

Sibson, R.H., 1989, Earthquake faulting as a structural process: Journal of Structural Geology, v. 11, p. 1–14.

Siders, M.A, and Shubat, M.A., 1986, Stratigraphy and structure of the northern Bull Valley Mountains and Antelope Range, Iron County, Utah, *in* Griffen, D.T., and Phillips, W.R., eds., Thrusting and extensional structures and mineralization in the Beaver Dam Mountains, southwestern Utah: Utah Geological Association Publication 15, p. 87–102.

Silberman, M.L., 1975, Ages and tectonic implications of the transition of calc-alkaline andesitic to basaltic volcanism in the western Great Basin and the Sierra Nevada: Geological Society of America Abstracts with Programs, v. 7, no. 3, p. 375.

Smith, R.L., 1979, Ash-flow magmatism, *in* Chapin, C.E., and Elston, W.E., eds, Ash-flow tuffs: Geological Society of America Special Paper 180, p. 5–27.

Smith, R.L., and Luedke, R.G., 1984, Potentially active volcanic lineaments and loci in western conterminous United States, *in* Explosive volcanism—Inception, evolution, and hazards: Washington, D.C., National Academy Press, Studies in Geophysics, p. 47–66.

Sparlin, M.A., Braile, L.W., and Smith, R.B., 1982, Crustal structure of the eastern Snake River Plain determined from ray trace modeling of seismic refraction data: Journal of Geophysical Research, v. 87, no. B4, p. 2619–2633.

Spurney, J.C., 1984, Geology of the Iron Peak intrusion, Iron County, Utah: Kent, Ohio, Kent State University M.S. thesis, 84 p.

Stern, T.A., 1985, A back-arc basin formed within continental lithosphere—The central volcanic region of New Zealand: Tectonophysics, v. 112, p. 385–409.

Steven, T.A., 1988, Mosaic faulting as a guide to mineral exploration in the Richfield 1°×2° quadrangle, western Utah, *in* Schindler, K.S., ed., USGS research on mineral resources—1989 program and abstracts: U.S. Geological Survey Circular 1035, p. 71.

Steven, T.A., Cunningham, C.G., Naeser, C.W., and Mehnert, H.H., 1979, Revised stratigraphy and radiometric ages of volcanic rocks and mineral deposits in the Marysvale area, west-central Utah: U.S. Geological Survey Bulletin 1469, 40 p.

Steven, T.A., and Morris, H.T., 1983, Geologic map of the Cove Fort quadrangle, west-central Utah: U.S. Geological Survey Miscellaneous Investigations Series Map I–1481, scale 1:50,000.

———1987, Summary mineral resource appraisal of the Richfield 1°×2° quadrangle, west-central Utah: U.S. Geological Survey Circular 916, 24 p.

Steven, T.A., Morris, H.T., and Rowley, P.D., 1990, Geologic map of the Richfield 1°×2° quadrangle, west-central Utah: U.S. Geological Survey Miscellaneous Investigations Series Map I–1901, scale 1:250,000.

Steven, T.A., Rowley, P.D., and Cunningham, C.G., 1984, Calderas of the Marysvale volcanic field, west central Utah: Journal of Geophysical Research, v. 89, no. B10, p. 8751–8764.

Stewart, J.H., 1971, Basin and range structure—A system of horsts and grabens produced by deep-seated extension: Geological Society of America Bulletin, v. 82, p. 1019–1044.

———1980, Regional tilt patterns of late Cenozoic basin-range fault blocks, Western United States: Geological Society of America Bulletin, Part I, v. 91, p. 460–464.

———1983, Cenozoic structure and tectonics of the northern Basin and Range province, California, Nevada, and Utah, *in* The role of heat in the development of energy and mineral resources in the northern Basin and Range province: Davis, Calif., Geothermal Resources Council, Special Report 13, p. 25–40.

Stewart, J.H., Moore, W.J., and Zietz, Isidore, 1977, East-west patterns of Cenozoic igneous rocks, aeromagnetic anomalies, and mineral deposits, Nevada and Utah: Geological Society of America Bulletin, v. 88, p. 67–77.

Stewart, J.H., and Roldan-Quintana, Jaime, 1994, Tilt domains and large-scale segmentation of the Basin and Range province with emphasis on new data from Sonora, Mexico: Geological Society of America Abstracts with Programs, v. 26, no. 2, p. 96.

Stewart, J.H., Walker, G.W., and Kleinhampl, F.J., 1975, Oregon-Nevada lineament: Geology, v. 3, p. 265–268.

Stoeser, D.B., 1993, Tertiary calderas and regional extension of the east-central part of the Tintic–Deep Creek mineral belt, eastern Great Basin, Utah, *in* Scott, R.W., Jr., Detra, P.S., and Berger, B.R., eds., Advances related to United States and international mineral resources—Developing frameworks and exploration technologies: U.S. Geological Survey Bulletin 2039, p. 5–23.

Stone, D.S., 1986, Wrench faulting and Rocky Mountains tectonics: The Mountain Geologist, v. 6, p. 67–79.

Sullivan, K.R., Kowallis, B.J., and Mehnert, H.H., 1991, Isotopic ages of igneous intrusions in southeastern Utah—Evidence for a mid-Cenozoic Reno–San Juan magmatic zone: Brigham Young University Geology Studies, v. 37, p. 139–144.

Swadley, W C, Page, W.R., Scott, R.B., and Pampeyan, E.H., 1994, Geologic map of the Delamar 3 SE quadrangle, Lincoln County, Nevada: U.S. Geological Survey Geologic Quadrangle Map GQ–1754, scale 1:24,000.

Tanaka, K.L., Shoemaker, E.M., Ulrich, G.E., and Wolfe, E.W., 1986, Migration of volcanism in the San Francisco volcanic field, Arizona: Geological Society of America Bulletin, v. 97, p. 129–141.

Taylor, W.J., and Bartley, J.M., 1992, Prevolcanic extensional Seaman breakaway fault and its geologic implications for eastern Nevada and western Utah: Geological Society of America Bulletin, v. 104, p. 255–266.

Taylor, W.J., Bartley, J.M., Lux, D.R., and Axen, G.J., 1989, Timing of Tertiary extension in the Railroad Valley–Pioche transect, Nevada—Constraints from $^{40}Ar/^{39}Ar$ ages of volcanic rocks: Journal of Geophysical Research, v. 94, no. B6, p. 7757–7774.

Thompson, G.A., and Burke, D.B., 1974, Regional geophysics of the Basin and Range province: Annual Review of Earth and Planetary Science, v. 2, p. 213–238.

Thompson, G.A., Parsons, Tom, and Smith, Richard, 1990, Examples of magma overpressure suppressing normal faulting and inhibiting seismicity—Snake River Plain, Idaho, Yucca Mountain, Nevada, and Mono Craters, California [abs.]: Eos, v. 71, no. 43, p. 1622.

Thompson, G.A., and Zoback, M.L., 1979, Regional geophysics of the Colorado Plateau: Tectonophysics, v. 61, p. 149–181.

Thompson, R.A., Milling, M.E., Jr., Fleck, R.J., Wright, L.A., Rogers, N.W., and Johnson, C.M., in press, Temporal, spatial, and compositional constraints on extension-related volcanism in central Death Valley, California: Journal of Geophysical Research.

Tobisch, O.T., Saleeby, J.B., and Fiske, R.S., 1986, Structural history of continental volcanic arc rocks, eastern Sierra Nevada, California—A case for extensional tectonics: Tectonics, v. 5, p. 65–94.

Van Kooten, G.K., 1988, Structure and hydrocarbon potential beneath the Iron Springs laccolith, southwestern Utah: Geological Society of America Bulletin, v. 100, no. 10, p. 1533–1540.

Von Tish, D.B., Allmendinger, R.W., and Sharp, J.W., 1985, History of Cenozoic extension in central Sevier Desert, west-central Utah, from COCORP seismic reflection data: American Association of Petroleum Geologists Bulletin, v. 69, p. 1077–1087.

Walker, J.A., and Mattox, S.R., 1989, The influences of subduction on mid-late Cenozoic volcanism in southwestern Utah:

Geological Society of America Abstracts with Programs, v. 21, no. 6, p. A57.

Wallace, R.E., 1984, Patterns and timing of late Quaternary faulting in the Great Basin Province and relation to some regional tectonic features: Journal of Geophysical Research, v. 89, no. B7, p. 5763–5769.

Ward, P.L., 1991, On plate tectonics and the geologic evolution of southwestern North America: Journal of Geophysical Research, v. 96, no. B7, p. 12479–12496.

Weiss, S.I., Noble, D.C., Worthington, J.E., IV, and McKee, E.H., 1993, Neogene tectonism from the southwestern Nevada volcanic field to the White Mountains, California—Part I, Miocene volcanic stratigraphy, paleotopography, extensional faulting and uplift between northern Death Valley and Pahute Mesa, *in* Lahren, M.M., Trexler, J.H., Jr., and Spinosa, Claude, Crustal evolution of the Great Basin and the Sierra Nevada; Geological Society of America Cordilleran/Rocky Mountain Sections Joint Meeting, Reno, Nev., 1993, Field Trip Guidebook: Reno, Nev., University of Nevada, Mackay School of Mines, p. 353–369.

Wells, M.L., Dallmeyer, R.D., and Allmendinger, R.W., 1990, Late Cretaceous extension in the hinterland of the Sevier thrust belt, northwestern Utah and southern Idaho: Geology, v. 18, p. 929–933.

Wernicke, Brian, 1981, Low-angle normal faults in the Basin and Range province—Nappe tectonics in an extending orogen: Nature, v. 291, p. 645–648.

———1985, Uniform-sense normal simple shear of the continental lithosphere: Canadian Journal of Earth Sciences, v. 22, p. 108–125.

———1992, Cenozoic extensional tectonics of the U.S. Cordillera, *in* Burchfiel, B.C., Lipman, P.W., and Zoback, M.L., eds., The Cordilleran orogen—Conterminous U.S.: Geological Society of America, The Geology of North America, v. G–3, p. 553–581.

Wernicke, Brian, and Axen, G.J., 1988, On the role of isostasy in the evolution of normal fault systems: Geology, v. 16, p. 848–851.

Wernicke, Brian, Axen, G.J., and Snow, J.K., 1988, Basin and Range extensional tectonics at the latitude of Las Vegas, Nevada: Geological Society of America Bulletin, v. 100, p. 1738–1757.

Wernicke, Brian, Walker, J.D., and Beaufait, M.S., 1985, Structural discordance between Neogene detachments and frontal Sevier thrusts, central Mormon Mountains, southern Nevada: Tectonics, v. 4, no. 2, p. 213–246.

White, R.S., Spence, G.D., Fowler, S.R., McKenzie, D.P., Westbrook, G.K., and Bowen, A.N., 1987, Magmatism at rifted continental margins: Nature, v. 330, p. 439–444.

Wiebe, R.A., 1993, Basaltic injections into floored silicic magma chambers [abs.]: Eos, v. 74, no. 1, p. 1–3.

Williams, P.L., and Hackman, R.J., 1971, Geology, structure, and uranium deposits of the Salina quadrangle, Utah: U.S. Geological Survey Miscellaneous Geologic Investigations Map I–591, scale 1:250,000.

Willis, G.C., 1985, Revisions to the geochronology and source areas of early Tertiary formations in the Salina area of Sevier Valley, central Utah: Geological Society of America Abstracts with Programs, v. 17, no. 4, p. 272.

Willis, G.F., and Tosdal, R.M., 1992, Formation of gold veins and breccias during dextral strike-slip faulting in the Mesquite mining district, southeastern California: Economic Geology, v. 87, p. 2002–2022.

Wilson, J.T., 1965, A new class of faults and their bearing on continental drift: Nature, v. 207, p. 343–347.

Wise, D.U., 1963, An outrageous hypothesis for the tectonic pattern of the North American Cordillera: Geological Society of America Bulletin, v. 74, p. 357–362.

Witkind, I.J., 1964, Geology of the Abajo Mountains area, San Juan County, Utah: U.S. Geological Survey Professional Paper 453, 110 p.

Wright, Lauren, 1976, Late Cenozoic fault patterns and stress fields in the Great Basin and westward displacement of the Sierra Nevada block: Geology, v. 4, p. 489–494.

Zietz, Isidore, Gilbert, F.P., and Kirby, J.R., 1978, Aeromagnetic map of Nevada—Color coded intensities: U.S. Geological Survey Geophysical Investigations Map GP–922, scale 1:1,000,000.

Zietz, Isidore, Shuey, Ralph, and Kirby, J.R., Jr., 1976, Aeromagnetic map of Utah: U.S. Geological Survey Geophysical Investigations Map GP–907, scale 1:1,000,000.

Zoback, M.L., 1989, State of stress and modern deformation of the northern Basin and Range province: Journal of Geophysical Research, v. 94, no. B6, p. 7105–7128.

Zoback, M.L., Anderson, R.E., and Thompson, G.A., 1981, Cainozoic evolution of the state of stress and style of tectonism in the Basin and Range province of the Western United States: Royal Society of London Philosophical Transactions, v. A300, p. 407–434.

Zoback, M.L., McKee, E.H., Blakely, R.J., and Thompson, G.A., 1994, The northern Nevada rift—Regional tectono-magmatic relations and the middle Miocene stress direction: Geological Society of America Bulletin, v. 106, p. 371–382.

Zoback, M.L., and Thompson, G.A., 1978, Basin and Range rifting in northern Nevada—Clues from a mid-Miocene rift and its subsequent offsets: Geology, v. 6, p. 111–116.

The Fate of the Colorado Plateau—A View from the Mantle

By Felix E. Mutschler,[1] Edwin E. Larson,[2] *and* David L. Gaskill[3]

CONTENTS

FIGURES

ABSTRACT

The Colorado Plateau is bordered by five passive hot spots: a southward extension of the Great Falls tectonic zone, the Colorado mineral belt, the northern Rio Grande Rift, the Great Basin regional gravity low, and the southern Basin and Range province. Each hot spot represents mantle upwelling induced by lithospheric extension related to plate-tectonic events. Manifestations of these hot spots include thin crust and lithosphere, hot low-density upper mantle, volcanism resulting from decompression melting of the mantle, and regional arching and rifting. As the hot spots developed and enlarged they progressively reduced the size of the stable cratonic block now represented by the Colorado Plateau.

PASSIVE HOT SPOTS BORDERING THE COLORADO PLATEAU

The Colorado Plateau is an isolated block of the Proterozoic craton which is being reduced in size by the lateral encroachment of a ring of Late Cretaceous to Holocene passive hot spots (fig. 1). Three features are characteristic of these hot spots: (1) Regional geophysical anomalies (figs. 2 and 3) indicative of thin crust, thin lithosphere, low-density upper mantle, and high heat flow. (2) Young and (or) active volcanism resulting from decompression melting of rising hot mantle. Volcanism tends to be younger outward from the apex of a static hot spot or along the trend of a migrating hot spot. (3) Regional doming or arching above a rising and expanding mantle bulge. Crustal extension and thinning causes axial rifting of the regional dome above the area of mantle upwelling.

Hot spots, in general, may be either (1) active, resulting from deep-seated asthenospheric mantle thermal plumes (fig. 4; Courtney and White, 1986), or (2) passive, resulting from subcrustal lithospheric thinning (fig. 5; Eaton, 1987). Assuming that active, deep-source mantle plumes tend to remain stationary over periods of tens of millions of years (Irvine, 1989), they should leave "volcanic tracks" on lithosphere plates that move across them, as did the Hawaiian hot spot on the Pacific plate (Clague, 1987). There is, however, little evidence of long-lived volcanic-chronologic tracks for the hot spots bordering the Colorado Plateau, suggesting that they are passive features. The loci of these hot spots appear to have remained essentially fixed to the southwestward-traveling North American plate for tens of millions of years, suggesting that they reside in the lithosphere or are mechanically coupled to it. This implies that if passive hot spots form at sites of significant subcrustal thinning, once they are initiated they may be self-sustaining and travel with the host lithospheric plate.

Various mechanisms have been suggested for large-scale thinning of the subcrustal continental lithosphere, including (1) differential shifting of lithospheric blocks resulting from plate movements (Mutschler and others, 1991), (2) isostatic rebound and gravitational collapse of tectonically thickened orogenic welts (Mutschler and others, 1987; Wernicke and others, 1987), (3) release of regional compressive stress upon termination of adjacent continental margin subduction (Scholz and others, 1971), (4) lithospheric erosion by asthenospheric advection (Eggler and others, 1988), (5) back-arc spreading (Thompson and Burke, 1974), (6) lithospheric delamination (Bird, 1979), (7) lithospheric weakening by mantle degassing (Bailey, 1970, 1978), and (8) lateral transfer of a "great wave" of lower crustal material from the coast to beneath a distant area, producing thickened crust (Bird, 1988). Whatever their ultimate cause, most of the Cordilleran passive hot spots we describe show initial magmatic crustal penetration controlled by regional crustal structures, including crustal province boundaries such as the Great Falls tectonic zone and ancient fault systems such as the Colorado mineral belt (fig. 6). As they evolve, however, these hot spots usually expand across crustal blocks and sutures (fig. 7), suggesting that their ultimate source resides at least as deep as the subcrustal lithosphere.

We will examine the magmatic, tectonic, and chronologic evolution of the five passive hot-spot loci marginal to, and encroaching on, the Colorado Plateau:

1. The Great Falls tectonic zone (GFTZ), active from ≈70 to 20(?) Ma.

---

[1] Petrophysics Crisis Center, Department of Geology, Eastern Washington University, Cheney, WA 99004.
[2] Department of Geological Sciences, University of Colorado, Boulder, CO 80309.
[3] 548 Pinesong Trail, Golden, CO 80401.

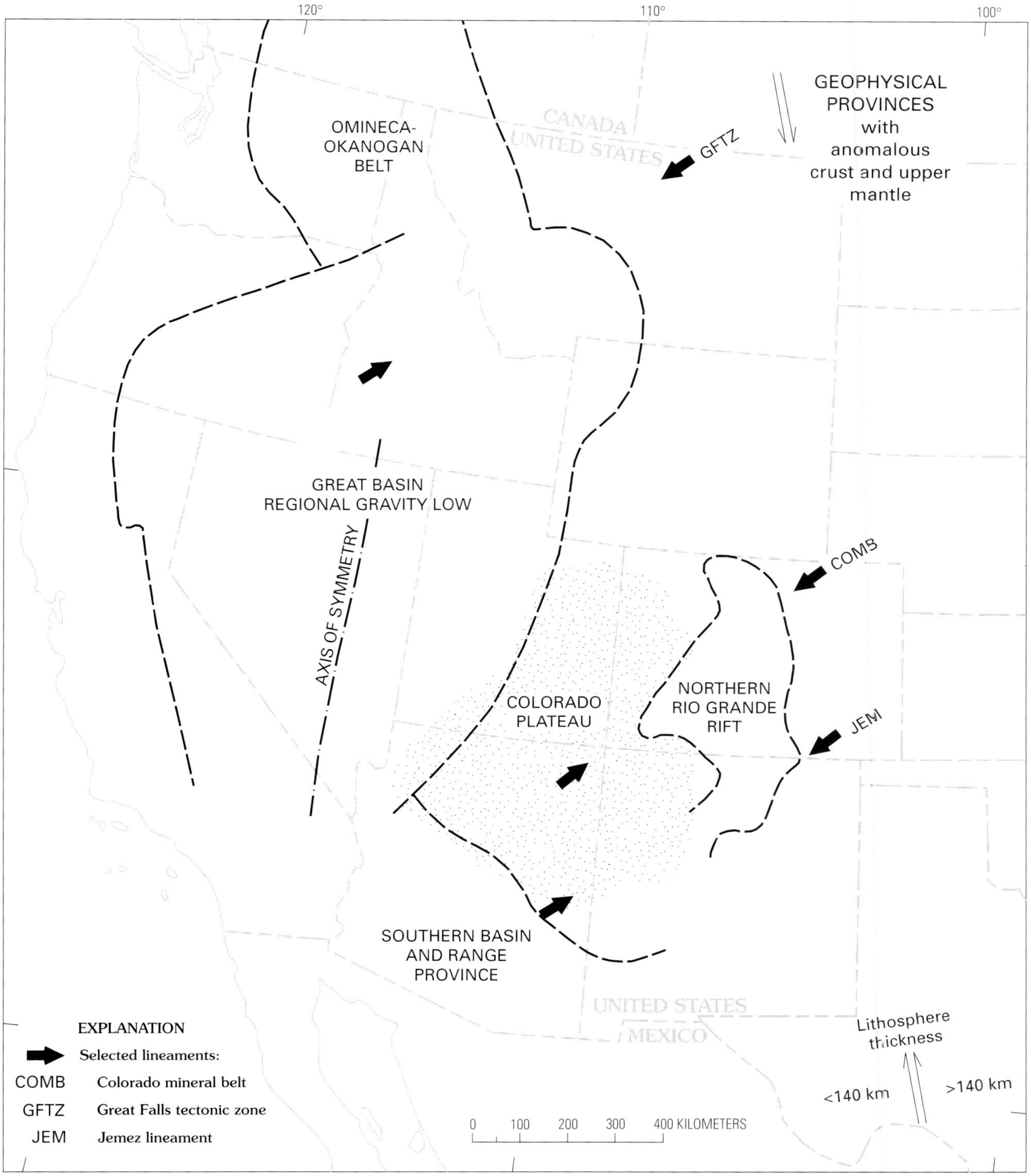

Figure 1. Relation of the Colorado Plateau to geophysical provinces characterized by crustal or upper mantle geophysical anomalies. Generalized axis of bilateral symmetry of observed Bouguer gravity and topography, in center of Great Basin regional gravity low, is from Eaton and others (1978, fig. 3–11–B). Colorado Plateaus physiographic province (stippled) modified from Bayer (1983).

2. The Colorado mineral belt (COMB), active from ≈75 to 17(?) Ma.
3. The northern Rio Grande Rift (NRGR), starting at ≈35–26 Ma and active from ≈17 to 0 Ma.
4. The Great Basin regional gravity low (GBRGL), active from ≈17 to 0 Ma.
5. The southern Basin and Range province (SBR), active from ≈40 to 0 Ma.

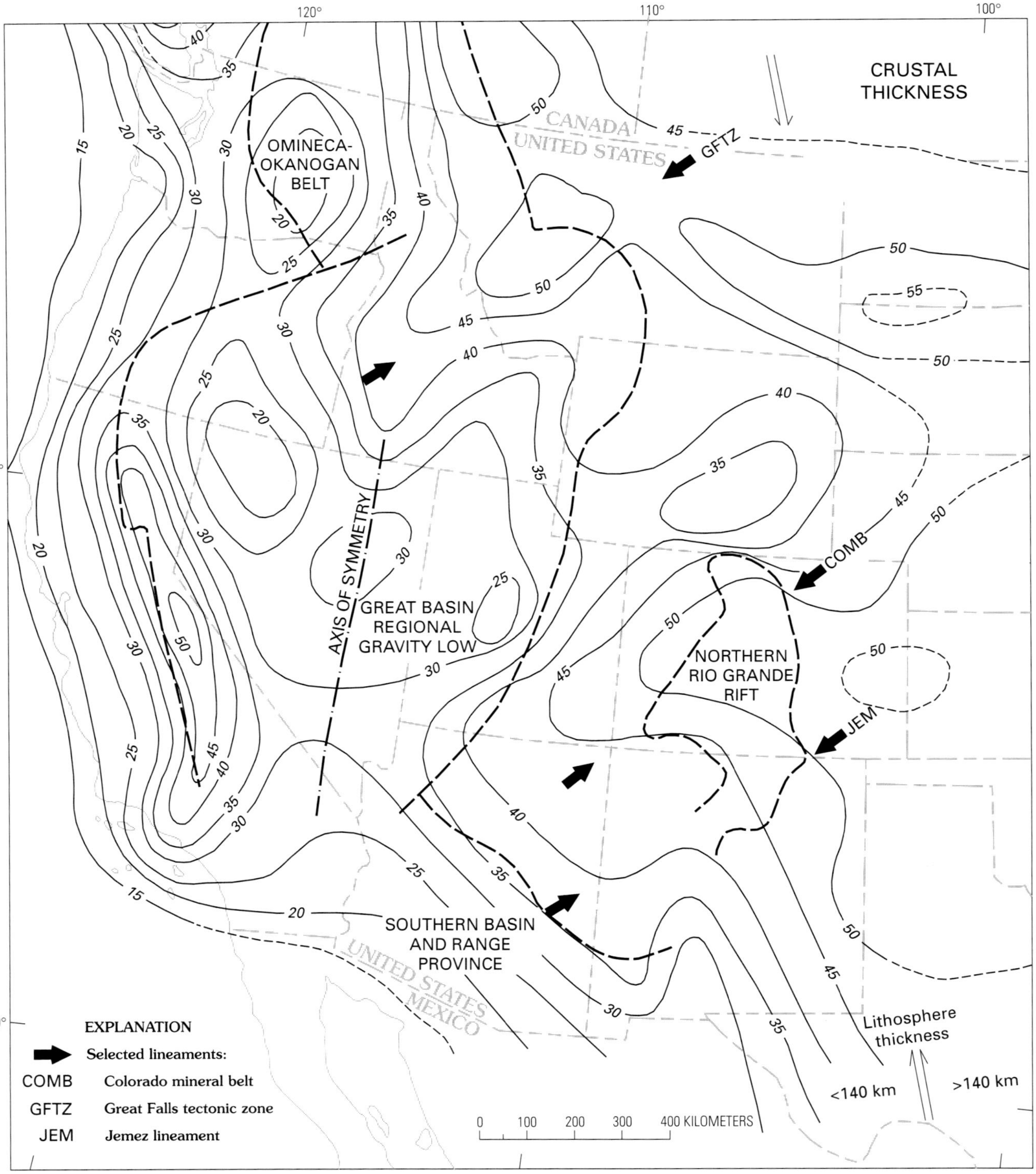

Figure 2. Crustal thickness in the Western United States. Contours show depth to reflection moho, in kilometers below sea level. From Allenby and Schnetzler (1983, fig. 2).

GREAT FALLS TECTONIC ZONE (GFTZ)

The Late Cretaceous–Eocene central Montana alkaline province and the Eocene Challis volcanic field lie along the northeast-trending Great Falls tectonic zone (GFTZ), an ancient, repeatedly reactivated crustal flaw (O'Neill and Lopez, 1985), which essentially coincides with the northwest side of the Archean Wyoming province cratonic block (fig. 7).

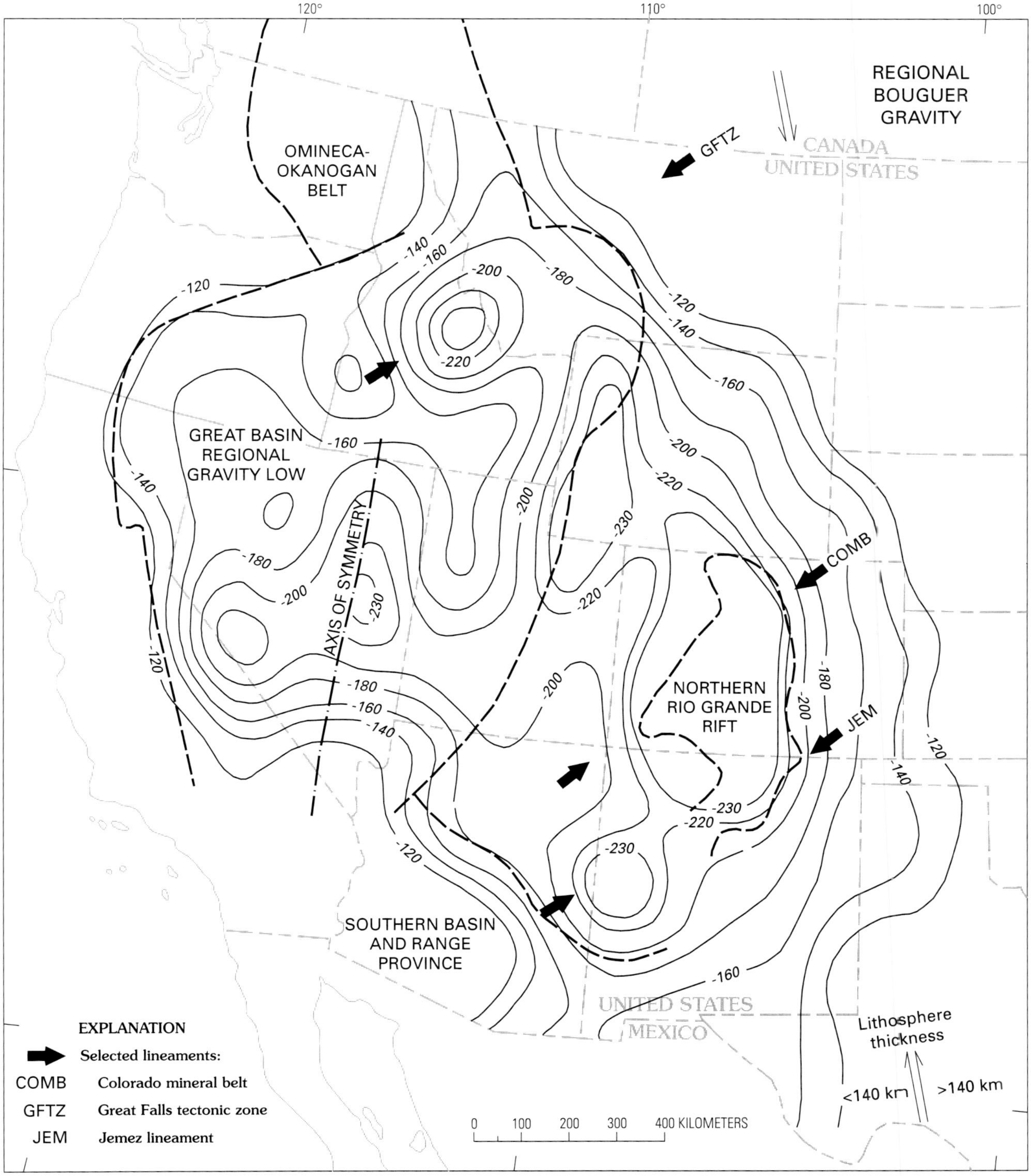

Figure 3. Regional Bouguer gravity of wavelengths greater than 250 km in the Western United States. Contours show gravity in milligals. A comparison of this map with 1,000-km-filtered regional gravity maps (Hildenbrand and others, 1982) suggests that the major negative anomalies shown here represent low-density material at depths extending from the crust-mantle boundary to >125 km.

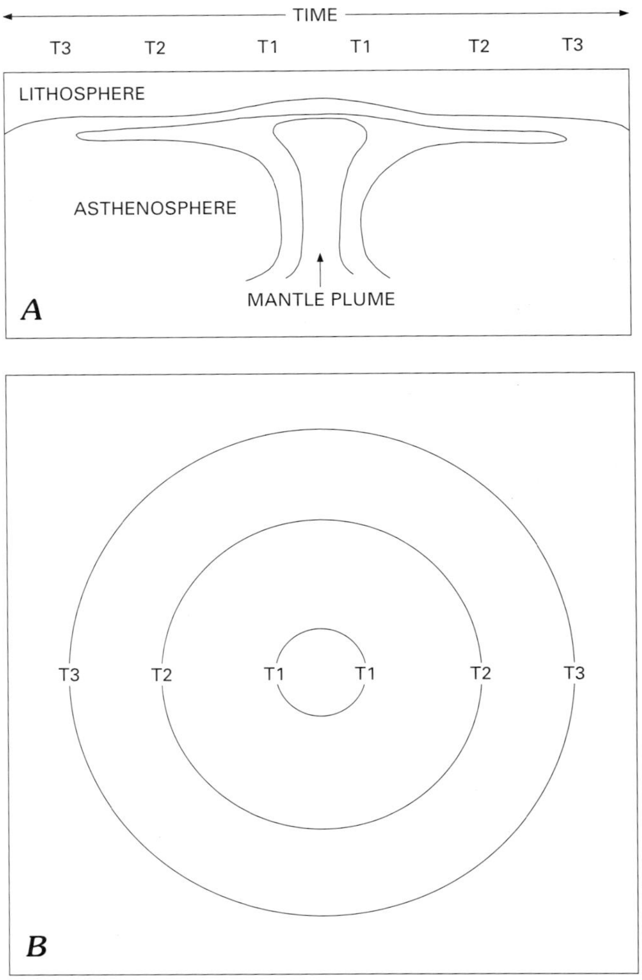

Figure 4. Development of an active hot spot over time. *A*, Generalized cross section showing temperature anomalies with respect to mean asthenosphere temperature in an axisymmetric convection model (after White and McKenzie, 1989, fig. 2). *B*, Map view showing isochrons with outward younging of inception of magmatism above an axisymmetric mantle plume. Similar isochron patterns may develop above passive hot spots.

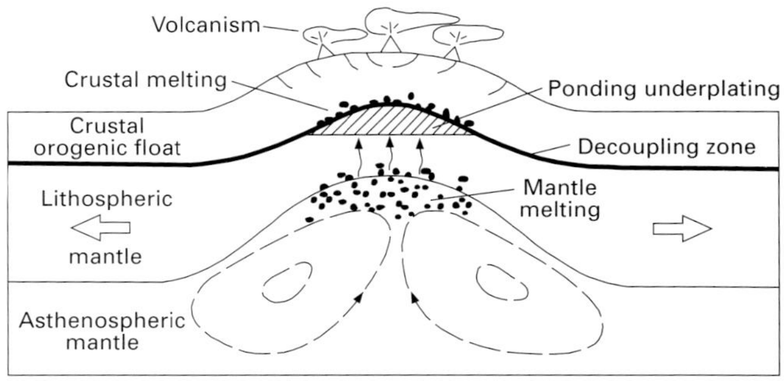

Figure 5. Cross section showing features of a passive hot spot resulting from subcrustal lithospheric thinning. Not to scale.

The Late Cretaceous–early Tertiary tectonic setting of the Montana alkaline province and Challis volcanic field included the following elements as shown on figure 6:

1. A regional northeast-trending Eocene topographic dome defined on the basis of paleobotanical studies by Axelrod (1968). The axis of the dome was essentially coincident with the GFTZ.

2. Extensive Eocene (≈50–44 Ma) mildly alkaline shoshonitic to calc-alkaline magmatism in the Challis volcanic field (Moye, 1988; Norman and Mertzman, 1991) on the crest of the dome, and Late Cretaceous–Eocene (≈76–46 Ma) alkaline-dominated magmatism on the flanks of the dome and along its northeast projection—the central Montana alkaline province described by Larsen (1940) and many subsequent workers. (See papers in Baker and Berg, 1991, for instance.)

3. Synvolcanic axial rifting along the crest of the dome indicated by recurrent movements on the trans-Challis fault system, northeast-trending dike swarms and volcano-tectonic grabens and calderas (Moye, 1988).

These features can be integrated into a generalized model in which decompression melting of rising mantle yielded mafic alkaline magmas, some of which parked in the crust. These accumulated mantle melts triggered partial crustal melting, generating the voluminous calc-alkaline magmas of the Challis volcanic field, which are the surface manifestations of large batholithic bodies (Mabey and Webring, 1985). Surface doming resulted from both emplacement of the granite batholiths at shallow levels and deep-level upward movement of thermally expanded mantle. The areal extent of the plutonic and volcanic loci and the topographic dome is comparable to that of similar features that surround recognized modern mantle hot spots.

Mutschler and others (1991) suggested that this passive hot spot developed in response to an offset in large-scale northwest-trending Cretaceous strike-slip zones that resulted from oblique convergence of the North American and Pacific plates (fig. 8). The mid-Cretaceous to Paleocene right-lateral transcurrent faults of the Columbia tectonic belt extend southeastward from British Columbia (Oldow and others, 1989) but do not continue south of the GFTZ. Similar Mesozoic right-lateral transcurrent faults, however, are present south of the projection of the GFTZ, in the Central tectonic belt of eastern California and western Nevada (Kistler, 1990; Oldow and others, 1989). In both the Columbia and Central tectonic belts, late Mesozoic movement on the transcurrent structures amounted to hundreds of kilometers. Thus, the GFTZ may have acted as a transtensional zone, or releasing bend, between the Columbia and Central tectonic belt transcurrent systems. This

Figure 6 (facing page). Laramide (75–42 Ma) igneous rocks and selected tectonic elements in the Western United States. Modified from Mutschler and others (1987, fig. 4).

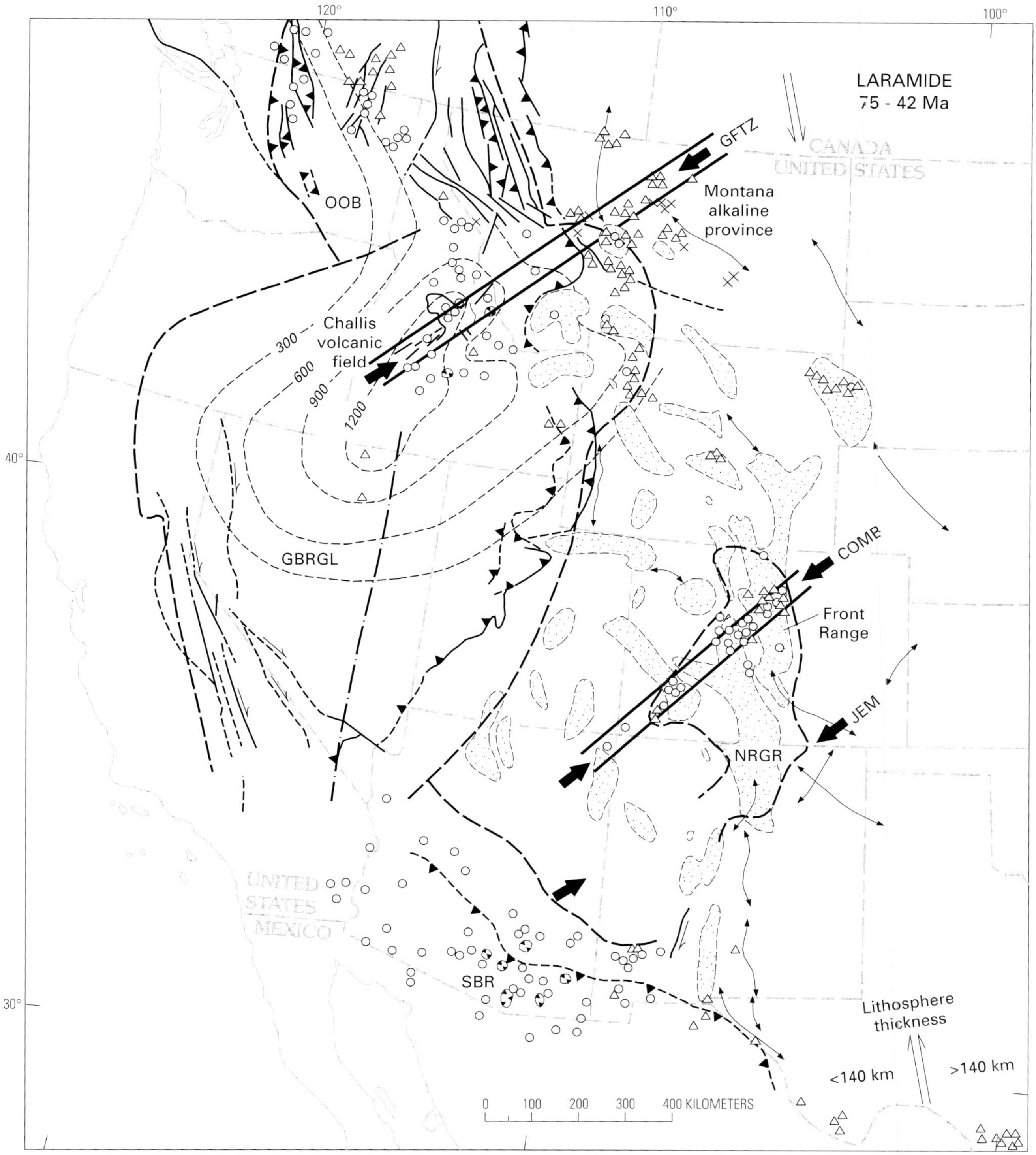

EXPLANATION

- Alkaline igneous center
- Lamproites, kimberlites, and other lamprophyric rocks
- Calc-alkaline igneous center
- Calc-alkaline caldera
- Rocky Mountain uplift
- Frontal Laramide thrust fault—Sawteeth on upper plate
- High-angle faults—Dashed where concealed or inferred
- --300-- Paleotopographic contours, in meters (modified from Axelrod, 1968)
- Major lineament
- COMB Colorado mineral belt
- GFTZ Great Falls tectonic zone
- JEM Jemez lineament
- GBRGL Great Basin regional gravity low
- NRGR Northern Rio Grande Rift
- OOB Omineca-Okanogan belt
- SBR Southern Basin and Range province

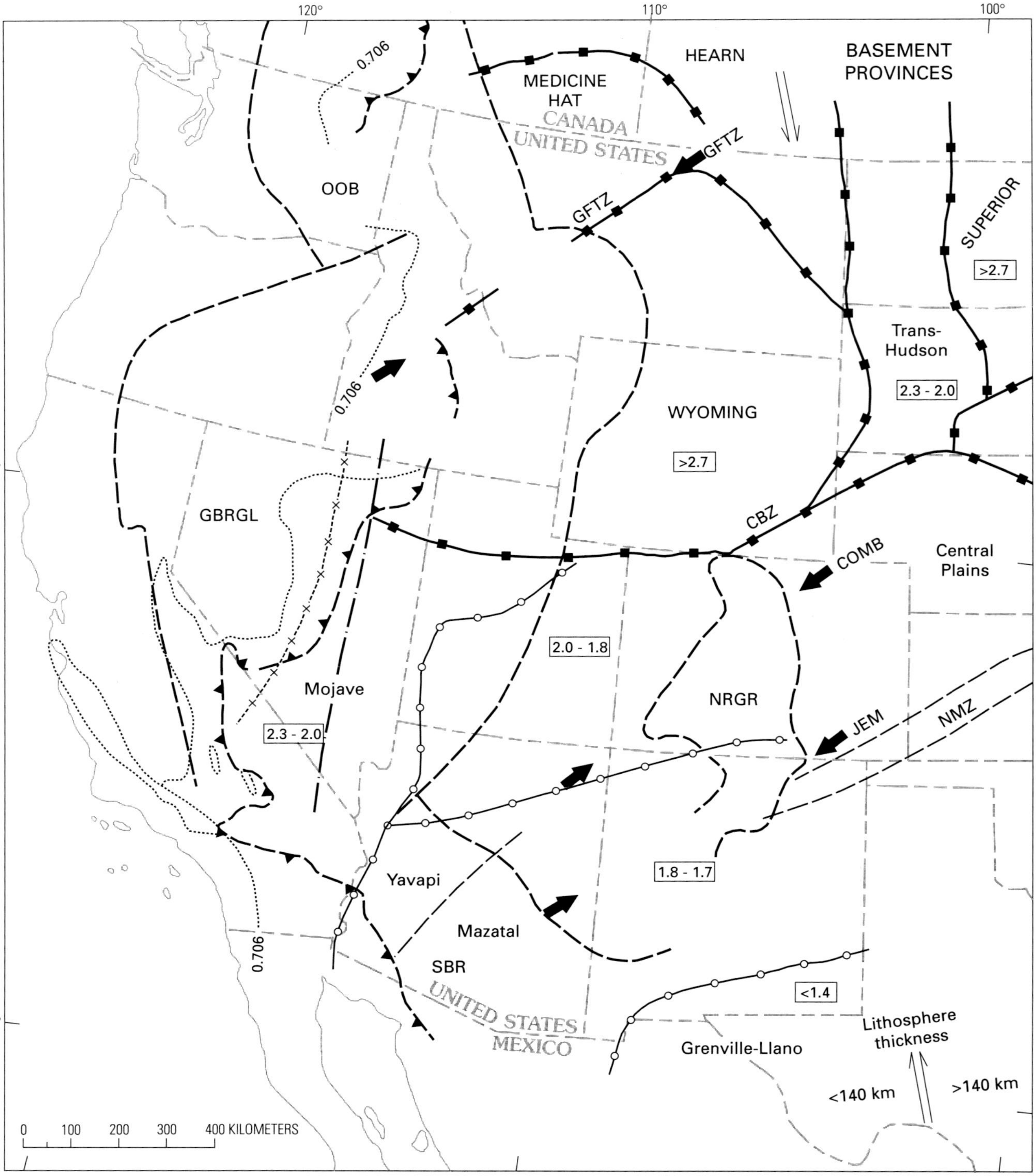

Figure 7 (above and facing page). Crustal provinces of the Western United States.

model is diagrammed in figure 8*B*, showing lithospheric mantle extension across the GFTZ axis beneath a decoupling zone. If the decoupling zone were fairly deep, evidence of the event in the crustal "orogenic float" could be sparse. Extension (shown in fig. 8*B* as occurring by pure shear) would have thinned the lithospheric mantle, resulting in upflow of hot deeper (asthenospheric) mantle. The influx of thermal energy, and perhaps magma, into the extended lithosphere would have set off the sequence of decompression melting, diapiric magma rise, local crustal

EXPLANATION

▲—▲	Edge of terranes accreted in the Phanerozoic—Sawteeth on outboard side
■—■	Major crustal province, or orogen, boundary (suture zone)
CBZ	Cheyenne belt suture zone (1.8–1.7 Ga)
GFTZ	Great Falls tectonic zone (≈1.8 Ga)
— — —	Second order crustal province boundary
NMZ	New Mexico-Michigan zone. Possible northern edge of 1.65 Ga accreted terranes. Based on filtered gravity data. Modified from van Schmus and others (1987)
···········	Initial $^{87}Sr/^{86}Sr$ isopleth = 0.706 for Mesozoic plutons. Approximates edge of Precambrian crust. Modified from Carlson and others (1991), Fleck and Criss (1985), and Kistler (1990)
--x-----x--	εNd = -7 and initial $^{87}Sr/^{86}Sr$ = 0.708 contour in Nevada and California which is interpreted as edge of Precambrian crust by Farmer and DePaolo (1983)
-o——o- >2.7	Crust-formation province age in Ga—Based on depleted mantle Nd-model ages from Bennett and DePaolo (1987)
WYOMING	Archean crustal province name
Mojave	Proterozoic crustal province name

ponding or penetration, and ultimately development of the paleotopographic and volcanic features recognized in the near-surface rock record of the Montana alkaline province and the Challis volcanic field. This model shares some features with the uplift and decompression scenarios suggested by Dudás (1991).

From its ≈50- to 45-Ma position beneath the Challis volcanic field, the magmatic focus of the GFTZ hot spot appears to have migrated southward during the ensuing 30 m.y. into central Nevada, as indicated by the successive 40-, 30-, and 20-Ma magmatic and caldera fronts shown in figure 9. The switch from northwest-directed extension across the GFTZ (with magmatism concentrated along the GFTZ) to east-northeast-directed extension (with southward-migrating magmatism) occurred at ≈48 Ma in east-central Idaho (Janecke, 1992). The Eocene-Miocene southward magmatic migration was essentially coeval with a southward sweep of upper crustal extensional domains (Seedorff, 1991). The ≈38- to 20-Ma ignimbrite flareup in the Great Basin (Best and others, 1989) resulted from the high-level emplacement of major calc-alkaline batholiths (fig. 10) during early, dominantly ductile, crustal extension (Gans and others, 1989). The southern limit of ≈30- to 20-Ma caldera-forming eruptions (fig. 9) approximately coincides with the east-trending Blue Ribbon–Warm Springs lineament (Rowley and others, 1978), possibly marking a major zone of transform accommodation between areas having different amounts of crustal extension (Eaton and others, 1978; Rowley and others, this volume).

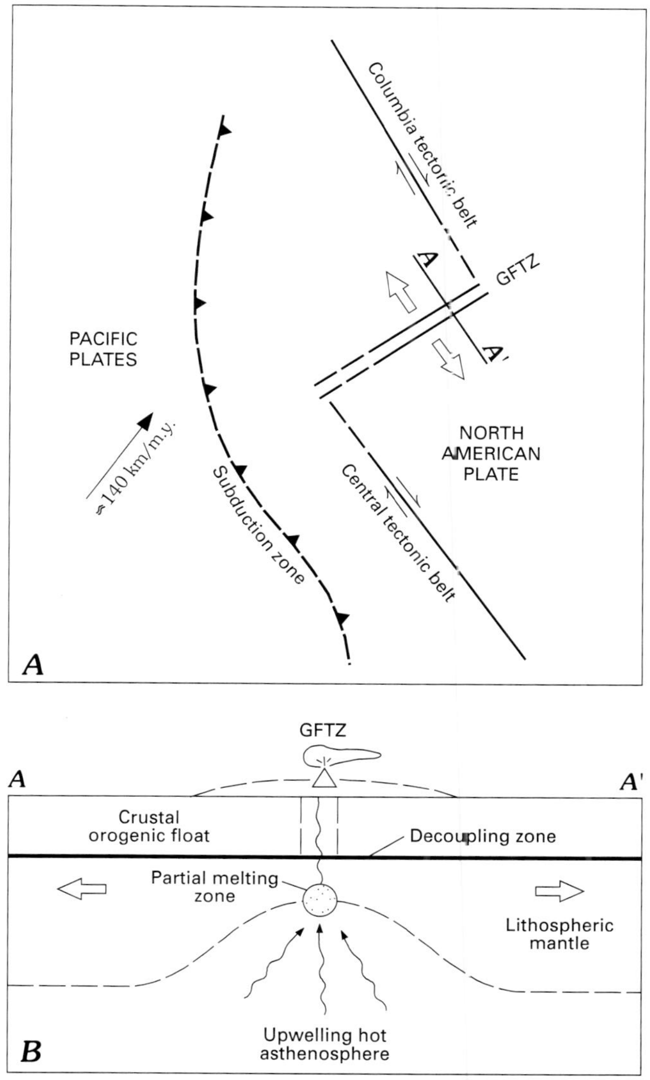

Figure 8. Diagrammatic map (*A*) and cross section (*B*) showing the Great Falls tectonic zone (GFTZ) as a Late Cretaceous–Eocene transtension zone between the Columbia and Central tectonic belt transcurrent fault systems (not to scale). From Mutschler and others (1991, fig. 2).

COLORADO MINERAL BELT (COMB)

The COMB hot spot initially developed along the Colorado mineral belt, a segment of a regional northeast-trending basement shear zone of Proterozoic origin (Tweto and Sims, 1963; Warner, 1980). During the Laramide orogeny, the COMB was oriented essentially parallel to the axis of maximum compression. Magmatism began shortly after the start of uplift of the Laramide ranges in Colorado (Mutschler and others, 1987) and was closely restricted to the axis of the COMB, which appears to have "unzipped" along a strike length of more than 500 km. The activity extended from the Carrizo Mountains, Ariz., in the Four Corners area, to the eastern edge of Colorado's Front Range (figs. 6, 11*A*)

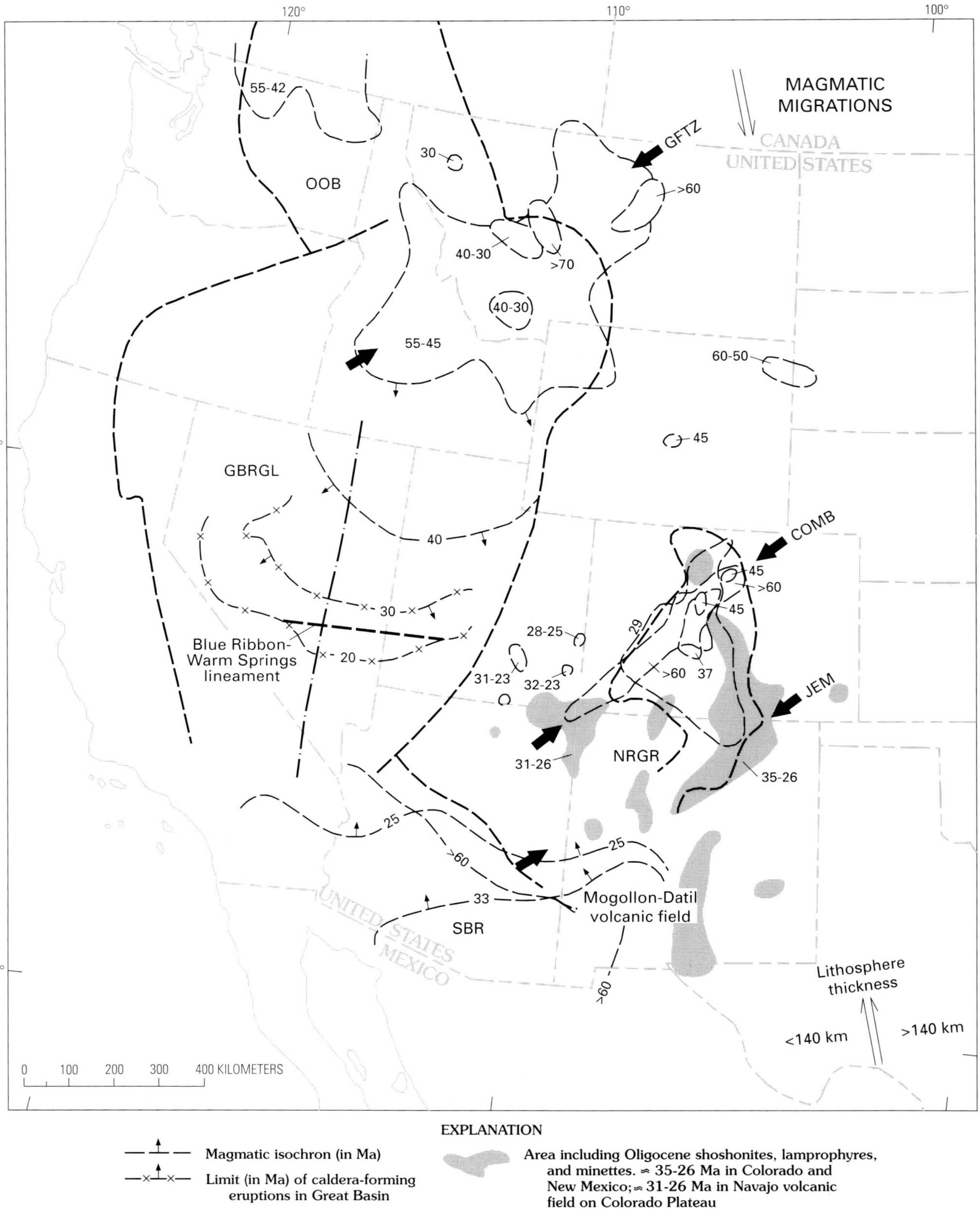

Figure 9. Magmatic migration patterns in the Western United States from about 75 to 20 Ma. See figure 1 for explanation of lines and symbols not explained here. Data from Best and others (1989), and Mutschler and others (1987). Arrows on isochrons show interpreted magmatic migration patterns.

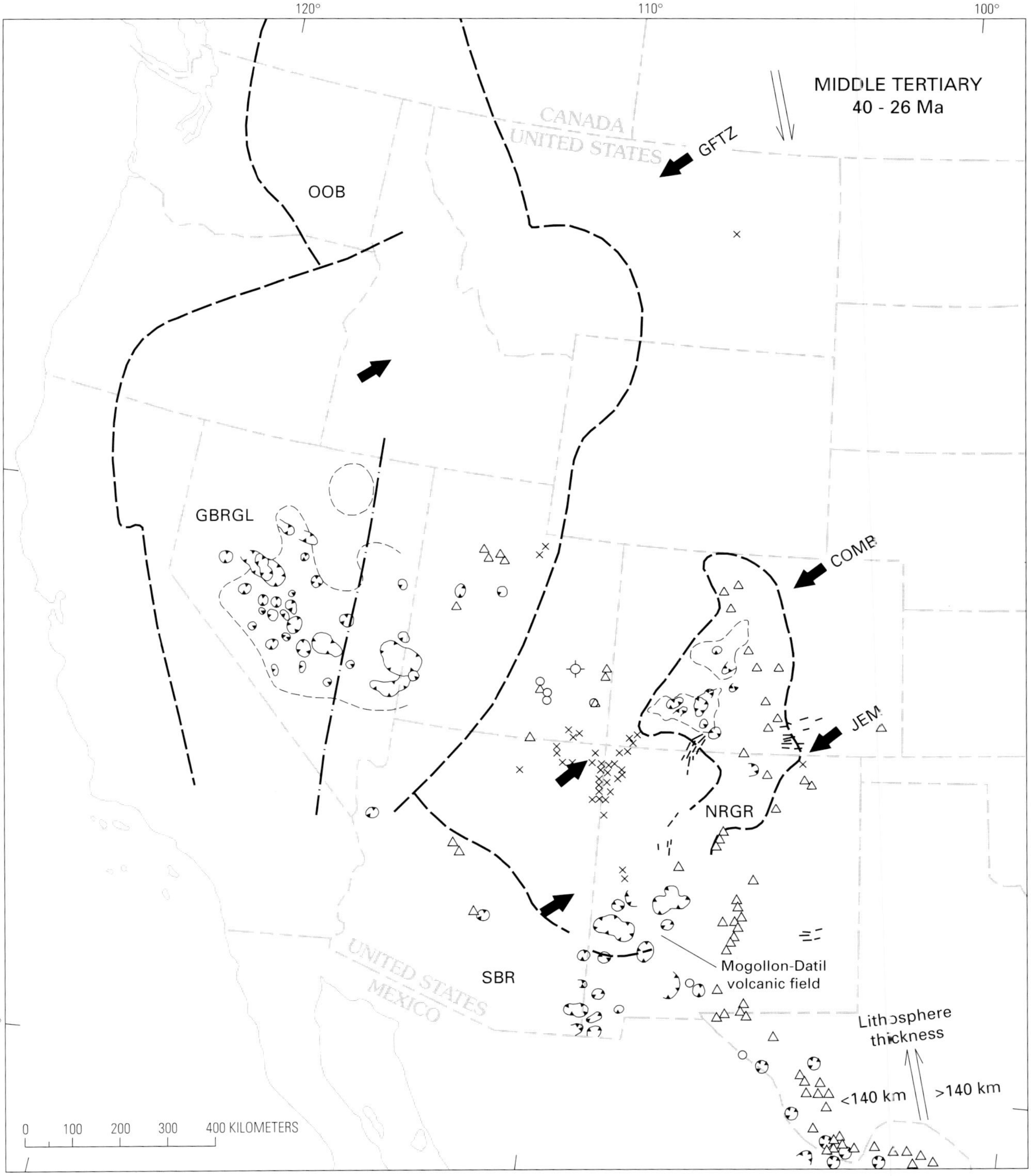

EXPLANATION

△	Alkaline igneous center	– – – – –	Extent of buried calc-alkaline batholith (based on gravity data)
⁄⁄⁄	Alkaline dike swarm	○	Calc-alkaline igneous center on Colorado Plateau
×	Minettes, lamproites, and other lamprophyric rocks	⌖	Upheaval Dome structure
	Calc-alkaline caldera		

Figure 10. Selected middle Tertiary (40–26 Ma) igneous features in the Southwestern United States. See figure 1 for explanation of lines and symbols not explained here. Modified from Mutschler and others (1987, fig. 10).

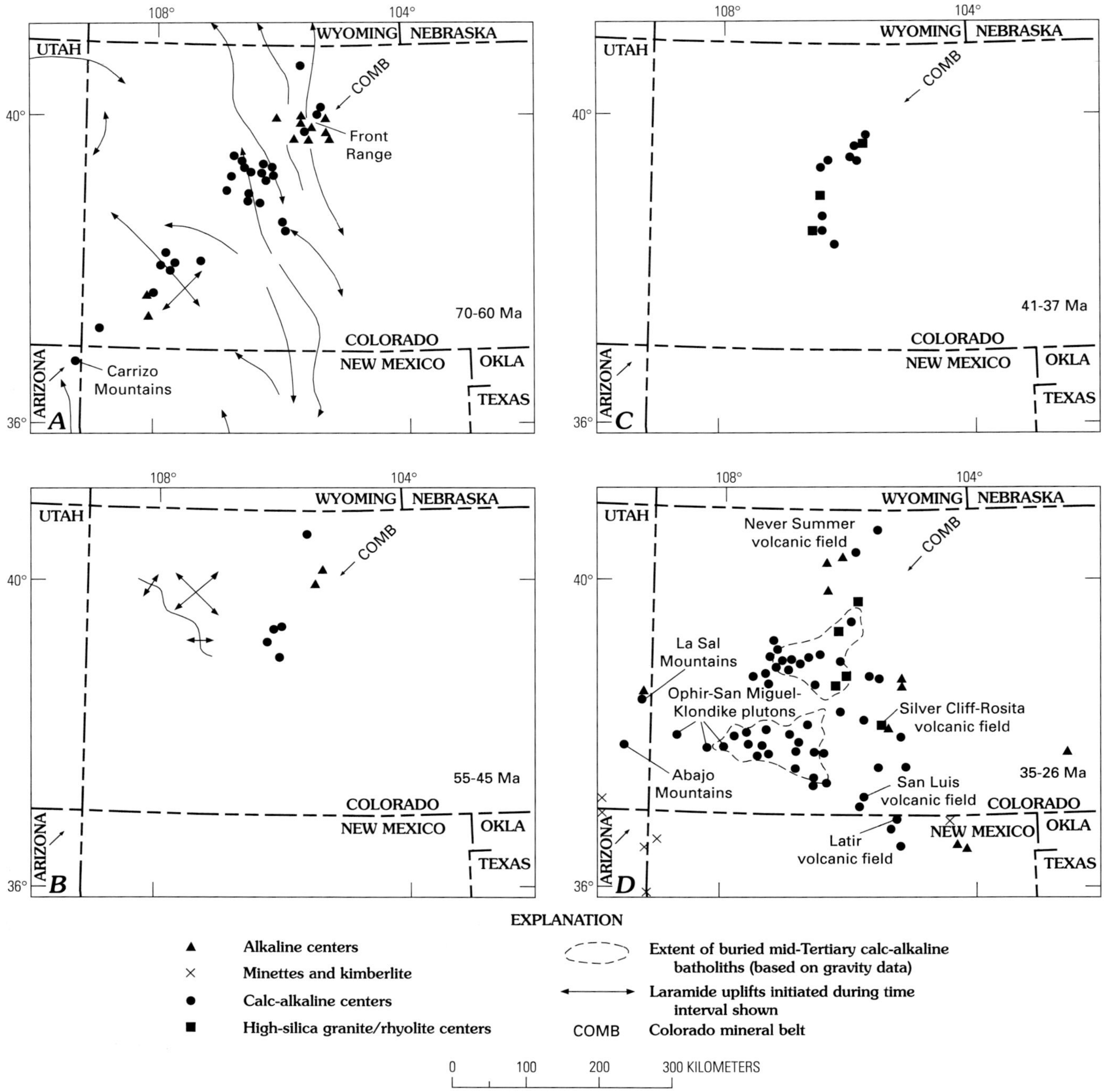

Figure 11. Laramide–middle Tertiary magmatic migration patterns, Colorado and environs. Modified from Mutschler and others (1987, fig. 11).

during the interval ≈74–64 Ma. No systematic age trends are apparent in rocks representing this time span along the COMB igneous belt, but mantle-derived alkaline rocks tend to be concentrated near the ends of the COMB, whereas calc-alkaline rocks containing significant crustal components predominate in the central part of the belt. By late Eocene time, igneous activity was restricted to the central and northeastern parts of the COMB (fig. 11*C*). The onset of regional crustal extension during middle Tertiary (Oligocene) time was marked by a rapidly enlarging ignimbrite flareup in central Colorado (fig. 11*D*), probably in response to massive basalt accumulation in or beneath the lower crust. This accumulation resulted in large-scale crustal melting, rise of the resulting calc-alkaline magmas to form shallow batholiths, and ignimbrite eruptions from at least 16 calderas during the period 36–27 Ma (Lipman, 1984; Steven and Lipman, 1976).

Small calc-alkaline centers—including the Abajo (32–23 Ma), Henry (31–23 Ma), and La Sal (28–25 Ma) Mountains, Utah; the Latir (26–19 Ma) volcanic field, New Mexico; and the San Luis (29–28 Ma), Silver Cliff–Rosita (33–27 Ma), Never Summer (29–28 Ma) volcanic fields, and the Ophir–San Miguel–Klondike (≈26 Ma) plutons, Colorado—developed outside of the central and southwestern Colorado batholithic area. Many of these peripheral centers began about 31–26 Ma, several million years after the onset of the voluminous mid-Tertiary batholithic magmatism along the COMB. Thus, from an early focus in central Colorado, the areas involved in middle Tertiary partial melting appear to have spread outward for about 10–12 m.y. (fig. 11).

Mutschler and others (1987) suggested that the COMB passive hot spot developed in response to decompression-triggered partial melting beneath isostatically rebounding crustal and lithospheric roots produced by Laramide compression. The model may be overly simplistic, especially as it failed to take into account possible regional lithospheric thinning resulting from differential subcrustal movements. Chapin (1983) documented a series of north-trending Eocene right-lateral faults and fault-bounded basins extending the length of the eastern Rocky Mountain uplifts of Colorado and New Mexico. Perhaps these crustal wrench structures reflect the thinning of partially decoupled lithosphere in a manner similar to that suggested for the GFTZ hot spot.

NORTHERN RIO GRANDE RIFT (NRGR)

The NRGR hot spot is in the north-central part of the Alvarado Ridge of Eaton (1986, 1987), which is a >1,200-km-long, north-trending, Neogene thermotectonic uplift (fig. 12). Eaton (1987) convincingly modeled the ridge crest as a feature that rose rapidly above the axis of a developing linear asthenospheric bulge beneath thinning lithospheric mantle. The model is supported by geophysical data (Eaton, 1987; Olsen and others, 1987; Cordell and others, 1991; Gibson and others, 1993) indicative of thinned crust and anomalously low-density mantle, and by regional heat-flow observations. Eaton (1986, 1987) suggested that the topographic ridge began to form at ≈17–12 Ma, and that uplift peaked between 7 and 4 Ma. The NRGR passive hot spot, however, may have a significant older history, including north-trending Precambrian shear zones (Cordell, 1978; Eaton, 1979; Tweto, 1979), which were reactivated in the Eocene wrenching event, and a magmatic episode of initial mantle melting between 35 and 26 Ma.[4] This magmatic precursor to the Miocene-Pliocene uplift event is represented by a north-trending belt of Oligocene mantle-derived shoshonitic plutons and lamprophyres extending from northern Colorado through New Mexico (fig. 9). In contrast, the Neogene period of rapid ridge uplift was characterized by bimodal basalt-rhyolite volcanism. Both tholeiitic and alkali basalts occur, representing lithospheric and asthenospheric mantle melting (Livaccari and Perry, 1993). The coeval high-silica rhyolites may represent melting of crustal granulites.

The differing locations and eruption times of Neogene lithosphere- and asthenosphere-derived basalts in different segments of the rift (Baldridge and others, 1984, 1991; Lipman, 1969; Perry and others, 1987, 1988) may result from local differences in the shear mechanisms involved in subcrustal lithospheric extension (fig. 13). Northeast-trending accommodation zones transverse to the Rio Grande Rift also appear to separate distinct tectonic and magmatic crustal blocks. The most striking accommodation zone is part of the Jemez lineament, which has acted as a >300-km-long locus for Neogene magmatism (Aldrich, 1986). The 15- to 0.001-Ma magmatism along the lineament marks a northwestward volcanic encroachment onto the Colorado Plateau (Aldrich and Laughlin, 1984; Baldridge and others, 1991).

GREAT BASIN REGIONAL GRAVITY LOW (GBRGL)

The Great Basin regional gravity low (GBRGL) of Eaton and others (1978) has long been recognized as a site of relatively rapid Neogene crustal extension. Lower crustal ductile extension of thickened Nevadan and Sevier lithosphere may have begun in the Cretaceous (Hodges and Walker, 1992), and significant normal faulting occurred during Eocene time (Gans and others, 1993). However, the majority of upper crustal brittle extension (basin-range faulting) did not begin until ≈20–17 Ma (Eaton and others, 1978) and it postdates the major part of the Oligocene great ignimbrite flareup. Upper crustal brittle extension is continuing today (Smith, 1978). As a result of this long-lived extension, the Great Basin is characterized by thin (≈30 km) crust underlain by anomalous mantle, high heat flow, regional doming, many calderas and voluminous ash-flow tuffs succeeded by modest amounts of bimodal (basalt-rhyolite) volcanics, and topography developed by basin-range faulting. Basaltic volcanism generally becomes more recent toward the Sierra Nevada and Wasatch transition zones bordering the Great Basin, and it has progressively overstepped these zones (Smith and Luedke, 1984; Stewart and Carlson, 1976). Neogene peralkaline rhyolites, probably derived from fractionation of trachybasalts, form an irregular ring around the periphery of the Great Basin (fig. 12). All of these features indicate regional mantle upwelling within a passive hot spot.

[4] Gregory and Chase (1992) used paleobotanical analysis to suggest that the Alvarado Ridge had reached essentially its present elevation by 35 Ma. This early uplift may be related to the ≈30-Ma low-angle normal faulting in the rift region discussed by Olsen and others (1987).

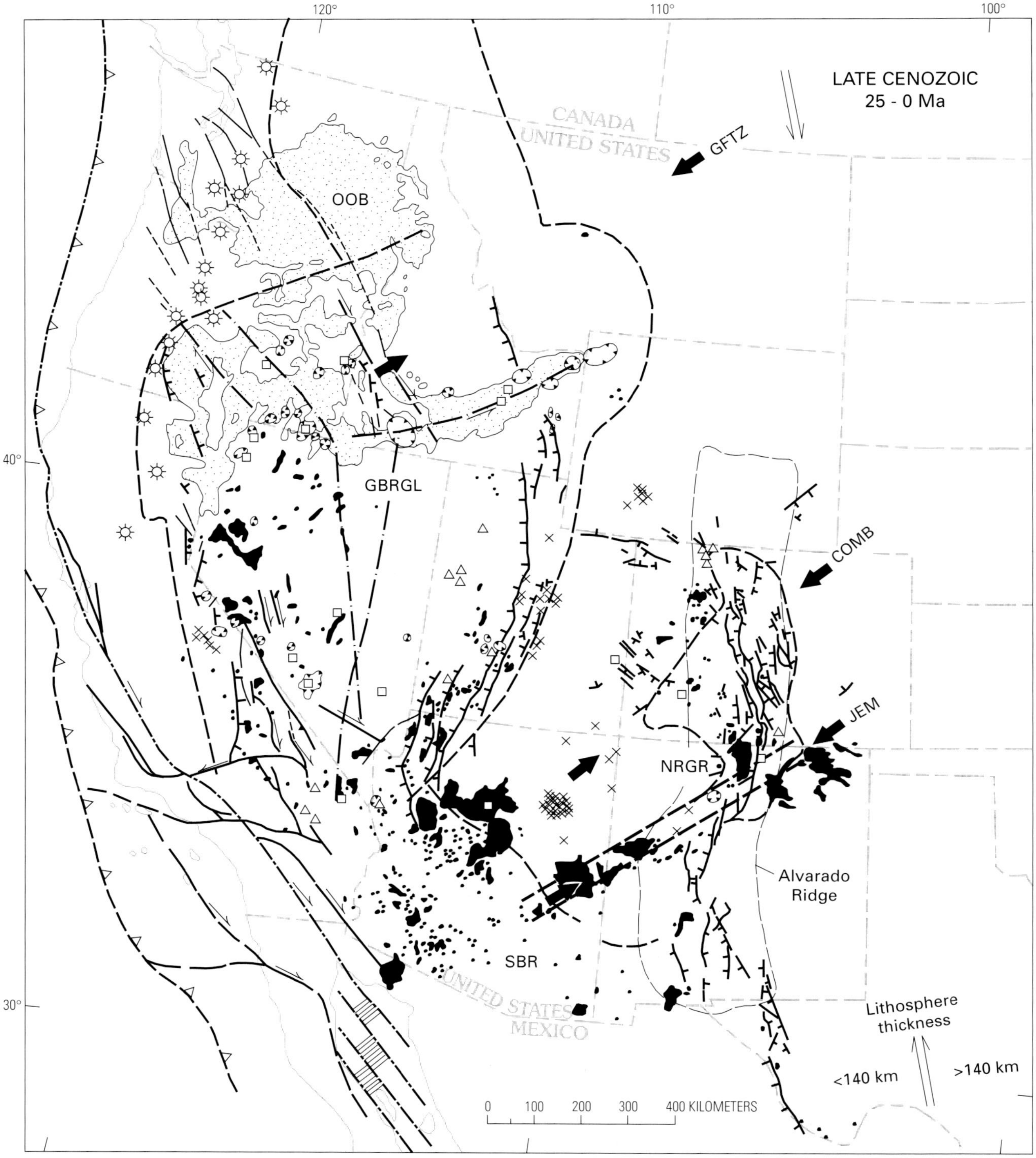

Figure 12 (above and facing page). Late Cenozoic (25–0 Ma) igneous rocks and selected tectonic features in the Western United States. Modified from Mutschler and others (1987, fig. 14).

The northern and southern borders of the Great Basin also show the effects of an evolving passive hot spot. The northwest-trending Brothers fault zone in Oregon (Lawrence, 1976) and the northeast-trending Snake River Plain in Idaho meet and form a "triple junction" with the south-southeast-trending northern Nevada or Oregon-Nevada magnetic lineament (Blakely, 1988; Stewart and others, 1975) near the common boundary of Oregon,

EXPLANATION

△ Alkaline igneous center

× Lamproites, lamprophyres

□ Peralkaline rhyolite/granite center (includes calderas)

Basalts, largely alkaline (south of 42°N.)—Locally includes small to moderate amounts of intermediate and silicic lavas

Flood basalts of Oregon Plateau, Columbia Plateau, and Snake River Plain

Major High Cascade stratovolcano

Caldera—Most result from rhyolitic eruptions

Extensional faults, shown only in Southern Rockies, Wasatch transition zone, and western edge of Great Basin

Major strike-slip fault

Other faults

Alvarado Ridge crestal province (Eaton, 1986, 1987)

Major lineament
JEM, Jemez lineament

Magnetic lineament in Nevada; transform accommodation zones in Oregon and Idaho

Active transform plate boundary

Active subduction plate boundary

Spreading ridge crust formed in last 1 million years

Nevada, and Idaho (fig. 14). This "triple junction" is similar to the radial rift geometry on a rising dome or on an inflating shield volcano, and may mark the crestal area of passive mantle upwelling. Neogene volcanism at the "triple junction" has an age of 17–16 Ma, but the volcanic features, especially silicic centers, become progressively younger outward on two of the three arms: northwestward along the Brothers fault zone to Newberry Crater (MacLeod and others, 1976) and northeastward along the Snake River Plain to Yellowstone (Christiansen, 1993; Christiansen and McKee, 1978). These two lineaments can be interpreted as "***diffuse (and very leaky) zones of transform accommodation between regions of greater and lesser cumulative tectonic (basin-range) extension to the south and north, respectively" (Hildreth and others, 1991, p. 65). The Quaternary Yellowstone Plateau volcanic field, therefore, is probably not the site of an active hot spot, but rather the northeast corner of a very large shield-shaped area of extended lithosphere located above the expanding passive hot spot that underlies the GBRGL. The southern end of the GBRGL can be interpreted in a similar fashion, with the Garlock fault (Davis and Burchfiel, 1973) and the Las Vegas shear zone serving as diffuse (but relatively dry) zones of transform accommodation.

SOUTHERN BASIN AND RANGE PROVINCE (SBR)

Laramide northeast-southwest compression destroyed a Cretaceous marine trough in southeastern Arizona and

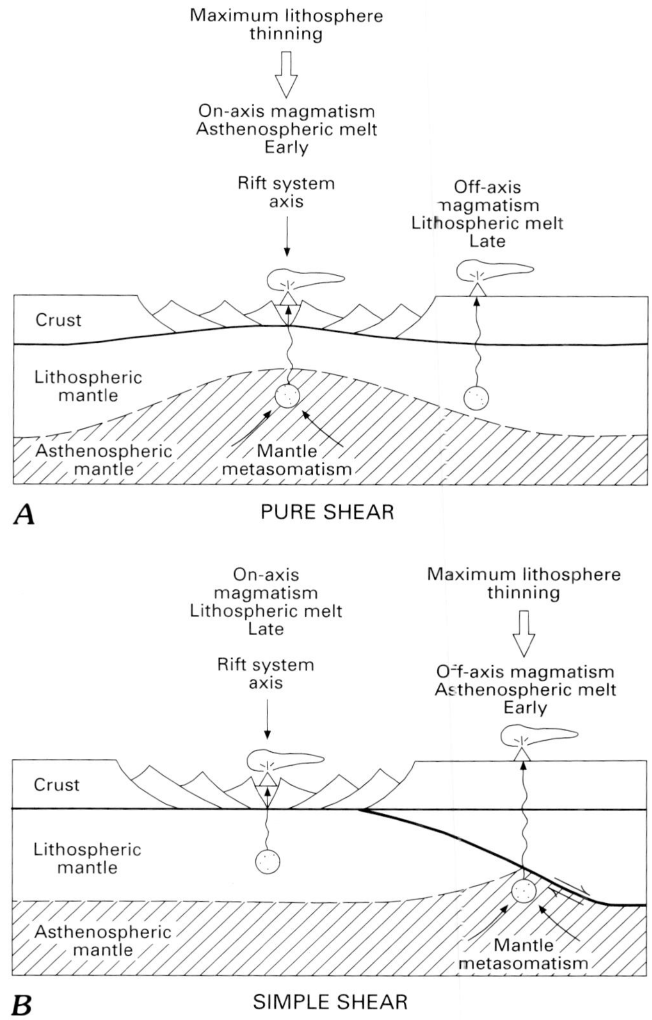

Figure 13. Hypothetical timing and distribution of mantle-derived magmatism and lithospheric thinning resulting from pure shear (*A*) and simple shear (*B*) modes of crustal extension. Modified from Farmer and others (1989, fig. 1).

southwestern New Mexico between ≈80 and 50 Ma. Deformation included uplift of basement welts and thrust faulting accompanied by extensive ≈75- to 50-Ma calc-alkaline plutonism and volcanism (Krantz, 1989). In southwestern Arizona, regional greenschist-facies metamorphism accompanied thrusting and plutonism. These Laramide events almost certainly resulted in significant crustal thickening. Yet today the southern Basin and Range province (SBR) is characterized by thin crust (fig. 2), evidence of large-scale lithospheric extension, high heat flow, and recent volcanism. These features, typical of passive hot spots, evolved during post-Laramide time.

Middle Tertiary ductile lithosphere extension involved development of major regional low-angle detachment faults and the isostatic uplift of metamorphic core complexes

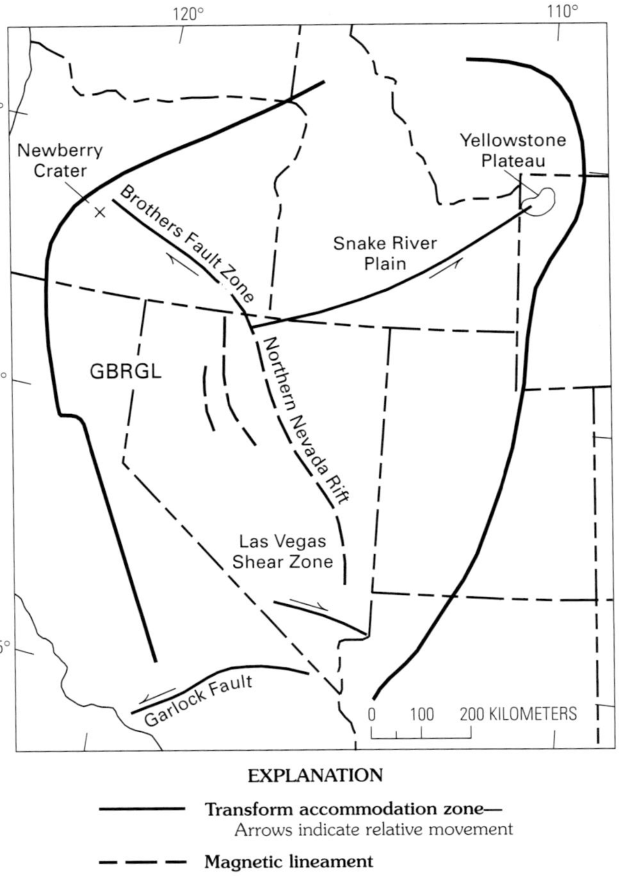

Figure 14. Late Cenozoic (≈17–0 Ma) major crustal tectonic elements that indicate an expanding passive hot spot beneath the Great Basin regional gravity low (GBRGL).

(Spencer and Reynolds, 1989). Middle Tertiary calc-alkaline-dominated magmatism shows a general westward and northwestward progression in the SBR, from a ≈40- to 36-Ma inception in the Mogollon-Datil volcanic field of southwestern New Mexico (Elston and Bornhorst, 1979; McIntosh and others, 1992) to an onset at ≈25 Ma in western Arizona (fig. 9). Major caldera formation and ignimbrite eruptions occurred between 36 and 24 Ma in the Mogollon-Datil field (fig. 10; McIntosh and others, 1992), and between ≈32 and 15 Ma in the Arizona part of the SBR (Nealey and Sheridan, 1989).

About 15–13 Ma, styles of deformation and magmatism changed significantly in the SBR (Menges and Pearthree, 1989). Brittle crustal extension began, in the form of high-angle normal (basin-range) faulting, and bimodal (basalt-rhyolite) magmatism became dominant. Bimodal volcanic features and, especially, rhyolitic centers show a northeastward migration from the SBR onto the Colorado Plateau since ≈15 Ma (Moyer and Nealey, 1989; Nealey and Sheridan, 1989).

The main pulse of basin-range faulting and magmatism ended at ≈5–2 Ma, although some normal faulting and seismic activity continue today, and at least four alkali basalt eruptions have occurred in and near Arizona in Holocene time (Lynch, 1989). Seismic reflection data suggest the presence of a horizontal basaltic magma body and solidified intrusions within the lower crust of the transition zone between the SBR and the Colorado Plateau (Parsons and others, 1992).

SCENARIO FOR THE EVOLUTION OF CONTINENTAL PASSIVE HOT SPOTS

A generalized model for the sequential development of the passive hot spots described is given herein. Some features of individual hot spots vary from this model.

1. Lithospheric thinning may be initiated by differential movements between lithospheric blocks, by back-arc spreading, or by gravitational collapse of an orogenic welt, all of which are common results of large-scale plate tectonic motions and reorganizations.

2. Thinning of the lithospheric mantle, which may be mechanically uncoupled from the crust, results in an upflow of the expanding asthenosphere, triggering decompression melting in the mantle. Early melts tend to be of two types: (a) Small volumes of mafic potassic magmas (such as alkaline lamprophyres or minettes) representing minimal mantle melting. These highly volatile-charged magmas generally transit through the lithosphere rapidly, with only minor fractionation en route. (b) Shoshonites, representing crustal-level fractionation and contamination of nepheline-normative alkaline basalts (Meen and Curtis, 1989). These may form moderate-size volcanic-plutonic complexes.

3. Continuing extension of the lithosphere causes increased mantle melting; the resulting basalt magmas rise and park at neutral buoyancy levels near the base of the crust (Glazner and Ussler, 1988) and (or) form distributed dike intrusion networks in the lower lithosphere and crust (Lachenbruch and Sass, 1978). Gentle regional crustal doming begins at this stage. Heat loss from gravitationally stalled basalts causes partial crustal melting, yielding calc-alkaline magmas which rise and collect at upper crustal neutral buoyancy levels, ultimately forming batholiths. Initial eruptions from the batholiths form intermediate-composition stratovolcano fields. These early andesites represent mixed mantle and crustal melts. As crustal melting continues, the bulk composition of the batholiths becomes increasingly silicic (dacitic to rhyolitic), and as the batholiths enlarge they contribute to crustal arching and thermally weaken the crust so that it extends ductilely (Armstrong and Ward, 1991; Gans and others, 1989). With time, as mantle melting spreads over a broader area, the zone of parked basalt in the lower crust spreads laterally, resulting in outward migration of the area of calc-alkaline magmatism as the zone of crustal melting widens and moves upward. (See figs. 9, 11.)

Eventually, many of the roofs of the fractionated calc-alkaline batholiths fail, producing multiple caldera eruptions and regional ignimbrite fields.

4. Finally, the thermally weakened upper crust may fail by listric (basin-range) faulting above the zone of ductile flow and distributed magmatic extension in the lower crust and uppermost mantle. Bimodal (basalt and rhyolite and (or) trachybasalt and peralkaline rhyolite) magmatism accompanies the basin-range faulting stage. These bimodal assemblages tend to be concentrated peripheral to earlier calc-alkaline batholiths, perhaps because the low-density batholiths inhibit the passage of mantle-derived magmas. The basalt and high-silica rhyolite suite probably represents limited crustal melting, inasmuch as the high-silica rhyolites have minimum melting compositions. The peralkaline rhyolites, which tend to be slightly older than the high-silica rhyolites, may represent fractionation of mantle-derived trachybasalts.

SUMMARY AND CONCLUSIONS

In the area we discuss, inboard passive hot-spot magmatism began in Late Cretaceous to Paleocene time with the development of the Great Falls tectonic zone and the Colorado mineral belt. Both these features started as linear volcanic-plutonic zones and expanded into large volcanic fields overlying calc-alkaline batholithic complexes.

Today, the Colorado Plateau is surrounded by Neogene passive hot spots, including the Great Basin regional gravity low to the west, the northern Rio Grande Rift to the east, and the southern Basin and Range province to the south and southwest. Magmatism in these areas is dominated by predominantly bimodal (alkali basalt-rhyolite) suites. Magmatic migration patterns (Nealey and Sheridan, 1989; Smith and Luedke, 1984) show that late Cenozoic magmatism is overstepping the plateau from all of these passive hot spots. Can basin-range faulting be far behind?

Armstrong and Ward (1991) and Ward (1991) have recently outlined and commented on many of the plate-tectonic scenarios invoked to explain Cordilleran magmatism. They emphasized, as have many other workers, a close spatial and temporal correlation between areas of crustal extension and inboard Cenozoic magmatism throughout the length of the Cordillera. Anderson (1992) has succinctly stated that the location of a hot spot is controlled by lithospheric conditions, and that even if the asthenosphere is relatively hot, a hot spot will not form unless the lithosphere is under extension. Lithospheric extension occurred at all the hot spots we describe, but different plate-motion phenomena were responsible for extension at different localities. For example, in the early Great Falls tectonic zone, oblique plate convergence produced intraplate transcurrent fault systems offset by a transtensional zone across which the lithosphere thinned. On the other hand, the late Cenozoic Great Basin regional gravity low, in Atwater's (1970) model, developed through gravitational collapse of an orogenic welt when a bounding plate margin changed from a subduction mode to a transform mode. In other cases (such as the Laramide to mid-Cenozoic Colorado mineral belt), it remains uncertain how plate interactions and motions relate to demonstrable inboard lithospheric extension and magmatism.

Since different plate-tectonic scenarios are involved in the development of different hot spots, it appears that direct involvement of a subducted oceanic slab is not a requisite for generation of the passive hot-spot magmatism we describe. Consequently, to term the igneous rocks of these inboard hot spots "subduction related" or "arc related" is perhaps misleading. Rather, the magmatism we describe can be considered to be "continental magmatism" in the sense of Ward (1991).

ACKNOWLEDGMENTS

Our subcrustal flights of fancy are largely based on interpretation of local and regional geological, geochemical, geochronological, and geophysical studies by the many enthusiastic workers who, for over a century, have contributed to the literature of Cordilleran geology. Space does not permit us to list all the references we used, or perhaps abused, but many of them are listed in Mutschler and others (1987, 1994). Our views were sharpened, and tempered, by discussions with the participants of the July 1992 U.S. Geological Survey Workshop on Laccolithic Complexes of Southeastern Utah convened by Jules Friedman and Curt Huffman. We especially thank Dave Nealey and Bill Steele, who reviewed the manuscript and made constructive and thoughtful suggestions to improve it.

REFERENCES CITED

Aldrich, M.J., Jr., 1986, Tectonics of the Jemez lineament in the Jemez Mountains and Rio Grande Rift: Journal of Geophysical Research, ser. B, v. 91, p. 1753–1762.

Aldrich, M.J., Jr., and Laughlin, A.W., 1984, A model for the tectonic development of the southeastern Colorado Plateau boundary: Journal of Geophysical Research, ser. B, v. 89, p. 10207–10218.

Allenby, R.J., and Schnetzler, C.C., 1983, United States crustal thickness: Tectonophysics, v. 93, p. 13–31.

Anderson, D.L., 1992, Mantle dynamics—The hotcell model [abs.]: Eos, v. 73, no. 14, Spring Meeting Supplement, p. 272.

Armstrong, R.L., and Ward, Peter, 1991, Evolving geographic patterns of Cenozoic magmatism in the North American Cordillera—The temporal and spatial association of magmatism and metamorphic core complexes: Journal of Geophysical Research, ser. B, v. 96, p. 13201–13224.

Atwater, Tanya, 1970, Implications of plate tectonics for the Cenozoic tectonic evolution of western North America: Geological Society of America Bulletin, v. 81, p. 3513–3536.

Axelrod, D.I., 1968, Tertiary floras and topographic history of the Snake River basin, Idaho: Geological Society of America Bulletin, v. 79, p. 713–734.

Bailey, D.K., 1970, Volatile flux, heat-focusing and the generation of magma: Geological Journal (London), Special Issue 2, p. 177–186.

———1978, Continental rifting and mantle degassing, *in* Neumann, E.-R., and Ramberg, I.B., eds., Petrology and geochemistry of continental rifts: Dordrecht, Holland, D. Reidel, p. 1–13.

Baker, D.W., and Berg, R.B., eds., 1991, Guidebook of the central Montana alkalic province—Geology, ore deposits and origin: Montana Bureau of Mines and Geology Special Publication 100, 201 p.

Baldridge, W.S., Olsen, K.H., and Callender, J.F., 1984, Rio Grande Rift—Problems and perspectives: New Mexico Geological Society Guidebook, 35th Field Conference, Rio Grande Rift—Northern New Mexico, p. 1–12.

Baldridge, W.S., Perry, F.V., Vaniman, D.T., Nealey, L.D., Laughlin, A.W., Kyle, P.R., Bartov, Yosef, Steinitz, Gideon, and Gladney, E.S., 1991, Middle to late Cenozoic magmatism of the southeastern Colorado Plateau and central Rio Grande Rift (New Mexico and Arizona, U.S.A.)—A model for continental rifting: Tectonophysics, v. 197, p. 327–354.

Bayer, K.C., 1983, Generalized structural, lithologic, and physiographic provinces in the fold and thrust belts of the United States: U.S. Geological Survey special map, 2 sheets, scale 1:2,500,000.

Bennett, V.C., and DePaolo, D.J., 1987, Proterozoic crustal history of the Western United States as determined by neodymium isotopic mapping: Geological Society of America Bulletin, v. 99, p. 674–685.

Best, M.G., Christiansen, E.H., Deino, A.L., Grommé, C.S., McKee, E.H., and Noble, D.C., 1989, Eocene through Miocene volcanism in the Great Basin of the Western United States: New Mexico Bureau of Mines and Mineral Resources Memoir 47, p. 91–133.

Bird, Peter, 1979, Continental delamination and the Colorado Plateau: Journal of Geophysical Research, v. 84, p. 7561–7571.

——— 1988, Formation of the Rocky Mountains, Western United States—A continuum computer model: Science, v. 239, p. 1501–1507.

Blakely, R.J., 1988, Curie temperature isotherm analysis and tectonic implications of aeromagnetic data from Nevada: Journal of Geophysical Research, ser. B, v. 93, p. 11817–11832.

Carlson, D.H., Fleck, R.J., Moye, F.J., and Fox, K.F., 1991, Geology, geochemistry and isotopic character of the Colville igneous complex, northeastern Washington: Journal of Geophysical Research, ser. B, v. 96, p. 13313–13333.

Chapin, C.E., 1983, An overview of Laramide wrench faulting in the Southern Rocky Mountains with emphasis on petroleum exploration, *in* Lowell, J.D., ed., Rocky Mountain foreland basins and uplifts: Denver, Colo., Rocky Mountain Association of Geologists, p. 169–179.

Christiansen, R.L., 1993, The Yellowstone hot spot—Deep mantle plume or upper mantle melting anomaly [abs.]: Eos, v. 74, no. 43, supplement, p. 602.

Christiansen, R.L., and McKee, E.H., 1978, Late Cenozoic volcanic and tectonic evolution of the Great Basin and Columbia Intermontane regions, *in* Smith, R.B., and Eaton, G.P., eds., Cenozoic tectonics and regional geophysics of the western Cordillera: Geological Society of America Memoir 152, p. 283–311.

Clague, D.A., 1987, Hawaiian alkaline volcanism, *in* Fitton, J.G., and Upton, B.G.J., eds., Alkaline igneous rocks: Geological Society of London Special Publication 30, p. 227–252.

Cordell, Lindrith, 1978, Regional geophysical setting of the Rio Grande Rift: Geological Society of America Bulletin, v. 89, p. 1073–1090.

Cordell, Lindrith, Zorin, Y.A., and Keller, G.R., 1991, The decompensative gravity anomaly and deep structure of the region of the Rio Grande Rift: Journal of Geophysical Research, ser. B, v. 96, p. 6557–6568.

Courtney, R.C., and White, R.S., 1986, Anomalous heat flow and geoid across the Cape Verde Rise—Evidence of dynamic support from a thermal plume in the mantle: Geophysical Journal of the Royal Astronomical Society, v. 87, p. 815–868.

Davis, G.A., and Burchfiel, B.C., 1973, Garlock Fault—An intracontinental transform structure, southern California: Geological Society of America Bulletin, v. 84, p. 1407–1422.

Dudás, F.Ö., 1991, Geochemistry of igneous rocks from the Crazy Mountains, Montana, and tectonic models for the Montana alkalic province: Journal of Geophysical Research, ser. B, v. 96, p. 13261–13277.

Eaton, G.P., 1979, A plate tectonic model for late Cenozoic crustal spreading in the Western United States, *in* Riecker, R.E., ed., Rio Grande Rift—Tectonics and magmatism: Washington, D.C., American Geophysical Union, p. 7–32.

———1986, A tectonic redefinition of the Southern Rocky Mountains: Tectonophysics, v. 132, p. 163–194.

———1987, Topography and origin of the Southern Rocky Mountains and Alvarado Ridge, *in* Coward, M.P., Dewey, J.F., and Hancock, P.L., eds., Continental extensional tectonics: Geological Society of London Special Publication 28, p. 355–369.

Eaton, G.P., Wahl, R.R., Prostka, H.J., Mabey, D.R., and Kleinkopf, M.D., 1978, Regional gravity and tectonic patterns—Their relation to late Cenozoic epeirogeny and lateral spreading in the western Cordillera, *in* Smith, R.B., and Eaton, G.P., eds., Cenozoic tectonics and regional geophysics of the western Cordillera: Geological Society of America Memoir 152, p. 51–91.

Eggler, D.H., Meen, J.K., Welt, Fred, Dudás, F.Ö., Furlong, K.P., McCallum, M.E., and Carlson, R.W., 1988, Tectonomagmatism of the Wyoming province: Colorado School of Mines Quarterly, v. 83, no. 2, p. 25–40.

Elston, W.E., and Bornhorst, T.J., 1979, The Rio Grande Rift in context of regional post-40 m.y. volcanic and tectonic events, *in* Riecker, R.E., ed., Rio Grande Rift—Tectonics and magmatism: Washington, D.C., American Geophysical Union, p. 416–438.

Farmer, G.L., and DePaolo, D.J., 1983, Origin of Mesozoic and Tertiary granite in the Western United States and implications for pre-Mesozoic crustal structure—1, Nd and Sr isotopic studies in the geocline of the northern Great Basin: Journal of Geophysical Research, ser. B, v. 88, p. 3379–3401.

Farmer, G.L., Perry, F.V., Semken, S., Crowe, B., Curtis, D., and DePaolo, D.J., 1989, Isotopic evidence on the structure and origin of subcontinental lithospheric mantle in southern

Nevada: Journal of Geophysical Research, ser. B, v. 94, p. 7885–7898.

Fleck, R.J., and Criss, R.E., 1985, Strontium and oxygen isotopic variations in Mesozoic and Tertiary plutons of central Idaho: Contributions to Mineralogy and Petrology, v. 90, p. 291–308.

Gans, P.B., Mahood, G.A., and Schermer, Elizabeth, 1989, Synextensional magmatism in the Basin and Range province; a case study from the eastern Great Basin: Geological Society of America Special Paper 233, 53 p.

Gans, P.B., Miller, E.L., and Lee, J., 1993, The Mesozoic-Cenozoic tectonic evolution of the eastern Great Basin—A revised history [abs.]: Eos, v. 74, no. 43, p. 608.

Gibson, S.A., Thompson, R.N., Leat, P.T., Morrison, M.A., Hendry, G.L., Dickin, A.P., and Mitchell, J.G., 1993, Ultrapotassic magmas along the flanks of the Oligo-Miocene Rio Grande Rift, USA—Monitors of a zone of lithospheric mantle extension and thinning beneath a continental rift: Journal of Petrology, v. 34, p. 187–228.

Glazner, A.F., and Ussler, William, III, 1988, Trapping of magma at midcrustal density discontinuities: Geophysical Research Letters, v. 15, p. 673–675.

Gregory, K.M., and Chase, C.G., 1992, Tectonic significance of paleobotanically estimated climate and altitude of the late Eocene erosion surface, Colorado: Geology, v. 20, p. 581–585.

Hildenbrand, T.G., Simpson, R.W., Godson, R.H., and Kane, M.F., 1982, Digital colored residual and regional Bouguer gravity maps of the conterminous United States with cut-off wavelengths of 250 km and 1000 km: U.S. Geological Survey Geophysical Investigations Map GP–953–A, scale 1:7,500,000.

Hildreth, Wes, Halliday, A.N., and Christiansen, R.L., 1991, Isotopic and chemical evidence concerning the genesis and contamination of basaltic and rhyolitic magma beneath the Yellowstone Plateau volcanic field: Journal of Petrology, v. 32, p. 63–138.

Hodges, K.V., and Walker, J.D., 1992, Extension in the Cretaceous Sevier orogen, North American Cordillera: Geological Society of America Bulletin, v. 104, p. 560–569.

Irvine, T.N., 1989, A global convection framework—Concepts of symmetry, stratification, and system in the Earth's dynamic structure: Economic Geology, v. 84, p. 2059–2114.

Janecke, S.U., 1992, Kinematics and timing of three superposed extensional systems, east-central Idaho—Evidence for an Eocene tectonic transition: Tectonics, v. 11, p. 1121–1138.

Kistler, R.W., 1990, Two different lithosphere types in the Sierra Nevada, California, *in* Anderson, J.L., ed., The nature of Cordilleran magmatism: Geological Society of America Memoir 174, p. 271–281.

Krantz, R.W., 1989, Laramide structures of Arizona, *in* Jenney, J.P., and Reynolds, S.J., eds., Geologic evolution of Arizona: Arizona Geological Society Digest, v. 17, p. 463–483.

Lachenbruch, A.H., and Sass, J.H., 1978, Models of an extending lithosphere and heat flow in the Basin and Range province, *in* Smith, R.B., and Eaton, G.P., eds., Cenozoic tectonics and regional geophysics of the western Cordillera: Geological Society of America Memoir 152, p. 209–250.

Larsen, E.S., 1940, Petrographic province of central Montana: Geological Society of America Bulletin, v. 51, p. 887–948.

Lawrence, R.D., 1976, Strike-slip faulting terminates the Basin and Range province in Oregon: Geological Society of America Bulletin, v. 87, p. 846–850.

Lipman, P.W., 1969, Alkalic and tholeiitic basaltic volcanism related to the Rio Grande depression, southern Colorado and northern New Mexico: Geological Society of America Bulletin, v. 80, p. 1343–1354.

———1984, The roots of ash flow calderas in western North America—Windows into the tops of granite batholiths: Journal of Geophysical Research, ser. B, v. 89, p. 8801–8841.

Livaccari, R.F., and Perry, F.V., 1993, Isotopic evidence for preservation of Cordilleran lithospheric mantle during the Sevier-Laramide orogeny, Western United States: Geology, v. 21, p. 719–722.

Lynch, D.J., 1989, Neogene volcanism in Arizona—The recognizable volcanoes, *in* Jenney, J.P., and Reynolds, S.J., eds., Geologic evolution of Arizona: Arizona Geological Society Digest, v. 17, p. 681–700.

Mabey, D.R., and Webring, M.W., 1985, Regional geophysical studies in the Challis quadrangle, chap. E *of* McIntyre, D.H., ed., Symposium on the geology and mineral deposits of the Challis 1°×2° quadrangle, Idaho: U.S. Geological Survey Bulletin 1658, p. 69–79.

MacLeod, N.S., Walker, G.W., and McKee, E.H., 1976, Geothermal significance of eastward increase in age of upper Cenozoic rhyolitic domes in southeastern Oregon, *in* Proceedings, Second United Nations Symposium on the Development and Use of Geothermal Resources: Washington, D.C., U.S. Government Printing Office, v. 1, p. 465–474.

McIntosh, W.C., Chapin, C.E., Ratté, J.C., and Sutter, J.F., 1992, Time-stratigraphic framework for the Eocene-Oligocene Mogollon-Datil volcanic field, southwest New Mexico: Geological Society of America Bulletin, v. 104, p. 851–871.

Meen, J.K., and Curtis, P.C., 1989, What are shoshonites? [abs.]: Eos, v. 70, p. 1421.

Menges, C.M., and Pearthree, P.A., 1989, Late Cenozoic tectonism in Arizona and its impact on regional landscape evolution, *in* Jenney, J.P., and Reynolds, S.J., eds., Geologic evolution of Arizona: Arizona Geological Society Digest, v. 17, p. 649–680.

Moye, F.J., 1988, Extensional control of Eocene volcanism and plutonism, Idaho and Washington: Geological Society of America Abstracts with Programs, v. 20, no. 6, p. 434.

Moyer, T.C., and Nealey, L.D., 1989, Regional compositional variations of late Tertiary bimodal rhyolite lavas across the Basin Range/Colorado Plateau boundary in western Arizona: Journal of Geophysical Research, ser. B, v. 94, p. 7799–7816.

Mutschler, F.E., Johnson, D.C., and Mooney, T.C., 1991, A speculative plate kinematic model for the central Montana alkalic province and related gold deposits, *in* Baker, D.W., and Berg, R.B., eds., Guidebook of the central Montana alkalic province—Geology, ore deposits and origin: Montana Bureau of Mines and Geology Special Publication 100 p. 121–124.

——— 1994, ALK.BIB—A selected bibliography of alkaline igneous rocks and related mineral deposits, with an emphasis on western North America: U.S. Geological Survey Open-File Report 94–624, 222 p.

Mutschler, F.E., Larson, E.E., and Bruce, R.M., 1987, Laramide and younger magmatism in Colorado—New petrologic and tectonic variations on old themes: Colorado School of Mines Quarterly, v. 82, no. 4, p. 1–47.

Nealey, L.D., and Sheridan, M.F., 1989, Post-Laramide volcanic rocks of Arizona and northern Sonora, Mexico, and their inclusions, *in* Jenney, J.P., and Reynolds, S.J., eds., Geologic evolution of Arizona: Arizona Geological Society Digest, v. 17, p. 609–647.

Norman, M.D., and Mertzman, S.A., 1991, Petrogenesis of Challis volcanics from central and southwestern Idaho—Trace element and Pb isotopic evidence: Journal of Geophysical Research, v. 96, p. 13279–13293.

Oldow, J.S., Bally, A.W., Avé Lallemant, H.G., and Leeman, W.P., 1989, Phanerozoic evolution of the North American Cordillera; United States and Canada, *in* Bally, A.W., and Palmer, A.R., eds., The geology of North America—An overview: Boulder, Colo., Geological Society of America, The Geology of North America, v. A, p. 139–232.

Olsen, K.H., Baldridge, W.S., and Callender, J.F., 1987, Rio Grande Rift—An overview: Tectonophysics, v. 143, p. 119–139.

O'Neill, J.M., and Lopez, D.A., 1985, Character and significance of Great Falls tectonic zone, east-central Idaho and west-central Montana: American Association of Petroleum Geologists Bulletin, v. 69, p. 437–447.

Parsons, Tom, Howie, J.M., and Thompson, G.A., 1992, Seismic constraints on the nature of lower crustal reflectors beneath the extending southern transition zone of the Colorado Plateau, Arizona: Journal of Geophysical Research, ser. B, v. 97, p. 12391–12407.

Perry, F.V., Baldridge, W.S., and DePaolo, D.J., 1987, Role of asthenosphere and lithosphere in the genesis of late Cenozoic basaltic rocks from the Rio Grande Rift and adjacent regions of the southwestern United States: Journal of Geophysical Research, ser. B, v. 92, p. 9193–9213.

———1988, Chemical and isotopic evidence for lithospheric thinning beneath the Rio Grande Rift: Nature, v. 332, p. 432–434.

Rowley, P.D., Lipman, P.W., Mehnert, H.H., Lindsey, D.A., and Anderson, J.J., 1978, Blue Ribbon lineament, an east-trending structural zone within the Pioche mineral belt of southwestern Utah and eastern Nevada: U.S. Geological Survey Journal of Research, v. 6, p. 175–192.

Scholz, C.H., Barazangi, Muawia, and Sbar, M.L., 1971, Late Cenozoic evolution of the Great Basin, Western United States, as an ensialic interarc basin: Geological Society of America Bulletin, v. 82, p. 2979–2990.

Seedorff, Eric, 1991, Magmatism, extension, and ore deposits of Eocene to Holocene age in the Great Basin—Mutual effects and preliminary proposed genetic relationships, *in* Raines, G.L., Lisle, R.E., Schafer, R.W., and Wilkinson, W.H., eds., Geology and ore deposits of the Great Basin: Reno, Nev., Geological Society of Nevada, p. 133–177.

Smith, R.B., 1978, Seismicity, crustal structure, and intraplate tectonics of the interior of the western Cordillera, *in* Smith, R.B., and Eaton, G.P., eds., Cenozoic tectonics and regional geophysics of the western Cordillera: Geological Society of America Memoir 152, p. 111–144.

Smith, R.L., and Luedke, R.G., 1984, Potentially active volcanic lineaments and loci in western conterminous United States, *in* Explosive volcanism—Inception, evolution, and hazards: Washington, D.C., National Academy Press, p. 47–66.

Spencer, J.E., and Reynolds, S.J., 1989, Middle Tertiary tectonics of Arizona and adjacent areas, *in* Jenney, J.P., and Reynolds, S.J., eds., Geologic evolution of Arizona: Arizona Geological Society Digest, v. 17, p. 539–574.

Steven, T.A., and Lipman, P.W., 1976, Calderas of the San Juan volcanic field, southwestern Colorado: U.S. Geological Survey Professional Paper 958, 35 p.

Stewart, J.H., and Carlson, J.E., 1976, Cenozoic rocks of Nevada: Nevada Bureau of Mines and Geology Map 52, 4 sheets, scale 1:1,000,000.

Stewart, J.H., Walker, G.W., and Kleinhampl, F.J., 1975, Oregon-Nevada lineament: Geology, v. 3, p. 265–268.

Thompson, G.A., and Burke, D.B., 1974, Regional geophysics of the Basin and Range province: Earth and Planetary Science Annual Reviews, v. 2, p. 213–238.

Tweto, Ogden, 1979, The Rio Grande Rift System in Colorado, *in* Riecker, R.E., ed., Rio Grande Rift—Tectonics and magmatism: Washington, D.C., American Geophysical Union, p. 33–56.

Tweto, Ogden, and Sims, P.K., 1963, Precambrian ancestry of the Colorado mineral belt: Geological Society of America Bulletin, v. 74, p. 991–1014.

Ward, P.L., 1991, On plate tectonics and the geological evolution of southwestern North America: Journal of Geophysical Research, ser. B, v. 96, p. 12479–12496.

Warner, L.A., 1980, The Colorado lineament, *in* Kent, H.C., and Porter, K.W., eds., Colorado geology: Denver, Colo., Rocky Mountain Association of Geologists, p. 11–21.

Wernicke, B.P., Christiansen, R.L., England, P.C., and Sonder, L.J., 1987, Tectonomagmatic evolution of Cenozoic extension in the North American Cordillera, *in* Coward, M.P., Dewey, J.F., and Hancock, P.L., eds., Continental extensional tectonics: Geological Society Special Publication 28, p. 203–221.

White, Robert, and McKenzie, Dan, 1989, Magmatism at rift zones—The generation of volcanic continental margins and flood basalts: Journal of Geophysical Research, ser. B, v. 94, p. 7685–7729.

Geochemistry of Volcanic Rocks in the Marysvale Volcanic Field, West-Central Utah

By Charles G. Cunningham, Daniel M. Unruh, Thomas A. Steven, Peter D. Rowley, Charles W. Naeser, Harald H. Mehnert, Carl E. Hedge, *and* Kenneth R. Ludwig[1]

CONTENTS

FIGURES

TABLE

ABSTRACT

Igneous activity in the Marysvale volcanic field of west-central Utah occurred in three episodes. The first took place 34–22 Ma and was dominated by calc-alkaline, intermediate composition rocks of the Bullion Canyon Volcanics and Mount Dutton Formations. The ε_{Nd} increased with time, reflecting decreasing crustal interaction with time. The second episode, 23–14 Ma, was dominated by an alkali-rhyolite and basalt (bimodal) assemblage. The similarity in isotopic ratios, together with a change to alkali rhyolites, is interpreted to be due to derivation of the nearly eutectic rhyolites by remelting the batholithic rocks that previously fed the Bullion Canyon volcanoes. Potassium-rich mafic rocks were erupted early in the second episode and were followed by more normal basalts. The third episode, 9–5 Ma, was also characterized by a bimodal assemblage of volcanic rocks. The highly variable isotopic ratios of the rhyolites indicate derivation from variable sources, reflecting episodic intervals of extensional tectonics.

[1]Berkeley Geochronology Center, 2455 Ridge Road, Berkeley, CA 94709.

Igneous activity and associated mineralization in the Marysvale volcanic field along the west side of the Colorado Plateau in west-central Utah occurred mainly during three episodes: ≈34–22 Ma, 23–14 Ma, and 9–5 Ma. In the first episode, during a regime of tectonic convergence, two contrasting suites of rocks were erupted concurrently: (1) poorly evolved, crystal-poor, pyroxene-plagioclase-bearing andesitic rocks, called the Mount Dutton Formation, from deep sources along the southern flank of the volcanic field, and (2) crystal-rich, biotite-hornblende-plagioclase-bearing dacitic rocks, called the Bullion Canyon Volcanics, from an east-northeast-trending shallow batholith under the middle of the field (Rowley and others, 1979, 1994; Steven and others, 1979, 1990; Steven, Rowley, and Cunningham, 1984; Cunningham and others, 1983; Steven and Morris, 1987).

Eruptions of crystal-rich ash-flow tuff from centers genetically related to the Bullion Canyon Volcanics (fig. 1) resulted in the formation of the Three Creeks caldera at 27.5 Ma (Steven, 1981), the Big John caldera at about 24 Ma (Steven, Cunningham, and Anderson, 1984), and the Monroe Peak caldera at 23 Ma (Rowley and others, 1988a,b). The Monroe Peak caldera is the largest caldera in the Marysvale volcanic field and was formed at the end of Bullion Canyon volcanism (Steven, Rowley, and Cunningham, 1984).

Isotopic data for Bullion Canyon Volcanics and related rocks (table 1, fig. 2) show a weak trend of increasing ε_{Nd} (–10 at 29 Ma to –5 at 21–22 Ma) with decreasing age. Initial strontium isotopic values are nearly constant ($^{87}Sr/^{86}Sr$=0.7052–0.7062), whereas Pb isotopic compositions are variable. With the exception of one sample obviously contaminated by upper crustal material (No. 78-958, initial $^{87}Sr/^{86}Sr$ = 0.716, $^{206}Pb/^{204}Pb$=19.8; table 1), Mount Dutton samples tend to have lower ε_{Nd} (–11) and $^{206}Pb/^{204}Pb$ (17.95–18.15), but similar initial $^{87}Sr/^{86}Sr$ to Bullion Canyon samples (fig. 2). Potassium, Rb, and U contents in samples from the Bullion Canyon Volcanics also increased with time, but remained relatively constant in Mount Dutton rocks.

At about 23–22 Ma, the composition of the igneous rocks changed to a bimodal suite of alkali rhyolite and basaltic rocks. These rocks are the products of the combined second and third igneous episodes, which continued to be erupted into the Quaternary. This change, from intermediate to bimodal volcanism at 23–22 Ma, is interpreted to be a result of the change from subduction to initial extension along the eastern margin of the Basin and Range province. The oldest basaltic lavas, erupted at 23–21 Ma (potassium-rich mafic lavas in table 1), contain 3.85 to 4.07 weight percent K_2O and have initial $^{87}Sr/^{86}Sr$ =0.7057–0.7061, ε_{Nd} = –6.7 to –8.5, and $^{206}Pb/^{204}Pb$=18.47–19.03. The high initial $^{87}Sr/^{86}Sr$ and low ε_{Nd} of these samples relative to similar age Bullion Canyon samples may reflect either crustal contamination or input from a subducted slab (Best and others, 1980; Walker and others, 1991).

Voluminous but locally erupted rhyolites were prevalent during the second and third episodes, from 22 to 14 Ma and from 9 to 5 Ma. The largest volume of alkali rhyolite,

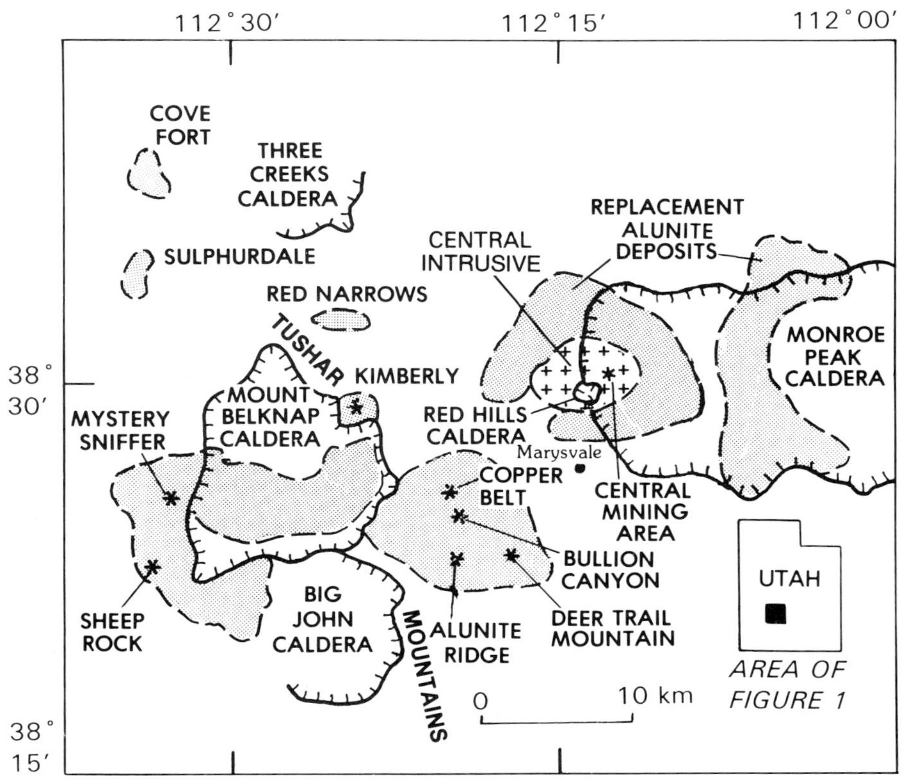

Figure 1. Location of calderas in central part of Marysvale volcanic field, Utah. Also shown are areas of altered rocks (shaded) and significant mineral deposits.

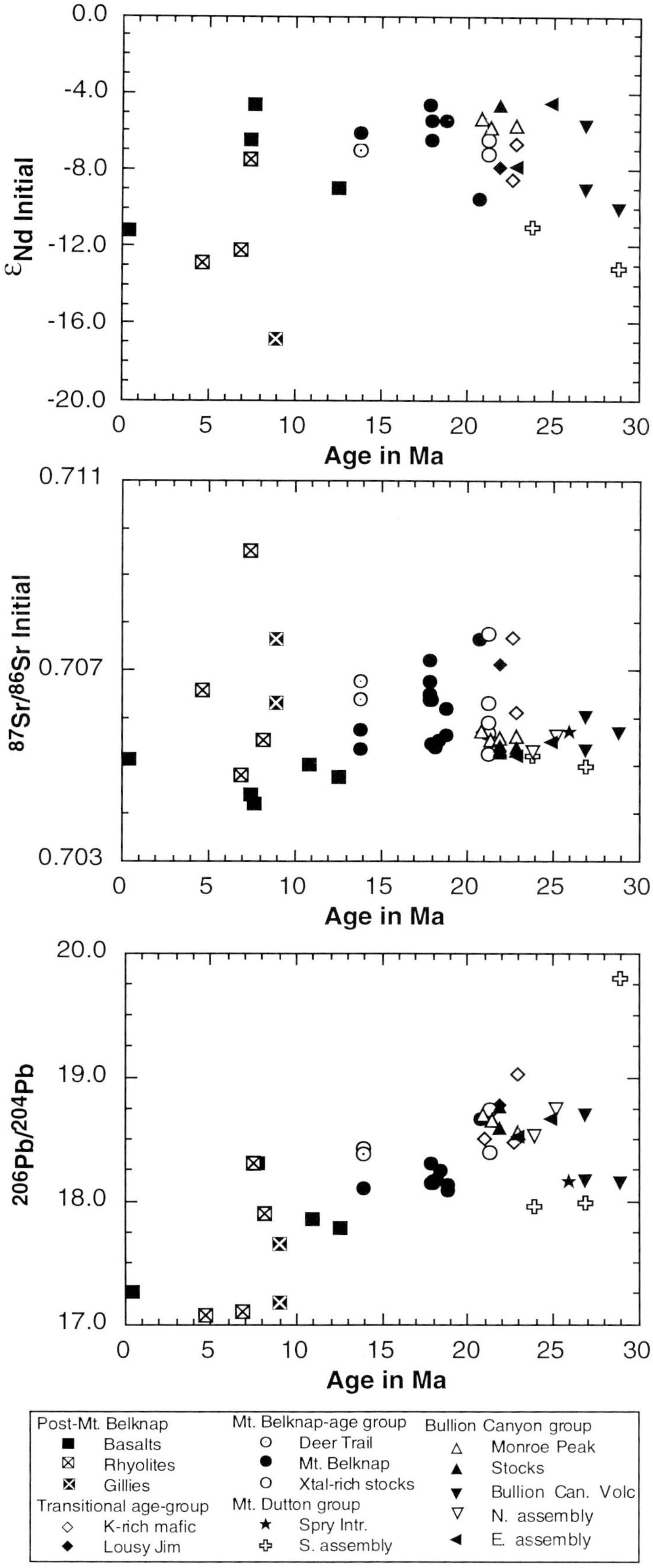

Figure 2. Plots of initial ε_{Nd}, $^{87}Sr/^{86}Sr$, and $^{206}Pb/^{204}Pb$ as functions of age for Marysvale samples.

called the Mount Belknap Volcanics (Cunningham and Steven, 1979; Budding and others, 1987), was erupted during the second episode (22–14 Ma, table 1). The Mount Belknap Volcanics was erupted from two areas (fig. 1): (1) the Mount Belknap caldera in the west, which formed at 19 Ma, and (2) a 12×12-km area, about 20 km east of the Mount Belknap caldera, that was active between 21 Ma at the northeast end of the area and 14 Ma at the southwest end (Cunningham and others, 1982).

The Mount Belknap Volcanics have higher silica contents (71.4–76.8 weight percent) than the older Bullion Canyon Volcanics (58.1–65.7 weight percent; table 1), as well as generally higher Rb contents (269–564 ppm) and lower Sr contents (23–100 ppm) than most of the Bullion Canyon Volcanics (69–142 ppm Rb, 473–928 ppm Sr).

The Mount Belknap Volcanics, in addition to having major-element compositions that contrast with the Bullion Canyon Volcanics and Mount Dutton Formation, have ε_{Nd} values similar to the latest Bullion Canyon Volcanics and among the highest ε_{Nd} values (–4.6 to –6.7, with one outlier; table 1, fig. 2) of any of the samples analyzed in this study. These data suggest that the processes involved in the production of these rhyolites involved comparatively little crustal contamination. Lead isotopic compositions of these samples (again with one exception) are tightly clustered with $^{206}Pb/^{204}Pb$=18.08–18.24 (table 1; fig. 2). In contrast, the initial $^{87}Sr/^{86}Sr$ values of the Mount Belknap Volcanics (0.7053–0.7072) are somewhat higher and more variable than those of the Bullion Canyon Volcanics (fig. 2). Because the Sr abundances of the Mount Belknap Volcanics are comparatively low, the Sr isotopic compositions of these samples were more susceptible to modification by relatively small amounts of crustal contamination than were those rocks of Bullion Canyon Volcanics, which have higher Sr contents.

During the third episode of igneous activity (9–5 Ma), sparse basaltic lavas had normal K_2O contents (1.82–2.97 weight percent), lower initial $^{87}Sr/^{86}Sr$ (0.7042–0.7051), highly variable ε_{Nd} (–4.7 to –11.2), and low but variable $^{206}Pb/^{204}Pb$ (17.3–18.3, table 1; fig. 2) when compared to the older, potassium-rich, basaltic lavas. Low initial Sr values in association with low ε_{Nd} and low $^{206}Pb/^{204}Pb$ have also been found among some late Tertiary basalt and basaltic andesite on the Markagunt Plateau approximately 80 km south of this study area (fig. 3), also along the western margin of the Colorado Plateau, by Nealey and others (in press). Although we have a rather limited data set, we note that the lowest $^{206}Pb/^{204}Pb$ and ε_{Nd} values are found in the sample (79-S-44) with the highest silica content, whereas the highest values are found in the only true basalt analyzed (78-164). This situation is analogous to that observed by

Table 1. Lead, Sr and Nd isotopic compositions of selected samples from the Marysvale Volcanic Field.

Sample	age (Ma)	SiO_2	$^{206}Pb/^{204}Pb$[1]	$^{207}Pb/^{204}Pb$[1]	$^{208}Pb/^{204}Pb$[1]	$^{87}Rb/^{86}Sr$	$^{87}Sr/^{86}Sr$[2]	$^{87}Sr/^{86}Sr$ Ini.	$^{147}Sm/^{144}Nd$	$^{143}Nd/^{144}Nd$[3]	$^{143}Nd/^{144}Nd$ Ini.	ε_{Nd} I
Basaltic lavas												
79-S-44	0.5	56.73	17.259	15.489	37.439	0.250	0.705122	0.70512	0.0998	0.512060	0.512060	-11.23
M848	7.6	55.12	18.302	15.589	37.434	0.257	0.704388	0.70436	0.1112	0.512297	0.512291	-6.53
78-164	7.8	48.65	18.297	15.568	37.907	0.230	0.704195	0.70417	0.1033	0.512390	0.512385	-4.71
M673	11	56.11	17.848	15.519	37.765	0.105	0.705026	0.70501				
M800	12.7	54.87	17.783	15.508	37.489	0.098	0.704748	0.70473	0.0914	0.512167	0.512159	-8.98
Rhyolites												
78-115	4.8	76.49	17.073	15.449	36.925	42.440	0.709433	0.7065	0.2048	0.511973	0.511967	-12.94
78-732	7	66.85	17.103	15.439	36.972	0.290	0.704789	0.70476	0.0985	0.512002	0.511997	-12.28
78-175	7.6	75.99	18.303	15.579	37.655	35.580	0.713320	0.7095	0.1134	0.512244	0.512238	-7.57
78-1695A	8.3	75.17	17.888	15.519	37.676	8.270	0.706495	0.70552				
Gillies Hill												
79-S-8A	9.1	69.61	17.171	15.479	37.652	0.919	0.707749	0.70763	0.0892	0.511762	0.511757	-16.92
79-S-9A	9.1	77.84	17.650	15.499	37.576	14.800	0.708223	0.70631				
MT. BELKNAP-AGE GROUP												
Deer Trail Rhyolite												
M945	14	73.2	18.417	15.586	38.537	0.628	0.706864	0.70674	0.0868	0.512296	0.512288	-6.43
Alunite												
M341	14	--	18.377	15.571	38.332	0.008	0.706362	0.70636	0.0503	0.512259	0.512254	-7.09
Mt. Belknap Volcanics												
M779A	14	76.15	18.097	15.568	38.080	24.120	0.710116	0.7053	0.1094	0.512315	0.512305	-6.11
M779B	14	75.31				23.300	0.710383	0.7058				
M687A	18	76.7	18.135	15.567	38.074	13.080	0.709814	0.70647				
M840A	18	76.54	18.303	15.596	38.371	45.230	0.717922	0.7064	0.1369	0.512391	0.512375	-4.64
M19A	18	75.64				24.340	0.712962	0.7067				
M672A	18.5	71.44	18.239	15.577	38.248	6.807	0.707308	0.70552				
M408A	18.1	76.76	18.143	15.567	38.234	1.945	0.706880	0.70638	0.0937	0.512290	0.512279	-6.51
M841	18	75.68				33.610	0.715782	0.7072				
M502A	18.1	76.04	18.148	15.567	38.149	15.350	0.709396	0.70545	0.1091	0.512344	0.512331	-5.49
M833	18.3	76.03	18.165	15.567	38.076	19.120	0.710339	0.70537				
M820A	19	75.77	18.123	15.566	38.085	18.750	0.710689	0.70563	0.0965	0.512343	0.512331	-5.47
M119A	19	76.41	18.075	15.577	38.107	20.730	0.711794	0.7062				
M50	20.9	71.5	18.662	15.638	38.999	5.180	0.709188	0.70765	0.1028	0.512131	0.512117	-9.60
Crystal-rich stocks												
M49A	21.4	74.45	18.380	15.605	38.494	2.293	0.705907	0.70521	0.0972	0.512251	0.512237	-7.24
M837A	21.4	74.76				11.060	0.709031	0.70567				
M837B	21.4	76.52				14.670	0.712219	0.70776				
M838A	21.4	74.26	18.730	15.623	38.910	5.280	0.707505	0.70590	0.1013	0.512287	0.512273	-6.55
M838B	21.4	76.62				17.730	0.711689	0.70630				
TRANSITIONAL-AGE GROUP												
K-rich mafic lavas												
M798	21.1	52.11	18.502	15.607	38.226	0.215	0.705754	0.70569				
M657	22.8	58.07	18.469	15.617	38.506	0.326	0.707785	0.70768	0.1171	0.512187	0.512170	-8.53
M856	23	53.39	19.027	15.667	39.001	0.220	0.706182	0.70611	0.1031	0.512278	0.512262	-6.71
Lousy Jim												
M822	22	69.4				1.991	0.707752	0.70713				
M823A	22	67.2	18.775	15.664	38.876	2.671	0.707955	0.70712	0.1041	0.512217	0.512202	-7.91
BULLION CANYON VOLCANICS AND RELATED ROCKS												
Monroe Peak												
M678	21	62.83	18.694	15.626	38.559	0.626	0.705887	0.70570	0.1052	0.512345	0.512331	-5.43
79-1664A	23	65.74	18.551	15.624	38.410	1.897	0.706210	0.70559	0.1016	0.512327	0.512312	-5.75
M569	21.5	64.87	18.640	15.626	38.446	1.448	0.705962	0.70552	0.0937	0.512319	0.512306	-5.90
M608	22	65.7				1.533	0.706049	0.70557				
Monzonite stocks												
M867	22	58.25	18.583	15.614	38.430	0.949	0.705566	0.70527				
M621	22	58.58	18.758	15.582	38.521	0.841	0.705583	0.70532	0.1060	0.512386	0.512308	-4.62
M845	22	59.58				1.238	0.705807	0.70542				
M851	22	58.69				0.941	0.705574	0.70528				
M869	23	58.1				0.833	0.705552	0.70528				
M870	23	58.67				0.945	0.705659	0.70535				
Bullion Canyon Volcanics												
M90	27	64.08	18.161	15.586	38.343	0.381	0.706156	0.70601	0.1026	0.512152	0.512134	-9.12
79-1666	27	60.43	18.685	15.655	38.366	0.557	0.705493	0.70528	0.1035	0.512322	0.512304	-5.80
79-S-1	29	61.89	18.146	15.577	38.088	0.190	0.705728	0.70565	0.1084	0.512100	0.512079	-10.14
Northern Assemblage												
79-S-3	24	55.91	18.513	15.617	38.154	0.302	0.705353	0.70525				
M739	25.3	61.75	18.736	15.616	38.163	0.830	0.705898	0.70560				
Eastern Assemblage												
79-1517	23	56.56	18.523	15.588	38.260	0.170	0.705255	0.70520	0.1052	0.512215	0.512199	-7.94
79-1518	25	58.74	18.661	15.616	38.232	0.350	0.705614	0.70549	0.1079	0.512388	0.512370	-4.55
MOUNT DUTTON FMN. AND RELATED ROCKS												
Mount Dutton Fmn. (southern assemblage)												
79-1665	24	56.55	17.953	15.558	38.324	0.151	0.705241	0.70519	0.1060	0.512056	0.512039	-11.04
78-698	27	59.33	17.987	15.548	38.171	0.343	0.705092	0.70496				
78-958	29	57.51	19.796	15.787	39.983	0.850	0.716530	0.71618	0.1216	0.511945	0.511922	-13.20
Spry Intrusion												
R2106	26.1	66.73	18.152	15.549	38.148	0.360	0.705833	0.70570				

[1] Corrected for mass fractionation during analyses of 0.12 ± 0.03% per mass based on analyses of Pb standard SRM-981. Uncertainties in the data generally correspond to the fractionation uncertainties.

[2] Measured uncertainties are generally 0.003% to 0.004% and are relative to a value of 0.710254 ± 15 (6 analyses) of Sr standard SRM-987.

[3] Measured uncertainties are generally 0.003% to 0.004% and are relative to a value of 0.511854 ± 12 (8 analyses) of the La Jolla Nd Standard.

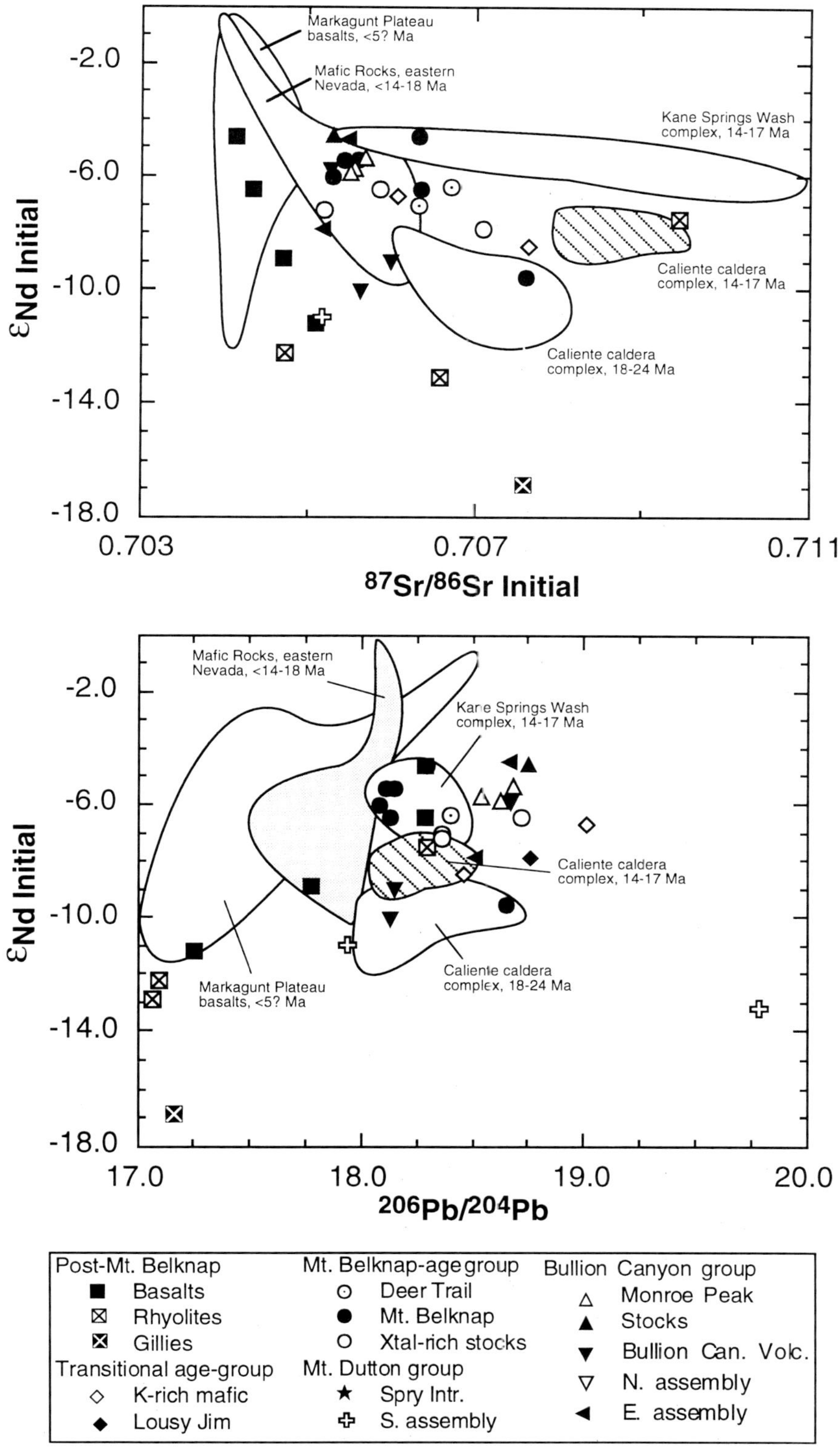

Figure 3. Plots of ε_{Nd} vs. $^{87}Sr/^{86}Sr$, and $^{206}Pb/^{204}Pb$ for Marysvale samples. Shown for reference are fields of data for the Kane Springs Wash and Caliente caldera complexes in southeastern Nevada and the Markagunt Plateau in southwestern Utah.

Nealey and others (in press), who ascribed this rather unusual isotopic signature to interaction of ascending mantle-derived magmas with the lower crust. This interpretation was based primarily on two factors: (1) a suite of deep-crustal xenoliths from the San Francisco volcanic field, central Arizona (along the southern margin of the Colorado Plateau) have isotopic characteristics that represent the end-member of this trend ($^{206}Pb/^{204}Pb \approx 16.0$, $^{87}Sr/^{86}Sr \approx 0.7025$, $\varepsilon_{Nd} \approx -20$; Nealey and Unruh, 1991), and (2) evidence for this isotopic signature is not found west of the approximate Colorado Plateau boundary, even among similar-aged basalts in the St. George basin immediately southwest of the Markagunt Plateau (Unruh and others, in press).

Rhyolites that were erupted during the third episode of igneous activity had major-element compositions similar to those of the 22–14 Ma group but had markedly different radiogenic-isotope characteristics. Lead, Sr, and Nd isotopic ratios all show much more variation than ratios observed among the other age groups (fig. 2), and the lowest ε_{Nd} values (–12 to –16) and $^{206}Pb/^{204}Pb$ (17.0–17.2) are found among these samples. However, initial $^{87}Sr/^{86}Sr$ values do not appear to be well correlated with Nd or Pb isotopic compositions (figs. 2 and 3), and rhyolites from the third episode show the widest variation in initial Sr and Nd values found within the entire sample suite (again with the exception of the highly contaminated Mount Dutton sample previously mentioned).

The isotopic trends exhibited by the Bullion Canyon and Mount Belknap Volcanics show similarities to those exhibited by the similar-age Caliente and Kane Springs Wash caldera complexes in southeastern Nevada (fig. 3; data from Scott and others, 1995; Unruh and others, 1995; D.M. Unruh, unpub. data, 1995). In both areas, ε_{Nd} values generally increase with decreasing age (at least from 29–18 Ma in Marysvale), whereas initial $^{87}Sr/^{86}Sr$ values tend to become more variable with decreasing age (figs. 2 and 3). In both areas these features may be attributed to different processes and sources involved during the change from subduction-related to extension-related volcanism. In the Caliente and Kane Springs Wash complexes, this transition is also reflected in the major-element chemistry of the rhyolite tuffs: Kane Springs Wash and post-17 Ma Caliente tuffs become metaluminous to weakly peralkaline, whereas the older tuffs are generally peraluminous (Rowley and others, 1995; Scott and others, 1995).

The Mount Belknap and Bullion Canyon Volcanics show a larger range of Pb isotopic compositions than do the tuffs of the Caliente and Kane Springs Wash complexes (figs. 3 and 4), although Pb isotopic data for the individual groups show fairly tight clusters (fig. 4). Lead isotopic data for the post-Mount Belknap samples are more similar to those for late Tertiary Markagunt Plateau basalts than to data for either Marysvale or Caliente–Kane Springs Wash samples.

All of the data on the $^{206}Pb/^{204}Pb$ vs. $^{207}Pb/^{204}Pb$ diagram plot significantly above the "northern hemisphere reference line" (NHRL; Hart, 1984), that represents the average trend of oceanic basalts in the northern hemisphere. This suggests that the Pb isotopic characteristics of these samples are derived primarily from lithospheric sources. Furthermore, Pb isotopic data for Marysvale samples in general show a pronounced average decrease in $^{206}Pb/^{204}Pb$ with decreasing age from about 22 Ma to the present (fig. 2), which may indicate a progressive change in the source region for these samples. A best-fit line through the $^{206}Pb/^{204}Pb$ vs. $^{207}Pb/^{204}Pb$ data in figure 4 yields an apparent age of 1.9±0.2 Ga (MSWD=13). However, in view of the evidence for multiple sources for these samples, this apparent age may have no rigorous significance.

The isotopic data presented here for Marysvale samples are consistent with an interpretation of having been derived from three primary sources: the lithospheric mantle (± a subduction-related component), lower crust, and upper crust. (In this context, the terms "upper" and "lower" crust are not meant to imply that these represent exactly two discreet physical entities; rather, these terms are used in the context of two end-member isotopic signatures.) The upper-crustal end-member, most evident in Mount Dutton sample 78-958 (table 1), is characterized by radiogenic Pb and Sr ($^{206}Pb/^{204}Pb>19.8$; $^{87}Sr/^{86}Sr>0.715$) and nonradiogenic Nd (ε_{Nd} <–13). The lower-crustal end-member, most evident in Gillies Hill sample 79-S8A, is also characterized by low ε_{Nd} (<–18) but perhaps only moderately radiogenic Sr ($^{87}Sr/^{86}Sr>0.708$) and nonradiogenic Pb ($^{206}Pb/^{204}Pb<17.0$). This lower-crustal end-member is similar to that inferred for the Markagunt Plateau except that Sr appears to have been more radiogenic in the Marysvale area.

The isotopic characteristics of the lithospheric mantle are probably most closely approximated by the only true basalt analyzed (sample 78-164, table 1; $\varepsilon_{Nd} \approx -4.7$, $^{206}Pb/^{204}Pb \approx 18.3$, and $^{87}Sr/^{86}Sr \approx 0.704$). These values are very similar to those estimated for the lithospheric mantle under the Kane Springs Wash and Caliente caldera complexes (Scott and others, 1995; Unruh and others, 1995), and surprisingly close to those estimated by Johnson and Thompson (1991; $\varepsilon_{Nd} \approx -4$, $^{206}Pb/^{204}Pb \approx 18.2$, $^{87}Sr/^{86}Sr \approx 0.705$) for the isotopic composition of the lithospheric mantle in the northern Rio Grande Rift on the opposite side of the Colorado Plateau. This isotopic component may represent the lithospheric mantle directly or the lithospheric mantle modified by a "subduction component" (Fitton and others, 1991) in the form of fluids or melts derived from the subducted slab. The fact that basalts with ε_{Nd} = –4 to 0 are relatively common among late Tertiary and Quaternary basalts on the Markagunt Plateau (Nealey and others, in press) suggests that the ε_{Nd} value of unmodified lithosphere in this area may be closer to zero (Farmer and others, 1989).

During eruption of the Mount Dutton Formation, the Bullion Canyon Volcanics, and the 20–18 Ma Mount Belknap Volcanics, crustal interaction progressively decreased, as evidenced by progressively increasing ε_{Nd} (fig. 2). The variation in Pb isotopic data during this interval suggests that both the Bullion Canyon and Mount Dutton magmas interacted with both upper and lower crust (figs. 2 and 3). The apparent progressive decrease in crustal interaction may reflect either thinning of the lithosphere during the onset of extension, providing a thinner column of crust for rising magmas to interact with, or production of progressively larger volumes of magma within the lithospheric mantle: all of these in effect dilute the crustal influence.

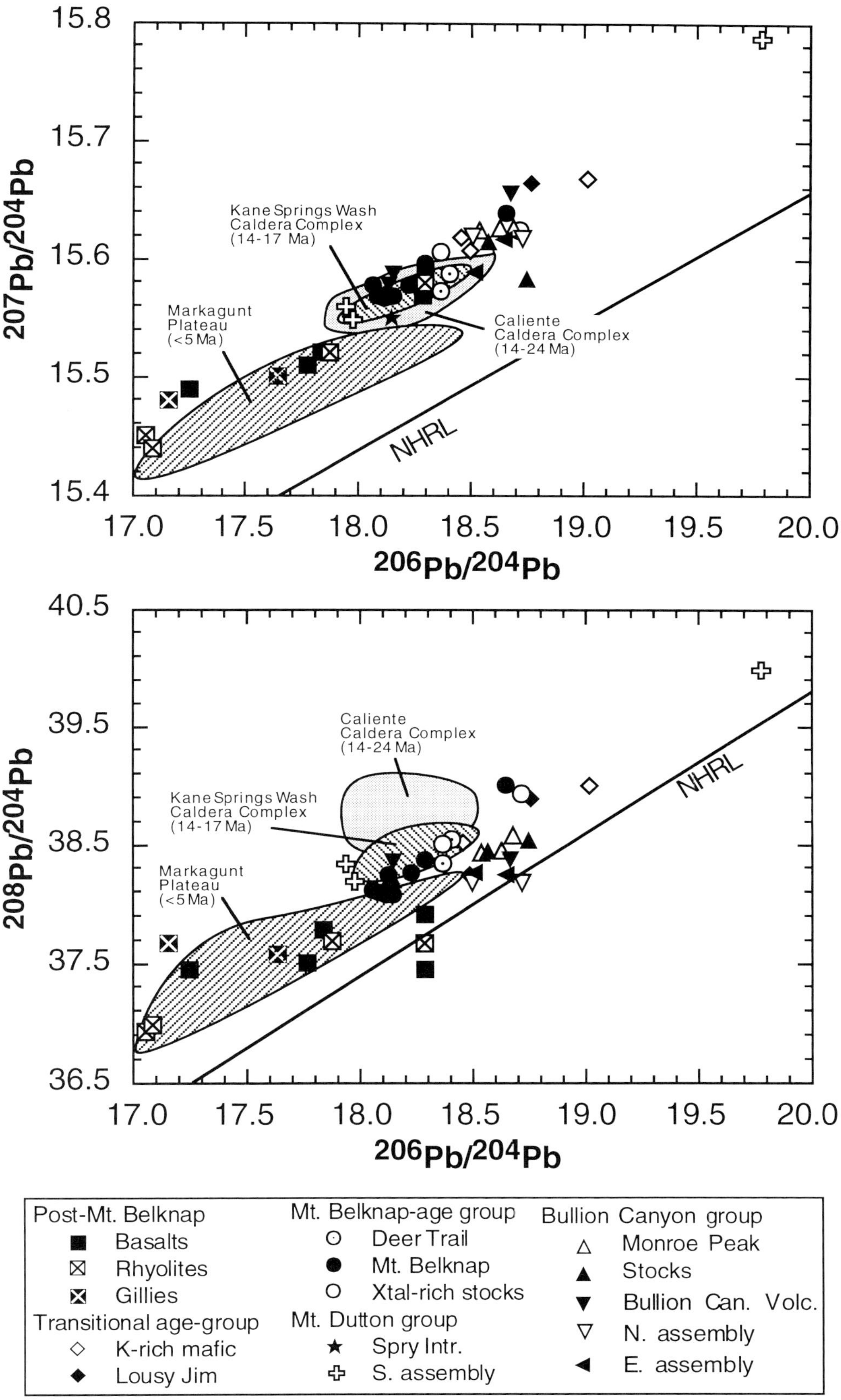

Figure 4. $^{207}Pb/^{204}Pb$ and $^{208}Pb/^{204}Pb$ vs. $^{206}Pb/^{204}Pb$ diagrams for Marysvale samples. Shown for reference are fields of data for the Kane Springs Wash and Caliente caldera complexes in southeastern Nevada and the Markagunt Plateau in southwestern Utah.

The similarity of neodymium, lead, and strontium isotopic values, together with the contrasting major element compositions, between the latest Bullion Canyon Volcanics (Monroe Peak caldera) and the spatially related early Mount

Belknap Volcanics (crystal-rich stocks), is consistent with the interpretation that the Mount Belknap Volcanics was derived from partial melting of Bullion Canyon batholithic rocks. This could be achieved as rising isotherms from the onset of extensional tectonics melted nearly eutectic magmas from the former magma chambers that fed Bullion Canyon volcanoes.

The Sr, Nd (fig. 3), and particularly the Pb isotopic compositions (fig. 4) of the samples formed during the third episode of volcanic activity appear to be strongly influenced by interaction with the lower crust (although rhyolites from this group have highly variable isotopic compositions). The fact that both mafic and silicic rocks have this signature may imply that the site of primary melt generation may have risen to the crust-mantle boundary during this period, particularly if this boundary had been modified by numerous earlier injections of mantle-derived magmas (Johnson and others, 1990). This interpretation is consistent with episodic intervals of increased extension and derivation of the late rhyolitic magmas from a variety of levels in the crust as isotherms rose in response to extension.

REFERENCES CITED

Best, M.G., McKee, E.H., and Damon, P.E., 1980, Space-time-composition patterns of late Cenozoic mafic volcanism, southwestern Utah and adjoining areas: American Journal of Science, v. 180, p. 1035–1050.

Budding, K.E., Cunningham, C.G., Zielinski, R.A., Steven, T.A., and Stern, C.R., 1987, Petrology and chemistry of the Joe Lott Tuff Member of the Mount Belknap Volcanics, Marysvale volcanic field, west-central Utah: U.S. Geological Survey Professional Paper 1354, 47 p.

Cunningham, C.G., Ludwig, K.R., Naeser, C.W., Weiland, E.K., Mehnert, H.H., Steven, T.A., and Rasmussen, J.D., 1982, Geochronology of hydrothermal uranium deposits and associated igneous rocks in the eastern source area of the Mount Belknap Volcanics, Marysvale, Utah: Economic Geology, v. 77, no. 2, p. 444–454.

Cunningham, C.G., and Steven, T.A., 1979, Mount Belknap and Red Hills calderas and associated rocks, Marysvale volcanic field, west-central Utah: U.S. Geological Survey Bulletin 1468, 34 p.

Cunningham, C.G., Steven, T.A., Rowley, P.D., Glassgold, L.B., and Anderson, J.J., 1983, Geologic map of the Tushar Mountains and adjoining areas, Marysvale volcanic field, Utah: U.S. Geological Survey Miscellaneous Investigations Series Map I–1430–A, scale 1:50,000.

Farmer, G.L., Perry, F.V., Semken, S., Crowe, B., Curtis, D., and DePaolo, D.J., 1989, Isotopic evidence on the structure and origin of subcontinental lithospheric mantle in southern Nevada: Journal of Geophysical Research, v. 94, p. 7885–7898.

Fitton, J.G., James, D., and Leeman, W.P., 1991, Basic magmatism associated with Late Cenozoic extension in the western United States—Compositional variations in space and time: Journal of Geophysical Research, v. 96, p. 13693–13711.

Hart, S.R., 1984, A large-scale isotope anomaly in the Southern Hemisphere mantle: Nature, v. 309, p. 753–757.

Johnson, C.M., and Thompson, R.A., 1991, Isotopic composition of Oligocene mafic volcanic rocks in the northern Rio Grande rift; evidence for contributions of intraplate and subduction magmatism to the evolution of the lithosphere: Journal of Geophysical Research, v. 96, p. 13593–13608.

Johnson, C.M., Lipman, P.W., and Czamanske, G.K., 1990, H, O, Sr, Nd, and Pb isotope geochemistry of the Latir volcanic field and cogenetic intrusions, New Mexico, and relations between evolution of a continental magmatic center and modifications of the lithosphere: Contributions to Mineralogy and Petrology, v. 104, p. 99–124.

Nealey, L.D., and Unruh, D.M., 1991, Geochemistry and isotopic characteristics of deep crustal xenoliths from Tule Tank, San Francisco Volcanic Field, northern Arizona: Arizona Geological Society Digest, v. 19, p. 153–163.

Nealey, L.D., Unruh, D.M., and Maldonado, F., in press, Sr, Nd, and Pb isotopic maps of Pleistocene and Holocene volcanic rocks of the Panguitch 30' × 60' quadrangle, southwest Utah: U.S. Geological Survey Open-File Report, scale 1:250,000.

Rowley, P.D., Cunningham, C.G., Steven, T.A., and Naeser, C.W., 1988a, Geologic map of the Marysvale quadrangle, Piute County, Utah: Utah Geological and Mineral Survey Map 105, 15 p., scale 1:24,000.

_____1988b, Geologic map of the Antelope Range quadrangle, Sevier and Piute Counties, Utah: Utah Geological and Mineral Survey Map 106, 14 p., scale 1:24,000.

Rowley, P.D., Mehnert, H.H., Naeser, C.W., Snee, L.W., Cunningham, C.G., Steven, T.A., Anderson, J.J., Sable, E.G., and Anderson, R.E., 1994, Isotopic ages and stratigraphy of Cenozoic rocks of the Marysvale volcanic field and adjacent areas, west-central Utah: U.S. Geological Survey Bulletin 2071, 35 p.

Rowley, P.D., Nealey, L.D., Unruh, D.M., Snee, L.W., Mehnert, H.H., Anderson, R.E., Grommé, S.C., and Siders, M.A., 1995, Stratigraphy of Oligocene and Miocene ash-flow tuffs in and near the Caliente caldera complex, southeastern Nevada and southwestern Utah, *in* Scott, R.B., and Swadley, W.C., eds., Geologic studies in the Basin and Range to Colorado Plateau transition in southeastern Nevada, southwestern Utah, and northwestern Arizona: U.S. Geological Survey Bulletin 2056–B, p. 47–88.

Rowley, P.D., Steven, T.A., Anderson, J.J., and Cunningham, C.G., 1979, Cenozoic stratigraphic and structural framework of southwestern Utah: U.S. Geological Survey Professional Paper 1149, 22 p.

Scott, R.B., Unruh, D.M., Snee, L.W., Harding, A.E., Blank, H.R. Jr., Budahn, J.R., and Mehnert, H.H., 1995, Relation of peralkaline magmatism to heterogeneous extension during the middle Miocene, southeastern Nevada: Journal of Geophysical Research, v. 100, p. 10381–10401.

Steven, T.A., 1981, Three Creeks caldera, southern Pavant Range, Utah: Brigham Young University Geology Studies, v. 28, pt. 3, p. 1–7.

Steven, T.A., Cunningham, C.G., and Anderson, J.J., 1984, Geologic history and uranium potential of the Big John caldera, southern Tushar Mountains, Utah, *in* Steven, T.A., ed., Igneous activity and related ore deposits in the western and southern Tushar Mountains, Marysvale volcanic field, west-central Utah: U.S. Geological Survey Professional Paper 1299–B, p. 23–33.

Steven, T.A., Cunningham, C.G., Naeser, C.W., and Mehnert, H.H., 1979, Revised stratigraphy and radiometric ages of volcanic rocks and mineral deposits in the Marysvale area, west-central Utah: U.S. Geological Survey Bulletin 1469, 40 p.

Steven, T.A., and Morris, H.T., 1987, Summary mineral resource appraisal of the Richfield 1° × 2° quadrangle, west-central Utah: U.S. Geological Survey Circular 916, 24 p.

Steven, T.A., Morris, H.T., and Rowley, P.D., 1990, Geologic map of the Richfield 1° × 2° quadrangle, west-central Utah: U.S. Geological Survey Miscellaneous Investigations Series Map I–1901, scale 1:250,000.

Steven, T.A., Rowley, P.D., and Cunningham, C.G., 1984, Calderas of the Marysvale volcanic field, west-central Utah: Journal of Geophysical Research, v. 89, no. B10, p. 8751–8764.

Unruh, D.M., Nealey, L.D., Rowley, P.D., Snee, L.W., Mehnert, H.H., and Anderson, R.E., 1995, Strontium and neodymium isotopic survey and ash-flow tuffs and related rocks from the Caliente caldera complex, southeastern Nevada and southwestern Utah, *in* Scott, R.B., and Swadley, W C, eds., Geologic studies in the Basin and Range to Colorado Plateau transition in southeastern Nevada, southwestern Utah, and northeastern Arizona: U.S. Geological Survey Bulletin 2056–D, p. 113–129.

Unruh, D.M., Nealey, L.D., and Maldonado, Florian, in press, Sr-Nd-Pb isotopic maps of late Cenozoic volcanic rocks of the St. George 30' × 60' quadrangle, southwest Utah: U.S. Geological Survey Open-File Report, scale 1:250,000.

Walker, J.A., Mattox, S.R., and Feigenson, M.D., 1991, Temporal changes in the source region of Cenozoic lavas of the Utah transition zone: Geological Society of America Abstracts with Programs, v. 23, no. 4, p. 102.

Potential for Alkaline Igneous Rock-Related Gold Deposits in the Colorado Plateau Laccolithic Centers

By Felix E. Mutschler,[1] Edwin E. Larson,[2] *and* Michael L. Ross[3]

CONTENTS

FIGURES

TABLES

ABSTRACT

Several types of productive gold deposits in the Rocky Mountains, ranging in age from ≥79 Ma to 26 Ma, show a close spatial, temporal, and genetic association with alkaline igneous rocks. Deposit types range from porphyry copper–precious metal systems characterized by Cu>Ag>Au or platinum-group elements, through transitional types, to epithermal precious-metal-only deposits commonly characterized by Au>Ag. The alkaline rocks associated with these deposits represent mantle melts which fractionated in crustal-level magma chambers. Coeval calc-alkaline igneous rocks formed by crustal melting and magma mixing occur with the alkaline rocks at many localities.

The igneous rocks of the Colorado Plateau laccolithic centers fall into two age groups: early Laramide (≈72–70 Ma) and middle Tertiary (33–23 Ma). Calc-alkaline diorite porphyries are the most voluminous igneous rocks in these centers. Essentially coeval alkaline syenite porphyries occur at Mount Pennell in the Henry Mountains, the North and Middle Mountain centers in the La Sal Mountains, and the Navajo Mountain center. Small volumes of late-stage peralkaline granite and rhyolite are also present at the Mount Pennell and North Mountain centers. The rock chemistry and alteration-mineralization assemblages of the Colorado Plateau laccolithic centers were compared to those of productive Rocky Mountain alkaline rock-related gold deposits. This comparison suggests a modest potential for discovery of gold deposits at several Colorado Plateau localities.

INTRODUCTION

A significant part of the gold production and reserves from Laramide and younger ore deposits in the Rocky Mountains comes from hypogene deposits associated with alkaline igneous rocks (table 1, fig. 1; Mutschler and others, 1990). In this report we compare the major-element chemistry of these productive alkaline rock suites with chemical data from the igneous rocks exposed in the laccolithic centers of the Colorado Plateau. The comparison suggests a possibility for discovery of alkaline rock-related gold deposits at several Colorado Plateau laccolithic centers, including the Henry and La Sal Mountains and Navajo Mountain, all in Utah.

Alkaline igneous rocks have been defined in many ways, and confusing nomenclature schemes based largely on variations in modal mineralogy abound. In this paper we use whole-rock major-element oxide analyses to define alkaline rocks as those igneous rocks that either (1) have weight percent $Na_2O+K_2O>0.3718$ (weight percent SiO_2) -14.5; or (2) have mol Na_2O + mol K_2O > mol Al_2O_3. Criterion 1 is from Macdonald and Katsura's (1964) alkalis versus silica plot for separating alkaline from subalkaline basalts (fig. 2). Criterion 2 defines peralkaline rocks in the sense of Shand (1951). Criteria 1 and 2 are independent; that is, peralkaline rocks as defined by criterion 2 need not satisfy criterion 1. Note that silica saturation (the presence or absence of either modal or normative feldspathoids) is *not* a criterion for alkaline rocks as used here. Alkaline rocks range in composition from relatively primitive kimberlites, lamproites, lamprophyres, and alkali basalts to highly evolved felsic syenites, phonolites, and peralkaline granites, rhyolites, and trachytes.

The worldwide association of a variety of types of gold deposits with alkaline igneous rocks (Mutschler and Mooney, 1993) suggests a genetic relationship. Various possibilities have been suggested to explain the relationship:

1. Parental alkaline magmas may be generated by partial mantle melting at sites where deeply penetrating fault systems extend through the crust (Cameron, 1990).
2. Gold may be transported from the deep mantle by mafic alkaline magmas (Rock and others, 1989).
3. The generally high volatile content of alkaline magmas (Bailey and Hampton, 1990; Webster and others, 1992) could provide ligands for gold acquisition, transport, and deposition (Cameron and Hattori, 1987; Mutschler and Mooney, 1993).

[1] Petrophysics Crisis Center, Department of Geology, Eastern Washington University, Cheney, WA 99004.

[2] Department of Geological Sciences, University of Colorado, Boulder, CO 80309.

[3] Utah Geological Survey, 2363 South Foothill Drive, Salt Lake City, UT 84109.

ACKNOWLEDGMENTS

Many colleagues in academia, industry, and government have helped us to compile data on the alkaline igneous rocks of the Cordillera and their associated mineral deposits. For providing us with unpublished material we especially thank James E. Elliott, Fess Foster, Bruce A. Geller, Stephen R. Mattox, Thomas C. Mooney, and Peter D. Rowley. Constructive reviews by Thomas Frost and Steve Ludington helped to clarify both our ideas and our expression.

ALKALINE ROCK-RELATED GOLD DEPOSITS OF THE ROCKY MOUNTAINS

Laramide and younger alkaline rock-related gold deposits in the Rocky Mountains are listed in table 1, and some typical ore-related rock assemblages are plotted on total alkali-silica (TAS) variation diagrams in figure 3. Many of these assemblages include relatively primitive mafic alkaline rocks together with highly evolved or fractionated rocks. This combination suggests that crustal level parking (perhaps at neutral buoyancy levels) and fractionation have been important processes in the evolution of these suites. Coeval calc-alkaline rocks are common at many Rocky Mountain alkaline rock localities (fig. 3*E–H*) and are predominant at some of them. In many cases the calc-alkaline magmas probably resulted from partial crustal melting by heat and volatiles from mantle-derived alkaline magmas that either ponded in or underplated the crust. In these situations mixing of calc-alkaline and alkaline magmas can produce a variety of hybrid magmas as at the Rosita–Silver Cliff volcanic centers, Colorado (fig. 3*H*).

Precious metal-bearing deposits associated with Rocky Mountain alkaline igneous centers can be divided into three

Table 1. Laramide and younger alkaline igneous rock-related gold deposits of the Rocky Mountains.

Locality	Igneous rocks	Age in Ma (method)[1]	Ore deposits	References[2]
Boulder County telluride camps (Eldora, Gold Hill, Jamestown, Magnolia, Sugarloaf, Sunset, Ward), Colorado	Alkaline to transitional sodic-syenite, quartz syenite, and bostonite stocks; bostonite and lamprophyre dikes. Coeval calc-alkaline plutons also present. An older (~74–60 Ma) alkaline and calc-alkaline suite of plutons is also present.	55 and 45 ($^{40}Ar/^{39}Ar$, FT)	Pyritic Au vein deposits. Au-Ag telluride vein and breccia deposits. Fluorspar-Ag-Pb vein and breccia deposits. Polymetallic Ag (Pb, Zn, Cu, Au) vein deposits. Epithermal W (wolframite) vein deposits. Stockwork Mo mineralization in quartz syenite stock. Production 1859–1990: Au, ~34,200 kg; +Ag, Cu, Pb, W, Zn.	Davis and Streufert (1990); Gable (1984) C; Jenkins (1979) C; Kane and others (1988) A; Kelly and Goddard (1969); F. E. Mutschler (unpub. data, 1990) C; Nash and Cunningham (1973); Phair and Jenkins (1975) C; Saunders (1991); Walter and Geller (1992) A.
Central City–Idaho Springs, Colorado	Alkaline to transitional sodic-syenite, quartz syenite, and bostonite stocks; bostonite dikes and breccia-pipes. Coeval calc-alkaline plutons also present.	65–58 (K-Ar, Rb-Sr)	Zoned district with central pyritic Au vein and breccia-pipe deposits and peripheral polymetallic vein deposits, both of which are probably related to calc-alkaline plutons. Late Au-Ag telluride-quartz-fluorite vein and breccia-pipe deposits, U vein deposits, and Mo breccia-pipe deposits related to alkaline plutons. District production 1859–1987: Au, ~171,070 kg; Ag, >2,278,000 kg; +Cu, Pb, Zn, U, W. Production from deposits directly related to alkaline rocks 1859–1984: Au, ~3,110 kg; +Ag, Pb, W.	Bastin and Hill (1917) C; Budge and Romberger (1983); Davis and Streufert (1990); Dickin and others (1986); F. E. Mutschler (unpub. data, 1990) C; Phair (1952) C; Phair and Jenkins (1975) C; Rice and others (1985) C; Rice and others (1982) A; Spurr and Garrey (1908) C; Wallace (1989).
Cripple Creek, Colorado	Large volcanic vent-diatreme subsidence basin complex includes alkaline phonolite, latite-phonolite, trachyphonolite, trachydolerite, trachyte, and syenite plutons, dikes, breccia pipes, and volcanic ejecta. Lamprophyre and foidal basalt dikes and breccia pipes common.	35–29 (K-Ar)	High-grade Au (Ag) telluride epithermal vein, stockwork, breccia-pipe, and disseminated deposits. Low-grade native Au (on pyrite) disseminated deposits. Production 1891–1990: Au, 653,175 kg; Ag, >7,300 kg. Reserves 1990: Au, >9,500 kg.	Cross (1895) C; Davis and Streufert (1990); Eriksson (1987); Fears and others (1986); Koschmann (1949); Lindgren and Ransome (1906) C; Loughlin (1927) C; Loughlin and Koschmann (1935); Marvin and others (1974) A; F. E. Mutschler (unpub. data, 1990) C; Mutschler and others (1985); Nelson (1989, 1990); Pontius (1991); Saunders (1988); Thompson and others (1985).
La Plata Mountains, Colorado	Early calc-alkaline to transitional diorite-monzonite porphyry sills, laccoliths, dikes, and stocks. Younger alkaline stocks include syenite (2), monzonite (1), and diorite (2). Lamprophyre and trachyte dikes, some of which are pre-mineralization, are the youngest igneous rocks.	70–67 (K-Ar, FT)	Porphyry Cu (Ag, PGE, Au) deposit in syenite stock. Au-Cu skarn and contact breccia deposits. Pyritic Au vein and replacement deposits. Ag-Au bearing polymetallic vein and replacement deposits. Au-Ag telluride vein and replacement deposits. Ruby silver vein deposits. Production 1878–1990: Au, ~6,850 kg; Ag, ~63,360 kg; +Cu, Pb, Zn. Reserves 1990: Au, ~6,875 kg; Ag, >170,000 kg.	Armstrong (1969) A; Cross and others (1899) C; Cunningham and others (1977) A; Eckel and others (1949); Lux (1977) C; McDowell (1971) A; Mutschler and others (1985); Saunders and May (1986); Werle and others (1984) C.
Rosita, Colorado	Volcanic-plutonic caldera system, includes alkaline syenogabbro diorite, monzonite, and trachyte plutons; lamprophyre dikes. Coeval high-silica rhyolites and other calc-alkaline volcanics and intrusions. Adjacent coeval Silver Cliff caldera system exposes high-silica rhyolites and intermediate calc-alkaline volcanics and intrusions.	33–27 (K-Ar, FT)	Au-Ag telluride-base metal sulfide and sulfosalt breccia-pipe deposit. Ag-Au bearing polymetallic vein deposits. Extensive acid-sulfate (alunite) alteration. REE-bearing apatite-magnetite veins and breccia-pipe deposits. Production 1872–1959: Au, ~3,110 kg; +Ag, Cu, Pb, Zn.	Cross (1896) C; Emmons (1896); Hildebrand and Conklin (1974); Kleinkopf and others (1979); McEwan (1985); F. E. Mutschler (unpub. data, 1990) C; Phair and Jenkins (1975) C; Sharp (1978) A; Sharp and Naeser (1986) A.
Caribou Mountain (Mount Pisgah), Idaho	Alkaline shonkinite-diorite subvolcanic plutonic complex.	52–50 (K-Ar)	Porphyry, skarn, and vein Cu (Au, Ag) deposits. Production 1870s–1959: Au, ≤1,860 kg; +Ag, Cu.	Anderson and Kirkham (1931); Hamilton (1961); Huntsman (1984, 1988) A.

Table 1. Laramide and younger alkaline igneous rock-related gold deposits of the Rocky Mountains—Continued.

Locality	Igneous rocks	Age in Ma (method)[1]	Ore deposits	References[2]
Goose Lake stock, Cooke City, Montana	Alkaline syenite-monzonite composite stock.	~76 (K-Ar)	Porphyry and pegmatitic Cu (Ag, PGE, Au) sulfide deposits. Major productive precious metal deposits of Cooke City district are related to younger (~40 Ma) felsic plutons and intrusive breccias.	Cordalis (1984); Elliott (1974, 1979) A; J. E. Elliott (unpub. data, 1983) C; Holt (1961); Lovering (1930); F. E. Mutschler (unpub. data, 1983) C; Mutschler and others (1985); Simons and others (1979); Size (1967).
Judith Mountains (Warm Spring, Maiden, Gilt Edge), Montana	Older calc-alkaline plutonic suite (69–67 Ma) includes quartz monzonite, monzonite, diorite, and rhyolite. Younger alkaline plutonic suite (65–62 Ma) includes syenite (with cumulate alkali gabbro and Cu-bearing sulfide inclusions), tinguaite (with cumulate ijolite inclusions), and peralkaline granite.	69–62 (K-Ar)	Au-Ag telluride and native gold vein, limestone replacement, and breccia deposits. Ag-Au bearing polymetallic vein deposits. Disseminated sulfide and skarn Au mineralization at Linster dome. Intense K-metasomatism, disseminated pyrite, and calcite-fluorite-quartz-sulfide-barite-aegirine-scapolite veining in Red Mountain area suggests possible buried carbonatite or alkaline rock-related porphyry metals system. Production 1880–1987: Au, 10,420 kg; Ag, 9,860 kg; +Cu, Pb. Reserves 1988: Au, >750 kg; Ag, 5,445 kg.	Giles (1983); Goddard (1988); Hall (1976) C; Kirchner (1982) A, C; Kohrt (1991); Lindsey and Fisher (1985); Marvin and others (1980) A; F. E. Mutschler (unpub. data, 1990) C; Priesmeyer (1986); Wallace (1953) C; Weed and Pirsson (1898) C; Zhang and Spry (1991).
Little Belt Mountains (Neihart), Montana	Alkaline shonkinite, syenite, monzonite plutons; lamprophyre dikes. Coeval calc-alkaline plutons include high-silica granites and rhyolites.	59; 55–46 (K-Ar)	Ag-Au bearing polymetallic vein and replacement deposits. Au skarn deposit. Fe skarn deposits. Au-Ag telluride vein and breccia mineralization. Gem sapphire deposits in hybrid mafic alkaline lamprophyre dike. Ag (Au) stockwork and disseminated deposit in rhyolite breccia-pipe. Stockwork Mo deposit in high-silica granite-rhyolite intrusive complex. Production 1881–1984: Au, ~2,085 kg; Ag, >9,420 kg; +Cu, Pb, Zn, gemstones.	Armstrong and others (1982) A; Baker and others (1991); Brownlow and Komorowski (1988); Clabaugh (1952) C; Dahy (1991); Embry (1987) C; Marvin and others (1973) A; Meyer and Mitchell (1988); Olmore (1991); Pirsson (1900) C; Rupp (1980) C; Schutz and others (1988); Walker (1991); Witkind (1970, 1973) C; Woodward (1991).
Little Rocky Mountains (Zortman-Landusky), Montana	Alkaline and transitional syenite plutons and trachyte dikes. Coeval calc-alkaline monzonite, quartz monzonite, and granite(?) plutons.	67–60 (K-Ar)	Au-Ag electrum and telluride stockwork, breccia, and vein deposits. Production 1884–1989: Au, 35,520 kg; Ag, >76,000 kg. Reserves 1990: Au, 23,850 kg; Ag, 267,500 kg.	Hastings (1988); Lindsey and Fisher (1985); Marvin and others (1980) A; F. E. Mutschler (unpub. data, 1990) C; Roemmel (1982) C; Russell (1991) C; Russell and Gabelman (1991); Ryzak (1990); Weed and Pirsson (1896) C; White and Lawless (1989); Wilson and Kyser (1988, 1989) C.
Moccasin Mountains (Kendall), Montana	Alkaline to transitional syenite, trachyte plutons and intrusion breccias. Coeval calc-alkaline rocks include rhyolite plutons and intrusion breccias.	66–64; 53 (K-Ar, FT)	Au-Ag telluride and electrum stratabound (limestone karst breccia) and intrusion breccia deposits. Production 1893–1987: Au, 14,000 kg; Ag, ~6,221 kg. Reserves 1989: Au, 6,800 kg.	Blixt (1933) C; Garverich (1991); Kurisoo (1991); Lindsey (1982, 1985) A, C; Lindsey and Fisher (1985); Lindsey and Naeser (1985) A; Marvin and others (1980) A; F. E. Mutschler (unpub. data, 1990) C.
Whitehall (Golden Sunlight), Montana	Calc-alkaline(?) latite sills; alkaline trachybasalt hypabyssal plutons. Post-mineralization lamprophyre dikes.	≥79 (K-Ar)	Epithermal stockwork and disseminated Au-Ag electrum and telluride deposit with Fe, Cu, Pb, Zn, Mo, As sulfides, etc., in hydrothermal breccia-pipe. Zones downward to porphyry-style Cu-Mo (Ag, Au) mineralization. Production 1890–1987: Au, 17,511 kg; Ag, 8,616 kg; +Cu, Pb, Zn. Reserves 1988: Au, 77,760 kg; Ag, ~77,760 kg.	Fess Foster, Golden Sunlight Mines, Inc. (written commun., 1991) A, C; Foster and Chadwick (1990); Porter and Ripley (1985).

Table 1. Laramide and younger alkaline igneous rock-related gold deposits of the Rocky Mountains—Continued.

Locality	Igneous rocks	Age in Ma (method)[1]	Ore deposits	References[2]
Cerrillos, New Mexico	Alkaline monzonite, syenite, diorite plutons; lamprophyre dikes. Coeval(?) latite-trachybasalt volcanics.	Oligocene >27 (K-Ar)	Cu (Ag, Au) porphyry system. Ag-Au bearing polymetallic vein deposits. Turquoise deposits. Production 1902–1956: Au, 43 kg; Ag, 3,890 kg; +Cu, Pb, Zn, gemstones.	Akright (1979); Aldrich and others (1986) A; Clarke (1915) C; Disbrow and Stoll (1957); Giles (1991); North and McLemore (1988); Sun and Baldwin (1958) C; Wargo (1964).
Jicarilla Mountains, New Mexico	Alkaline syenogabbro, monzonite, and syenite plutons.	38 (K-Ar)	Au-Ag disseminated and vein mineralization in altered syenite and syenogabbro plutons. Fe skarn mineralization. Production 1905–1968: Au, ~260 kg; +Ag. Reserves 1985: large but low-grade Au-Ag resource.	Allen and Foord (1991) A, C; North and McLemore (1988); Segerstrom and Ryberg (1974) A, C; Segerstrom and others (1979).
Nogal (Sierra Blanca), New Mexico	Two pulses of alkaline magmatism. Older (~37 Ma) event includes trachybasalt to trachyphonolite flows; essexite and nepheline syenite plutons. Younger (30–27 Ma) event includes trachyte, quartz trachyte, and rhyodacite volcanics; syeno-diorite, peralkaline syenite, and peralkaline granite plutons.	~37; 30–27 (K-Ar)	Au-Ag-Cu breccia-hosted deposits. Polymetallic vein mineralization. Stockwork Mo-Cu mineralization. Production 1902–1953: Au, ~470 kg; Ag, ~620 kg; +Cu. Reserves 1991: Au, ~3,300 kg; +Ag.	Allen and Foord (1991) A, C; Allen and McLemore (1991) C; Black (1977) C; Cepeda (1990); Fulp and Woodward (1991); Giles and Thompson (1972) A, C; North and McLemore (1988); Thompson (1968; 1972 A, C; 1991).
Ortiz (Old Placers), New Mexico	Older (~34 Ma) calc-alkaline suite includes latite-andesite and gra-nodiorite plutons. Younger (30–26 Ma) alkaline suite includes nepheline monzonite plutons, latite plutons and dike swarms, and a diatreme vent breccia.	~34; 30–26 (K-Ar)	Cu-Ag skarn deposit. Au-Ag (Cu, W) breccia-pipe deposit. Porphyry Cu (Au, Ag) deposit. Au-Ag polymetallic vein deposits. Production 1833–1987: Au, ~10,890 kg; +Ag, Cu, Pb, W, Zn. Reserves 1990: Au, 18,560 kg; +Ag, Cu.	Kay (1986) C; Maynard and others (1990) A; Maynard and others (1991) A; Ogilvie (1908) C; Wright (1983).
San Pedro (New Placers), New Mexico	Alkaline–calc-alkaline syenite, monzonite plutons.	Oligocene	Au-Cu-W skarn deposit. Ag-Pb-Zn limestone replacement deposit. Production 1909–1938: Au, 3,640 kg; Ag, 9,490 kg; +Cu, W, Zn.	Atkinson (1961); Elston (1967); Lindgren and others (1910); North and McLemore (1988).
White Oaks, New Mexico	Large breccia-pipe complex intruded by alkaline syenogabbro to syenite plutons and lamprophyre plugs and dikes.	34 (K-Ar)	Au-Ag (W) vein, stockwork, and breccia-hosted epithermal deposits. Production 1879–1984: Au, 5,070 kg; +Ag, Cu, W.	Allen and Foord (1991); Griswold (1959); Lindgren and others (1910); North and McLemore (1988); Ronkos (1991) A.
Northern Black Hills, South Dakota	Alkaline, peralkaline, and calc-alkaline plutons include nephe-line syenite, syenite, peralkaline syenite, phonolite, tinguaite, trachyte, monzonite, quartz monzonite, peralkaline rhyolite, rhyolite, and lamprophyres.	~60–53 (K-Ar)	Au-Ag (Pb-W) vein, stockwork, replace-ment, breccia-pipe, and disseminated deposits in intrusions and sedimentary rocks. Production 1875–1989: Au, ~87,090 kg; +Ag, Cu, Pb, W, Zn. Reserves: 1989: Au, 121,200 kg.	DeWitt and others (1986); Grun-wald (1970) C; Irving (1899) C; Kirchener (1971) C; Larsen (1977) C; Loomis and Alexander (1990); Noble (1948) C; Norton (1989); Paterson (1990); Paterson and others (1989); Pirsson (1894) C; Sharwood (1911) C; Shearer (1990).

[1] FT in age method indicates fission track age.
[2] A or C following date indicates that reference contains age date (A), or major-element chemical analyses (C).

types, or deposit models. Attributes of the two end-member models are summarized in table 2. The third, transitional, model can show features of both end-member models.

Porphyry copper–precious metal deposits.—These occur in or adjacent to shoshonitic syenite stocks and are characterized by precious metals contained in copper sulfides occurring in stockworks, disseminations, veins, pegmatite dikes and segregations, endoskarns, exoskarns, and local immiscible sulfide concentrations; by relatively high sulfur abundance; and by Cu>Ag>Au or PGE (platinum-group elements). Examples include the Allard stock, La Plata Mountains, Colo. (Werle and others, 1984); the Goose Lake stock, Cooke City, Mont. (Elliott, 1972, 1974; Lovering, 1930); and the Cerrillos district, New Mexico

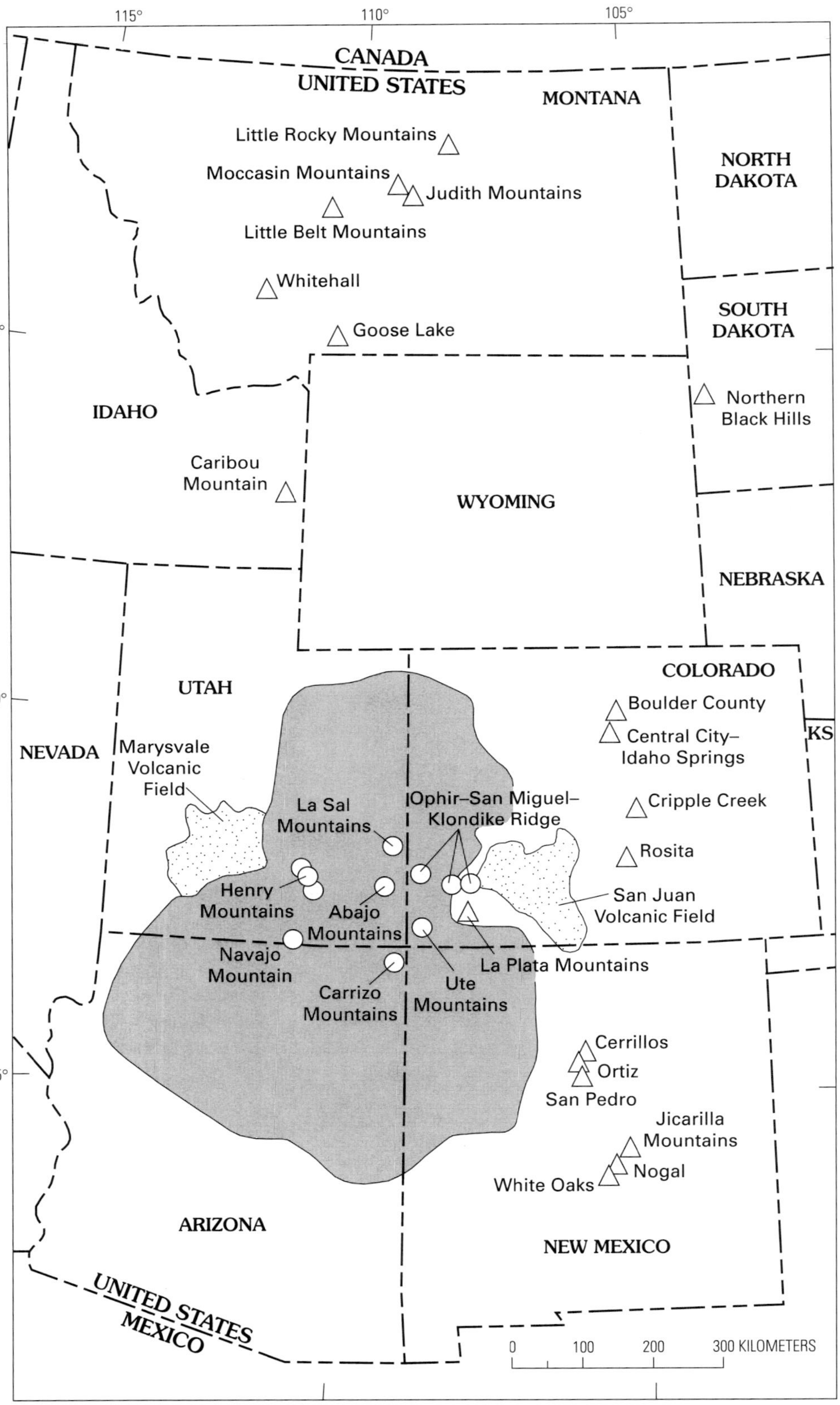

Figure 1. Rocky Mountain and Colorado Plateau localities discussed in text. Circles are Colorado Plateau laccolithic centers; triangles are Laramide and younger alkaline igneous rock-related gold deposits.

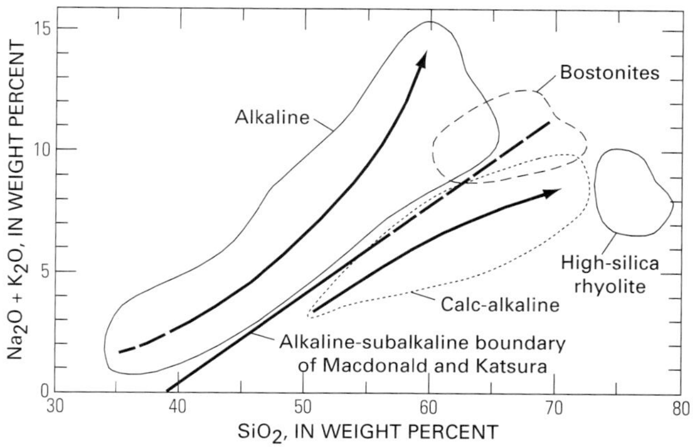

Figure 2. Total alkali-silica plot showing Macdonald and Katsura's (1964) boundary for separating alkaline and subalkaline rocks, and compositional fields for alkaline rocks related to precious metal deposits and selected other rock types. Arrows show generalized fractionation trends.

(Giles, 1991). In alkaline rock porphyry copper deposits, the precious metals constitute byproducts or coproducts of copper production. Deposits of this type in Mesozoic accreted terranes are being actively mined in British Columbia (McMillan, 1991; Schroeter and others, 1989).

Epithermal gold deposits.—These are generally associated with syenites, trachytes, phonolites, and lamprophyres, and they occur in a variety of settings including volcanic vent complexes, breccia pipes, hot-spring and geyser systems, bonanza veins, and replacements and disseminations in sedimentary and igneous rocks. They are characterized by Au (Ag)-telluride and native gold mineralization, relatively low sulfur abundance, and commonly by Au>Ag. Examples of bonanza epithermal vein deposits include the Cripple Creek district (Loughlin and Koschmann, 1935), the Boulder County telluride camps (Saunders, 1991), and the Bessie G mine in the La Plata Mountains (Saunders and May, 1986), all in Colorado.

Table 2. Attributes of alkaline rock-related precious-metal systems.

[PGE=platinum-group elements]

Attribute	Epithermal Au systems (example, Cripple Creek, Colorado)	Porphyry copper–precious metal systems (example, Allard stock, Colorado)
Recoverable metals	Au>Ag (precious metals usually sole products or major products).	Cu>Ag>PGE or Au (precious metals usually byproducts or coproducts).
Mineralization	Largely open-space vein filling	Disseminated, pegmatitic, vein, local immiscible sulfides, skarn.
Au-bearing species	Tellurides, native Au	Native Au (largely in sulfides)
Hydrothermal alteration	Vein envelope and/or pervasive	Pervasive.
Facies	Propylitic, K-metasomatism, carbonatic, redox (phyllic).	Propylitic, K-metasomatism, carbonatic, redox (phyllic, solfataric).
S (as S^{-2}) abundance	Relatively low	High.
Volatile-element concentrations:		
Te	Very high	Moderate.
Tl	High	Low.
Hg	Generally high	Low.
As, Sb	Generally high	Variable.
Ore stage fluids:		
Temperature	≤210 °C	~300–800 °C.
Salinity	Low (≥5 wt. % NaCl)	High.
CO_2 content	High	High (often saturated).
H_2O	Meteoric dominated	Magmatic dominated.
Pressure	1– >350 bars	~350–1000 bars.
Source of precious metals	Au from bisulfide and/or ditelluride complexes; carbonyl complexes(?).	Ag and PGE from chloride complexes; at least some Au late in sequence from bisulfide complexes(?).

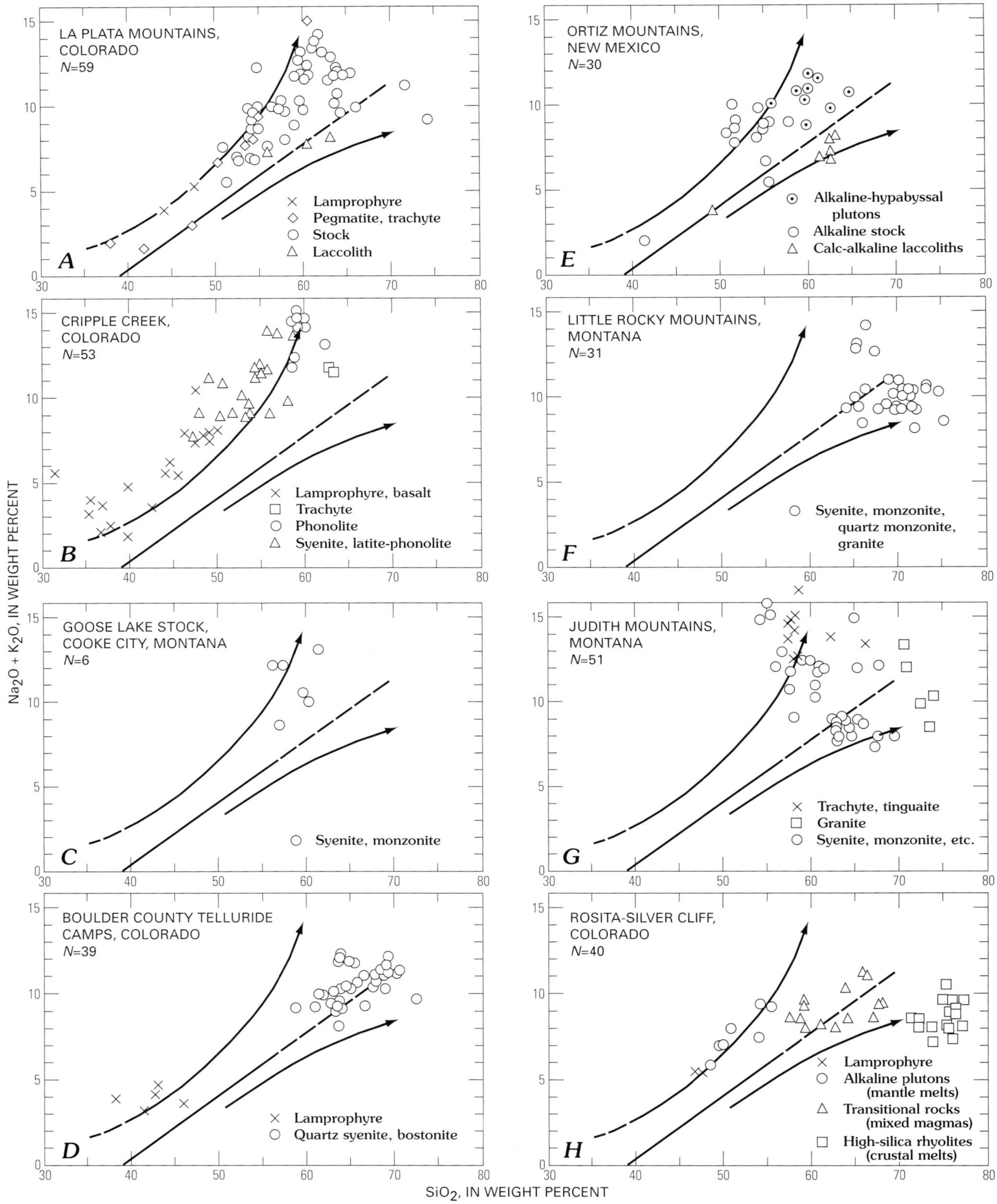

Figure 3. Total alkali-silica plots for selected alkaline rock suites related to gold deposits in the Rocky Mountains. Data from references listed in table 1. Boundary between alkaline and calc-alkaline rocks shown by inclined line. Curved arrows show generalized fractionation trends: upper arrow for ore-related alkaline rocks; lower arrow for calc-alkaline rocks. *N* indicates number of samples.

Noteworthy low-grade bulk-tonnage epithermal deposits include those of the Little Rocky (Russell, 1991), Moccasin (Kurisoo, 1991), and Judith (Giles, 1983) Mountains, Mont.

Transitional epithermal-mesothermal gold-silver deposits.—These include disseminated, breccia-pipe, skarn and vein Au-Ag (Cu, Pb, Zn, W) deposits related to syenite-monzonite-diorite plutons. The disseminated deposits range from Au-only porphyry to sedimentary-rock-hosted micron-size Au. Gold mineralization in the porphyry and skarn deposits typically appears to be late in the paragenetic sequence (post-base metal) and to be accompanied by retrograde alteration events. Examples include the Ortiz Mountains (Maynard and others, 1991), Jicarilla Mountains (Allen and Foord, 1991), and perhaps the White Oaks district (Ronkos, 1991), New Mexico; the Red Mountain area, Judith Mountains, Mont. (Hall, 1976); and some of the Tertiary districts in the northern Black Hills, S. Dak. (Paterson and others, 1988).

These three deposit models may represent a vertical (and perhaps short-term temporal) progression. All three types are associated with chemically similar alkaline rocks (see fig. 3) and show similar hydrothermal alteration assemblages (table 2). The ore fluids for both epithermal and porphyry copper–precious metal systems were CO_2-rich and relatively oxidized. Fluid inclusions in vein and rock minerals are CO_2-rich. Alteration assemblages, both pervasive ones and those found as envelopes around veins, feature carbonate minerals and hematite. Ore-stage gangue minerals commonly include carbonates and sulfates, and negative $\delta^{34}S$ values in sulfides are common. The two end-member deposit types differ, however, in Au, Ag, PGE, Cu, and S abundances, in volatile-element concentrations, and in ore-fluid pressure, temperature, and composition (table 2), suggesting that they were deposited from separate fluids. Cameron and Hattori (1987) proposed a scheme for the essentially simultaneous development of two chemically distinct fluids in an oxidized (high f_{O_2}), CO_2-rich magma chamber, which Mutschler and Mooney (1993) modified to explain the formation of epithermal "gold-only" deposits above and (or) peripheral to alkaline rock-related porphyry copper–precious metal systems. This model is shown diagrammatically in figure 4. A fractionated, oxidized, CO_2-saturated alkaline magma chamber exsolves a CO_2-rich, highly saline, metal-bearing aqueous fluid, which then unmixes into two immiscible phases: (1) a high-salinity fluid into which Cu, Fe, Ag, and PGE are partitioned as Cl complexes, which can form porphyry copper mineralization with high Ag and (or) PGE values; and (2) a low-salinity $H_2O–CO_2$ fluid into which Au (±As, Hg, Sb) is partitioned as S and (or) Te complexes which can form epithermal Au-dominated mineralization.

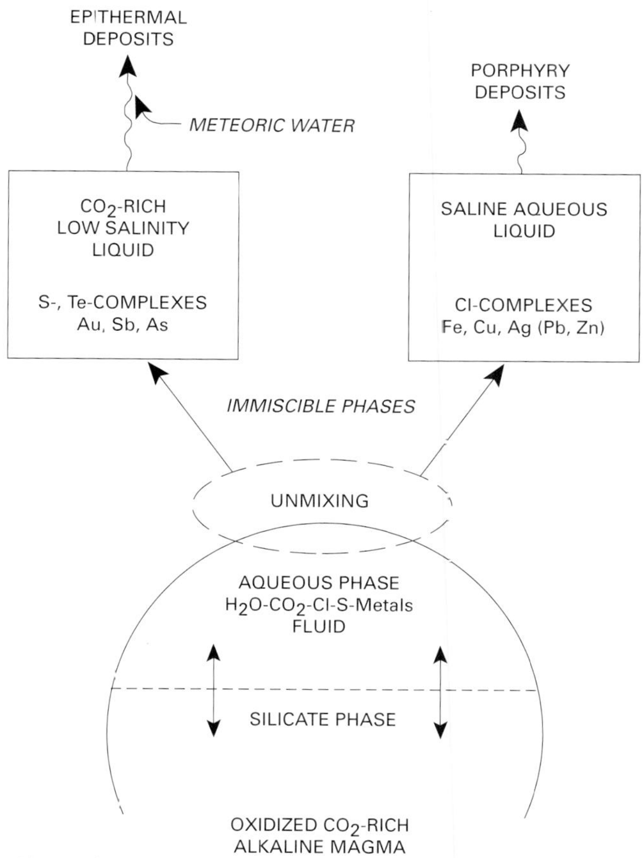

Figure 4. Schematic model for the evolution of two ore fluids from an oxidized CO_2-rich alkaline magma chamber. From Mutschler and Mooney (1993).

PROSPECTING GUIDES

A variety of gold-bearing deposits are associated with alkaline rocks; consequently various techniques may be useful in prospecting for different types of deposits (Mutschler and Mooney, 1993). Some useful indicators are as follows:

1. The source-host alkaline rocks for both porphyry and epithermal mineralization show evidence of significant crustal-level fractionation; thus chemically diverse suites of alkaline (±calc–alkaline) rocks are typically associated with mineralization.

2. Both porphyry and epithermal deposits are accompanied by one or more of the following pervasive hydrothermal alteration assemblages (diagnostic characteristics in parentheses): K-metasomatism (whole-rock $K_2O>Na_2O$; hydrothermal K-feldspar and (or) biotite), carbonatic (whole-rock $CO_2>0.5$ weight percent; hydrothermal carbonate minerals; CO_2-rich fluid inclusions), redox (whole-rock $Fe_2O_3>1.5$ FeO; hydrothermal hematite), $\delta^{34}S$ evidence that sulfide S equilibrated with sulfate S, or sulfidization (hydrothermal pyritization and (or) sulfate minerals).

3. Concealed porphyry and skarn deposits, which have relatively high sulfide concentrations, may be recognized by induced polarization surveys.

Table 3. Colorado Plateau laccolithic centers.

Locality	Igneous rocks	Age in Ma (method)[1]	Mineralization and alteration	References[2]
Carrizo Mountains, Arizona	Calc-alkaline diorite porphyry laccoliths, sills, and dikes. Younger (Oligocene) minette plugs and dikes of Navajo volcanic field are also present.	~70 (K-Ar)	Near Pastora Peak, brecciated sandstone cemented by veinlets of quartz overlies fractured diorite porphyry containing sparse pyrite and rare chalcopyrite.	Armstrong (1969) A; Cross (1894) C; Lux (1977) C; O'Sullivan and Beikman (1963); Strobell (1956).
Ophir–San Miguel–Klondike Ridge, Colorado	West-trending belt of four stocks and many laccoliths, sills and dikes extends from Ophir on the west flank of the San Juan dome to Glade Mountain and Klondike Ridge. Calc-alkaline plutons range from microdiorite to porphyritic adamellite. Oil test wells have intersected buried sills of similar composition in the Paradox Member of the Hermosa Formation. A small andesite vent(?) crops out at Glade Mountain. Post-mineralization lamprophyre dikes occur in the Mount Wilson district.	33–26 (K-Ar)	Two types of productive fissure veins occur in the Iron Springs (Ophir or Ames) and Mount Wilson districts at the east end of the porphyry belt: (1) Gold veins—narrow high-grade pyrite-arsenopyrite-quartz (±chalcopyrite, carbonate, barite) veins. (2) Silver-base metal veins—galena-sphalerite-pyrite-chalcopyrite-tetrahedrite-quartz-carbonate-barite veins. Vein-envelope and pervasive alteration includes propylitic, quartz-sericite, and pyritic assemblages. Estimated total production since 1882: Au, 3,888 kg; Ag, 28,250 kg; Cu, Pb, Zn, W.	Bromfield (1967) C; Lipman and others (1976) A; McDowell (1971) A; Shawe and others (1968) C; Steenland (1962); Vogel (1960).
Sleeping Ute Mountain (or "Ute Mountains"), Colorado	Calc-alkaline diorite porphyry forms three stocks, three bysmaliths, and many laccoliths, sills, and dikes. Small dikes and sills of lamprophyre (spessartite) are younger than the diorite porphyry.	~72 (K-Ar, FT)	Minor shear-zone-controlled pyrite-chalcopyrite-quartz-carbonate-barite veins occur in sandstone. A large part of the Marble Mountain bysmalith has been pyritized, sericitized, and propylitized; small showings of oxide copper minerals are present. Several small travertine-cemented breccia pipes are younger than the lamprophyres and may represent post-Laramide hot-spring activity.	Cross (1894) C; Cunningham and others (1977) A; Ekren and Houser (1965) C.
Abajo Mountains, Utah	Calc-alkaline diorite porphyry forms two stocks surrounded by shatter zones (which include intrusion breccias), many laccoliths, sills, and dikes. Two additional stocks may be present in the subsurface.	32–23 (FT, K-Ar)	Disseminated, stockwork, and vein Au-bearing pyrite-quartz-carbonate (±chalcopyrite, sphalerite, galena, ruby, silver?) deposits occur in shear zones in both stocks and at the margins of several laccoliths. Pervasive to shear-zone-controlled propylitic alteration occurs in both stocks and several laccoliths. Alkali metasomatism altered some East Mountain stock diorite porphyry to alkaline compositions. Minor, undocumented gold production from a few small lode mines and placers.	Armstrong (1969) A; Nelson and others (1992) A; Sullivan and others (1991) A; Witkind (1964) C.
Henry Mountains, Utah	Calc-alkaline to mildly alkaline diorite porphyry occurs in five distinct laccolithic centers at Mounts Ellen, Pennell, Hillers, Holmes, and Ellsworth. Each center consists of a central stock, in part surrounded by a zone of shattered indurated sedimentary rocks and diorite porphyry cut by irregular diorite porphyry intrusions; numerous bysmaliths, laccoliths, sills, and dikes. At the youngest center (Mount Pennell, ~25 Ma), the diorite porphyry stock is cut by a complex alkaline syenite porphyry stock, which in turn is intruded by small peralkaline granite porphyry plugs and dikes. The most primitive igneous rock recognized in the Henry Mountains is a basalt porphyry occurring in the shatter zone and as small sills at the Mount Ellen center.	31–23 ($^{40}Ar/^{39}Ar$, FT)	Vein and shear-zone-controlled stockwork and disseminated Cu-Au-Ag deposits have been prospected at all the central stocks and shatter zones. They are most extensively developed at the Mount Pennell and Mount Ellen centers. The hypogene vein assemblage includes pyrite, chalcopyrite, bornite, molybdenite, magnetite, hematite, quartz, and carbonates. Pervasive propylitic alteration occurs in many plutons; pervasive alkali metasomatism and pyritization have affected central stocks and shatter zones, some of which also show epidote-amphibole-garnet skarn development. Silicification is present in mineralized shear zones at the Mount Pennell center. Reported production since 1889 from Mount Ellen (Bromide Basin): Au, 21 kg; Ag, 93 kg; Cu, 8.2 Mg. Mount Pennell lode production since 1925: Au<1 kg. Small amounts of placer gold are reported to have been recovered from streams draining Mounts Ellen and Pennell.	Butler and others (1920); Cross (1894) C; Doelling (1980); Dubiel and others (1988a, b; 1990) A; Engel (1959) C; Hunt and others (1953) C; Hunt (1988) C; Kilinc (1979) C; Nelson (1991) C; Nelson and Davidson (1993) C; Nelson and others (1992) A; Sullivan and others (1991) A.

Table 3. Colorado Plateau laccolithic centers—Continued.

Locality	Igneous rocks	Age in Ma (method)[1]	Mineralization and alteration	References[2]
La Sal Mountains, Utah	Calc-alkaline to mildly alkaline diorite porphyry occurs in three laccolithic centers at North, Middle, and South Mountains. Each center has a central stock, several peripheral laccoliths and/or bysmaliths, and dikes and sills of diorite porphyry. At the North Mountain center the diorite porphyry is cut by a small stock and dikes of alkaline syenite porphyries. Emplacement of the syenites was accompanied by hydrothermal alteration and mineralization. A single syenite porphyry laccolith/sill is present at the Middle Mountain center. A group of explosion breccia pipes associated with small peralkaline granite and rhyolite porphyry plutons formed during or shortly after the emplacement of the alkaline syenite porphyries. Coeval peralkaline rhyolite and nosean trachyte porphyry dikes mark the final magmatic event at North Mountain.	28–25 ($^{40}Ar/^{39}Ar$, FT)	Fissure vein and fracture-zone-controlled stockwork and disseminated Cu-Au-Ag deposits have been prospected at all three centers. Alteration and mineralization are most extensively developed at North Mountain, where they accompanied emplacement of the syenite porphyries. Here pervasive and vein envelope alkali metasomatism, pyritization, carbonatic, and redox assemblages are widespread. Hypogene vein minerals include pyrite, chalcopyrite, bornite, hematite, quartz, carbonates, and fluorite. A second stage of carbonate, hematite, and silica alteration and minor sulfide mineralization accompanied formation of the explosion breccia pipes at North Mountain. Small ore shipments were made from several North Mountain mines, but production figures are not available. Traces of placer gold occur in pre-Wisconsin gravels derived from North and Middle Mountains.	Armstrong (1969) A; Butler and others (1920); Cross (1894) C; Hunt (1958) C; Irwin (1973) C; Nelson (1991) C; Nelson and others (1992) A; Stern and others (1965) A; Sullivan and others (1991) A
Navajo Mountain, Utah	Small alkaline quartz-bearing syenite porphyry intrusion exposed on southwest side of ~10-km-diameter dome probably uplifted by emplacement of buried pluton.	No data	Primary hematite in syenite indicates crystallization at high-P_{O_2} conditions. Most of the sedimentary rocks exposed on the top of the dome are highly silicified.	Condie (1964)

[1] FT = fission-track.
[2] A or C following date indicates that reference contains age date (A), or major-element chemical analyses (C).

4. Low-sulfide "invisible," or micron- (micrometer) size Au deposits may be recognized by geochemical anomalies in Au (>10 ppb) and some of the following: Ag, As, Bi, Ce, F, Hg, Sb, Se, Te, Tl, U, V, W, and high Ba and Sr in Ba:Sr:Rb ratios. Gold, commonly at levels in the parts-per-billion range, is the only universal geochemical guide to ore; concentrations and dispersion halos of other "pathfinder" elements can vary widely from deposit to deposit, even within a single district (Mutschler and others, 1985).

ALKALINE ROCKS AND MINERALIZATION IN THE COLORADO PLATEAU LACCOLITHIC CENTERS

Depending on how the borders of the Colorado Plateau are defined, laccolithic centers on the plateau can be enumerated differently. We have somewhat arbitrarily considered that the Laramide-age La Plata, Ouray, and Rico laccolithic centers in Colorado are Rocky Mountain features, whereas we have included the Laramide-age Sleeping Ute Mountain (or "Ute Mountains"), Colo., and Carrizo Mountains, Ariz., as Colorado Plateau features. In a similar fashion, we have excluded the middle Tertiary West Elk Mountains, Colo., laccoliths from the Colorado Plateau.

Data on the lithology, form, age, and associated mineralization of the igneous rocks in the Colorado Plateau laccolithic centers (fig. 1) are summarized in table 3. Total alkali-silica (TAS) plots for those laccolithic centers for which whole-rock chemical analyses are available are shown in figure 5. Most of the igneous rocks fall into two groups:

1. Dominantly calc-alkaline intermediate (55–65 weight percent SiO_2) rocks, here collectively termed *diorite porphyry*. Different investigators have applied various names to these rocks, including diorite-monzonite porphyry, diorite porphyry, granodiorite, granodiorite porphyry, microgabbro, microgranogabbro, monzonite porphyry, plagioclase-hornblende porphyry, porphyritic adamellite, quartz diorite porphyry, and quartz monzonite porphyry. Diorite porphyry forms concordant plutons, including laccoliths and sills, and discordant stocks, dikes, and bysmaliths. It constitutes more than 90 percent of the igneous rocks exposed in the major laccolithic centers of the Colorado Plateau (Hunt, 1956; Hunt and others, 1953) and adjacent areas. On TAS plots most of the diorite porphyries fall within the compositional fields of the mid-Tertiary

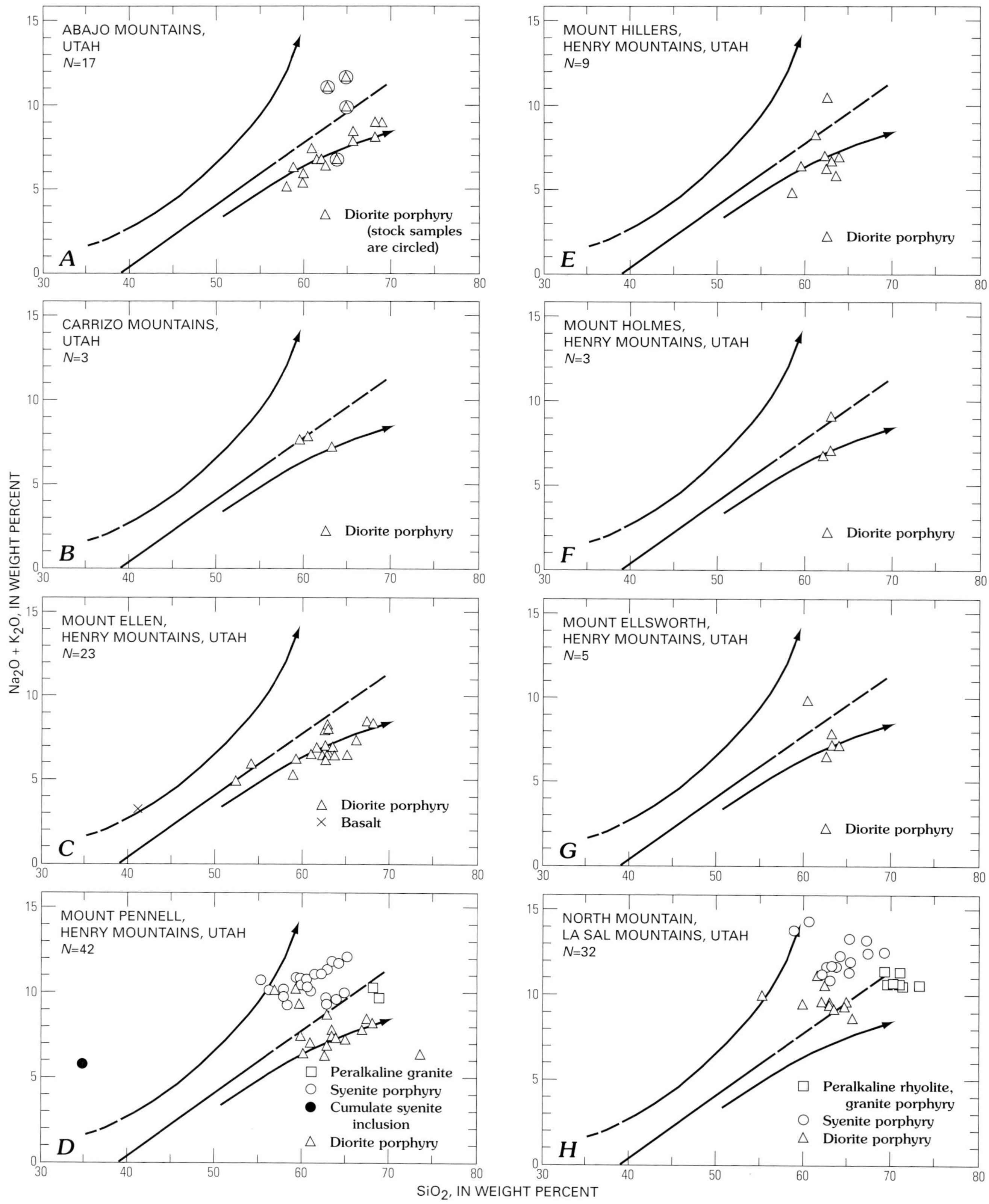

Figure 5 (above and facing page). Total alkali-silica plots for Colorado Plateau laccolithic centers. Data from references listed in table 3. Boundary between alkaline and calc-alkaline rocks shown by inclined line. Curved arrows show generalized fractionation trends: upper arrow for ore-related alkaline rocks; lower arrow for calc-alkaline rocks. *N*, number of samples.

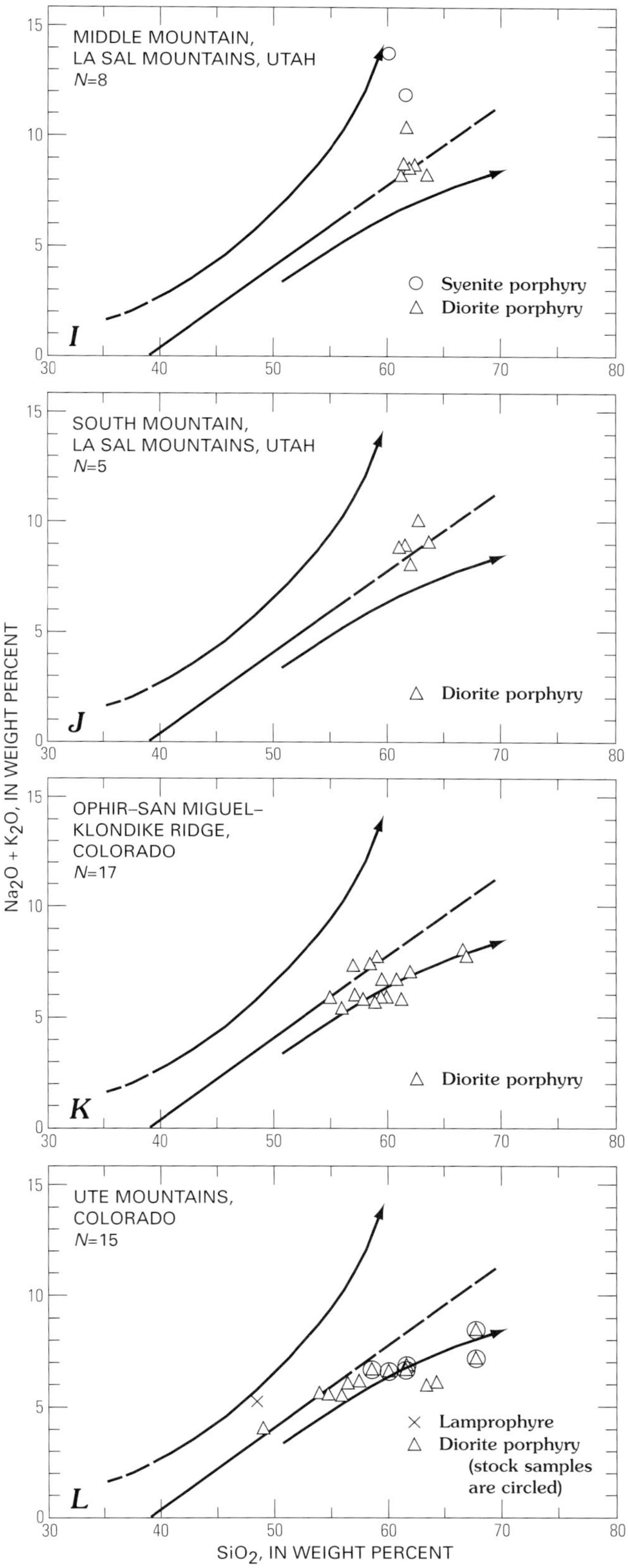

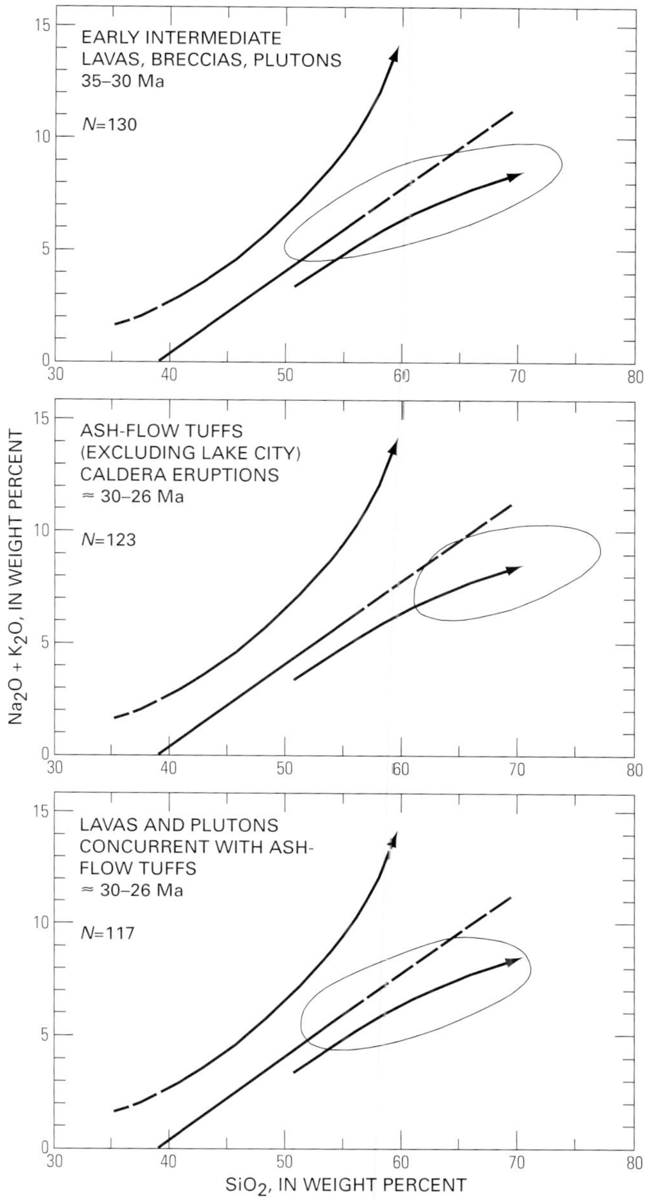

Figure 6. Total alkali-silica plots for middle Tertiary (35–26 Ma) volcanic rocks and plutons of the San Juan volcanic field, Colorado. Data from Mutschler and others (1981). Boundary between alkaline and calc-alkaline rocks shown by inclined line. Curved arrows show generalized fractionation trends: upper arrow for ore-related alkaline rocks; lower arrow for calc-alkaline rocks. *N*, number of samples.

calc-alkaline volcanics and plutons of the San Juan volcanic field, Colorado, as shown in figure 6. Many of the diorite porphyry analyses that plot significantly above the alkaline-subalkaline rock divider in figure 5*A*, *E*, *G*, and *H* reflect post-crystallization alkali metasomatism, including albitization (Nelson and Davidson, 1993) and potassic alteration (represented by secondary potassium feldspar and (or) biotite).

2. Alkaline intermediate (55–65 weight percent SiO_2) rocks, here collectively termed *syenite porphyry*. These rocks occur at the Mount Pennell center in the Henry Mountains and at the North and Middle Mountain centers

in the La Sal Mountains, Utah, forming stocks, sills, irregular plutons, and dikes, all of which are younger than the diorite porphyry plutons. In addition to these well-known localities, Condie (1964) has described a small quartz-bearing, highly oxidized, syenite porphyry pluton on the southwest flank of the Navajo Mountain dome, Utah. Both nepheline- and quartz-normative syenite porphyries are present in the La Sal and Henry Mountains. Although most of the La Sal syenite porphyries are peralkaline, none of the analyzed Henry Mountain syenite porphyries are. The syenite porphyries of the Henry and La Sal Mountains are chemically distinct from, and have no counterpart in, the voluminous coeval mid-Tertiary rocks of the San Juan volcanic field of southwestern Colorado, as shown by comparing the TAS plots of figure 5*D*, *H*, and *I* with figure 6. In fact, in both the San Juan, Colo. (Mutschler and others, 1987), and Marysvale, Utah (Mattox, 1992; Rowley and others, this volume), volcanic fields, alkaline magmatism is essentially restricted to bimodal (alkali basalt/trachybasalt-rhyolite) suites, which began to erupt during the onset of extensional faulting at ≈23 Ma.

Nelson and Davidson (1993) have presented isotopic evidence that both the diorite porphyry and syenite porphyry magmas of the Henry Mountains were derived from the same mantle source, but that they probably represent different degrees, or depths, of partial melting. According to Nelson and Davidson the parental magmas for the two suites reacted with, and assimilated, crustal rocks at different parking levels.

Pervasive hydrothermal alteration and local base- and precious-metal mineralization followed emplacement of the syenite porphyry stocks at both the North Mountain center in the La Sals and the Mount Pennell center in the Henry Mountains (Hunt, 1958; Hunt, 1988; Irwin, 1973; Nelson and others, 1992). Propylitic (chlorite-calcite-epidote-magnetite-hematite) alteration generally extends farthest from the stocks. Other, generally more proximal, pervasive alteration assemblages include hornfels skarn (aegirine-augite, riebeckite, glaucophane, hedenbergite, actinolite, biotite, magnetite, hematite, garnet), alkali metasomatism (albitization and subordinate potassium feldspathization), carbonatic alteration (calcite, siderite?), redox alteration (hematitization), and sulfidization (pyrite±base metal sulfides). Narrow breccia zones and small fissure veins occur locally, commonly following sheeted zones. Hypogene vein minerals include quartz, carbonates, fluorite, Au-bearing pyrite, chalcopyrite, bornite, galena, sphalerite, molybdenite, magnetite, and hematite. Similar but less intense alteration and mineralization occurred in and adjacent to some diorite porphyry plutons at most of the other Colorado Plateau laccolithic centers. (See table 3.)

At the North Mountain center in the La Sals the alteration and mineralization caused by the syenite porphyry were accompanied and also followed by formation of the breccia pipes (volcanic vents?) and coeval small peralkaline rhyolite porphyry plutons and dikes (Ross, 1992). Several of these pipes exceed 500 m in maximum horizontal dimension. A second alteration and mineralization event is represented by disseminated hematite, minor vein chalcopyrite and pyrite, and vug and cone sheet-fracture filling by calcite and quartz in the pipes. At North Mountain, the final intrusive phases are late-stage to postmineralization peralkaline rhyolite porphyry and nosean trachyte porphyry dikes that crosscut most intrusive phases, breccia pipes, and pervasively altered areas (Ross, this volume). Peralkaline granite also occurs as the youngest intrusive phase in the Mount Pennell center of the Henry Mountains, where Hunt (1988) suggested it preceded or accompanied mineralization.

EXPLORATION POTENTIAL FOR GOLD IN THE COLORADO PLATEAU LACCOLITHIC CENTERS

Three features that are characteristic of productive alkaline rock-related gold deposits in the Rocky Mountains (tables 1, 2) are also present in the Colorado Plateau laccolithic centers:

1. The centers include highly fractionated alkaline plutons such as the syenites of Mount Pennell in the Henry Mountains, North Mountain in the La Sals, and Navajo Mountain. Evolved peralkaline granites and rhyolites are also present at Mount Pennell and North Mountain.

2. Pervasive hydrothermal alteration, including alkali metasomatism and propylitic, carbonatic, redox, pyritic, and silicic alteration, has been reported at most of the laccolithic centers.

3. Subeconomic occurrences of hypogene gold have been recognized in the Abajo, Henry, and La Sal Mountains, and significant historic gold-silver production took place in the Ophir–San Miguel centers.

In the last decade, several major mining companies have evaluated the precious metal potential of the Colorado Plateau laccolithic centers with geochemical surveys, and, in the La Sal Mountains, with drilling programs. Although some geochemical anomalies and limited mineralized drill intercepts have been found, no economic ore deposits were discovered. However, relatively large areas of geologically suitable terrain remain untested, especially for low-sulfide, low-grade gold deposits. Inasmuch as such "invisible" deposits are still being discovered in and adjacent to well-studied mining camps (see Tooker, 1990, for example), we conclude that a modest potential exists for finding economic alkaline rock-related gold deposits in the laccolithic centers of the Colorado Plateau.

REFERENCES CITED

Akright, R.L., 1979, Geology and mineralogy of the Cerrillos copper deposit, Santa Fe County, New Mexico, *in* Ingersoll, R.V., Woodward, L.A., and James, H.L., eds., Guidebook of Santa Fe Country: New Mexico Geological Society Annual Field Conference, 30th, Santa Fe, N. Mex., October 4–6, 1979, p. 257–260.

Aldrich, M.J., Jr., Chapin, C.E., and Laughlin, A.W., 1986, Stress history and tectonic development of the Rio Grande rift, New Mexico: Journal of Geophysical Research, v. 91, p. 6199–6211.

Allen, M.S., and Foord, E.E., 1991, Geological, geochemical and isotopic characteristics of the Lincoln County porphyry belt, New Mexico—Implications for regional tectonics and mineral deposits, *in* Barker, J.M., Kues, B.S., Austin, G.S., and Lucas, S.G., eds., Guidebook of the Sierra Blanca, Sacramento and Capitan Ranges, New Mexico: New Mexico Geological Society Annual Field Conference, 42d, p. 97–113.

Allen, M.S., and McLemore, V.T., 1991, The geology and petrogenesis of the Capitan pluton, New Mexico, *in* Barker, J.M., Kues, B.S., Austin, G.S., and Lucas, S.G., eds., Guidebook of the Sierra Blanca, Sacramento and Capitan Ranges, New Mexico: New Mexico Geological Society Annual Field Conference, 42d, p. 115–127.

Anderson, A.L., and Kirkham, V.R.D., 1931, Alkaline rocks of the Highwood type in southeastern Idaho: American Journal of Science, 5th series, v. 22, p. 51–68.

Armstrong, R.L., 1969, K-Ar dating of laccolithic centers of the Colorado Plateau and vicinity: Geological Society of America Bulletin, v. 80, p. 2081–2086.

Armstrong, R.L., Harakal, J.E., and Hollister, V.F., 1982, Eocene mineralization at Mount Tolman (Keller), Washington, and Silver Dyke, Montana: Isochron/West, no. 33, p. 9–10.

Atkinson, W.W., 1961, Geology of the San Pedro Mountains, Santa Fe County, New Mexico: New Mexico Bureau of Mines and Mineral Resources Bulletin 77, 49 p.

Bailey, D.K., and Hampton, C.M., 1990, Volatiles in alkaline magmatism: Lithos, v. 26, p. 157–165.

Baker, D.W., McBride, Gwen, and Dahy, James, 1991, Field guide, Little Belt Mountains, *in* Baker, D.W., and Berg, R.B., eds., Guidebook of the Central Montana alkalic province: Montana Bureau of Mines and Geology Special Publication 100, p. 144–162.

Bastin, E.S., and Hill, J.M., 1917, Economic geology of Gilpin County and adjacent parts of Clear Creek and Boulder Counties, Colorado: U.S. Geological Survey Professional Paper 94, 379 p.

Black, K.D., 1977, Petrology, alteration, and mineralization of two Tertiary intrusives, Sierra Blanca igneous complex, New Mexico: Fort Collins, Colo., Colorado State University M.S. thesis, 124 p.

Blixt, J.E., 1933, Geology and gold deposits of the North Moccasin Mountains, Fergus County, Montana: Montana Bureau of Mines and Geology Memoir 8, 25 p.

Bromfield, C.S., 1967, Geology of the Mount Wilson quadrangle, western San Juan Mountains, Colorado: U.S. Geological Survey Bulletin 1227, 100 p.

Brownlow, A.H., and Komorowski, J.-C., 1988, Geology and origin of the Yogo sapphire deposit, Montana: Economic Geology, v. 83, p. 875–880.

Budge, Suzanne, and Romberger, S.B., 1983, Trace element distribution around precious/base metal veins, Idaho Springs dist., CO: Geological Society of America Abstracts with Programs, v. 15, p. 398.

Butler, B.S., Loughlin, G.F., Heikes, V.C., and others, 1920, The ore deposits of Utah: U.S. Geological Survey Professional Paper 111, 672 p.

Cameron, E.M., 1990, Alkaline magmatism at Kirkland Lake, Ontario—Product of strike slip orogenesis: Canada Geological Survey Paper 90–1C, p. 261–269.

Cameron, E.M., and Hattori, Keiko, 1987, Archean gold mineralization and oxidized hydrothermal fluids: Economic Geology, v. 82, p. 1177–1191.

Cepeda, J.C., 1990, The Capitan dike swarm, Lincoln County, New Mexico—A Tertiary alkalic magma system: Geological Society of America Abstracts with Programs, v. 22, p. 5.

Clabaugh, S.E., 1952, Corundum deposits of Montana: U.S. Geological Survey Bulletin 983, 100 p.

Clarke, F.W., 1915, Analyses of rocks and minerals from the laboratory of the United States Geological Survey, 1880–1914: U.S. Geological Survey Bulletin 591, 376 p.

Condie, K.C., 1964, Crystallization Po_2 of syenite porphyry from Navajo Mountain, southern Utah: Geological Society of America Bulletin, v. 75, p. 359–362.

Cordalis, Charles, 1984, Mineralogy and petrology of the Copper King platinum prospect, Park County, Montana: Butte, Mont., Montana College of Mineral Science and Technology M.S. thesis, 62 p.

Cross, Whitman, 1894, The laccolithic mountain groups of Colorado, Utah, and Arizona: U.S. Geological Survey 14th Annual Report, pt. 2, p. 165–241.

———1895, General geology of the district, part 1 *of* Geology and mining industries of the Cripple Creek district, Colorado: U.S. Geological Survey 16th Annual Report, pt. 2, p. 13–109.

———1896, Geology of Silver Cliff and the Rosita Hills, Colorado: U.S. Geological Survey 17th Annual Report, pt. 2, p. 263–403.

Cross, Whitman, Spencer, A.C., and Purington, C.W., 1899, La Plata folio, Colorado: U.S. Geological Survey Geological Atlas Folio 60, 14 p.

Cunningham, C.G., Naeser, C.W., and Marvin, R.F., 1977, New ages for intrusive rocks in the Colorado mineral belt: U.S. Geological Survey Open-File Report 77–573, 7 p.

Dahy, J.P., 1991, Geology and igneous rocks of the Yogo sapphire deposit, Little Belt Mountains, Montana, *in* Baker, D.W., and Berg, R.B., eds., Guidebook of the Central Montana alkalic province: Montana Bureau of Mines and Geology Special Publication 100, p. 45–54.

Davis, M.W., and Streufert, R.K., 1990, Gold occurrences of Colorado: Colorado Geological Survey Resource Series, no. 28, 101 p.

DeWitt, Ed, Redden, J.A., Wilson, A.B., Buscher, David, and Dersch, J.S., 1986, Mineral resource potential and geology of the Black Hills National Forest, South Dakota and Wyoming: U.S. Geological Survey Bulletin 1580, 135 p.

Dickin, A.P., Rice, C.M., and Harmon, R.S., 1986, A strontium and oxygen isotope study of Laramide magmatic and hydrothermal activity near Central City, Colorado: Economic Geology, v. 81, p. 904–914.

Disbrow, A.E., and Stoll, W.C., 1957, Geology of the Cerrillos area, Santa Fe County, New Mexico: New Mexico Bureau of Mines and Mineral Resources Bulletin 48, 73 p.

Doelling, H.H., 1980, Mineral deposits of the Henry Mountains, *in* Picard, M.D., ed., Henry Mountains Symposium: Utah Geological Association Publication 8, p. 287–296.

Dubiel, R.F., Bromfield, C.S., Church, S.E., Kemp, W.M., Larson, M.J., Peterson, Fred, and Neubert, J.T., 1988a, Mineral resources of the Bull Mountain Wilderness Study Area, Garfield and Wayne Counties, Utah: U.S. Geological Survey Bulletin 1751–B, p. B1–B12.

———1988b, Mineral resources of the Mt. Hillers Wilderness Study Area, Garfield County, Utah: U.S. Geological Survey Bulletin 1751–C, p. C1–C14.

Dubiel, R.F., Bromfield, C.S., Church, S.E., Kemp, W.M., Larson, M.J., Peterson, Fred, and Pierson, C.T., 1990, Mineral resources of the Mount Pennell Wilderness Study Area, Garfield County, Utah: U.S. Geological Survey Bulletin 1751–D, p. D1–D17.

Eckel, E.B., Williams, J.S., Galbraith, F.W., and others, 1949, Geology and ore deposits of the La Plata district, Colorado: U.S. Geological Survey Professional Paper 219, 179 p.

Ekren, E.B., and Houser, F.N., 1965, Geology and petrology of the Ute Mountains area, Colorado: U.S. Geological Survey Professional Paper 481, 74 p.

Elliott, J.E., 1972, Platinum at Cooke City, Mont.: U.S. Geological Survey Professional Paper 750–A, p. A36.

———1974, The geology of the Cooke City area, Montana and Wyoming, *in* Voight, Barry, ed., Rock mechanics, the American Northwest: University Park, Pa., The Pennsylvania State University, p. 102–107.

———1979, Geologic map of the southwest part of the Cooke City quadrangle, Montana and Wyoming: U.S. Geological Survey Miscellaneous Investigations Series Map I–1084, scale 1:24,000.

Elston, W.E., 1967, Summary of the mineral resources of Bernalillo, Sandoval, and Santa Fe Counties, New Mexico: New Mexico Bureau of Mines and Mineral Resources Bulletin 81, 81 p.

Embry, P.A., 1987, Petrogenesis of the Yogo Peak stock, Little Belt Mountains, Montana: Missoula, Mont., University of Montana M.S. thesis, 94 p.

Emmons, S.F., 1896, The mines of Custer County, Colorado: U.S. Geological Survey 17th Annual Report, pt. 2, p. 405–472.

Engel, C.G., 1959, Igneous rocks and constituent hornblendes of the Henry Mountains, Utah: Geological Society of America Bulletin, v. 70, p. 951–980.

Eriksson, C.L., 1987, Petrology of the alkalic hypabyssal and volcanic rocks at Cripple Creek, Colorado: Golden, Colo., Colorado School of Mines M.S. thesis, 114 p.

Fears, D.W., Mutschler, F.E., and Larson, E.E., 1986, Cripple Creek, Colorado—A petrogenetic model: Geological Society of America Abstracts with Programs, v. 18, p. 599.

Foster, Fess, and Chadwick, Tom, 1990, Relationship of the Golden Sunlight mine to the Great Falls tectonic zone: Tobacco Root Geological Society 15th Annual Field Conference Guidebook, p. 77–81.

Fulp, M.S., and Woodward, L.A., 1991, Mineral deposits of the New Mexico alkalic province: Geological Society of America Abstracts with Programs, v. 23, no. 4, p. 23.

Gable, D.J., 1984, Geologic setting and petrochemistry of the Late Cretaceous–Early Tertiary intrusives in the northern Front Range mineral belt, Colorado: U.S. Geological Survey Professional Paper 1280, 33 p.

Garverich, M.R., 1991, Field guide, North Moccasin and Judith Mountains, *in* Baker, D.W., and Berg, R.B., eds., Guidebook of the Central Montana alkalic province: Montana Bureau of Mines and Geology Special Publication 100, p. 175–184.

Giles, D.L., 1983, Gold mineralization in the laccolithic complexes of central Montana, *in* Babcock, J.W., chairperson, The genesis of Rocky Mountain ore deposits—Changes with time and tectonics: Wheat Ridge, Colo., Denver Region Exploration Geologists Society, p. 157–162.

———1991, Gold mineralization in the Cerrillos district, New Mexico: Geological Society of America Abstracts with Programs, v. 23, no. 4, p. 24.

Giles, D.L., and Thompson, T.B., 1972, Petrology and mineralization of a molybdenum-bearing alkalic stock, Sierra Blanca, New Mexico: Geological Society of America Bulletin, v. 83, p. 2129–2148.

Goddard, E.N., 1988, Geologic map of the Judith Mountains, Fergus County, Montana: U.S. Geological Survey Miscellaneous Investigations Series Map I–1729, scale 1:31,680.

Griswold, G.B., 1959, Mineral deposits of Lincoln County, New Mexico: New Mexico Bureau of Mines and Mineral Resources Bulletin 67, 117 p.

Grunwald, R.R., 1970, Geology and mineral deposits of the Galena mining district, Black Hills, South Dakota: Rapid City, S.Dak., South Dakota School of Mines and Technology Ph. D. thesis, 323 p.

Hall, R.J., 1976, Petrology of diamond drill core from Judith Peak–Red Mountain area, Fergus County, Montana: Cheney, Wash., Eastern Washington University M.S. thesis, 38 p.

Hamilton, J.A., 1961, Geology of the Caribou Mountain area, Bonneville and Caribou Counties, Idaho: Moscow, Idaho, University of Idaho M.S. thesis, 59 p.

Hastings, J.S., 1988, Gold deposits of Zortman-Landusky, Little Rocky Mountains, Montana, *in* Schafer, R.W., Cooper, J.J., and Vikre, P.G., eds., Bulk mineable precious metal deposits of the western United States: Reno, Nev., Geological Society of Nevada, p. 187–205.

Hildebrand, F.A., and Conklin, N.M., 1974, A breccia dike containing rare-earth-bearing apatite, molybdenite and magnetite at Iron Hill, Custer County, Colorado: Economic Geology, v. 69, p. 508–515.

Holt, R.E., 1961, Copper King project, Goose Lake district, Park County, Montana: Denver, Colo., Bear Creek Mining Company unpublished report, 21 p.

Hunt, C.B., 1956, Cenozoic geology of the Colorado Plateau: U.S. Geological Survey Professional Paper 279, 99 p.

———1958, Structural and igneous geology of the La Sal Mountains, Utah: U.S. Geological Survey Professional Paper 294–I, p. 305–364.

Hunt, C.B., Averitt, Paul, and Miller, R.L., 1953, Geology and geography of the Henry Mountains region, Utah: U.S. Geological Survey Professional Paper 228, 234 p.

Hunt, G.L., 1988, Petrology of the Mt. Pennell central stock, Henry Mountains, Utah: Brigham Young University Geology Studies, v. 35, p. 81–100.

Huntsman, J.R., 1984, Caribou Mountain—A porphyry copper deposit in southeastern Idaho: Economic Geology, v. 79, p. 748–754.

———1988, Structural controls and age of the Caribou Mountain intrusion, southeastern Idaho: Geological Society of America Abstracts with Programs, v. 20, p. 421.

Irving, J.D., 1899, A contribution to the geology of the northern Black Hills: New York Academy of Science Annals, v. 12, p. 187–340.

Irwin, T.D., 1973, The petrologic evolution of the North Mountain stock, La Sal Mountains, Utah: Tucson, Ariz., University of Arizona M.S. thesis, 71 p.

Jenkins, R.E., II, 1979, Geology, geochemistry and origin of mineral deposits in the Hill Gulch area, Jamestown, Colorado: Golden, Colo., Colorado School of Mines Ph. D. thesis, 220 p.

Kane, W.T., Atkinson, W.W., Jr., and Snee, L.W., 1988, Trace element geochemistry and Ar/Ar geochronology of epithermal mineralization, Four Mile Canyon, Boulder County, Colorado: Geological Society of America Abstracts with Programs, v. 20, p. A353.

Kay, B.D., 1986, Vein and breccia gold mineralization and associated igneous rocks at the Ortiz mine, New Mexico, U.S.A.: Golden, Colo., Colorado School of Mines M.S. thesis, 179 p.

Kelly, W.C., and Goddard, E.M., 1969, Telluride ores of Boulder County, Colorado: Geological Society of America Memoir 109, 237 p.

Kilinc, I.A., 1979, Melting relations in the quartz diorite-H_2O and quartz diorite-H_2O-CO_2 systems: Neues Jahrbuch für Mineralogie, 1979, no. 2, p. 67–72.

Kirchener, J.G., 1971, Petrology and petrography of the phonolite porphyry intrusions of the northern Black Hills, South Dakota: Iowa City, Iowa, University of Iowa Ph. D. thesis, 199 p.

Kirchner, G.L., 1982, Field relations, petrology, and mineralization of the Linster Peak dome, Fergus County, Montana: Missoula, Mont., University of Montana M.S. thesis, 115 p.

Kleinkopf, M.D., Peterson, D.L., and Mattick, R.E., 1979, Gravity, magnetic, and seismic studies of the Silver Cliff and Rosita Hills volcanic area, Colorado: U.S. Geological Survey Professional Paper 726–F, p. F1–F17.

Kohrt, P.B., 1991, Alkalic rocks of the Judith Mountains, central Montana, *in* Baker, D.W., and Berg, R.B., eds., Guidebook of the Central Montana alkalic province: Montana Bureau of Mines and Geology Special Publication 100, p. 77–85.

Koschmann, A.H., 1949, Structural control of the gold deposits of the Cripple Creek district, Teller County, Colorado: U.S. Geological Survey Bulletin 955–B, p. 19–60.

Kurisoo, Peter, 1991, Gold deposits of the Kendall mining district, Fergus County, Montana, *in* Baker, D.W., and Berg, R.B., eds., Guidebook of the Central Montana alkalic province: Montana Bureau of Mines and Geology Special Publication 100, p. 39–44.

Larsen, R.K., 1977, Geology, alteration and mineralization of the northern Cutting stock, Lawrence County, South Dakota: Rapid City, S.Dak., South Dakota School of Mines and Technology M.S. thesis, 147 p.

Lindgren, Waldemar, Graton, L.C., and Gordon, C.H., 1910, The ore deposits of New Mexico: U.S. Geological Survey Professional Paper 68, 361 p.

Lindgren, Waldemar, and Ransome, F.L., 1906, Geology and gold deposits of the Cripple Creek district, Colorado: U.S. Geological Survey Professional Paper 54, 516 p.

Lindsey, D.A., 1982, Geologic map and discussion of selected mineral resources of the North and South Moccasin Mountains, Fergus County, Montana: U.S. Geological Survey Miscellaneous Investigations Series Map I–1362, scale 1:24,000.

———1985, A gold-mineralized breccia zone at Kendall, North Moccasin Mountains, Fergus County, Montana: U.S. Geological Survey Professional Paper 1301–C, p. 43–56.

Lindsey, D.A., and Fisher, F.S., 1985, Mineralized breccias in intrusive complexes of Late Cretaceous and Paleocene age, north-central Montana: U.S. Geological Survey Professional Paper 1301–A, p. 1–34.

Lindsey, D.A., and Naeser, C.W., 1985, Relation between igneous intrusion and gold mineralization in the North Moccasin Mountains, Fergus County, Montana: U.S. Geological Survey Professional Paper 1301–B, p. 35–41.

Lipman, P.W., Fisher, F.S., Mehnert, H.H., Naeser, C.W., Luedke, R.G., and Steven, T.A., 1976, Multiple ages of mid-Tertiary mineralization and alteration in the western San Juan Mountains, Colorado: Economic Geology, v. 71, p. 571–588.

Loomis, T.A., and Alexander, D.L., 1990, The geology of the Foley Ridge gold mine, *in* Fourth Western Regional Conference on Precious Metals and the Environment, Lead, S. Dak., September 19–22, 1990, Proceedings: AIME, Society for Mining, Metallurgy, and Exploration, Black Hills Section, p. 77–87.

Loughlin, G.F., 1927, Ore at deep levels in the Cripple Creek district, Colorado: American Institute of Mining and Metallurgical Engineers Technical Publication 13, 32 p.

Loughlin, G.F., and Koschmann, A.H., 1935, Geology and ore deposits of the Cripple Creek district, Colorado: Colorado Scientific Society Proceedings, v. 13, no. 5, p. 217–435.

Lovering, T.S., 1930, The New World or Cooke City mining district, Park County, Montana: U.S. Geological Survey Bulletin 811–A, p. 1–87.

Lux, D.R., 1977, A major element geochemical study of Laramide igneous rocks of the Colorado mineral belt: Houston, Texas, Rice University M.S. thesis, 77 p.

Macdonald, G.A., and Katsura, T., 1964, Chemical composition of Hawaiian lavas: Journal of Petrology, v. 5, p. 82–133.

Marvin, R.F., Hearn, B.C., Jr., Mehnert, H.H., Naeser, C.W., Zartman, R.E., and Lindsey, D.A., 1980, Late Cretaceous-Paleocene-Eocene igneous activity in north central Montana: Isochron/West, no. 29, p. 5–25.

Marvin, R.F., Witkind, I.J., Keefer, W.R., and Mehnert, H.H., 1973, Radiometric ages of intrusive rocks in the Little Belt Mountains, Montana: Geological Society of America Bulletin, v. 84, p. 1977–1986.

Marvin, R.F., Young, E.J., Mehnert, H.H., and Naeser, C.W., 1974, Summary of radiometric age determinations on Mesozoic and Cenozoic igneous rocks and uranium and base metal deposits in Colorado: Isochron/West, no. 11, 41 p.

Mattox, S.R., 1992, Geochemistry, origin, and tectonics of mid-Tertiary and late Tertiary and Quaternary volcanic rocks, southern Marysvale volcanic field, Utah: De Kalb, Ill., Northern Illinois University Ph. D. thesis, 239 p.

Maynard, S.R., Nelsen, C.J., Martin, K.W., and Schultz, J.L., 1990, Geology and gold mineralization of the Ortiz Mountains, Santa Fe County, New Mexico: Mining Engineering, v. 42, no. 8 (August), p. 1007–1011.

Maynard, S.R., Woodward, L.A., and Giles, D.L., 1991, Tectonics, intrusive rocks, and mineralization of the San Pedro–Ortiz porphyry belt, north-central New Mexico: New Mexico Bureau of Mines and Mineral Resources Bulletin 137, p. 57–69.

McDowell, F.W., 1971, K-Ar ages of igneous rocks from the western United States: Isochron/West, no. 2, p. 1–16.

McEwan, C.J.A., 1986, Geological, stable isotope and fluid inclusion evidence for concealed epithermal precious metal mineralisation, Rosita Hills, Colorado: Geological Society of America Abstracts with Programs, v. 16, p. 688.

McMillan, W.J., 1991, Porphyry deposits in the Canadian Cordillera: British Columbia Ministry of Energy, Mines and Petroleum Resources Paper 1991–4, p. 253–276.

Meyer, H.O.A., and Mitchell, R.H., 1988, Sapphire bearing ultramafic lamprophyre from Yogo, Montana—An ouachitite: Canadian Mineralogist, v. 26, p. 81–88.

Mutschler, F.E., Griffin, M.E., Stevens, D.S., and Shannon, S.S., Jr., 1985, Precious metal deposits related to alkaline rocks in the North American Cordillera—An interpretive review: Geological Society of South Africa Transactions, v. 88, p. 355–377.

Mutschler, F.E., Larson, E.E., and Bruce, R.M., 1987, Laramide and younger magmatism in Colorado—New petrologic and tectonic variations on old themes: Colorado School of Mines Quarterly, v. 82, no. 4, p. 1–47.

Mutschler, F.E., and Mooney, T.C., 1993, Precious metal deposits related to alkaline igneous rocks—Provisional classification, grade-tonnage data, and exploration frontiers, *in* Kirkham, R.V., Sinclair, W.D., Thorpe, R.I., and Duke, J.M., eds., Mineral deposit modeling: Geological Association of Canada Special Paper 40.

Mutschler, F.E., Mooney, T.C., and Johnson, D.C., 1990, Precious metal deposits related to alkaline igneous rocks—A space-time trip through the Cordillera, *in* Fourth Western Regional Conference on Precious Metals and the Environment, Lead, S. Dak., September 19–22, 1990, Proceedings: AIME, Society for Mining, Metallurgy, and Exploration, Black Hills Section, p. 43–59.

Mutschler, F.E., Rougon, D.J., Lavin, O.P., and Hughes, R.D., 1981, PETROS—A data bank of major-element chemical analyses of igneous rocks (version 6–1): Boulder, Colo., NOAA National Geophysical Data Center (magnetic tape).

Nash, J.T., and Cunningham, C.G., Jr., 1973, Fluid-inclusion studies of the fluorspar and gold deposits, Jamestown district, Colorado: Economic Geology, v. 68, p. 1247–1262.

Nelson, S.E., 1989, Geology, alteration, and mineral deposits of the Cresson diatreme, Cripple Creek district, Colorado: Fort Collins, Colo., Colorado State University M.S. thesis, 147 p.

———1990, Geology, alteration and Au-Ag mineralization of the Cresson diatreme, Cripple Creek district, Colorado: Geological Society of America Abstracts with Programs, v. 22, no. 6, p. 40–41.

Nelson, S.T., 1991, Mid-Tertiary magmatism of the Colorado Plateau—The Henry and La Sal Mountains, Utah: Los Angeles, Calif., University of California Ph. D. thesis, 318 p.

Nelson, S.T., and Davidson, J.P., 1993, Interactions between mantle-derived magmas and mafic crust, Henry Mountains, Utah: Journal of Geophysical Research, v. 98, p. 1837–1852.

Nelson, S.T., Sullivan, K.R., and Davidson, J.P., 1992, New age determinations of central Colorado Plateau laccoliths, Utah—Recognizing disturbed K-Ar systematics and re-evaluating tectonomagmatic relationships: Geological Society of America Bulletin, v. 104, p. 1547–1560.

Noble, J.A., 1948, High-potash dikes in the Homestake mine, Lead, South Dakota: Geological Society of America Bulletin, v. 59, p. 927–940.

North, R.M., and McLemore, V.T., 1988, A classification of the precious metal deposits of New Mexico, *in* Schafer, R.W., Cooper, J.J., and Vikre, P.G., eds., Bulk mineable precious metal deposits of the western United States: Reno, Nev., Nevada Geological Society, p. 625–659.

Norton, J.J., 1989, Bald Mountain gold mining region, northern Black Hills, South Dakota: U.S. Geological Survey Bulletin 1857–C, p. C1–C13.

Ogilvie, I.H., 1908, Some igneous rocks from the Ortiz Mountains, New Mexico: Journal of Geology, v. 16, p. 230–238.

Olmore, S.D., 1991, Geologic summary of the Big Ben molybdenum deposit near Neihart, Montana, *in* Baker, D.W., and Berg, R.B., eds., Guidebook of the Central Montana alkalic province: Montana Bureau of Mines and Geology Special Publication 100, p. 124–127.

O'Sullivan, R.B., and Beikman, H.M., 1963, Geology, structure, and uranium deposits of the Shiprock quadrangle, New Mexico and Arizona: U.S. Geological Survey Miscellaneous Geologic Investigations Map I–345, scale 1:250,000.

Paterson, C.J., 1990, Magmatic-hydrothermal model for epithermal-mesothermal Au-Ag deposits in the northern Black Hills, *in* Fourth Western Regional Conference on Precious Metals and the Environment, Lead, S. Dak., September 19–22, 1990, Proceedings: AIME, Society for Mining, Metallurgy, and Exploration, Black Hills Section, p. 89–102.

Paterson, C.J., Lisenbee, A.L., and Redden, J.A., 1988, Gold deposits in the Black Hills, South Dakota, *in* Diedrich, R.P., Dyka, M.A.K., and Miller, W.R., eds., Thirty-Ninth Field Conference guidebook—Eastern Powder River Basin–Black Hills: Wyoming Geological Association Field Conference, 39th, Casper, Wyo., Sept. 9–11, 1988, p. 295–304.

Paterson, C.J., Uzunlar, Nuri, Groff, J., and Longstaffe, F.J., 1989, A view through an epithermal-mesothermal precious metal system in the northern Black Hills, South Dakota—A magmatic origin for the ore-forming fluids, *in* Keays, R.R., Ramsay, W.R.H., and Groves, D.I., eds., The geology of gold deposits—The perspective in 1988: Economic Geology Monograph 6, p. 564–570.

Phair, George, 1952, Radioactive Tertiary porphyries in the Central City district, Colorado, and their bearing upon pitchblende deposition: U.S. Geological Survey Trace-Element Investigations Report TEI–247, 53 p.

Phair, George, and Jenkins, L.B., 1975, Tabulation of uranium and thorium data on the Mesozoic-Cenozoic intrusive rocks of known chemical composition in Colorado: U.S. Geological Survey Open-File Report 75–501, 57 p.

Pirsson, L.V., 1894, On some phonolite rocks from the Black Hills: American Journal of Science, 3rd series, v. 47, p. 341–346.

———1900, Petrography of igneous rocks of the Little Belt Mountains, Montana: U.S. Geological Survey 20th Annual Report, pt. 3, p. 463–581.

Pontius, J.A., 1991, Geology and gold mineralization of the Cresson deposit, Cripple Creek, Colorado: Northwest Mining Association 97th Annual Convention, Dec. 1–6, 1991, Spokane, Wash., Abstract Booklet, p. 26.

Porter, E.W., and Ripley, Edward, 1985, Petrologic and stable isotope study of the gold-bearing breccia pipe at the Golden Sunlight deposit, Montana: Economic Geology, v. 80, p. 1689–1706.

Priesmeyer, S.T., 1986, Geology and gold mineralization on the New Year mine and North Giltedge properties, Fergus County, Montana: Moscow, Idaho, University of Idaho M.S. thesis, 84 p.

Rice, C.M., Harmon, R.S., and Shepherd, T.J., 1985, Central City, Colorado—The upper part of an alkaline porphyry molybdenum system: Economic Geology, v. 80, p. 1769–1796.

Rice, C.M., Lux, D.R., and Macintyre, R.M., 1982, Timing of mineralization and related intrusive activity near Central City, Colorado: Economic Geology, v. 77, p. 1655–1666.

Rock, N.M.S., Groves, D.I., Perring, C.S., and Golding, S.D., 1989, Gold, lamprophyres, and porphyries—What does their association mean?, *in* Keays, R.R., Ramsay, W.R.H., and Groves, D.I., eds., The geology of gold deposits—The perspective in 1988: Economic Geology Monograph 6, p. 609–625.

Roemmel, J.S., 1982, A petrographic and economic evaluation of the White Cow intrusion, Little Rocky Mountains, Montana: Missoula, Mont., University of Montana M.S. thesis, 51 p.

Ronkos, C.J., 1991, Geology, alteration and gold mineralization in the White Oaks mining district, Lincoln County, New Mexico: Geological Society of America Abstracts with Programs, v. 23, no. 4, p. 88.

Ross, M.L., 1992, Geology and petrology of the Oligocene North Mountain intrusive center, La Sal Mountains, Utah: Geological Society of America Abstracts with Programs, v. 24, no. 6, p. 59.

Rupp, J.A., 1980, Tertiary rhyolite dikes and plutons of the northern Little Belt Mountains, Montana: Cheney, Wash., Eastern Washington University M.S. thesis, 136 p.

Russell, C.W., 1991, Gold mineralization in the Little Rocky Mountains, Montana, *in* Baker, D.W., and Berg, R.B., eds., Guidebook of the Central Montana alkalic province: Montana Bureau of Mines and Geology Special Publication 100, p. 1–18.

Russell, C.W., and Gabelman, Joan, 1991, Field guide, Little Rocky Mountains, *in* Baker, D.W., and Berg, R.B., eds., Guidebook of the Central Montana alkalic province: Montana Bureau of Mines and Geology Special Publication 100, p. 185–194.

Ryzak, David, 1990, Gold deposits of active mining areas, Little Rocky Mountains, Montana, *in* Fourth Western Regional Conference on Precious Metals and the Environment, Lead, S. Dak., September 19–22, 1990, Proceedings: AIME, Society for Mining, Metallurgy, and Exploration, Black Hills Section, p. 61–75.

Saunders, J.A., 1988, Textural and geochemical characteristics of gold mineralization from the Cresson mine, Cripple Creek district, Colorado, U.S.A.: Institution of Mining and Metallurgy Transactions, section B, v. 97, p. B36–B39.

———1991, Gold deposits of the Boulder County gold district, Colorado: U.S. Geological Survey Bulletin 1857–I, p. I37–I48.

Saunders, J.A., and May, E.R., 1986, Bessie G—A high-grade epithermal gold telluride deposit, La Plata County, Colorado, U.S.A., *in* Macdonald, A.J., ed., Gold 86—An International Symposium on the Geology of Gold Deposits, Toronto, 1986, Proceedings: Willowdale, Ontario, Konsult International, p. 436–444.

Schroeter, T.G., Lund, C., and Carter, G., 1989, Gold production and reserves in British Columbia: British Columbia Ministry of Energy, Mines and Petroleum Resources Open File 1989–22, 86 p.

Schutz, J.L., Fulp, M.S., Suchomel, B.J., and Woodward, L.A., 1988, Geology and precious-metal mineralization at the Lenny mine, Little Belt Mountains, central Montana, *in* Schafer, R.W., Cooper, J.J., and Vikre, P.G., eds., Bulk mineable precious metal deposits of the western United States: Reno, Nev., Geological Society of Nevada, p. 747.

Segerstrom, Kenneth, and Ryberg, G.E., 1974, Geology and placer gold deposits of the Jicarilla Mountains, Lincoln County, New Mexico: U.S. Geological Survey Bulletin 1308, 25 p.

Segerstrom, Kenneth, Stotelmeyer, R.B., and Williams, F.E., 1979, Mineral resources of the White Mountain Wilderness and adjacent areas, Lincoln County, New Mexico: U.S. Geological Survey Bulletin 1453, 135 p.

Shand, H.S., 1951, Eruptive rocks, 4th edition: New York, John Wiley, 488 p.

Sharp, W.N., 1978, Geologic map of the Silver Cliff and Rosita volcanic centers, Custer County, Colorado: U.S. Geological Survey Miscellaneous Investigations Series Map I–1081, scale 1:24,000.

Sharp, W.N., and Naeser, C.W., 1986, Apatite fission-track age for the Bull Domingo boulder pipe, Custer County, Colorado, chap. H *of* Peterman, Z.E., and Schnabel, D.C., eds., Shorter contributions to isotope research: U.S. Geological Survey Bulletin 1622, p. 77–80.

Sharwood, W.J., 1911, Analyses of some rocks and minerals from the Homestake mine, Lead, South Dakota: Economic Geology, v. 6, p. 729–789.

Shawe, D.R., Simmons, G.C., and Archbold, N.L., 1968, Stratigraphy of Slick Rock district and vicinity, San Miguel and Dolores Counties, Colorado: U.S. Geological Survey Professional Paper 576–A, p. A1–A108.

Shearer, C.K., 1990, Geochemistry of Laramide igneous intrusions, northern Black Hills, with implications for petrogenetic relationships and magma sources, *in* Peterson, C.J., and Lisenbee, A.L., eds., Metallogeny of gold in the Black Hills, South Dakota: Society of Economic Geologists Guidebook Series, v. 7, p. 35–42.

Simons, F.S., Armbrustmacher, T.J., van Noy, R.M., Zilka, N.T., Federspiel, F.E., and Ridenour, James, 1979, Mineral resources of the Beartooth Primitive area and vicinity, Carbon, Park, Stillwater, and Sweet Grass Counties, Montana, and Park County, Wyoming: U.S. Geological Survey Bulletin 1391–F, p. F1–F125.

Size, W.B., 1967, The petrology of the Goose Lake Tertiary intrusive complex, Park County, Montana: De Kalb, Ill., Northern Illinois University M.S. thesis, 96 p

Spurr, J.E., and Garrey, G.H., 1908, Economic geology of the Georgetown quadrangle (together with the Empire district), Colorado: U.S. Geological Survey Professional Paper 63, 422 p.

Steenland, N.C., 1962, Gravity and aeromagnetic exploration in the Paradox basin: Geophysics, v. 27, p. 73–89.

Stern, T.W., Newell, M.F., Kistler, R.W., and Shawe, D.R., 1965, Zircon uranium-lead and thorium-lead ages and mineral potassium-argon ages of La Sal Mountains rocks, Utah: Journal of Geophysical Research, v. 70, p. 1503–1507.

Strobell, J.D., Jr., 1956, Geology of the Carrizo Mountains area in northeastern Arizona and northwestern New Mexico: U.S. Geological Survey Oil and Gas Investigations Map OM–160, scale 1:48,000.

Sullivan, K.R., Kowallis, B.J., and Mehnert, H.H., 1991, Isotopic ages of igneous intrusions in southeastern Utah—Evidence for a Mid-Cenozoic Reno–San Juan magmatic zone: Brigham Young University Geology Studies, v. 37, p. 139–144.

Sun, M.-S., and Baldwin, Brewster, 1958, Volcanic rocks of the Cienega area, Santa Fe County, New Mexico: New Mexico Bureau of Mines and Mineral Resources Bulletin 54, 80 p.

Thompson, T.B., 1968, Hydrothermal alteration and mineralization of the Rialto stock, Lincoln County, New Mexico: Economic Geology, v. 63, p. 943–949.

———1972, Sierra Blanca igneous complex, New Mexico: Geological Society of America Bulletin, v. 83, p. 2341–2356.

———1991, The Bonito-Nogal district, Lincoln County, New Mexico: Geological Society of America Abstracts with Programs, v. 23, no. 4, p. 99.

Thompson, T.B., Trippel, A.D., and Dwelley, P.C., 1985, Mineralized veins and breccias of the Cripple Creek district, Colorado: Economic Geology, v. 80, p. 1669–1688.

Tooker, E.W., 1990, Gold in the Bingham district, Utah: U.S. Geological Survey Bulletin 1857–E, p. E1–E16.

Vogel, J.D., 1960, Geology and ore deposits of the Klondike Ridge area, Colorado: U.S. Geological Survey Open-File Report, 206 p.

Walker, G.E., 1991, Geology of the Barker mining district, Judith Basin and Cascade Counties, Montana, *in* Baker, D.W., and Berg, R.B., eds., Guidebook of the Central Montana alkalic province: Montana Bureau of Mines and Geology Special Publication 100, p. 28–37.

Wallace, A.R., 1989, Gold in the Central City mining district, Colorado: U.S. Geological Survey Bulletin 1857–C, p. C38–C47.

Wallace, S.R., 1953, The petrology of the Judith Mountains, Fergus County, Montana: Ann Arbor, Mich., University of Michigan Ph. D. thesis, 245 p.

Walter, R.C., and Geller, B.A., 1992, Age of telluride-stage mineralization from single-crystal $^{40}Ar/^{39}Ar$ dating of vein-deposited adularia [abs.]: Eos, v. 73, no. 14, Spring Meeting Supplement, p. 362.

Wargo, J.G., 1964, Geology of a disseminated copper deposit near Cerrillos, New Mexico: Economic Geology, v. 59, p. 1525–1538.

Webster, I.D., Taylor, R.P., and Bean, C., 1992, Water-rich pantellerite magma—Analysis of glass inclusions from Fantale volcano, Ethiopia [abs]: Eos, v. 73, no. 14, Spring Meeting Supplement, p. 368.

Weed, W.H., and Pirsson, L.V., 1896, Geology of the Little Rocky Mountains: Journal of Geology, v. 4, p. 399–428.

———1898, Geology and mineral resources of the Judith Mountains of Montana: U.S. Geological Survey 18th Annual Report, pt. 3, p. 437–616.

Werle, J.L., Ikramuddin, Mohammed, and Mutschler, F.E., 1984, Allard stock, La Plata Mountains, Colorado—An alkaline rock-hosted porphyry copper–precious metal deposit: Canadian Journal of Earth Sciences, v. 21, p. 630–641.

White, P.J., and Lawless, J.V., 1989, Geochemistry of porphyry-hosted Au-Ag deposits in the Little Rocky Mountains, Montana—A discussion: Economic Geology, v. 84, p. 970–971.

Wilson, M.R., and Kyser, T.K., 1988, Geochemistry of porphyry-hosted Au-Ag deposits in the Little Rocky Mountains, Montana: Economic Geology, v. 83, p. 1329–1346.

———1989, Geochemistry of porphyry-hosted Au-Ag deposits in the Little Rocky Mountains, Montana—A reply: Economic Geology, v. 84, p. 971–973.

Witkind, I.J., 1964, Geology of the Abajo Mountains area, San Juan County, Utah: U.S. Geological Survey Professional Paper 453, 110 p.

———1970, Composite dikes in the Little Belt Mountains, central Montana: U.S. Geological Survey Professional Paper 700–C, p. C82–C88.

———1973, Igneous rocks and related mineral deposits of the Barker quadrangle, Little Belt Mountains, Montana: U.S. Geological Survey Professional Paper 752, 58 p.

Woodward, L.A., 1991, Metallic mineralization in the Yogo and Running Wolf mining districts, Little Belt Mountains, Montana, *in* Baker, D.W., and Berg, R.B., eds., Guidebook of the Central Montana alkalic province: Montana Bureau of Mines and Geology Special Publication 100, p. 19–27.

Wright, A., 1983, The Ortiz gold deposit (Cunningham Hill)—Geology and exploration: Nevada Bureau of Mines and Geology Report 36, p. 42–51.

Zhang, Xiaomao, and Spry, P.G., 1991, Mineralogical and fluid inclusion characteristics of the epithermal Gies gold-silver telluride deposit, Judith Mountains, Montana—A preliminary study, *in* Baker, D.W., and Berg, R.B., eds., Guidebook of the Central Montana alkalic province: Montana Bureau of Mines and Geology Special Publication 100, p. 63–76.

Laccolithic Complexes of Southeastern Utah: Time of Emplacement and Tectonic Setting—Recommendations for Additional Studies

By Felix E. Mutschler[1]

1. An electronic database of major- and trace-element, isotopic, geochronologic, and petrographic analyses should be compiled for the igneous rocks on and adjacent to the Colorado Plateau. Such a database would be a powerful tool to characterize and compare igneous suites (Nealey and Sheridan, 1989, among others); evaluate metallic mineral exploration potential (for example, Mutschler and others, 1985); and develop and test models relating igneous rock chemistry to magma genesis and tectonic setting (for example, Fitton and others, 1991). To maximize the potential for interfacing the igneous rock information with geophysical and other data bases through GIS systems, accurate latitude and longitude or UTM coordinates for each sample location should be incorporated in the database.

2. Additional geochronological studies should be directed to date both oldest and youngest plutons in the laccolithic complexes on and adjacent to the Colorado Plateau. The following should be dated: The syenite pluton at Navajo Mountain, Utah (Condie, 1964); the Klondike Hills, Colo., plutons (Shawe and others, 1968); the syenodiorite and monzonite plutons of the Levan, Utah area (Witkind and Weiss, 1991 and references therein). Likewise, the following have not been dated: in the Henry Mountains, Utah, the youngest plutons (granite porphyry of Hunt, 1988); in the La Sal Mountains, Utah, the young granite porphyry and soda rhyolite porphyry (Hunt and Waters, 1958). The only dates available for the Carrizo Mountains, Ariz., are two K-Ar dates of ≈70 Ma (Armstrong, 1969), which may reflect excess argon. The Carrizo Mountains may well be Laramide, but further dating is in order.

3. Further characterization of the physical, chemical, and age attributes of the crystalline basement rocks underlying the Colorado Plateau is in order. Cuttings and core from petroleum test wells that reached basement are largely undescribed in terms of detailed lithology, age, density, and magnetic properties. These studies would allow us to refine and better constrain geophysical and geological models.

[1]Petrophysics Crisis Center, Department of Geology, Eastern Washington University, Cheney, WA 99004.

4. Evaluation of the possibility of unconventional metal deposits on the Colorado Plateau should be given a high priority. Possible deposit models include Ni-Mo-PGE-Au ores in black shales (Coveney and Nansheng, 1991) in the Paradox basin; PGE-Au mineralization associated with roll-front uranium deposits; and PGE-Au mineralization associated with subtle alteration zones and (or) "redbed" Cu-Ag mineralization in Lisbon Valley and Bull Canyon, Utah.

REFERENCES CITED

Armstrong, R.L., 1969, K-Ar dating of laccolithic centers of the Colorado Plateau and vicinity: Geological Society of America Bulletin, v. 80, p. 2081–2086.

Condie, K.C., 1964, Crystallization Po_2 of syenite porphyry from Navajo Mountain, southern Utah: Geological Society of America Bulletin, v. 75, p. 359–362.

Coveney, R.M., Jr., and Nansheng, C., 1991, Ni-Mo-PGE-Au-rich ores in Chinese black shales and speculations on possible analogues in the United States: Mineralium Deposita, v. 26, p. 83–88.

Fitton, J.G., James, Dodie, and Leeman, W.P., 1991, Basic magmatism associated with late Cenozoic extension in the western United States: compositional variations in space and time: Journal of Geophysical Research, v. 96, p. 13693–13711.

Hunt, C.B., and Waters, A.C., 1958, Structural and igneous geology of the La Sal Mountains, Utah: U.S. Geological Survey Professional Paper 294–I, p. 305–364.

Hunt, G.L., 1988, Petrology of the Mt. Pennell central stock, Henry Mountains, Utah: Brigham Young University Geology Studies, v. 35, p. 81–100.

Mutschler, F.E., Griffin, M.E., Stevens, D.S., and Shannon, S.S., Jr., 1985, Precious metal deposits related to alkaline rocks in the North American Cordillera—An interpretive review: Geological Society of South Africa Transactions, v. 88, p. 355–377.

Nealey, L.D., and Sheridan, M.F., 1989, Post-Laramide volcanic rocks of Arizona and northern Sonora, Mexico, and their inclusions, *in* Jenney, J.P., and Reynolds, S.J., eds., Geologic evolution of Arizona: Tucson, Arizona Geological Society Digest 17, p. 609–647.

Shawe, D.R., Simmons, G.C., and Archbold, N.L., 1968, Stratigraphy of Slick Rock district and vicinity, San Miguel and Dolores Counties, Colorado: U.S. Geological Survey Professional Paper 576–A, p. A1–A108.

Witkind, I.J., and Weiss, M.P., 1991, Geologic map of the Nephi 30'×60' quadrangle, Carbon, Emery, Juab, Sanpete, Utah, and Wasatch Counties, Utah: U.S. Geological Survey Miscellaneous Investigations Series Map I–1937, scale 1:100,000.

Appendix 1

Around the Henry Mountains with Charlie Hanks, Some Recollections

By Charles B. Hunt[1]

It was my good fortune in 1935 to be assigned chief of a U.S. Geological Survey field party studying and mapping the geology of the Henry Mountains, Utah (fig. 1). Geologically the area is of great interest because of the classic work done there in 1876 by G.K. Gilbert for the Powell Survey. In the 1930's the area still was frontier (fig. 2)—a long distance from railroads, paved roads, telephones, stores, or medical services. It was the heart of an area the size of New York State without a railroad, and a third of that area was without any kind of a road. This was not Marlboro country; it was Bull Durham country. The geological work had to be done by pack train (fig. 3); it was about the last of the big pack-train surveys in the West—the end of an era.

It was my good fortune also to obtain the services of a veteran horseman who knew that country, Charles R. Hanks of Green River, Utah (fig. 4). Charlie served as packer during each of our five field seasons there. He had played a leading role in the history of the area; the town of Hanksville was named in 1885 for his father when a post office was established there. Charlie, an old-time cow puncher, had spent more hours in saddles than he had in chairs, and he had slept more nights on the ground under the stars than he had in bed under a roof. Most of his meals had been before an open fire on the range. He knew that country, both its good features and its hazards. He knew and understood horses and mules and knew how to travel and live comfortably in the desert and its mountains. And he had learned about geologists.

Charlie had worked for the Geological Survey in the summers of 1930 and 1931 as a packer for Art Baker's field party mapping the Green River Desert. Art helped me get started in 1935 and arranged for Charlie Hanks to join us as a packer. We also employed Lou Christensen, one of Charlie's sons-in-law, as our cook. The geologists on the party in 1935 were Paul Averitt, Jack Hirsch, and Ralph Miller.

Figure 1. C.B. Hunt, in 1936, examines the contact between a porphyry intrusion and overlying altered beds of the Mancos Shale, near Mount Ellen in the Henry Mountains.

When that field project was started, I was 29; Charlie was a quarter century farther along life's road. Always,

[1] The Johns Hopkins University, Emeritus. Deceased, September 3, 1997.

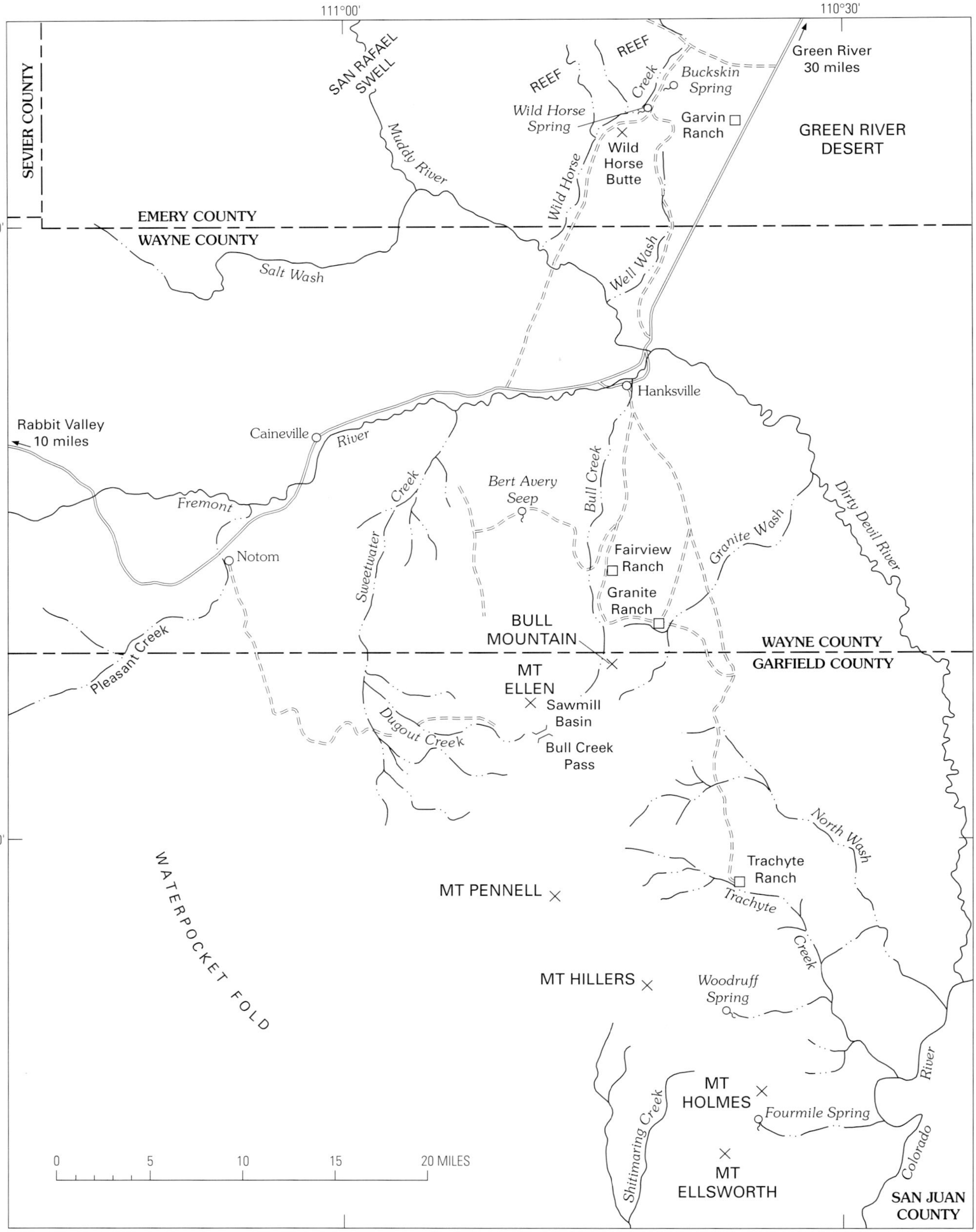

Figure 2. The Henry Mountains, Utah, and vicinity in the late 1930's.

Figure 3. Going to the field. A USGS pack train carries supplies for a spike camp at Fourmile Spring, April 1936.

Figure 4. Charlie Hanks, veteran packer, takes a well-deserved rest near the triangulation flag at the top of Mount Ellen.

though, he was cooperative and helpful to *the kid* in charge. The other geologists were young too, and all of us were city lads. But despite the inadequacies of our western ways, the group won Charlie's respect by being conscientious, hard working, and good natured. The group enjoyed Charlie Hanks and he enjoyed us.

Our first camp was at Buckskin Spring, about 4 miles northeast of Wild Horse Butte. While most of the party established camp and set triangulation flags, Charlie Hanks and Art Baker took me on a conducted tour of the area around camp. We took a car and drove cross country; if there had been a road along the route we took, it had not been used for many years. We drove down Well Wash, battling loose sand all the way, and then back up some tracks over the red hills, northeast of what now is the Hanksville airport. As we started up the grade, Charlie commented to Art, "Now we'll show Charlie Hunt what we mean by a sandy sunnovabich." We'll just say it was sandy, very sandy, but we did get to the top by two of us pushing the car for 5 miles.

Most of the work was by horseback, but not all. There are places horses cannot go. Mapping the Reef of the San Rafael Swell, for example, involved a rocky climb to the top of the Reef and a stadia traverse along the crest. Wherever a canyon crosses the Reef, we had to climb down into it and back up the other side. Charlie would bring our spike camp outfit along the outside of the Reef and meet us for supper and overnight camp. The rendezvous could not be planned in advance because we did not know how many canyons there were or how they were spaced. Charlie had to guess how many miles of crest we could traverse and how many canyons we could cross in the day and make camp accordingly. We took nearly a week traversing the 20 miles to the Muddy River, but every evening Charlie had camp made at the canyon we descended. He came to know the capacity of his geologists as well as that of his horses.

After a few weeks at Buckskin Spring we moved our base camp to Wild Horse Creek and, after a while there, moved it on to the Muddy River where it issues from the San Rafael Swell (fig. 5). On that move Charlie came with the trucks because he knew how to get them to the place where we wanted to camp. Lou and Ralph brought the horses. They spent the night camping in the Swell, and that night one of the horses strayed and was lost.

Next morning Charlie took camping gear and a pack animal to go find the stray. It might have taken days—but no, he was back with the missing horse by middle of the afternoon. It turns out that he had jogged 12 miles to where Wild Horse Creek issues from the Swell and there found the horse's tracks heading down valley toward our former camp. Instead of following those tracks he took a short cut to the trail from Wild Horse Spring to Buckskin Spring. As he approached that trail, there was the missing horse ambling back toward Buckskin Spring, where we had first camped. Asked about his maneuvers, Charlie said, "Oh, I just figured what I would do if I was the horse."

Managing a large pack string poses logistics problems. A riding horse in the desert must have alternate days to rest, so each of us had two animals. A working horse consumes about 100 pounds of oats monthly, and a pack animal can carry about 300 pounds. So the third animal barely managed to carry enough feed for itself and the other two.

At our Muddy River camp we were left with several extra animals that were not needed when Jack Hirsch was taken ill with appendicitis and had to leave our party. We had rented horses from Sam Adams, who lived in Green River, and Charlie Hanks went there to find out from Sam where we could leave his horses. Charlie returned to camp reporting he had missed Sam who had left that morning by wagon for Rabbit Valley, a 300-mile round trip that would take about a month. What to do? Charlie began reasoning: "Sam will spend tonight at San Rafael and tomorrow he'll get to Garvin's. Wednesday evening he'll pull into Hanksville and he'll stop at Andrew's place. He won't get away very early Thursday, and in town he'll stop to see Les and Nelus. You know, I think if I go down to the road here about the middle of the afternoon next Thursday, I'll find Sam." And Charlie did just that. In fact he had to wait only 30 minutes for Sam's wagon to pass en route to Rabbit Valley, 100 miles on its way from Green River.

When Art Baker returned to Washington after spending a couple of weeks helping the party get started, Charlie Hanks and I took him to Green River, where he caught a train. In those days all trains stopped at Green River, where the locomotives took on coal and water. We spent the evening visiting on the porch of Charlie's home, and in the course of the evening Art asked Charlie how many grandchildren he had. Charlie replied, "I reckon about a couple of dozen." Mrs. Hanks expostulated, "Charlie Hanks, you know how many grandchildren you have!" So they started

Figure 5. Base camp along the Muddy River, 1935. Base camps were established only in areas accessible by truck, and were almost luxurious, compared to the single-tent "spike camps" (see fig. 8) used for overnight stays in the back country.

counting, and the number proved to be 25. Charlie resumed his rocking and smilingly said, "I figured I was close."

In those days our supplies—for horses and men—were obtained at a general store owned by Mr. Asimus. The dry cereal boxes then were carrying stories for children, and each morning Paul Averitt would solemnly read us the story. We consumed a good many boxes of cereal and a good many stories that summer, but there was one we could not get—a story called "Two Knights of Red Castle." Paul insisted that Charlie shop for the box having that story; it must have been quite a scene at the store and the subject of many pointed remarks as Asimus and Charlie opened cartons of corn flakes searching for "Two Knights of Red Castle." Again, though, Charlie was successful, and he delivered the box with appropriate ceremony to Paul.

The geologists, all city bred, learned to ride their horses in a fashion that was less than good Western form. One reason, though, was the handicap of the surveying equipment that had to be carried—a plane-table board strapped to one's back, an alidade slung over the right shoulder and held under the left arm, and a tripod carried on the right shoulder (fig. 6). Getting aboard and staying aboard the horse had to be learned by practice, with Charlie's help.

At times Charlie would tease about our horsemanship, but Paul Averitt managed to even scores with Charlie during an automobile trip to town. It began to rain. This was the period when automobiles were changing from manually operated windshield wipers to mechanically operated ones. Charlie didn't know how to start the blasted thing, so Paul ceremoniously showed him how, and *never* let him forget it.

In 1935 the road between Hanksville and Green River was little more than a pair of tracks (fig. 7), and the position of the tracks shifted with every storm. The 15 miles from Green River to the San Rafael River was across shale—good in dry weather but slippery as grease and hub-deep in mud in wet weather. The 40 miles from the San Rafael to Hanksville was in sand—good in wet weather but loose and tractionless in dry weather. These conditions served to separate the optimists from the pessimists: to the optimists part of the road always was good, and to the pessimists part always was bad. When possible we took two cars and carried 50 gallons of water. The 55-mile trip was considered easy if made in less than a full day, and many is the time we had to borrow part of the second day to get through. During our second season, Charlie and Lou went to town as a storm broke and were unable to return for more than a week.

Three of us started the 1936 field season—Marcus Goldman, Charlie Hanks, and I. We spent April and May mapping Mounts Holmes and Ellsworth from a camp at Fourmile Spring (fig. 8). To get there we could drive our trucks to Trachyte Ranch and pack in from there. The pack trip took two days; the night on the trail was spent at Woodruff Spring. To obtain supplies was a 6-day trip for Charlie Hanks. With the horses not loaded he could get back to Trachyte Ranch in a day, and the next day he might get to

Figure 6. Have tripod, will travel. Ralph Miller shows some of the equipment a working geologist needed to carry on horseback. Normally, Ralph would also have an 18- by 24-inch plane-table board strapped to his back.

Green River. A day was spent there, and a fourth day returning to Trachyte Ranch. The 5th and 6th days were on the trail to Fourmile Spring. During that period I learned one of the verities of the West: It was never the horse packing the oats or canned goods that rolled over or fell off the trails—oh no, it was always the horse with the eggs or the jug of wine.

I had a favorite type of raw leather field shoe to wear in the desert, and that season I had brought some new ones obtained at an army surplus store. They had been quite a bargain, and I proudly showed them to Charlie. He shook his head. "They ain't gonna' last you," he said, "them's belly leather." And again he was right.

Our spike camps, like the one at Fourmile Spring (fig. 8), consisted of a single tent with a rectangular pit 15 inches deep at one end. A sheep stove rested at one side of the pit. Sacks of oats around the pit served as backrests, and we could sit on the edge of the pit around the stove—a sort of Indian pithouse. In cold or wet weather this was comfortably cozy; in warm weather we might move outside.

Figure 7. A thoroughfare. Utah State Highway 24, between Green River and Hanksville, in 1935.

G.K. Gilbert's report on the Henry Mountains warns that the "Little Rockies," as Mounts Holmes and Ellsworth are known, are exceedingly rough and can be scaled only on foot. Working there was strenuous. One evening at supper Marcus, Charlie, and I were seated around the pit. Suddenly Marcus rose, stepped across the pit, picked up the salt, and returned to his seat. "Why didn't you ask, Marcus?" Charlie said. "We would have passed the salt." Wearily, Marcus replied, "I was just too tired to ask."

In June, Marcus returned to Washington, and Charlie and I were joined by three geologists—Ralph Miller, who was returning for his second summer, Edgar "Jerry" Bowles, and W.W. Simmons. Lou Christensen returned as cook (fig. 9). Our first base camp was at Bert Avery Seep. Later we moved to Granite Ranch, which had been abandoned, and finally to Dugout Creek at the west foot of Mount Ellen. From these base camps we had brief "spike trips" to the more remote parts of the areas—into the mountains and along the canyon rims.

That summer my wife, Alice, visited our camp. While Jerry Bowles and I were mapping Bull Mountain, she would ride out with us and then, when we started our instrument setups, she would return to camp at Granite Ranch. She expressed apprehension about finding her way back. Charlie reassured her, "You just give Dollie the rein and she'll bring you back to Granite the shortest route. That horse knows where to find her oats." Alice didn't get lost, but admits there were anxious moments when she was sure Dollie was making a wrong turn.

Alice was with us at our spike camp at Log Flat in Sawmill Basin. The day we were to move camp to Bull Creek Pass, Jerry and I had gone on ahead to continue the mapping; Charlie Hanks and Alice were to break camp and move it. While Charlie was loading the pack animals, one of the horses, "Moose," started straying away. Alice, trying to be helpful, went to fetch Moose, but Moose would eye her suspiciously, wait until Alice was near, and then jog a bit farther up the mountain side. Alice spoke softly and sweetly, but Moose kept moving toward the summit. Chagrined, she returned and told Charlie that Moose now was far up the mountainside. When the pack string finally was loaded,

Figure 8. The spike camp at Fourmile Spring, April 1936. Charlie Hanks, at right, is preparing to lead two pack horses down to Trachyte Ranch for supplies.

Figure 9. The 1936 field party ready for business. From left to right are Charlie Hanks, C.B. Hunt, W.W. Simmons, Jerry Bowles, the cook Lou Christensen, and Ralph Miller.

Charlie took off on his horse, and addressing himself in no uncertain terms to the errant Moose, soon had her back in camp. Moose carried an extra load that day.

Charlie Hanks and I set the triangulation flag on the summit of Mt. Ellen. We cut a tall tree a thousand feet below the peak, and Charlie with his horse dragged it up to the saddleback ridge south of the peak. From there he and I had to drag it to the peak (fig. 10); the ground is too bouldery for horses. Setting that flag on the peak was the original flag-raising ceremony. The flag pole stood for a year, which is no mean feat considering the gale winds at the 11,000-foot altitude.

We were well acquainted with folks in Hanksville. We were made to feel welcome there and they regularly visited our camp when riding the range. Informality prevailed. One day Charlie Hanks was fetching our mail at the post office; there was mail for everyone except Ralph Miller. Charlie commented there was no mail for Ralph, and Mrs. MacDougal, the Postmistress, looked again. Presently she returned with a bundle of mail for Ralph, exclaiming, "Somebody had put it under the M's."

Another experience with Mrs. MacDougal concerns the name of the Dirty Devil River. The name had been applied by John Wesley Powell when he went down the Colorado River—the tributary in question proved not to be a trout stream, it was a dirty devil. The name stuck, and became the antithesis of the better known "Bright Angel" in Grand Canyon. But "Dirty Devil" was unpopular with some early settlers, and when the Board on Geographic Names, about 1890, wrote to the post offices inquiring about local usage, the Hanksville Post Office replied, "The nicer people call it Fremont." Charlie Hanks referred to the river as Dirty Devil and so did everyone else in Hanksville. I was anxious to restore the original name because of its historical flavor. Knowing the Board on Geographic Names would consult the Hanksville Post Office, then managed by Mrs. MacDougal, I asked her the name of the river. She replied, "Oh you mean the Dirty Devil?" I then told her that her predecessor had said that the nicer people call it Fremont. She only smiled, and said, "Oh, the nicer people moved away from Hanksville a long time ago."

At the end of the 1936 field season the project was visited by my boss from Washington, Hugh D. Miser. First I showed him the mapping north of Mount Ellen, and then we convened at Fairview Ranch for a pack trip onto the north flank of Mount Ellen. Charlie Hanks took the pack string to Log Flat while I showed Miser the pediments along Bull Creek, the well-exposed contact between the Bull Mountain bysmalith and the wall rock, the floor of the Horseshoe Ridge laccolith, and the roof of the Bull Creek laccolith. It was a day showing off spectacular geology spectacularly exposed.

When we reached camp, Charlie was not there, so we unsaddled and relaxed with a cup of wine. Presently Charlie came in, with a deer. That evening, after a feast—including some of Charlie Hanks' sourdough biscuits—Miser said to him, "Charlie, they ought to make this into a national park." It wasn't just the good camping weather, the feast, the wine, or the geology; it was the delightful blend of all four. Clearly the boss was pleased.

Figure 10. All in a day's work. C.B. Hunt and Charlie Hanks drag a heavy log of Engelmann spruce toward the summit of Mount Ellen, to be a pole for a triangulation flag. June 1936.

In 1937, Paul Averitt, Jerry Bowles, and I extended the mapping southward to Mount Hillers. George Wolgamot, owner of Trachyte Ranch, joined Charlie Hanks in helping to manage our camps and horses. That summer we worked mostly from temporary spike camps.

Mount Pennell had been so overgrazed that Charlie had trouble finding a place where he could pasture our horses. The mountains at that time were in bad shape because of drought and overgrazing. We rode out together one morning, he looking for pasturage and I to set triangulation flags. En route we came on a sheep herd and stopped for a visit with the herder. Under the boulder where Charlie was sitting, I noticed a blade of grass, the first of the day. I commented on it; the herder almost apologetically said, "Yeh, yu see, we've only been over this part once."

That summer we had to map the bare ledges of the country between Trachyte Creek and North Wash and between North Wash and the Dirty Devil. There is no water there, so we waited for rain to fill the waterpockets, or tanks. When the rains came Charlie and I took off. He made camp in North Wash and I was to join him there after spending the day mapping along the rim.

When it came time to quit, I took off following some fresh tracks that I discovered too late were not headed where Charlie should be. Some Ekkers[2] had crossed by that day and I was following them. It was too late to find Charlie, so I returned to Trachyte Ranch, had late supper and went to bed. About midnight I was awakened by horse's hooves; there came Charlie Hanks. Fearing an accident, he was wasting no time beginning a search. That was the only connection we failed to make in the five field seasons.

Next day he and I again headed for North Wash and from there onto the rocky ledges towards the rim of the Dirty Devil. By noon on the second day we had reached the rim and were having lunch together looking across that magnificent canyon. The Dirty Devil is not as deep as the Grand Canyon, but 3,000 feet is deep enough, and the colors surpass those of the Grand Canyon. We ate lunch in quiet; I was drinking in the scenery and thanking my lucky stars for the privilege of being there. Charlie Hanks finally broke the silence, "You know, there just has to be minerals in those rocks over there; no piece of country could be so gosh-blamed worthless."

At one of our camps that summer, on the west side of Mount Pennell, Charlie Hanks and I were joined by George Wolgamot and Paul Averitt. We were to move camp to a spring that Paul and I had found in a trailless part of the southwest side of the mountain. I told George he could follow our tracks. At supper next evening I asked George if he had any trouble finding the place; there had been no trouble, he had followed the tracks of a wild cow. I said, "George, do you mean you would rather follow the tracks of a wild cow than of a geologist?" "Yeh," George replied softly. "You see, the wild cow knew where he was going." Charlie Hanks sure appreciated that one.

[2] Editor's note: The large Ekker family had lived in the Hanksville area for generations and often provided useful information during Hunt's investigations (Hunt, written commun., 1995).

Figure 11. VIP inspection tour on east side of Mount Ellen, September 1938. Paul Averitt and Wilbur Burbank are at left, C.B. Hunt stands in front of Chief Geologist G.F. Loughlin at center, and Herbert E. Gregory is at right.

In 1938 mapping was completed southward along the Colorado River to Halls Crossing and westward to the Waterpocket Fold. Paul Averitt, Ralph Miller, and Robert E. Bates assisted that summer, and again we had the good services of George Wolgamot and Charlie Hanks. We operated from two camps and periodically I would visit the other camp to see what progress was being made.

One day Charlie Hanks and I took off across the desert in the general direction where we thought the other camp would be located. On the desert, Charlie always could spot a cow or horse on a hillside too far away for me to make out the animal. Furthermore, he not only could see the animal before I could, he could tell whether it was a horse, cow, or man on horseback. I thought his eyesight remarkable, until that day. About noon, far off on a hilltop, I saw Ralph Miller at a plane table. I not only knew it was a geologist at a plane table, I knew it was Ralph. Charlie couldn't see him, but presently spotted Ralph's horse. It was clear that each of us could see and identify a tiny speck if it was the kind we knew.

The deserts south of Mount Hillers include the dry (usually) wash known locally as Shitimaring. This was another of the colorful local names I wished retained, but knew it had no hope in the Board on Geographic Names as it was constituted in the 1930's. (The viewpoint later changed, as witness the 1952–53 Mt. Hillers and Mt. Ellsworth Quadrangles of the U.S. Geological Survey.) We decided to try the name "Shootaring." One of Charlie Hanks' heartiest laughs came one day when some of the Ekkers visited our camp and, looking at our maps, shouted, "Shootaring! Shootaring! Yu damn sissies."

At the end of that field season we were to be inspected by a considerable group of VIP's from Washington (fig. 11). Miser came again, and with him G.F. Loughlin, Chief Geologist, Herbert E. Gregory, who had begun work on the Colorado Plateau about the time I was born, and Wilbur Burbank, who had experience with similar geology in the San Juan Mountains. Bert Loper, veteran Colorado River boatman, helped Charlie and George with the camps during the field conference.

Our only bad accidents in five field seasons occurred during that conference. The first day out Gregory's horse fell, and Gregory had to be returned to town with cracked ribs. The second day, Loughlin's horse fell, and Loughlin had to be taken to Price with a dislocated shoulder. The third day, Miser's horse fell and had to be led off the mountain. It was decided to terminate the field conference.

At the end of the 1939 field season we were host to a group from the Carnegie Geophysical Laboratory, including N.L. (Ham) Bowen, J.W. Greig, Earl Ingerson, E.F. Osborn, and Frank Schairer. We convened at Notom, where I had arranged to rent a string of horses from the ranch, owned by George Durfey, a contemporary and old friend of Charlie Hanks. Charlie and George selected the horses for the five new arrivals, and George would explain to each one the attributes of the particular horse he was being given, and why that horse was selected. In turn, Joe Greig, Earl Ingerson, Ossie Osborn, and Frank Schairer were given their horses, and finally George brought out one for Ham, who obviously was a dignified elder statesman in the midst of a group of youngsters. Said George, "This is just the horse for you—gentle as they come, sure footed, neck reins easily, easy gaited, and keeps moving. All the kids around here love to ride him. Just one thing, though—he is a sunnovabich to jump sideways." Needless to say that became the password for that trip.

Finally came the end of the Henry Mountains project. Four of the horses owned by the Survey still survived and I asked Charlie Hanks what he thought they might be worth. "What do you mean," he replied, "'What are they worth?' Do you want to buy 'em or do you want to sell 'em?"

On our many revisits to the Henry Mountains, Alice and I would stop to visit with Charlie in Green River. He seemed to age as little as anyone possibly could. His humor and quick response continued with him. In 1967, in connection with some additional work in the Henry Mountains with some graduate students, we needed to be reminded about certain features that could be seen in a 1938 photograph. In the picture was a person I thought was Charlie. He looked at the picture, but quickly passed it back saying, "That ain't me; he's wearing galluses and I never did."

It is ironical that this veteran of the saddle and the open range should meet his end in an automobile accident in a city, but that is how it happened—in Salt Lake City about 1970. It was an unhappily violent end. But Charlie Hanks had mostly a quiet life. His skills with the horses and in camp were those of an artist/expert, and those of us who were with him as he practiced his arts and expertise appreciated and respected his ability. We are richer for the experience of having been associated with him.

APPENDIX 2. INDEX OF CITATIONS—SELECTED REFERENCES PERTAINING TO GEOCHRONOLOGY AND TECTONICS OF THE COLORADO PLATEAU

Abrams, Gerda, 1993, Complete Bouguer gravity anomaly map of the state of Colorado: U.S. Geological Survey Miscellaneous Field Studies Map MF–2263, scale 1:500,000.

Agrell, S.O., Charnley, N.R., and Rowley, P.D., 1986, The occurrence of hibonite, perovskite, zirconolite, pseudobrookites and other minerals in a metamorphosed hydrothermal system at Pine Canyon, Piute County, Utah, U.S.A. [abs.]: Mineralogical Society (London) Bulletin, no. 72, p. 4.

Aiken, C.L.V., 1975, Residual Bouguer gravity anomaly map of Arizona: Arizona University Geophysical Lab, Tucson, Ariz., scale 1:1,000,000.

Akers, J.P., Irwin, J.H., Stevens, P.R., and McClymonds, N.E., 1962, Geology of the Cameron quadrangle, Arizona: U.S. Geological Survey Geological Quadrangle Map GQ–162, 1 sheet, scale 1:62,500.

Akright, R.L., 1979, Geology and mineralogy of the Cerrillos copper deposit, Santa Fe County, New Mexico, *in* Ingersoll, R.V., Woodward, L.A., and James, H.L., eds., Guidebook of Santa Fe Country: New Mexico Geological Society Annual Field Conference, 30th, Oct. 4–6, 1979, p. 257–260.

Aldrich, M.J., Jr., 1986, Tectonics of the Jemez lineament in the Jemez Mountains and Rio Grande rift: Journal of Geophysical Research, v. 91, p. 1753–1762.

Aldrich, M.J., Jr., Chapin, C.E., and Laughlin, A.W., 1986, Stress history and tectonic development of the Rio Grande rift, New Mexico: Journal of Geophysical Research, v. 91, p. 6199–6211.

Aldrich, M.J., Jr., and Laughlin, A.W., 1984, A model for the tectonic development of the southeastern Colorado Plateau boundary: Journal of Geophysical Research, v. 89, p. 10207–10218.

Allen, J.C., and Boettcher, A.L., 1978, Amphiboles in andesite and basalt—II, Stability as a function of P–T–fH20–fO2: American Mineralogist, v. 63, p. 1074–1087.

Allen, M.S., and Foord, E.E., 1991, Geologic, geochemical and isotopic characteristics of the Lincoln County porphyry belt, New Mexico—Implications for regional tectonics and mineral deposits, *in* Barker, J.M., Kues, B.S., Austin, G.S., and Lucas, S.G., eds., Guidebook of the Sierra Blanca, Sacramento and Capitan Ranges, New Mexico: New Mexico Geological Society Guidebook, 42nd Field Conference, p. 97–113.

Allen, M.S., and McLemore, V.T., 1991, The geology and petrogenesis of the Capitan pluton, New Mexico: New Mexico Geological Society Guidebook, 42nd Field Conference, p. 115–127.

Allenby, R.J., and Schnetzler, C.C., 1983, United States crustal thickness: Tectonophysics, v. 93, p. 13–31. (map of depth to reflection M)

Allmendinger, R.W., Farmer, Harlow, Hauser, E.C., Sharp, James, Von Tish, Douglas, Oliver, Jack, and Kaufman, Sidney, 1986, Phanerozoic tectonics of the Basin and Range–Colorado Plateau transition from COCORP data and geologic data, *in* Barazangi, M., and Brown, L., eds., Reflection seismology—The continental crust: American Geophysical Union Geodynamics Series, v. 14, p. 257–268.

Allmendinger, R.W., Hauge, T.A., Hauser, E.C., Potter, D.J., Klemperer, S.L., Nelson, K.D., Kneupfer, P. and Oliver, J., 1987, Overview of the COCORP 40° N transect, western United States—The fabric of an orogenic belt: Geological Society of America Bulletin, v. 98, p. 308–319.

Allmendinger, R.W., and Jordan, T.E., 1984, Mesozoic structure of the Newfoundland Mountains, Utah—Horizontal shortening and subsequent extension in the hinterland of the Sevier orogenic belt: Geological Society of America Bulletin, v. 95, p. 1280–1292.

Allmendinger, R.W., Sharp, J.W., Von Tish, Douglas, Serpa, Laura, Brown, Larry, Kaufman, Sidney, Oliver, Jack, and Smith, R.B., 1983, Cenozoic and Mesozoic structure of the eastern Basin and Range province, Utah, from COCORP seismic-reflection data: Geology, v. 11, p. 532–536.

Anderson, A.L., and Kirkham, V.R.D., 1931, Alkaline rocks of the Highwood type in southeastern Idaho: American Journal of Science, 5th series, v. 22, p. 51–68.

Anderson, D.L., 1992, Mantle dynamics; the hotcell model [abs.]: Eos, v. 73, no. 14, Spring Meeting Supplement, p. 272.

———1982, Hotspots, polar wander, Mesozoic convection, and the geoid: Nature, v. 297, p. 391–393.

———1994, Superplumes or supercontinents?: Geology, v. 22, p. 39–42. —Plumes *not necessary* to account for large low-velocity-mantle region with hotspots and ridges, as in Pacific and Indian Oceans—extensive plateau and continental flood-basalt provinces. Instead, invoke global plate reorganizations.

Anderson, D.L., Tanimoto, T., and Zhang, Y.-S., 1992, Plate tectonics and hotspots: Science, v. 256, p. 1645–1651.

Anderson, E.M., 1951, The dynamics of faulting and dyke formation with applications to Britain: London, Oliver and Boyd, 206 p.

Anderson, J.J., 1971, Geology of the southwestern High Plateaus of Utah—Bear Valley Formation, an Oligocene-Miocene volcanic arenite: Geological Society of America Bulletin, v. 82, p. 1179–1205.

———1985, Mid-Tertiary block faulting along west and northwest trends, southern High Plateaus, Utah: Geological Society of America Abstracts with Programs, v. 17, no. 7, p. 513.

———1987, Late Cenozoic drainage history of the northern Markagunt Plateau, Utah, *in* Kopp, R.S., and Cohenour, R.E., eds., Cenozoic geology of western Utah—Sites for precious metal and hydrocarbon accumulations: Utah Geological Association Publication 16, p. 271–278.

———1988, Pre-basin-range block faulting along west and northwest trends, southeastern Great Basin and southern High Plateaus: Geological Society of America Abstracts with Programs, v. 20, no. 3, p. 139.

———1993, The Markagunt megabreccia—Large Miocene gravity slide mantling the northern Markagunt Plateau, southwestern Utah: Utah Geological Survey Miscellaneous Publication 93-2,37 p.

Anderson, J.J., and Rowley, P.D., 1975, Cenozoic stratigraphy of the southwestern High Plateaus of Utah, *in* Anderson, J.J., Rowley, P.D., Fleck, R.J., and Nairn, A.E.M., Cenozoic geology of southwestern High Plateaus of Utah: Geological Society of America Special Paper 160, p. 1–52.

Anderson, J.J., Rowley, P.D., Blackman, J.T., Mehnert, H.H., and Grant, T.C., 1990, Geologic map of the Circleville Canyon area, southern Tushar Mountains and northern Markagunt Plateau, Beaver, Garfield, Iron, and Piute Counties, Utah: U.S. Geological Survey Miscellaneous Investigations Series Map I–2000, scale 1:50,000.

Anderson, J.J., Rowley, P.D., Machette, M.N., Decatur, S.H., and Mehnert, H.H., 1990, Geologic map of the Nevershine Hollow area, eastern Black Mountains, southern Tushar Mountains, and northern Markagunt Plateau, Beaver and Iron Counties, Utah: U.S. Geological Survey Miscellaneous Investigations Series Map I–1999, scale 1:50,000.

Anderson, R.E., 1971, Thin skin distension in Tertiary rocks of southeastern Nevada: Geological Society of America Bulletin, v. 82, p. 43–58.

———1973, Large-magnitude late Tertiary strike-slip faulting north of Lake Mead, Nevada: U.S. Geological Survey Professional Paper 794, 18 p.

———1978, Chemistry of Tertiary volcanic rocks in the Eldorado Mountains, Clark County, Nevada, and comparisons with rocks from some nearby areas: U.S. Geological Survey Journal of Research, v. 6, p. 409–424.

———1984, Strike-slip faults associated with extension in and adjacent to the Great Basin: Geological Society of America Abstracts with Programs, v. 16, no. 6, p. 429.

———1986, Coeval mixed-mode dip-slip and strike-slip faulting in and adjacent to the Basin and Range, Utah-Nevada: Geological Society of America Abstracts with Programs, v. 18, no. 5, p. 338.

———1987, Neogene geologic history of the Nevada-Utah border area at and near latitude 37° 30' N.: Geological Society of America Abstracts with Programs, v. 19, no. 7, p. 572.

———1988, Hazard implications of joint-controlled basaltic volcanism in southwestern Utah: Geological Society of America Abstracts with Programs, v. 20, no. 7, p. A115.

———1989, Tectonic evolution of the Intermontane System, Basin and Range, Colorado Plateau, and High Lava Plains, Chapter 10 *in* Pakiser, L.C., and Mooney, W.D., eds., Geophysical framework of the continental United States: Geological Society of America Memoir 172, p. 163–176.

———1990, West-directed tectonic escape between the northern and southern sectors of the Basin and Range province: Geological Society of America Abstracts with Programs, v. 22, no. 3, p. 2.

———1993, Tectonic significance of a Miocene dike swarm and its post-emplacement vertical and meridional collapse, Lake Mead area, Nevada, Arizona: Geological Society of America Abstracts with Programs, v. 25, no. 5, p. 3.

Anderson, R.E., and Barnhard, T.P., 1991, Relationship between Miocene plutonism, uplift, and extension, Lake Mead area, northernmost Arizona and adjacent Nevada: Geological Society of America Abstracts with Programs, v. 23, no. 5, p. 245.

———1992, Neotectonic framework of the central Sevier Valley area, Utah, and its relationship to seismicity, *in* Gori, P.L., and Hays, W.W., eds., Assessment of regional earthquake hazards and risk along the Wasatch front, Utah: U.S. Geological Survey Professional Paper 1500–F, 47 p.

———1993a, Heterogeneous Neogene strain and its bearing on horizontal extension and horizontal and vertical contraction at the margin of the extensional orogen, Mormon Mountains area, Nevada and Utah: U.S. Geological Survey Bulletin 2011, 43 p.

———1993b, Aspects of three-dimensional strain at the margin of the extensional orogen, Virgin River depression area, Nevada, Utah, and Arizona: Geological Society of America Bulletin, v. 105, p. 1019–1052.

Anderson, R.E., Barnhard, T.P., and Snee, L.W., 1994, Roles of plutonism, mid-crustal flow, tectonic rafting, and horizontal collapse in shaping the Miocene strain field of the Lake Mead area, Nevada and Arizona: Tectonics, v. 13, p. 1381–1410.

Anderson, R.E., and Bohannon, R.G., 1993, Three-dimensional aspects of the Neogene strain field, Nevada-Utah-Arizona tricorner area, *in* Lahren, M.M., Trexler, J.H., Jr., and Spinosa, Claude, eds., Crustal evolution of the Great Basin and the Sierra Nevada; Geological Society of America Cordilleran/Rocky Mountain Sections Joint Meeting, Reno, Nev., 1993, Field Trip Guidebook: Reno, Nev., University of Nevada, Mackay School of Mines, p. 167–196.

Anderson, R.E., Bucknam, R.C., and Hamblin, Kenneth, 1978, Road log to the Quaternary tectonics of the Intermontane Seismic Belt between Provo and Cedar City, Utah: Geological Society of America Rocky Mountain Section Annual Meeting, field trip 8, 50 p.

Anderson, R.E., and Ekren, E.B., 1977, Comment *on* Late Cenozoic fault patterns and stress fields in the Great Basin and westward displacement of the Sierra Nevada block (Wright, 1976): Geology, v. 5, p. 388–389.

Anderson, R.E., Longwell, C.R., Armstrong, R.L., and Marvin, R.F., 1972, Significance of K-Ar ages of Tertiary rocks from the Lake Mead region, Nevada-Arizona: Geological Society of America Bulletin, v. 83, p. 273–288.

Anderson, R.E., and Mehnert, H.H., 1979, Reinterpretation of the history of the Hurricane fault in Utah, *in* Newman, G.W., and Goode, H.D., eds., Basin and Range symposium and Great Basin Field Conference: Rocky Mountain Association of Geologists and Utah Geological Association, p. 145–165.

Anderson, R.E., Zoback, M.L., and Thompson, G.A., 1983, Implications of selected subsurface data on the structural form and evolution of some basins in the northern Basin and Range province, Nevada and Utah: Geological Society of America Bulletin, v. 94, p. 1055–1072.

Anderson, R.N., and Noltimier, H.C., 1973, A model for the horst and graben structure of midocean ridge crests based upon spreading velocity and basalt delivery to the oceanic crust: Geophysics Journal of the Royal Astronomical Society, v. 34, p. 137–147.

Angelier, Jacques, Coletta, Bernard, and Anderson, R.E., 1985, Neogene paleostress changes in the Basin and Range—A case study at Hoover Dam, Nevada-Arizona: Geological Society of America Bulletin, v. 96, p. 347–361.

Armstrong, R.L., 1969, K-Ar dating of laccolithic centers of the Colorado Plateau and vicinity: Geological Society of America Bulletin, v. 80, p. 2081–2086.

———1990, Cenozoic magmatism in the North America Cordillera and the origin of metamorphic core complexes: Geological Society of America Abstracts with Programs, v. 22, no. 3, p. 4.

Armstrong, R.L., Ekren, E.B., McKee, E.H., and Noble, E.C., 1969, Space-time relations of Cenozoic silicic volcanism in the Great Basin of the Western United States: American Journal of Science, v. 267, p. 478–490.

Armstrong, R.L., Harakal, J.E., and Hollister, V.F., 1982, Eocene mineralization at Mount Tolman (Keller), Washington, and Silver Dyke, Montana: Isochron/West, no. 33, p. 9.

Armstrong, R.L., and Ward, Peter, 1991, Evolving geographic patterns of Cenozoic magmatism in the North American cordillera—The temporal and spatial association of magmatism and metamorphic core complexes: Journal of Geophysical Research, v. 96, no. B8, p. 13201–13224.

Atkinson, W.W., 1961, Geology of the San Pedro Mountains, Santa Fe County, New Mexico: New Mexico Bureau of Mines and Mineral Resources Bulletin 77, 49 p.

Atwater, Tanya, 1970, Implications of plate tectonic evolution of western North America: Geological Society of America Bulletin, v. 81, p. 3513–3536.

———1989, Plate tectonic history of the northeast Pacific and western North America, *in* Winterer, E.L., Hussong, D.M., and Decker, R.W., eds., The Geology of North America; DNAG Series, v. N, The Eastern Pacific Ocean and Hawaii: Boulder, Colo., Geological Society of America, p. 27–72.

Axelrod, D.I., 1968, Tertiary floras and topographic history of the Snake River basin, Idaho: Geological Society of America Bulletin, v. 79, p. 713–734.

Axen, G.J., 1991, Tertiary extension, magmatism, and thrust reactivation in the southern Great Basin, and a mechanical model for detachment faulting: Cambridge, Mass., Harvard University Ph. D. dissertation, 235 p.

Axen, G.J., Lewis, P.R., Burke, K.J., Sleeper, K.G., and Fletcher, J.M., 1988, Tertiary extension in the Pioche area, Lincoln County, Nevada, *in* Weide, D.L., and Faber, M.L., eds., This extended land—Geological journeys in the southern Basin and Range: Geological Society of America, Cordilleran section, Field trip guidebook, p. 3–5.

Axen, G.J., Taylor, W.J., and Bartley, J.M., 1993, Space-time patterns and tectonic controls of Tertiary extension and magmatism in the Great Basin of the western United States: Geological Society of America Bulletin, v. 105, p. 56–76.

Aydin, Atilla, 1978, Small faults as deformation bands in sandstone, *in* Byerlee, J.D., and Wyss, M., eds., Rock friction and earthquake prediction: Pure and Applied Geophysics, v. 116, no. 4–5, p. 913–930.

Aydin, Attila, and Nur, Amos, 1982, Evolution of pull-apart basins and their scale independence: Tectonics, v. 1, no. 1, p. 91–105.

Baars, D.L., 1966, Pre-Pennsylvanian paleotectonics—Key to basin evolution and petroleum occurrences in Paradox Basin, Utah and Colorado: American Association of Petroleum Geologists Bulletin, v. 50, no. 10, p. 2082–2111.

Baars, D.L., and Stevenson, G.M., 1981, Tectonic evolution of the Paradox Basin, Utah and Colorado, *in* Wiegand, D.L., ed., Geology of the Paradox Basin: Denver, Colo., Rocky Mountain Association of Geologists, Field Conference, 1981, Guidebook, p. 23–31.

Babenroth, D.L., and Strahler, A.N., 1945, Geomorphology and structure of the East Kaibab monocline, Arizona and Utah: Geological Society of America Bulletin, v. 56, no. 2, p. 107–150.

Bacon, C.R., 1982, Time-predictable bimodal volcanism in the Coso Range, California: Geology, v. 10, p. 65–69.

Bagby, W.C., 1989, Patterns of gold mineralization in Nevada and Utah, *in* Shawe, D.R., Ashley, R.P., and Carter, L.M.H., eds., United States gold terranes—Part I, Chapter B: U.S. Geological Survey Bulletin 1857, p. B11–B21.

Bailey, D.K., 1970, Volatile flux, heat-focusing and the generation of magma: Geological Journal (London) Special Issue 2, p. 177–186.

———1978, Continental rifting and mantle degassing, *in* Neumann, E.R., and Ramberg, I.B., eds., Petrology and geochemistry of continental rifts: Dordrecht, Holland, D. Reidel, p. 1–13.

Bailey, D.K., and Hampton, C.M., 1990, Volatiles in alkaline magmatism: Lithos, v. 26, p. 157–165.

Bailey, R.A., 1989, Geologic map of the Long Valley caldera, Mono–Inyo Craters volcanic chain and vicinity, eastern California: U.S. Geological Survey Miscellaneous Investigations Series Map I–1933.

Baker, D.W., and Berg, R.B., eds., 1991, Guidebook of the central Montana alkalic province: Montana Bureau of Mines and Geology Special Publication 100, 201 p.

Baker, D.W., McBride, Gwen, and Dahy, James, 1991, Field guide, Little Belt Mountains, *in* Baker, D.W., and Berg, R.B., eds., Guidebook of the central Montana alkalic province: Montana Bureau of Mines and Geology Special Publication 100, p. 144–162.

Baldridge, W.S., Olsen, K.H., and Callender, J.F., 1984, Rio Grande rift—Problems and perspectives: New Mexico Geological Society Guidebook, 35th Field Conference, Rio Grande rift, northern New Mexico, p. 1–12.

Baldridge, W.S., Perry, F.V., Vaniman, D.T., Nealey, L.D., Laughlin, A.W., Kyle, P.R., Bartov, Yosef, Steinitz, Gideon, and Gladney, E.S., 1991, Middle to late Cenozoic magmatism of the southeastern Colorado Plateau and central Rio Grande Rift (New Mexico and Arizona, U.S.A.); a model for continental rifting: Tectonophysics, v. 197, p. 327–354.

Barker, D.S., 1991, Quartz monzonite and associated iron deposits, Iron Springs, Utah: Geological Society of America Abstracts with Programs, v. 23, no. 5, p. A388.

Barnes, C.W., 1974, Interference and gravity tectonics in the Gray Mountain area, Arizona, *in* Karlstrom, T.N.V., Swann, G.A., and Eastwood, R.L., eds., Geology of Northern Arizona, Part I—Regional studies: Geological Society of America, Rocky Mountain Section meeting [guidebook], p. 442–453.

———1987, Geology of the Gray Mountain area, Arizona: Geological Society of America Centennial Field Guide, Rocky Mountain Section, p. 379–384.

Barnhard, T.P., and Anderson, R.E., 1991, Tectonic significance of dike orientations in the Lake Mead area of Nevada and Arizona: Geological Society of America Abstracts with Programs, v. 23, no. 5, p. 233.

Bartley, J.M., 1989, Changing Tertiary extension directions in the Dry Lake Valley area, Nevada, and a possible dynamic model, *in* Garside, L.J., and Shaddrick, D.R., eds., Compressional and extensional structural styles in the northern Basin and Range province—Seminar Proceedings: Reno, Nev., Nevada Petroleum Society and Geological Society of Nevada, p. 35–39.

_____1990, Hypothetical mid-Tertiary synvolcanic stress rotation in the southeastern Great Basin: Geological Society of America Abstracts with Programs, v. 22, no. 3, p. 6.

Bartley, J.M., Axen, G.J., Taylor, W.J., and Fryxell, J.E. 1988, Cenozoic tectonics of a transect through eastern Nevada near 38° N latitude, *in* Weide, D.L., and Faber, M.L., eds., This extended land—Geological journeys in the southern Basin and Range: Geological Society of America, Cordilleran section, Field trip guidebook, p. 1–20.

Bartley, J.M., Friedrich, A.M., Walker, J.D., Coleman, D.S., and Price, D.E., 1994, Post-Sevier belt normal faulting in southwestern Utah: Geological Society of America Abstracts with Programs, v. 26, no. 2, p. 37.

Bartley, J.M., Taylor, W.J., and Lux, D.R., 1992, Blue Ribbon volcanic rift in southeastern Nevada and its effects on Basin and Range fault segmentation: Geological Society of America Abstracts with Programs, v. 24, no. 6, p. 2.

Bartsch-Winkler, Susan, Barton, H.N., Cady, J.W., Cook, K.L., and Martin, C.M., 1988, Mineral resources of the Fifty Mile Mountain Wilderness Study Area, Kane County, Utah: United States Geological Survey Bulletin 1747–A, 20 p. (p. 11 — Cady: "almost certainly the short wavelength magnetic anomalies [over the Kaiparowits coal field] are caused by clinkers") [See also A. McCafferty, Idaho U.S.A. reports.]

Bastin, E.S., and Hill, J.M., 1917, Economic geology of Gilpin County and adjacent parts of Clear Creek and Boulder Counties, Colorado: U.S. Geological Survey Professional Paper 94, 379 p.

Bayer, K.C., 1983, Generalized structural, lithologic, and physiographic provinces in the fold and thrust belts of the United States (exclusion of Alaska and Hawaii): U.S. Geological Survey special map, map scale 1:2,500,000 (3 sheets).

Bayley, R.W., and Muehlberger, W.R., 1968, Basement rock map of the United States exclusive of Alaska and Hawaii: Washington, D.C., U.S. Geological Survey, scale 1:2,500,000 (2 sheets).

Beard, J.S., and Lofgren, G.E., 1991, Dehydration melting and water-saturated melting of basaltic and andesitic greenstones and amphibolites: Journal of Petrology, v. 32, p. 365–401.

Beaty, D.W., Cunningham, C.G., Rye, R.O., Steven, T.A., and Gonzales-Urien, E., 1986, Geology and genesis of the Deer Trail Pb-Zn-Ag-Au manto deposits, Marysvale district, west-central Utah: Economic Geology, v. 81, no. 8, p. 1932–1952.

Beghoul, Noureddine, and Barazangi, Muawia, 1989, Mapping high P_n velocity beneath the Colorado Plateau constrains uplift models: Journal of Geophysical Research 94, v. B6, p. 7083–7104. —Mantle lid thickness decreases 20 km beneath B-R wrt CP.

Bennett, V.C., and DePaolo, D.J., 1987, Proterozoic crustal history of the western United States as determined by neodymium isotopic mapping: Geological Society of America Bulletin, v. 99, p. 674–685.

Bergerat, Francoise, Bouroz-Weil, Catherine, and Angelier, Jacques, 1991, Paleostresses inferred from macrofractures, Colorado Plateau, western U.S.A.: Tectonophysics, v. 206, p. 219–243.

Best, M.G., 1988, Early Miocene change in direction of least principal stress, southwestern United States—Conflicting inferences from dikes and metamorphic core-detachment fault terranes: Tectonics, v. 7, p. 249–259.

Best, M.G., and Brimhall, W.H., 1974, Late Cenozoic alkalic basaltic magmas in the western Colorado Plateaus and the Basin and Range transition zone, U.S.A., and their bearing on mantle dynamics: Geological Society of America Bulletin, v. 85, p. 1677–1690.

Best, M.G., and Christiansen, E.H., 1991, Limited extension during peak Tertiary volcanism, Great Basin of Nevada and Utah: Journal of Geophysical Research, v. 96, no. B8, p. 13509–13528.

Best, M.G., Christiansen, E.H., and Blank, H.R.., Jr., 1989, Oligocene caldera complex and calc-alkaline tuffs and lavas of the Indian Peak volcanic field, Nevada and Utah: Geological Society of America Bulletin, v. 101, p. 1076–1090.

Best, M.G., Christiansen, E.H., Deino, A.L., Grommé, C.S., McKee, E.H., and Noble, D.C., 1989, Eocene through Miocene volcanism in the Great Basin of the western United States: New Mexico Bureau of Mines and Mineral Resources Memoir 47, p. 91–133.

Best, M.G., McKee, E.H., and Damon, P.E., 1980, Space-time-composition patterns of late Cenozoic mafic volcanism, southwestern Utah and adjoining areas: American Journal of Science, v. 180, p. 1035–1050.

Best, M.G., Scott, R.B., Rowley, P.D., Swadley, WC, Anderson, R.E., Gromme, C.S., Harding, A.E., Deino, A.L., Christiansen, E.H., Tingey, D.G., and Sullivan, K.R., 1993, Oligocene-Miocene caldera complexes, ash-flow sheets, and tectonism in the central and southeastern Great Basin, *in* Lahren, M.M., Trexler, J.H., Jr., and Spinosa, Claude, Crustal evolution of the Great Basin and the Sierra Nevada; Geological Society of America Cordilleran/Rocky Mountain Sections Joint Meeting, Field Trip Guidebook: Reno, Nev., University of Nevada, Mackay School of Mines, p. 285–311.

Billingsley, G.H., Barnes, C.W., and Ulrich, G.E., 1985, Geologic map of the Coconino Point and Grandview Point Quadrangles, Coconino County, Arizona: U.S. Geological Survey Miscellaneous Investigations Map I–1644, scale 1:62,500.

Billingsley, G.H., and Beus, S.S., 1985, The Surprise Canyon Formation—An upper Mississippian and lower Pennsylvanian(?) rock unit in the Grand Canyon, Arizona: U.S. Geological Survey Bulletin 1605–A, p. A27–A33.

Billingsley, G.H., Wenrich, K.J., and Huntoon, P.W., 1986, Breccia pipe and geologic map of the southeastern Hualapai Indian Reservation and vicinity, Arizona: U.S. Geological Survey Open-File Report 86–458–B, 1:48,000, 2 plates, with 26 p. text.

Billingsley, G.H., Wenrich, K.J., Huntoon, P.W., and Young, R.A., 1990, Breccia pipe and geologic map of the southwestern Hualapai Indian Reservation and vicinity, Arizona: U.S. Geological Survey Open-File Report 86–458–D, 1:48,000, 2 plates, with 33 p. text.

———in press, Breccia pipe and geologic map of the southwestern Hualapai Indian Reservation and vicinity, Arizona: U.S. Geological Survey Miscellaneous Investigations Series Map I–2554, 33 p., scale 1:48,000, 2 plates.

Bingler, E.C., 1971, Major east-west lineament in west-central Nevada: Geological Society of America Abstracts with Programs, v. 3, no. 2, p. 83.

Bird, Peter, 1979, Continental delamination and the Colorado Plateau: Journal of Geophysical Research, v. 84, no. B13, p. 7561–7571.

———1984, Laramide crustal thickening event in the Rocky Mountain foreland and Great Plains: Tectonics 3, p. 741–758.

———1988, Formation of the Rocky Mountains, western United States—A continuum computer model: Science, v. 239, p. 1501–1507.

Black, K.D., 1977, Petrology, alteration, and mineralization of two Tertiary intrusives, Sierra Blanca igneous complex, New Mexico: Fort Collins, Colo., Colorado State University M.S. thesis, 124 p.

Blakely, R.J., 1988, Curie temperature isotherm analysis and tectonic implications of aeromagnetic data from Nevada: Journal of Geophysical Research, v. 93, no. B10, p. 11817–11832.

Blakely, R.J., and Jachens, R.C., 1991, Regional study of mineral resources in Nevada—Insights from three-dimensional analysis of gravity and magnetic anomalies: Geological Society of America Bulletin, v. 103, p. 795–803.

Blakely, R.J., and Simpson, R.W., 1986, Approximating edges of source bodies from magnetic or gravity anomalies: Geophysics, v. 51, no. 7, p. 1494–1498.

Blanchet, P.H., 1957, Development of fracture analysis as exploration method: American Association of Petroleum Geologists Bulletin, v. 41, no. 8, p. 1748–1759.

Blank, H.R., 1994, Rafting and magmatism in E. Central Nevada—Insights from gravity and magnetic data: Geological Society of America Abstracts with Programs, v. 26, no. 2, p. 39.

Blank, H.R., Jr., and Kucks, R.P., 1989, Preliminary aeromagnetic, gravity, and generalized geologic maps of the USGS Basin and Range–Colorado Plateau transition zone study area in southwestern Utah, southeastern Nevada, and northwestern Arizona (the "BARCO" project): U.S. Geological Survey Open-File Report 89–432, 16 p.

Blank, H.R., Jr., and Mackin, J.H., 1967, Geologic interpretation of an aeromagnetic survey of the Iron Springs district, Utah: U.S. Geological Survey Professional Paper 516–B, 14 p.

Blank, H.R., Jr., Rowley, P.D., and Hacker, D.B., 1992, Miocene monzonitic intrusions and associated megabreccias of the Iron Axis region, southwestern Utah, *in* Wilson, J.R., ed., Field guide to geologic excursions in Utah and adjacent areas of Nevada, Idaho, and Wyoming, Geological Society of America Rocky Mountain Section Meeting: Utah Geological Survey Miscellaneous Publication 92–3, p. 399–420.

Blixt, J.E., 1933, Geology and gold deposits of the North Moccasin Mountains, Fergus County, Montana: Montana Bureau of Mines and Geology Memoir 8, 25 p.

Boardman, S.J., and Condie, K.C., 1986, Early Proterozoic bimodal volcanic rocks in central Colorado, U.S.A.—Part R, Geochemistry, petrogenesis and tectonic setting: Precambrian Research, v. 34, p. 37–68.

Bohannon, R.G., 1984, Nonmarine sedimentary rocks of Tertiary age in the Lake Mead region, southeastern Nevada and northwestern Arizona: U.S. Geological Survey Professional Paper 1259, 69 p.

Bowers, W.E., 1990, Geologic map of Bryce Canyon National Park and vicinity, southwestern Utah: U.S. Geological Survey Miscellaneous Investigations Series Map I–2108, scale 1:24,000.

Bowring, S.A., and Karlstrom, K.E., 1990, Growth, stabilization and reactivation of Proterozoic lithosphere in the southwestern United States: Geology, v. 18, p. 1203–1206. —Proterozoic basement terrane boundaries (west Yavapai, 1.7–1.8 Ga; Mazatzal, 1.6–1.7 Ga=E), boundary trends NE through Four Corners.

Bredehoeft, J.D., Wolff, R.G., Keys, W.S., and Shuter, Eugene, 1976, Hydraulic fracturing to determine the regional in situ stress field, Piceance basin, Colorado: Geological Society of America Bulletin, v. 87, no. 2, p. 250–258.

Bromfield, C.S., 1967, Geology of the Mount Wilson quadrangle, western San Juan Mountains, Colorado: U.S. Geological Survey Bulletin 1227, 100 p.

Brown, C.L., and Bartley, J.M., 1994, Multiple-phase extension in the Sheeprock and Simpson Mountains of west-central Utah: Geological Society of America Abstracts with Programs, v. 26, no. 2, p. 41.

Brownlow, A.H., and Komorowski, J.C., 1988, Geology and origin of the Yogo sapphire deposit, Montana: Economic Geology, v. 83, p. 875–880.

Bryant, Bruce, Naeser, C.W., Marvin R.F., and Mehnert, H.H., 1989, Upper Cretaceous and Paleogene sedimentary rocks and isotopic ages of Paleogene tuffs, Uinta Basin, Utah: U.S. Geological Survey Bulletin 1787–J, p. J1–J22.

Bryant, Bruce, and Nichols, D.J., 1988, Late Mesozoic and Early Tertiary reactivation of an ancient crustal boundary along the Uinta trend and its interaction with the Sevier orogenic belt, *in* Schmidt, C.J., and Perry, W.J., Jr., eds., Interaction of the Rocky Mountains Foreland and the Cordilleran Thrust Belt: Geological Society of America Memoir 171, p. 411–430. Between Archaean and Proterozoic crust, geologic map, p. 414; see additional notes.

Budding, K.E., Cunningham, C.G., Zielinski, R.A., Steven, T.A., and Stern, C.R., 1987, Petrology and chemistry of the Joe Lott Tuff Member of the Mount Belknap Volcanics, Marysvale volcanic field, west-central Utah: U.S. Geological Survey Professional Paper 1354, 47 p.

Budge, Suzanne, and Romberger, S.B., 1983, Trace element distribution around precious/base metal veins, Idaho Springs dist., CO: Geological Society of America Abstracts with Programs, v. 15, p. 398.

Burford, A.E., and Dixon, J.M., 1977, Systematic fracturing in clay and its relation to bedrock joints and drainage segments: Geological Society of America Abstracts with Programs, v. 9, no. 7, p. 916.

———1978, Systematic fracturing in young clay of the Cuyahoga River valley, Ohio and its relation to bedrock jointing and drainage segments: Geotechnique, v. 28, p. 201–206.

Burke, D.B., and McKee, E.H., 1979, Mid-Cenozoic volcano-tectonic troughs in central Nevada: Geological Society of America Bulletin, v. 90, p. 181–184.

Burke, K.J., 1991, Tertiary extensional tectonism in the northern Chief and southernmost Highland Ranges, Lincoln County, Nevada: Flagstaff, Ariz., Northern Arizona University M.S. thesis, 91 p.

Bursik, Marcus, and Sieh, Kerry, 1989, Range front faulting and volcanism in the Mono Basin, eastern California: Journal of Geophysical Research, v. 94, no. B11, p. 15587–15609.

Butler, B.S., 1991, Elevation data for the Precambrian surface in the central and southern Colorado Plateau and vicinity: U.S. Geological Survey Open-File Report 92–194, 75 p., with map, scale 1:1,250,000.

Butler, B.S., Loughlin, G.F., Heikes, V.C., and others, 1920, The ore deposits of Utah: U.S. Geological Survey Professional Paper 111, 672 p.

Butler, W.C., 1991, Elevation data for the Precambrian surface in the central and southern Colorado Plateau and vicinity: U.S. Geological Survey Open-File Report 92–194, 75 p., with map, scale 1:1,250,000.

Butler, W.C., and Kirkpatrick, J.C., in press, Topographic contour map on top of Precambrian age rocks, central and southern Colorado Plateau and vicinity: U.S. Geological Survey Miscellaneous Field Studies Map MF–2310.

Callaghan, Eugene, 1938, Preliminary report on the alunite deposits of the Marysvale region, Utah: U.S. Geological Survey Bulletin 886–D, p. 91–134.

———1939, Volcanic sequence in the Marysvale region in southwest-central Utah: American Geophysical Union Transactions, 20th Annual Meeting, Washington, D.C., pt. 3, p. 438–452.

Cameron, E.M., 1990, Alkaline magmatism at Kirkland Lake, Ontario—Product of strike slip orogenesis: Canada Geological Survey Paper 90–1C, p. 261–269.

Cameron, E.M., and Hattori, K., 1987, Archean gold mineralization and oxidized hydrothermal fluids: Economic Geology, v. 82, p. 1177–1191.

Campbell, D.L., Cunningham, C.G., Steven, T.A., Rowley, P.D., Glassgold, L.B., and Anderson, J.J., 1984, Aeromagnetic map on a geologic base map of the Tushar Mountains and adjoining areas, Marysvale volcanic field, Utah: U.S. Geological Survey Miscellaneous Investigations Series Map I–1430–D, scale 1:50,000.

Carlson, D.H., Fleck, R.J., Moye, F.J., and Fox, K.F., 1991, Geology, geochemistry and isotopic character of the Colville igneous complex, northeastern Washington: Journal of Geophysical Research, ser. B, v. 96, p. 13313–13333.

Carter, W.D., and Gualtieri, J.L., 1957, Preliminary geologic map of the Mount Peale 1 SW quadrangle, San Juan County, Utah: U.S. Geological Survey Mineral Investigations Field Studies Map MF–124, scale 1:24,000.

———1958, Preliminary geologic map of the Mount Peale 1NW quadrangle, San Juan County, Utah: U.S. Geological Survey Mineral Investigations Field Studies Map MF–140, scale 1:24,000.

———1965, Geology and uranium-vanadium deposits of the La Sal quadrangle, San Juan County, Utah, and Montrose County, Colorado: U.S. Geological Survey Professional Paper 508, 82 p.

Case, J.E., 1965, Gravitational evidence for a batholithic mass of low density along a segment of the Colorado Mineral Belt: Geological Society of America Special Paper 82, Abstracts for 1964, p. 26.

———1966, Geophysical anomalies over Precambrian rocks, northwestern Uncompahgre Plateau, Utah and Colorado: American Association of Petroleum Geologists Bulletin, v. 50, no. 7, p. 1423–1443.

———1991, Geologic map of the northwestern part of the Uncompahgre uplift, Grand County, Utah, and Mesa County, Colorado, with emphasis on Proterozoic rocks: U.S. Geological Survey Miscellaneous Investigations, Map I–2088, scale 1:24,000.

Case, J.E., and Joesting, H.R., 1972, Regional geophysical investigations in the central Colorado Plateau: U.S. Geological Survey Professional Paper 736, 31 p.

Case, J.E., Joesting, H.R., and Byerly, P.C., 1963, Regional geophysical investigations in the La Sal Mountains area, Utah and Colorado: U.S. Geological Survey Professional Paper 316–F, p. 91–116.

Cashion, W.B., 1973, Geologic and structure map of the Grand Junction Quadrangle, Colorado and Utah: U.S. Geological Survey Miscellaneous Investigations Series Map I–736, scale 1:250,000.

Cashion, W.B., Kilburn, J.E., Barton, H.N., Kelley, K.D., Kulik, D.M., and McDonnell, J.R., Jr., 1990, Mineral resources of the Desolation Canyon, Turtle Canyon, and Floy Canyon Wilderness Study Areas, Carbon, Emery, and Grand Counties, Utah: U.S Geological Survey Bulletin 1753, p. B18.

Catchings, R.D., 1992, A relation among geology, tectonics, and velocity structure, western to central Nevada Basin and Range: Geological Society of America Bulletin, v. 104, p. 1178–1192.

Catchings, R.D., and Mooney, W.D., 1993, Geophysical indicators of magmatism and tectonism in the Basin and Range: Geological Society of America Abstracts with Programs, v. 25, no. 5, p. 19.

Cater, F.W., Jr., 1955, The salt anticlines of southwestern Colorado and southeastern Utah, *in* Geology of parts of the Paradox, Black Mesa, and San Juan basins, Guidebook: Four Corners Geological Society, lst, p. 125–131.

———1970, Geology of the salt anticline region in southwestern Colorado: U.S. Geological Survey Professional Paper 637, 80 p.

Cater, F.W., Jr., and Elston, D.P., 1963, Structural development of salt anticlines of Colorado and Utah, *in* Backbone of the Americas: American Association of Petroleum Geologists Memoir 2, p. 152–159.

Catchings, R.D., 1992, A relation among geology, tectonics, and velocity structure, western to central Nevada Basin and Range: Geological Society of America Bulletin, v. 104, p. 1178–1192.

Cather, S.M., 1990, Stress and volcanism in the northern Mogollon-Datil volcanic field, New Mexico—Effects of the post-Laramide tectonic transition: Geological Society of America Bulletin, v. 102, p. 1447–1458.

Cepeda, J.C., 1990, The Capitan dike swarm, Lincoln County, New Mexico—A Tertiary alkalic magma system: Geological Society of America Abstracts with Programs, v. 22, p. 5.

Chamberlin, R.M., 1983, Cenozoic domino-style crustal extension in the Lemitar Mountains, New Mexico—A summary: New Mexico Geological Society Guidebook 34, p. 111–118.

Chapin, C.E., 1983, An overview of Laramide wrench faulting in the Southern Rocky Mountains with emphasis on petroleum exploration, *in* Lowell, J.D., ed., Rocky Mountain foreland basins and uplifts: Denver, Colo., Rocky Mountain Association of Geologists, p. 169–179.

———1988, Axial basins of the northern and central Rio Grande rift, *in* Sloss, L.L., ed., Sedimentary cover–North American craton: U.S. Geological Society of America, Decade of North American Geology, v. D–2, Chap. 8, Rocky Mountain Region, p. 165–170.

Chapin, C.E., Chamberlin, R.M., Osburn, G.R., White, D.W., and Sanford, A.R., 1978, Exploration framework of the Socorro geothermal area, New Mexico, p. 115–130 *in* Chapin, C.E., and Elston, W.E., eds., Field guide to selected cauldrons and mining districts of the Datil-Mogollon volcanic field, New Mexico: New Mexico Geological Society Special Paper 7. —Jemez lineament; Capitan lineament.

Chenoweth, W.L., and Blakemore, P.P., 1961, The Riverview mine, Coconino County, Arizona: Plateau, v. 33, no. 4, p. 112–114.

Christiansen, R.L., 1993, The Yellowstone hot spot—Deep mantle plume or upper mantle melting anomaly [abs.]: Eos, v. 74, no. 43, supplement, p. 602.

Christiansen, R.L., and Lipman, P.W., 1972, Cenozoic volcanism and plate-tectonic evolution of the western United States—II, Late

Cenozoic: Philosophical Transactions of Royal Society of London, v. A271, p. 249–284.

Christiansen, R.L., and McKee, E.H., 1978, Late Cenozoic volcanic and tectonic evolution of the Great Basin and Columbia Intermontane regions, *in* Smith, R.B., and Eaton, G.P., eds., Cenozoic tectonics and regional geophysics of the western Cordillera: Geological Society of America Memoir 152, p. 283–311.

Christiansen, R.L., and Yeats, R.S., 1992, Post-Laramide geology of the U.S. Cordilleran region, *in* Burchfiel, B.C., Lipman, P.W., and Zoback, M.L., eds., The Cordilleran orogen—Conterminous U.S.: Geological Society of America, The Geology of North America, v. B-3, p. 261–406.

Clabaugh, S.E., 1952, Corundum deposits of Montana: U.S. Geological Survey Bulletin 983, 100 p.

Clague, D.A., 1987, Hawaiian alkaline volcanism, *in* Fitton, J.G., and Upton, B.G.J., eds., Alkaline igneous rocks: Geological Society of London Special Publication 30, p. 227–252.

Clarke, F.W., 1915, Analyses of rocks and minerals from the laboratory of the United States Geological Survey, 1880–1914: U.S. Geological Survey Bulletin 591, 376 p.

Coleman, D.S., and Walker, J.D., 1992, Evidence for the generation of juvenile granitic crust during continental extension, Mineral Mountains batholith, Utah: Journal of Geophysical Research, v. 97, no. B7, p. 11011–11024.

Colman, S.M., 1983, Influence of the Onion Creek diapir on the late Cenozoic history of Fisher Valley, southeastern Utah: Geology, v. 11, p. 240–243.

Colucci, M.T., Dungan, M.A., Ferguson, K.M., Lipman, P.W., and Moorbath, Stephen, 1991, Precaldera lavas of the southeast San Juan volcanic field—Parent magmas and crustal interactions: Journal of Geophysical Research, v. 96, p. 13413–13434.

Committee for the Magnetic Anomaly Map of North America, 1987, Magnetic anomaly map of North America: Geological Society of America, 4 sheets, scale 1:5,000,000.

———1993, Magnetic anomaly map of part of the conterminous United States and some adjacent parts of Canada, scale 1:5,000,000, plate 3 *in* Reed, J.C., Jr., Bickford, M.E., Houston, R.S., Link, P.K., Rankin, D.W., Sims, P.K., and Van Schmus, W.R., eds., DNAG Series, The Geology of North America, vol. C–2, Precambrian; Conterminous U.S.: Geological Society of America, 657 p.

Condie, K.C., 1964, Crystallization Po_2 of syenite porphyry from Navajo Mountain, southern Utah: Geological Society of America Bulletin, v. 75, p. 359–362.

———1978, Geochemistry of Proterozoic granitic plutons from New Mexico, U.S.A.: Chemical Geology, v. 21, p. 131–149.

———1982, Plate tectonics model for Proterozoic continental accretion in the southwestern United States: Geology, v. 10, p. 37–42.

Condit, C.D., Crumpler, L.S., Aubele, J.C., and Elston, W.E., 1989, Patterns of volcanism along the southern margin of the Colorado Plateau—The Springerville field: Journal of Geophysical Research, v. 94, no. B6, p. 7975–7986.

Coney, P., 1976, Plate tectonics and the Laramide orogeny: New Mexico Geological Society Special Publication 6, p. 5–10.

Coney, P.J., and Harms, T.A., 1984, Cordilleran metamorphic core complexes—Cenozoic extensional relics of Mesozoic compression: Geology, v. 12, p. 550–554.

Coney, P.J., and Reynolds, S.J., 1977, Cordilleran benioff zones: Nature, v. 270, p. 403–406.

Cook, E.F., 1957, Geology of the Pine Valley Mountains, Utah: Utah Geological and Mineralogical Survey Bulletin 58, 111 p.

Cook, K.L., Bankey, Viki, and Mabey, D.R., 1990, Patterns of geologic structure determined from new gravity map of Utah: Society of Exploration Geophysicists Annual Meeting, p. 670–672.

Cook, K.L., Bankey, Viki, Mabey, D.R., and DePangher, Michael, 1989, Complete Bouguer gravity anomaly map of Utah: Utah Geological and Mineral Survey Map 122, scale 1:1,000,000.

Cook, K.L., Halliday, M.E., Cunningham, C.G., Steven, T.A., Rowley, P.D., Glassgold, L.B., Anderson, J.J., and Coles, L.L., 1984, Complete Bouguer gravity anomaly map on a geologic base map of the Tushar Mountains and adjoining areas, Marysvale volcanic field, Utah: U.S. Geological Survey Miscellaneous Investigations Series Map I–1430–C, scale 1:50,000.

Cordalis, Charles, 1984, Mineralogy and petrology of the Copper King platinum prospect, Park County, Montana: Butte, Mont., Montana College of Mineral Science and Technology M.S. thesis, 62 p.

Cordell, Lindrith, 1978, Regional geophysical setting of the Rio Grande rift: Geological Society of America Bulletin, v. 89, p. 1073–1090.

———1984, Composite residual total intensity aeromagnetic map of New Mexico, Overlay 2 *in* Geothermal resources of New Mexico: Scientific Map Series, scale 1:500,000 (National Geophysical Data Center, Boulder, Colorado, no. 84–TGB–05).

Cordell, Lindrith, and Keller, G.R., 1984, Regional structural trends inferred from gravity and aeromagnetic data in the New Mexico–Colorado border region: New Mexico Geological Society Guidebook 35, p. 21–23.

Cordell, Lindrith, Keller, G.R., and Hildenbrand, T.G., 1982, Bouguer gravity anomaly map of the Rio Grande rift, Colorado, New Mexico, and Texas: U.S. Geological Survey Geophysical Investigations Map GP–949, scale 1:1,000,000.

Cordell, Lindrith, Zorin, Y.A., and Keller, G.R., 1991, The dĕcompensative gravity anomaly and deep structure of the region of the Rio Grande rift: Journal of Geophysical Research, v. 96, p. 6557–6568.

Corry, C.E., 1988, Laccoliths—Mechanics of emplacement and growth: Geological Society of America Special Paper 220, 110 p.

Cosca, M.A., Essene, E.J., Geissman, J.W., Simmons, W.B., and Coates, D.A., 1989, Pyrometamorphism of rocks associated with natural burns in coal seams, Powder River Basin: American Mineralogist, v. 74, p. 85–100.

Courtney, R.C., and White, R.S., 1986, Anomalous heat flow and geoid across the Cape Verde Rise—Evidence of dynamic support from a thermal plume in the mantle: Geophysical Journal of the Royal Astronomical Society, v. 87, p. 815–868.

Coveney, R.M., Jr., and Nansheng, C., 1991, Ni–Mo–PGE–Au-rich ores in Chinese black shales and speculations on possible analogues in the United States: Mineralium Deposita, v. 26, p. 83 ff.

Crittenden, M.D., Jr., Coney, P.J., and Davis, G.H., eds., 1980, Cordilleran metamorphic core complexes: Geological Society of America Memoir 153, 490 p.

Cross, C.W., 1895, General geology of the district—Geology and mining industries of the Cripple Creek district, Colorado: U.S. Geological Survey 16th Annual Report, part 2, p. 13–109.

Cross, T.A., and Pilger, R.H., 1978, Constraints on absolute motion and plate interaction inferred from Cenozoic igneous activity in the western United States: American Journal of Science, v. 278, p. 865–902.

Cross, Whitman, 1894, The laccolithic mountain groups of Colorado, Utah, and Arizona: U.S. Geological Survey 14th Annual Report, part 2, p. 165–241.

———1895, General geology of the district, part 1 *of* Geology and mining industries of the Cripple Creek district, Colorado: U.S. Geological Survey 16th Annual Report, pt. 2, p. 13–109.

———1896, Geology of Silver Cliff and the Rosita Hills, Colorado: U.S. Geological Survey 17th Annual Report, part 2, p. 263–403.

Cross, Whitman, Spencer, A.C., and Purington, C.W., 1899, La Plata folio, Colorado: U.S. Geological Survey Geological Atlas Folio 60, 14 p.

Crough, T.S., 1983, Hotspot swells, p. 165–194 *in* Wetherill, G.W., Arden, A.L., and Stehli, F.G., eds., Annual Review of Earth and Planetary Sciences, v. 11, 1983: Palo Alto, Calif., Annual Reviews, Inc., 509 p.

Cruikshank, K.M., and Aydin, Atilla, 1995, Unweaving the joints in Entrada Sandstone, Arches National Park, Utah, U.S.A.: Journal of Structural Geology, v. 17, no. 3, p. 409–422.

Cruikshank, K.M., Zhao, Guozhu, and Johnson, A.M., 1991, Analysis of minor fractures associated with joints and faulted joints: Journal of Structural Geology, v. 13, no. 8, p. 865–886.

Cunningham, C.G., Ludwig, K.R., Naeser, C.W., Weiland, E.K., Mehnert, H.H., Steven, T.A., and Rasmussen, J.D., 1982, Geochronology of hydrothermal uranium deposits and associated igneous rocks in the eastern source area of the Mount Belknap Volcanics, Marysvale, west-central Utah: Economic Geology, v. 77, no. 2, p. 453–463.

Cunningham, C.G., Naeser, C.W., and Marvin, R.F., 1977, New ages for intrusive rocks in the Colorado mineral belt: U.S. Geological Survey Open-File Report 77–573, 7 p.

Cunningham, C.G., Naeser, C.W., Marvin, R.F., Luedke, R.G., and Wallace, A.R., 1994, Ages of selected intrusive rocks and associated ore deposits in the Colorado mineral belt: U.S. Geological Survey Bulletin 2109, 31 p.

Cunningham, C.G., Rye, R.O., Steven, T.A., and Mehnert, H.H., 1984, Origins and exploration significance of replacement and vein-type alunite deposits in the Marysvale volcanic field, west-central Utah: Economic Geology, v. 79, p. 50–71.

Cunningham, C.G., and Steven, T.A., 1979a, Mount Belknap and Red Hills calderas and associated rocks, Marysvale volcanic field, west-central Utah: U.S. Geological Survey Bulletin 1468, 34 p.

———1979b, Uranium in the central mining area, Marysvale district, west-central Utah: U.S. Geological Survey Miscellaneous Investigations Series Map I–1177, scale 1:24,000.

———1979c, Geologic map of the Deer Trail Mountain–Alunite Ridge mining area, west-central Utah: U.S. Geological Survey Miscellaneous Investigations Series Map I–1230, scale 1:24,000.

Cunningham, C.G., Steven, T.A., Rowley, P.D., Glassgold, L.B., and Anderson, J.J., 1983, Geologic map of the Tushar Mountains and adjoining areas, Marysvale volcanic field, Utah: U.S. Geological Survey Miscellaneous Investigations Series Map I–1430–A, scale 1:50,000.

Czamanske, G.K., Foland, K.A., Kubacher, F.A., and Allen, J.C., 1990, The ^{40}Ar/^{39}Ar chronology of caldera formation, intrusive activity and Mo-ore deposition near Questa, New Mexico: New Mexico Geological Society Guidebook 41, p. 355–358.

Dahy, J.P., 1991, Geology and igneous rocks of the Yogo sapphire deposit, Little Belt Mountains, Montana, *in* Baker, D.W., and Berg, R.B., eds., Guidebook of the central Montana alkalic province: Montana Bureau of Mines and Geology Special Publication 100.

Dalrymple, G.B., 1979, Critical tables for conversion of K-Ar ages from old to new constants: Geology, v. 7, p. 558–560.

Damon, P.E., Shafiqullah, Muhammad, and Leventhal, J.S., 1974, K-Ar chronology for the San Francisco volcanic field and rate of erosion of the Little Colorado River, *in* Karlstrom, T.N.V., Swann, G.A., and Eastwood, R.L., eds., Geology of Northern Arizona, with notes on archaeology and paleoclimate—Part I, Regional studies: Geological Society of America, Rocky Mountain Section Meeting (Guidebook for Field Trips, No. 27), p. 221–235.

Davis, G.A., and Burchfield, B.C., 1973, Garlock fault—An intracontinental transform structure, southern California: Geological Society of America Bulletin, v. 84, p. 1407–1422.

Davis, G.A., and Lister, G.S., 1988, Detachment faulting in continental extension—Perspectives from the southwestern U.S. Cordillera: Geological Society of America Special Paper 218, p. 133–159.

Davis, G.H., 1978, Monoclinal fold pattern of the Colorado Plateau, *in* Matthews, Vincent, III, ed., Laramide folding associated with basement block faulting in the western United States: Geological Society of America Memoir 151, p. 215–234.

Davis, G.H., and Krantz, R.W., 1986, Post- "Laramide" thrust faults in the Claron Formation, Bryce Canyon National Park, Utah: Geological Society of America Abstracts with Programs, v. 18, no. 5, p. 98.

Davis, G.H., and Rowley, P.D., 1993, Miocene thrusting, gravity sliding, and near-surface batholithic emplacement, Marysvale volcanic field, southwestern Utah [abs.]: Eos, v. 74, no. 43, p. 647.

Davis, M.W., and Streufert, R.K., 1990, Gold occurrences of Colorado: Colorado Geological Survey Resource Series 28, 101 p.

Delaney, P.T., and Pollard, D.D., 1981, Deformation of host rocks and flow of magma during growth of minette dikes and breccia-bearing intrusions near Ship Rock, New Mexico: U.S. Geological Survey Professional Paper 1201, 61 p.

Delaney, P.T., Pollard, D.D., Ziony, F.I., and McKee, E.H., 1986, Field relations between dikes and joints—Emplacement processes and paleostress analysis: Journal of Geophysical Research, v. 91, no. B5, p. 4920–4938.

DePaolo, D.J., 1981, Trace element and isotopic effects of combined wall-rock assimilation and fractional crystallization: Earth and Planetary Science Letters, v. 53, p. 189–202.

DeWitt, Ed, Redden, J.A., Wilson, A.B., Buscher, David, and Dersch, J.S., 1986, Mineral resource potential and geology of the Black Hills National Forest, South Dakota and Wyoming: U.S. Geological Survey Bulletin 1580, 135 p.

Dickin, A.P., Rice, C.M., and Harmon, R.S., 1986, A strontium and oxygen isotope study of Laramide magmatic and hydrothermal activity near Central City, Colorado: Economic Geology, v. 81, p. 904–914.

Dickinson, W.R., 1981, Plate tectonic evolution of the southern Cordillera, *in* Dickinson, W.R., and Payne, W.D., eds., Relations of tectonics to ore deposits in the southern Cordillera: Arizona Geological Society Digest, v. 14, p. 135–172.

Dickinson, W.R., and Snyder, W.S., 1979, Geometry of subducted slabs related to San Andreas transform: Journal of Geology, v. 87, p. 609–627.

Dilles, J.H., and Gans, P.B., 1995, The chronology of Cenozoic volcanism and deformation in the Yerington area, western Basin and Range and Walker Lake: Geological Society of America Bulletin,v. 107, p. 474–486.

Disbrow, A.E., and Stoll, W.C., 1957, Geology of the Cerrillos area, Santa Fe County, New Mexico: New Mexico Bureau of Mines and Mineral Resources Bulletin 48, 73 p.

Dodge, C.N., and Bartleson, B.L., 1986, The Eagle basin—A new exploration frontier, *in* Stone, D.S., ed., New interpretations of northwest Colorado geology: Denver, Colo., Rocky Mountain Association of Geologists, p. 113–121.

Doelling, H.H., 1980, Mineral deposits of the Henry Mountains, *in* Picard, M.D., ed., Henry Mountains Symposium: Utah Geological Association Publication 8, p. 287–296.

———1985, Geology of Arches National Park: Utah Geological and Mineral Survey Map 74, 15 p.

———1988, Geology of Salt Valley anticline and Arches National Park, Grand County, Utah, *in* Salt deformation in the Paradox region: Utah Geological and Mineral Survey Bulletin 122, p. 1–58.

Doelling, H.H., and Ross, M.L., 1993, Geologic map of the Big Bend quadrangle, Grand County, Utah: Utah Geological Survey Open-File Report 285, 108 p., scale 1:24,000.

Dubiel, R.F., Bromfield, C.S., Church, S.E., Kemp, W.M., Larson, M.J., Peterson, Fred, and Neubert, J.T., 1988a, Mineral resources of the Bull Mountain wilderness study area, Garfield and Wayne Counties, Utah: U. S. Geological Survey Bulletin 1751–B, p. B1–B12.

Dubiel, R.F., Bromfield, C.S., Church, S.E., Kemp, W.M., Larson, M.J., Peterson, Fred, and Neubert, J.T., 1988b, Mineral Resources of the Mt. Hillers wilderness study area, Garfield County,Utah: U. S. Geological Survey Bulletin 1751–C, p. C1–C14.

Dubiel, R.F., Bromfield, C.S., Church, S.E., Kemp, W.M., Larson, M.J., Peterson, Fred, and Pierson, C.T., 1990, Mineral resources of the Mount Pennell wilderness study area, Garfield County, Utah: U. S. Geological Survey Bulletin 1751–D, p. D1–D17.

Dudás, F.Ö., 1991, Geochemistry of igneous rocks from the Crazy Mountains, Montana, and tectonic models for the Montana alkalic province: Journal of Geophysical Research, v. 96, p. 13261–13277.

Duebendorfer, E.M., and Black, R.A., 1992, Kinematic role of transverse structures in continental extension—An example from the Las Vegas Valley shear zone, Nevada: Geology, v. 20, p. 1107–1110.

Duebendorfer, E.M., and Simpson, D.A., 1994, Kinematics and timing of Tertiary extension in the western Lake Mead region, Nevada: Geological Society of America Bulletin, v. 106, p. 1057–1073.

Duebendorfer, E.M., and Wallin, E.T., 1991, Basin development and syntectonic sedimentation associated with kinematically coupled strike-slip and detachment faulting, southern Nevada: Geology, v. 19, p. 87–90.

Dueker, K.G., and Humphreys, E.D., 1990, Upper-mantle velocity structure of the Great Basin: Geophysical Research Letters, v. 17, p. 1327–1330.

Dumitru, T.A., Duddy, I.R., and Green, P.F., 1994, Mesozoic-Cenozoic burial, uplift, and erosion history of the west-central Colorado Plateau: Geology, v. 22, no. 6, p. 499–502.

Duncan, R.A., and Richards, M.A., 1991, Hotspots, mantle plumes, flood basalts, and true polar wander: Reviews of Geophysics, v. 29, p. 31–50. —Relative motion of hotspots ≤3–5 mm/yr.

Dungan, M.A., Lipman, P.W., Colucci, M., Ferguson, K., and Blasley, S., 1989, Southeastern (Platoro) caldera complex, *in* Chapin, C.E., and Zidek, J., eds., Field excursions to volcanic terranes in the western United States—Volume I, Southern Rocky Mountain Region: New Mexico Bureau of Mines and Mineral Resources Memoir 46, p. 305–329.

Dyer, J.R., 1988, Using joint interactions to estimate paleostress ratios: Journal of Structural Geology, v. 10, no. 7, p. 685–699.

Eastwood, R.L., 1974, Cenozoic volcanism and tectonism of the southern Marble Plateau, *in* Karlstrom, T.N.V., Swann, G.A., and Eastwood, R.L., eds., Geology of northern Arizona, with notes on archaeology and paleoclimate—Part I, Regional studies: Geological Society of America, Rocky Mountain Section Meeting [Guidebook for Field Trips, No. 27], p. 236–256.

Eaton, G.P., 1979a, A plate-tectonic model for late Cenozoic crustal spreading in the western United States, *in* Riecker, R.E., ed., 1979, Rio Grande Rift—Tectonics and magmatism: American Geophysical Union, Washington, D.C., p. 7–32.

———1979b, Regional geophysics, Cenozoic tectonics, and geologic resources of the Basin and Range province and adjoining regions, *in* Newman, G.W., and Goode, H.D., eds., Basin and Range Symposium and Great Basin Field Conference: Denver, Colo., Rocky Mountain Association of Geologists, p. 11–39.

———1980, Geophysical and geological characteristics of the crust of the Basin and Range province, *in* Continental tectonics, studies in geophysics: Washington, D.C., National Academy of Sciences, p. 96–113.

———1982, The Basin and Range province—Origin and tectonic significance: Annual Reviews of Earth and Planetary Sciences, v. 10, p. 409–440.

———1984, The Miocene Great Basin of western North America as an extending back-arc region: Tectonophysics, v. 102, p. 275–295.

———1986, A tectonic redefinition of the southern Rocky Mountains: Tectonophysics 132, p. 163–194.

———1987, Topography and origin of the southern Rocky Mountains and Alvarado Ridge, p. 355–369 *in* Coward, M.P., Dewey, J.F., and Hancock, P.L., eds., Continental Extension Tectonics: Geological Society of London Special Publication 28, 637 p. (passive hot spots, due to sub-crustal lithospheric thinning) (E.-W. profile across the Colorado Plateau, Rio Grande rift-southern Rockies, and Great Plains, showing that crest (axis of symmetry) of 2,000 km wide swell is Rio Grande Rift, with max uplift = 3 km since 17 Ma)

Eaton, G.P., Christiansen, R.L., Iyer, H.M., Pitt, A.M., Mabey, D.R., Blank, H.R., Jr., Zietz, Isidore, and Gettings, M.E., 1975, Magma beneath Yellowstone National Park: Science, v. 188, p. 787–796.

Eaton, G.P., Wahl, R.R., Prostka, H.J., Mabey, D.R., and Kleinkopf, M.D., 1978, Regional gravity and tectonic patterns—Their relation to late Cenozoic epeirogeny and lateral spreading in the western Cordillera, *in* Smith, R.B., and Eaton, G.P., eds., Cenozoic tectonics and regional geophysics of the western Cordillera: Geological Society of America Memoir 152, p. 51–91.

Eckel, E.B., Williams, J.S., Galbraith, F.W., and others, 1949, Geology and ore deposits of the La Plata district, Colorado: U.S. Geological Survey Professional Paper 219, 179 p.

Eggler, D.H., Meen, J.K., Welt, Fred, Dudas, F.O., Furlong, K.P., McCallum, M.E., and Carlson, R.W., 1988, Tectonomagmatism of the Wyoming province: Colorado School of Mines Quarterly, v. 83, no. 2, p. 25–40.

Ekren, E.B., Anderson, R.E., Rogers, C.L., and Noble, D.C., 1971, Geology of northern Nellis Air Force Base Bombing and Gunnery Range, Nye County, Nevada: U.S. Geological Survey Professional Paper 651, 91 p.

Ekren, E.B., Bucknam, R.C., Carr, W.J., Dixon, G.L., and Quinlivan, W.D., 1976, East-trending structural lineaments in central Nevada: U.S. Geological Survey Professional Paper 986, 16 p.

Ekren, E.B., and Houser, F.N., 1965, Geology and petrology of the Ute Mountains area, Colorado: U.S. Geological Survey Professional Paper 481, 74 p.

Ekren, E.B., McIntyre, D.H., and Bennett, E.H., 1984, High-temperature, large-volume, lavalike ash-flow tuffs without calderas in southwestern Idaho: U.S. Geological Survey Professional Paper 1272, 76 p.

Ekren, E.B., Orkild, P.P., Sargent, K.A., and Dixon, G.L., 1977, Geologic map of Tertiary rocks, Lincoln County, Nevada: U S. Geological Survey Miscellaneous Investigations Series Map I–1041, scale 1:250,000.

Ekren, E.B., Rogers, C.L., Anderson, R.E., and Orkild, P.P., 1968, Age of Basin and Range normal faults in Nevada Test Site and Nellis Air Force Range, Nevada, *in* Eckel, E.B., ed., Nevada Test Site: Geological Society of America Memoir 110, p. 247–250.

Elliott, J.E., 1972, Platinum at Cooke City, Montana: U.S. Geological Survey Professional Paper 750–A, p. A36.

———1974, The geology of the Cooke City area, Montana and Wyoming, *in* Voight, Barry, ed., Rock mechanics, the American Northwest: University Park, Pa., Pennsylvania State University, p. 102–107.

———1979, Geologic map of the southwest part of the Cooke City quadrangle, Montana and Wyoming: U.S. Geological Survey Miscellaneous Investigations Series Map I–1084, scale 1:24,000.

Ellis, Michael, and King, Geoffrey, 1991, Structural control of flank volcanism in continental rifts: Science, v. 254, p. 839–842.

Elston, D.P., Shoemaker, E.M., and Landis, E.R., 1962, Uncompahgre front and salt anticline region of Paradox basin, Colorado and Utah: American Association of Petroleum Geologists Bulletin, v. 46, p. 1857–1858.

Elston, W.E., 1967, Summary of the mineral resources of Bernalillo, Sandoval, and Santa Fe Counties, New Mexico: New Mexico Bureau of Mines and Mineral Resources Bulletin 81, 81 p.

———1984, Subduction of young oceanic lithosphere and extensional orogeny in southwestern North America during mid-Tertiary time: Tectonics, v. 3, no. 2, p. 229–250.

Elston, W.E., and Bornhorst, T.J., 1979, The Rio Grande rift in context of regional post-40 m.y. volcanic and tectonic events, *in* Riecker, R.E., ed., Rio Grande rift—Tectonics and magmatism: Washington, D.C., American Geophysical Union, p. 416–438.

Embry, P.A., 1987, Petrogenesis of the Yogo Peak stock, Little Belt Mountains, Montana: Missoula, Mont., University of Montana M.S. thesis, 94 p.

Emmons, S.F., 1896, The mines of Custer County, Colorado: U.S. Geological Survey 17th Annual Report, part 2, p. 405–472.

Engebretson, D.C., Cox, Allan, and Thompson, G.A., 1984, Correlation of plate motions with continental tectonics—Laramide to Basin and Range: Tectonics, v. 3, no. 2, p. 115–119.

Engel, C.G., 1959, Igneous rocks and constituent hornblendes of the Henry Mountains, Utah: Geological Society of America Bulletin, v. 70, p. 951–980.

Engelder, Terry, 1982, Is there a genetic relationship between selected regional joints and contemporary stress within the lithosphere of North America?: Tectonics, v. 1, no. 2, p. 161–177.

———1982, Reply to a comment on "Is there a genetic relationship between selected regional joints and contemporary stress within the lithosphere of North America," by A.E. Scheidegger: Tectonics, v. 1, no. 5, p. 465–470.

Eriksson, C.L., 1987, Petrology of the alkalic hypabyssal and volcanic rocks at Cripple Creek, Colorado: Golden, Colo., Colorado School of Mines M.S. thesis, 114 p.

Erslev, E.A., and Sutter, J.F., 1990, Evidence for Proterozoic mylonitization in the northwestern Wyoming province: Geological Society of America Bulletin, v. 102, p. 1681–1694.

Evans, S.H., Jr., and Steven, T.A., 1982, Rhyolites in the Gillies Hill–Woodtick Hill area, Beaver County, Utah: Geological Society of America Bulletin, v. 93, p. 1131–1141.

Farmer, G.L., and DePaolo, D.J., 1983, Origin of Mesozoic and Tertiary granite in the western United States and implications for pre-Mesozoic crustal structure—1, Nd and Sr isotopic studies in the geocline of the northern Great Basin: Journal of Geophysical Research, v. 88, p. 3379–3401.

Farmer, G.L., Perry, F.V., Semken, S., Crowe, B., Curtis, D., and DePaolo, D.J., 1989, Isotopic evidence on the structure and origin of subcontinental lithospheric mantle in southern Nevada: Journal of Geophysical Research, v. 94, p. 7885–7898.

Faulds, J.E., 1992, From accommodation zones to metamorphic core complexes—Tracking the progressive development of major normal fault systems: Geological Society of America Abstracts with Programs, v. 24, no. 7, p. A158.

———1993, The Mt. Perkins block, northwestern Arizona—An exposed cross section of a synextensional volcano in highly extended terrane: Geological Society of America Abstracts with Programs, v. 25, no. 5, p. 36.

———1994, New insights on the geometry and kinematics of the Black Mountains–Highland Spring Range accommodation zone (BHZ), Arizona and Nevada: Geological Society of America Abstracts with Programs, v. 26, no. 2, p. 51.

Faulds, J.E., Gans, P.B., and Smith, E.I., 1994, Spatial and temporal patterns of extension in the northern Colorado River extensional corridor, northwestern Arizona and southern Nevada: Geological Society of America Abstracts with Programs, v. 26, no. 2, p. 51.

Faulds, J.E., Geissman, J.W., and Mawer, C.K., 1990, Structural development of a major extensional accommodation zone in the Basin and Range province, northeastern Arizona and southern Nevada—Implications for kinematic models of continental extension, *in* Wernicke, Brian, ed., Basin and Range extensional tectonics near the latitude of Las Vegas, Nevada: Geological Society of America Memoir 176, p. 37–76.

Faulds, J.E., Olson, E.L., and Littrell, R.L., 1994, Magmatic origin of rupture barriers and accommodation zones in extensional orogens—Analogies between continental and oceanic rifts [abs.]: Eos, v. 75, no. 44, p. 678.

Fears, D.W., Mutschler, F.E., and Larson, E.E., 1986, Cripple Creek, Colorado—A petrogenetic model: Geological Society of America Abstracts with Programs, v. 18, p. 599.

Fenneman, N.M., 1931, Physiography of Western United States: New York, McGraw-Hill Book Company, Inc., 534 p.

Ferguson, C.A., 1990, Localized high extension related to emplacement of a shallow synvolcanic pluton, Rio Grande Rift, New Mexico: Geological Society of America Abstracts with Programs, v. 22, no. 3, p. 22.

———1991, Stratigraphic and structural studies in the Mt. Withington caldera, Grassy Lookout quadrangle, Socorro County, New Mexico: New Mexico Geology, v. 13, no. 3, p. 50–54.

Finley, S.J., and Lorenz, J.C., 1989, Characterization and significance of natural fractures in Mesaverde reservoirs at the Multiwell Experiment site: Society of Petroleum Engineers Joint Rocky Mountain Regional/Low Permeability Reservoir Symposium and Exhibition, Denver, SPE Paper 19007, p. 721–728.

Fitton, J.G., James, Dodie, Kempton, P.D., Ormerod, D.S., and Leeman, W.P., 1988, The role of lithospheric mantle in the generation of late Cenozoic basic magmas in the western United States: Journal of Petrology Special Lithosphere Issue, p. 331–349.

Fitton, J.G., James, Dodie, and Leeman, W.P., 1991, Basic magmatism associated with late Cenozoic extension in the western United States—Compositional variations in space and time: Journal of Geophysical Research, v. 96, p. 13693–13711.

Fitzgerald, P.G., Fryxell, J.E., and Wernicke, B.P., 1991, Miocene crustal extension and uplift in southeastern Nevada—Constraints from fission track analysis: Geology, v. 19, p. 1013–1016.

Fleck, R.J., 1970, Age and possible origin of the Las Vegas Valley shear zone, Clark and Nye Counties, Nevada: Geological Society of America Abstracts with Programs, v. 2, no. 5, p. 333.

Fleck, R.J., Anderson, J.J., and Rowley, P.D., 1975, Chronology of mid-Tertiary volcanism in High Plateaus region of Utah, *in* Anderson, J.J., Rowley, P.D., Fleck, R.J., and Nairn, A.E.M., Cenozoic geology of southwestern High Plateaus of Utah: Geological Society of America Special Paper 160, p. 53–62.

Fleck, R.J., and Criss, R.E., 1985, Strontium and oxygen isotopic variations in Mesozoic and Tertiary plutons of central Idaho: Contributions to Mineralogy and Petrology, v. 90, p. 291–308.

Foster, Fess, and Chadwick, Tom, 1990, Relationship of the Golden Sunlight mine to the Great Falls tectonic zone: Tobacco Root Geological Society 15th Annual Field Conference Guidebook p. 77–81.

Fountain, D.M., and Christiansen, N.I., 1989, Composition of the continental crust and upper mantle—A review, *in* Pakiser, L.C., and Mooney, W.D., eds., Geophysical framework of the continental United States: Geological Society of America Memoir 172, p. 711–742.

Frahme, C.W., and Vaughn, E.B., 1983, Paleozoic geology and seismic stratigraphy of the northern Uncompahgre front, Grant County, Utah, *in* Lowell, J.D., and Gries, Robbie, eds., Rocky Mountain foreland basins and uplifts: Denver, Colo., Rocky Mountain Association of Geologists Guidebook, p. 201–211.

Franczyk, K.J., Fouch, T.D., Johnson, R.C., Molenaar, C.M., and Cobban, W.A., 1992, Cretaceous and Tertiary paleogeographic reconstructions for the Uinta-Piceance basin study area, Colorado and Utah: U.S. Geological Survey Bulletin 1787–Q, 37 p.

Fridich, C.J., and Mahood, G.H., 1984, Reverse zoning in the resurgent intrusions of the Grizzly Peak cauldron, Sawatch Range, Colorado: Geological Society of America Bulletin, v. 95, p. 779–787.

Friedman, J.D., Case, J.E., and Simpson, S.L., 1994, Tectonic trends of the northern Paradox Basin, Utah and Colorado, as derived from Landsat multispectral scanner imaging and geophysical and geologic mapping: U.S. Geological Survey Bulletin 2000–C, 30 p.

Friedman, J.D., and Simpson, S.L., 1980, Lineaments and structure of the northern Paradox basin, Colorado and Utah: U.S. Geological Survey Miscellaneous Field Studies Map MF–1221, scale 1:250,000.

Frizzell, V.A., Jr., and Zoback, M.L., 1987, Stress orientation determined from fault slip data in Hampel Wash area, Nevada, and its relation to contemporary regional stress field: Tectonics, v. 6, no. 2, p. 89–98.

Fuller, M.D., 1964, Expression of E-W fractures in magnetic surveys in parts of the U.S.A.: Geophysics, v. 29, p. 602–622.

Fulp, M.S., and Woodward, L.A., 1991, Mineral deposits of the New Mexico alkalic province: Geological Society of America Abstracts with Programs, v. 23, no. 4, p. 23.

Gable, D.J., 1984, Geologic setting and petrochemistry of the Late Cretaceous–Early Tertiary intrusives in the northern Front Range mineral belt, Colorado: U.S. Geological Survey Professional Paper 1280, 33 p.

Gans, P.B., 1987, An open-system, two-layer crustal stretching model for the eastern Great Basin: Tectonics, v. 6, p. 1–12.

Gans, P.B., Mahood, G.A., and Schermer, Elizabeth, 1989, Synextensional magmatism in the Basin and Range province—A case study from the eastern Great Basin: Geological Society of America Special Paper 233, 53 p.

Gans, P.B., Miller, E.L., and Lee, J., 1993, The Mesozoic-Cenozoic tectonic evolution of the eastern Great Basin—A revised history [abs.]: Eos, v. 74, no. 43, p. 608.

Gans, P.B., Miller, E.L., McCarthy, J., and Ouldcott, M.L., 1985, Tertiary extensional faulting and evolving ductile-brittle transition zones in the northern Snake Range and vicinity—New insights from seismic data: Geology, v. 13, p. 189–193.

Garverich, M.R., 1991, Field guide, North Moccasin and Judith Mountains, *in* Baker, D.W., and Berg, R.B., eds., Guidebook of the central Montana alkalic province: Montana Bureau of Mines and Geology Special Publication 100, p. 175–184.

Gibson, S.A., Thompson, R.N., Leat, P.T., Morrison, M.A., Hendry, G.L., Dickin, A.P., and Mitchell, J.G., 1993, Ultrapotassic magmas along the flanks of the Oligo-Miocene Rio Grande rift, U.S.A.—Monitors of a zone of lithospheric mantle extension and thinning beneath a continental rift: Journal of Petrology, v. 34, p. 187–228.

Gilbert, G.K., 1877, Report on the geology of the Henry Mountains: U.S. Geographical and Geological Survey, Rocky Mountains Region (Powell), 160 p.

———1882, Post-glacial joints: American Journal of Science, Third series, v. 23, no. 133, p. 25–27.

———1928, Basin Range faulting along the Oquirrh Range, Utah: Geological Society of America Bulletin, v. 39, p. 1103–1130.

Giles, D.L., 1983, Gold mineralization in the laccolithic complexes of central Montana, *in* The genesis of Rocky Mountain ore deposits—Changes with time and tectonics: Wheat Ridge, Colo., Denver Region Exploration Geologists Society, p. 157–162.

———1991, Gold mineralization in the Cerrillos district, New Mexico: Geological Society of America Abstracts with Programs, v. 23, no. 4, p. 24.

Giles, D.L., and Thompson, T.B., 1972, Petrology and mineralization of a molybdenum-bearing alkalic stock, Sierra Blanca, New Mexico: Geological Society of America Bulletin, v. 83, p. 2129–2148.

Gill, J.B., 1981, Orogenic andesites and plate tectonics: New York/Berlin, Springer Verlag, 390 p.

Glazner, A.F., and Bartley, J.M., 1984, Timing and tectonic setting of Tertiary low-angle normal faulting and associated magmatism in the southwestern United States: Tectonics, v. 3, p. 385–396.

Glazner, A.F., and Ussler, William, III, 1988, Trapping of magma at mid-crustal density discontinuities: Geophysical Research Letters, v. 15, p. 673–675.

———1989, Crustal extension, crustal density, and the evolution of Cenozoic magmatism in the Basin and Range of the Western United States: Journal of Geophysical Research, v. 94, no. B6, p. 7952–7960.

Goddard, E.N., 1988, Geologic map of the Judith Mountains, Fergus County, Montana: U.S. Geological Survey Miscellaneous Investigations Series Map I–1729, scale 1:31,680.

Gol'braikh, I.G., Zabaluyev, V.V., Lastochkin, A.M., Mirkin, G.R., and Reinin, I.V., 1968, Morphostructural methods for the tectonic study of covered platform oil and gas regions: Leningrad, NEDRA Publishing House, Leningrad Division, 136 p.

Gould, L.M., 1926, The geology of the La Sal Mountains of Utah: Michigan Academy of Science Papers, v. 7, p. 55–106.

Goydas, M.J., 1990, Provisional geologic map of the Fisher Valley quadrangle, Grand County, Utah: Utah Geological Survey Open-File Report 167, 48 p., scale 1:24,000.

Grand, S.P., 1994, Mantle shear structure beneath the Americas and surrounding oceans: Journal of Geophysical Research, v. 99, no. B6, p. 11591–11621.

Grant, S.K., and Proctor, P.D., 1988, Geologic map of the Antelope Peak quadrangle, Iron County, Utah: Utah Geological and Mineral Survey Open-File Report 130, 20 p., scale 1:20,000.

Gregory, K.M., and Chase, C.G., 1992, Tectonic significance of paleobotanically estimated climate and altitude of the late Eocene erosion surface, Colorado: Geology, v. 20, p. 581–585.

Gries, Robbie, 1983, North-south compression of Rocky Mountain structures, *in* Lowell, J.D., ed., Rocky Mountain foreland basins and uplifts: Denver, Colo., Rocky Mountain Association of Geologists, p. 9–32.

Griswold, G.B., 1959, Mineral deposits of Lincoln County, New Mexico: New Mexico Bureau of Mines and Mineral Resources Bulletin 67, 117 p.

Gross, M.R., 1993, The origin and spacing of cross joints—Examples from the Monterey Formation, Santa Barbara coastline, California: Journal of Structural Geology, v. 15, no. 6, p. 737–751.

Grout, M.A., 1991, Cleats in coal beds of the southern Piceance basin, Colorado—Correlation with regional and local fracture sets in associated clastic rocks, *in* Schwochow, S.D., Murray, D.K., and Fahy, M.F., eds., Coalbed methane of western North America: Rocky Mountain Association of Geologists, p. 35–47.

Grout, M.A., Abrams, G.A., Tang, R.L., Hainsworth, T.J., and Verbeek, E.R., 1991, Late Laramide thrust-related and evaporite-domed anticlines in the southern Piceance basin, northeastern Colorado Plateau: American Association of Petroleum Geologists Bulletin, v. 75, no. 2, p. 205–218.

Grout, M.A., and Verbeek, E.R., 1982, Intrusion of clastic dikes during diagenesis and fracture of oil shales: Geological Society of America Abstracts with Programs, v. 14, no. 7, p. 503.

———1983, Field studies of joints—Insufficiencies and solutions, with examples from the Piceance basin, Colorado, *in* Gary, J.H., ed., 16th Oil Shale Symposium, Proceedings: Golden, Colo., Colorado School of Mines Press, p. 68–80.

———1985, Fracture history of the Plateau Creek and adjacent Colorado River valleys, southern Piceance basin—Implications for predicting joint patterns at depth: U.S. Geological Survey Open-File Report 85–744, 17 p.

———1992, Fracture history of the Divide Creek and Wolf Creek anticlines and its relation to Laramide basin-margin tectonism, southern Piceance basin, northwestern Colorado: U.S. Geological Survey Bulletin 1787–Z, 32 p.

———1992, Joint-history and orientation data for Upper Cretaceous sandstones, Rawlins and Rock Springs uplifts, Washakie basin, southern Wyoming: U.S. Geological Survey Open-File Report 92–388, 30 p.

Grunwald, R.R., 1970, Geology and mineral deposits of the Galena mining district, Black Hills, South Dakota: Rapid City, S. Dak., South Dakota School of Mines and Technology Ph. D. thesis, 323 p.

Hacker, B.R., and Christie, J.M., 1990, Brittle/ductile and plastic/cataclastic transition in experimentally deformed and metamorphosed amphibolite, *in* Duba, A.G., Durham, W.B., Handin, J.W., and Wang, H.F., eds., The brittle-ductile transition in rocks: Geophysical Monograph 56, 243 p.

Hagstrum, J.T., and Lipman, P.W., 1986, Paleomagnetism of the structurally deformed Latir volcanic field, northern New Mexico—Relations to formation of the Questa caldera and development of the Rio Grande rift: Journal of Geophysical Research, v. 91, p. 7383–7402.

Hall, R.J., 1976, Petrology of diamond drill core from Judith Peak–Red Mountain area, Fergus County, Montana: Cheney, Wash., Eastern Washington University M.S. thesis, 38 p.

Haman, P.J., 1961, Lineament analysis on aerial photographs exemplified in the North Sturgeon Lake area, Alberta: West Canadian Research Publications of Geology and Related Sciences, series 2, no. 1, 20 p.

Hamilton, J.A., 1961, Geology of the Caribou Mountain area, Bonneville and Caribou Counties, Idaho: Moscow, Idaho, University of Idaho M.S. thesis, 59 p.

Hamilton, W.B., 1981, Plate-tectonic mechanism of Laramide deformation, *in* Boyd, D.W., and Lillegraven, J.A., eds., Rocky Mountain foreland basement tectonics: University of Wyoming Contributions to Geology, v. 19, p. 87–92. — First introduces CW rotation of CP.

———1985, Subduction, magmatic arcs, and foreland deformation, *in* Howell, D.G., ed., Tectonostratigraphic terranes of the circum-Pacific region: Houston, Texas, Circumpacific Council for Energy and Mineral Resources, p. 259–262.

———1987, Crustal extension in the Basin and Range province, southwestern United States, *in* Coward, M.P., Dewey, J.F., and Hancock, P.L., eds., Continental extensional tectonics: Geological Society Special Publication 28, p. 155–176.

———1988a, Tectonic setting and variations with depth of some Cretaceous and Cenozoic structural and magmatic systems of the western United States, *in* Ernst, W.G., ed., Metamorphism and crustal evolution of the western United States: Englewood Cliffs, New Jersey, Prentice Hall, p. 1–40.

———1988b, Detachment faulting in the Death Valley region, California and Nevada, *in* Carr, M.D., and Yount, J.C., eds., Geologic and hydrologic investigations of a potential nuclear waste disposal site at Yucca Mountain, southern Nevada: U.S. Geological Survey Bulletin 1790, p. 51–86.

———1988c, Plate tectonics and island arcs: Geological Society of America Bulletin, v. 100, p. 1503–1527.

———1988d, Laramide crustal shortening, *in* Schmidt, C.J., and Perry, W.J., eds., Interaction of the Rocky Mountain foreland and the Cordilleran thrust belt: Geological Society of America Memoir 171, p. 27–39.

———1989, Crustal geologic processes of the United States, *in* Pakiser, L.C., and Mooney, W.D., eds., Geophysical framework of the continental United States: Geological Society of America Memoir 172, p. 743–781.

———1995, Subduction systems and magmatism, *in* Smellie, J.L., ed., Volcanism associated with extension at consuming plate margins: Geological Society Special Publication 81, p. 3–28.

Hamilton, W.B., and Myers, W.B., 1966, Cenozoic tectonics of the Western United States: Reviews of Geophysics, v. 4, p. 509–549.

Hansen, W.R., 1986, History of faulting in the eastern Uinta Mountains, Colorado and Utah, *in* Stone, D.S., New Interpretations of Northwestern Colorado Geology: Denver, Colo., Rocky Mountain Association of Geologists, p.5–17.

Harden, D.R., Biggar, N.E., and Gillam, M.L., 1985, Quaternary deposits and soils in and around Spanish Valley, Utah, *in* Weide, D.L., ed., Soils and Quaternary geology of the southwestern United States: Geological Society of America Special Paper 203, p. 43–64.

Hardyman, R.F., and Oldow, J.S., 1991, Tertiary tectonic framework and Cenozoic history of the central Walker Lane, Nevada, *in* Raines, G.L., Lisle, R.E., Schafer, R.W., and Wilkinson, W.H., eds., Geology and ore deposits of the Great Basin: Geological Society of Nevada Symposium Proceedings, v. 1, p. 279–301.

Harrison, T.S., 1927, Colorado-Utah salt domes: American Association of Petroleum Geologists Bulletin, v. 11, p. 111–133.

Hasbrouck, W.P., and Hadsell, F.A., 1978, Geophysical techniques for coal exploration and development, *in* Hodgson, H.E., ed., Proceedings of the Second Symposium on the Geology of Rocky Mountain Coal—1977: Denver, Colo., Colorado Geological Survey, Department of Resources, p. 187–218.

Hastings, J.S., 1988, Gold deposits of Zortman-Landusky, Little Rocky Mountains, Montana, *in* Schafer, R.W., Cooper, J.J., and Vikre, P.G., eds., Bulk mineable precious metal deposits of the western United States: Reno, Nev., Geological Society of Nevada, p. 187–205.

Hauser, E.C., and Lundy, J., 1989, COCORP deep reflections—Moho at 50 km (165) beneath the Colorado Plateau: Journal of Geophysical Research, v. 94, no. B6, p. 7071–7081.

Haxby, W.F., and Turcotte, D.L., 1976, Stress induced by the addition or removal of overburden and associated thermal effects: Geology, v. 4, p. 181–185.

Haxel, G.B., Simmons, A.M., and McCarthy, J., 1990, Synextensional dioritic magmatism within and beneath metamorphic core complexes, southern Arizona and southeastern California—Integration of structural, petrologic, and seismic refraction data: Geological Society of America Abstracts with Programs, v. 22, no. 3, p. 28–29.

Haynes, D.D., and Hackman, R.J., 1978, Geology, structure, and uranium deposits of the Marble Canyon 1°×2° quadrangle, Arizona: U.S. Geological Survey Miscellaneous Investigations Series Map I–1003, 2 sheets, 1:250,000 scale.

Haynes, D.D., Vogel, J.D., and Wyant, D.G., 1972, Geology, structure, and uranium deposits of the Cortez quadrangle, Colorado and Utah: U.S. Geological Survey Miscellaneous Investigations Series Map I–629, sheet 1, scale 1:250,000.

Hedge, C.E., Houston, R.S., Tweto, O.L., Peterman, Z.E., Harrison, J.E., and Reid, R.R., 1986, The Precambrian of the Rocky Mountain region: U.S. Geological Survey Professional Paper 1241–D, 17 p.

Heizler, M.T., and Harrison, T.M., 1988, Multiple trapped argon isotope components revealed by $^{40}Ar/^{39}Ar$ isochron analysis: Geochimica et Cosmochemica Acta, v. 52, p. 1295–1303.

Helmstaedt, H.H., and Schulze, D.J., 1991, Early to mid-Tertiary inverted metamorphic gradient under the Colorado Plateau—Evidence from eclogite xenoliths in ultramafic microbreccias, Navajo volcanic field: Journal of Geophysical Research, v. 96, no. B8, p. 13225–13235.

Helz, R.T., 1976, Phase relations of basalt in their melting ranges at P_{H_2O}=5 kb—Part II, Melt compositions: Journal of Petrology, v. 17, p. 139–193.

Henderson, L.J., Gordon, R.G., and Engebretson, D.C., 1984, Mesozoic aseismic ridges on the Farallon plate and southward migration of shallow subduction during the Laramide orogeny: Tectonics, v. 3, p. 121–132.

Henley, R.W., and Adams, D.P.M., 1992, Strike-slip fault reactivation as a control on epithermal vein-style gold mineralization: Geology, v. 20, p. 443–446.

Henry, C.D., and Aranda-Gomez, J.J., 1992, The real southern Basin and Range—Mid- to late Cenozoic extension in Mexico: Geology, v. 20, p. 701–704.

Henry, C.D., Muehlberger, W.R., and Price, J.G., 1991, Igneous and structural evolution of the Solitario laccocaldera, Trans-Pecos Texas: Geological Society of America Abstracts with Programs, v. 23, no. 5, p. A451.

Henry, C.D., and Price, J.G., 1988, Laccocaldera—A new caldera type from Trans-Pecos Texas: Geological Society of America Abstracts with Programs, v. 20, no. 7, p. 113.

Henry, C.D., Price, J.G., Parker, D.F., and Wolff, J.A., 1989, Excursion 9A: Mid-Tertiary silicic alkalic magmatism of Trans-Pecos Texas—Rheomorphic tuffs and extensive silicic lavas, *in* Chapin, C.E., and Zidek, J., eds., Field excursions to volcanic terranes in the western United States—Volume I, Southern Rocky Mountain region: New Mexico Bureau of Mines and Mineral Resources Memoir 46, p. 231–274.

Hetz, R.T., 1976, Phase relations of basalts in their melting ranges at P_{H_2O} = 5 kb—Part II, Melt compositions: Journal of Petrology, v. 17, p. 139–193.

Herzberg, C.T., 1987, Magma density at high pressure—Part 2, A test of the olivine flotation hypothesis, *in* Mysen, B.C., ed., Magmatic processes—Physicochemical principles: Geochemical Society Special Publication 1, p. 47–58.

Heyl, A.V., 1972, The 38th Parallel lineament and its relation to ore deposits: Economic Geology, v. 67, p. 879–894.

———1983, Some major lineaments reflecting deep-seated fracture zones in the central United States, and mineral districts related to the zones: Global Tectonics and Metallogeny, v. 2, nos. 1 & 2, p. 75–89.

Heyman, O.G., 1983, Distribution and structural geometry of faults and folds along the northwestern Uncompahgre uplift, western Colorado and eastern Utah, *in* Averett, W.R., editor, Northern Paradox basin–Uncompahgre Uplift: Grand Junction Geological Society Guidebook, p. 45–57.

Heyman, O.G., Huntoon, P.W., and White-Heyman, M.A., 1986, Laramide deformation on the Uncompahgre Plateau—Geometry and mechanisms, *in* Stone, D.S., and Johnson, K.S., eds., New interpretations of Northwest Colorado geology: Denver, Colo., Rocky Mountain Association of Geologists, p. 65–76.

Hildebrand, F.A., and Conklin, N.M., 1974, A breccia dike containing rare-earth-bearing apatite, molybdenite and magnetite at Iron Hill, Custer County, Colorado: Economic Geology, v. 69, p. 508–515.

Hildenbrand, T.G., and Kucks, R.P., 1988a, Total intensity magnetic anomaly map of Nevada: Nevada Bureau of Mines and Geology Map 93A, scale 1:750,000.

———1988b, Filtered magnetic anomaly maps of Nevada: Nevada Bureau of Mines and Geology Map 93B, scale 1:1,000,000 and 1:2,500,000 (5 sheets).

Hildenbrand, T.G., Simpson, R.W., Godson, R.H., and Kane, M.F., 1982, Digital colored residual and regional Bouguer gravity maps of the conterminous United States with cut-off wavelengths of 250 km and 1000 km: U.S. Geological Survey Geophysical Investigations Map GP–953–A, scale 1:7,500,000 (2 sheets).

Hildreth, Wes, Halliday, A.N., and Christiansen, R.L., 1991, Isotopic and chemical evidence concerning the genesis and contamination of basaltic and rhyolitic magma beneath the Yellowstone Plateau volcanic field: Journal of Petrology, v. 32, p. 63–138.

Hilpert, L.S., and Roberts, R.J., 1964, Geology—Economic geology, in U.S. Geological Survey, Mineral and water resources of Utah: U.S. 88th Congress, 2d sess., p. 28–38.

Hintze, L.F., 1975, Geological highway map of Utah: Provo, Utah, Brigham Young University, scale 1:1,000,000.

Hintze, L.F., compiler, 1980, Geologic map of Utah: Salt Lake City, Utah, Utah Geological and Mineral Survey, scale 1:500,000, 2 sheets.

Hintze, L.F., 1988, Geologic history of Utah: Brigham Young University Geology Studies Special Publication 7, 202 p.

Hite, R.J., 1961, Potash-bearing evaporite cycles in the salt anticlines of the Paradox basin, Colorado and Utah, *in* Short papers in the geologic and hydrologic sciences: U.S. Geological Survey Professional Paper 424–D, p. D135–D138.

———1975, An unusual northeast-trending fracture zone and its relation to basement wrench faulting in northern Paradox basin, *in* Fassett, J.E., ed., Canyonlands country: Four Corners Geological Society Field Conference, 8th, Guidebook, p. 217–223.

Hite, R.J., and Lohman, S.W., 1973, Geologic appraisal of Paradox basin salt deposits for waste emplacement: U.S. Geological Survey Open-File Report 73–114, 75 p.

Hodges, K.V., and Walker, J.D., 1992, Extension in the Cretaceous Sevier orogen, North American Cordillera: Geological Society of America Bulletin, v. 104, p. 560–569.

Hodgson, R.A., 1961a, Regional study of jointing in Comb Ridge–Navajo Mountain area, Arizona and Utah: American Association of Petroleum Geologists Bulletin, v. 45, no. 1, p. 1–38.

———1961b, Reconnaissance of jointing in Bright Angel area, Grand Canyon, Arizona: American Association of Petroleum Geologists Bulletin, v. 45, no. 1, p. 95–97.

Hodgson, R.A., Gay, S.P., Jr., and Benjamins, J.Y., 1974, First international conference on the new basement tectonics, proceedings: Utah Geological Association Publication 5, 636 p.

Holm, R.F., 1987, Geomorphic evidence for youngest dated faults in north-central Arizona: The Mountain Geologist, v. 24, no. 1, p. 19–25.

Holt, R.E., 1961, Copper King project, Goose Lake district, Park County, Montana: Bear Creek Mining Company unpublished report, 21 p.

Hudson, M.R., and Rosenbaum, J.G., 1994, Broad transfer zones—Examples of vertical-axis rotation and discontinuous faulting in the Great Basin: Geological Society of America Abstracts with Programs, v. 26, no. 3, p. 61.

Hudson, M.R., Rosenbaum, J.G., Scott, R.B., Rowley, P.D., and Gromme, C.S., 1993, Paleomagnetic evidence for counterclockwise rotation in an extensional transfer zone, southeastern Great Basin [abs.]: Eos, v. 74, no. 43, p. 207–208.

———1995, The western Delamar–Iron Springs belt—A broad zone of sinistral shear in the Basin and Range province, U.S.A. [abs.]: International Union of Geodesy and Geophysics, 21st Assembly, Boulder, Colo., week B, p. B185.

Huffman, A.C., Jr., and Taylor, D.J., 1994, Pennsylvanian thrust faulting along the northeastern margin of the Paradox Basin, Colorado and Utah: Geological Society of America Abstracts with Programs, v. 26, no. 6, p. 19.

Humphreys, E.D., and Dueker, K.G., 1994a, Western United States upper mantle structure: Journal of Geophysical Research, v. 99, no. B5, p. 9615–9634.

———1994b, Physical state of the western U.S. upper mantle: Journal of Geophysical Research, v. 99, no. B5, p. 9635–9650.

Hunt, C.B., 1953, Geology and geography of the Henry Mountains region, Utah: U.S. Geological Survey Professional Paper 228, v. 50, 234 p.

———1956, Cenozoic geology of the Colorado Plateau: U.S. Geological Survey Professional Paper 279, 99 p. —Includes structural contour map of entire plateau for top of Kaibab limestone or its equivalent.

———1958, Structural and igneous geology of the La Sal Mountains, Utah: U.S. Geological Survey Professional Paper 294–I, p. 305–364.

———1969, Geologic history of the Colorado River: U.S. Geological Survey Professional Paper 669–C, p. C59–C130.

———1980, Structural and igneous geology of the Henry Mountains, Utah, *in* Picard, M.D., ed., Henry Mountains symposium: Utah Geological Association Publication 8, p. 25–106.

———1988, The laccolith-stock controversy: New results from the southern Henry Mountains, Utah—Discussion and reply: Geological Society of America Bulletin, v. 100, p. 1657–1659.

Hunt, C.B., Averitt, Paul, and Miller, R.L., 1953, Geology and geography of the Henry Mountains region, Utah: U.S. Geological Survey Professional Paper 228, 234 p.

Hunt, C.B., and Waters, A.C., 1958, Structural and igneous geology of the La Sal Mountains, Utah: U.S. Geological Survey Professional Paper 294–I, p. 305–364.

Hunt, G.L., 1988, Petrology of the Mt. Pennell central stock, Henry Mountains, Utah: Brigham Young University Geology Studies, v. 35, p. 81–100.

Huntoon, P.W., 1970, Hydro-mechanics of the groundwater system in the southern portion of the Kaibab Plateau, Arizona: Tucson, Ariz., University of Arizona Ph. D. dissertation, 251 p.

———1974, The post-Paleozoic structural geology of the eastern Grand Canyon, Arizona, *in* Breed, W.J. and Roat, E.C., eds., Geology of the Grand Canyon: Museum of Northern Arizona and Grand Canyon Natural History Association, p. 82–115.

———1993, Influence of inherited Precambrian basement structure on the localization and form of Laramide monoclines, Grand Canyon, Arizona, *in* Schmidt, C.J., Chase, R.B., and Erslev, E.A., eds., Laramide basement deformation in the Rocky Mountain foreland of the western United States: Geological Society of America Special Paper 280, p. 243–256.

Huntoon, P.W., Billingsley, G.H., Jr., and Breed, W.B., 1982, Geologic map of Canyonlands National Park and vicinity, Utah: Moab, Utah, The Canyonlands Natural History Association, scale 1:62,500.

Huntoon, P.W., Billingsley, G.H., Breed, W.J., Sears, J.W., Ford, T.D., Clark, M.D., Babcock, R.S., and Brown, E.H., 1976, Geologic map of the Grand Canyon National Park, Arizona: Grand Canyon Natural History Association and Museum of Northern Arizona, 1 sheet, scale 1:62,500.

Huntoon, P.W., and Sears, J.W., 1975, Bright Angel and Eminence faults, eastern Grand Canyon, Arizona: Geological Society of America Bulletin, v. 86, p. 465–472.

Huntsman, J.R., 1984, Caribou Mountain—A porphyry copper deposit in southeastern Idaho: Economic Geology, v. 79, p. 748–754.

———1988, Structural controls and age of the Caribou Mountain intrusion, southeastern Idaho: Geological Society of America Abstracts with Programs, v. 20, p. 421.

Hurtubise, D.O., 1989, Stratigraphy and structure of the Seaman Range and Fox Mountain area, Lincoln and Nye Counties, Nevada, with an emphasis on the Devonian System: Golden, Colo., Colorado School of Mines Ph. D. dissertation, 443 p.

———1994, Silver King lineament, the missing link of a 500 km east-trending structure in the southern Great Basin, *in* Dobbs, S.W., and Taylor, W.J., eds., Structural and stratigraphic investigations and petroleum potential of Nevada, with special emphasis south of the Railroad Valley producing trend: Nevada Petroleum Society, Conference Volume 2, p. 127–139.

Hutton, D.H.W., 1988, Granite emplacement mechanisms and tectonic controls—Inferences from deformational studies: Transactions of the Royal Society of Edinburgh, Earth Sciences, v. 79, p. 245–255.

Irvine, T.N., 1989, A global convection framework—Concepts of symmetry, stratification, and system in the Earth's dynamic structure: Economic Geology, v. 84, p. 2059–2114.

Irving, E., and Irving, G.A., 1982, Apparent polar wander paths Carboniferous through Cenozoic and the assembly of Gondwana: Geophysical Surveys, v. 5, p. 141–188.

Irving, J.D., 1899, A contribution to the geology of the northern Black Hills: New York Academy of Science Annals, v. 12, p. 187–340.

Irwin, T.D., 1973, The petrologic evolution of the North Mountain stock, La Sal Mountains, Utah: Tucson, Ariz., University of Arizona M.S. thesis, 71 p.

Isacks, B.L., and Barazangi, Muawia, 1977, Geometry of Benioff zones—Lateral segmentation and downward bending of the subducted lithosphere, *in* Talwani, Manik, and Pitman, W.C., III, eds., Island arcs, deep sea trenches and back-arc basins: Washington, D.C., American Geophysical Union, p. 99–111.

Izett, G.A., 1975, Late Cenozoic sedimentation and deformation in northern Colorado and adjoining areas, *in* Curtis, B.F., ed., Cenozoic history of the Southern Rocky Mountains: Geological Society of America Memoir 144, p. 179–209.

Izett, G.A., Honey, J.G., and Brownfield, M.E., 1985, Geologic map of the Citadel Plateau quadrangle, Moffat County, Colorado: U.S. Geological Survey Miscellaneous Investigations Series Map I–1532, 1 sheet, scale 1:48,000.

Jachens, R.C., Simpson, R.W., Blakely, R.J., and Saltus, R.W., 1993, Isostatic residual gravity map of the conterminous United States and some adjacent parts of Canada, scale 1:5,000,000, plate 4, *in* Reed, J.C., Jr., Bickford, M.E., Houston, R.S., Link, P.K., Rankin, D.W., Sims, P.K., and Van Schmus, W.R., eds., The Geology of North America, volume C–2, Precambrian—Conterminous U.S.: Boulder, Colo., Geological Society of America, 657 p.

Jackson, M.D., and Champion, D.E., 1987, Paleomagnetism of rotated sills at Mount Hillers, Henry Mountains, Utah—Implications for the early stages of growth of near surface igneous domes: Geological Society of America Abstracts with Programs, v. 19, no. 6, p. 391.

Jackson, M.D., and Pollard, D.D., 1988a, The laccolith-stock controversy—New results from the southern Henry Mountains, Utah: Geological Society of America Bulletin, v. 100, p. 117–139.

———1988b, The laccolith-stock controversy—New results from the southern Henry Mountains, Utah—Discussion and reply: Geological Society of America Bulletin, v. 100, p. 1657–1659.

———1990, Flexure and faulting of sedimentary host rocks during growth of igneous domes, Henry Mountains, Utah: Journal of Structural Geology, v. 12, p. 185–206.

Jackson, M.P.A., and Talbot, C.J., 1991, A glossary of salt tectonics: Texas Bureau of Economic Geology Geological Circular 91–4, 44 p.

Jamison, W.R., and Stearns, D.W., 1982, Tectonic deformation of Wingate Sandstone, Colorado National Monument: American Association of Petroleum Geologists Bulletin, v. 66, p. 2584–2608.

Janecke, S.U., 1992, Kinematics and timing of three superposed extensional systems, east-central Idaho—Evidence for an Eocene tectonic transition: Tectonics, v. 11, p. 1121–1138.

Jenkins, R.E., II, 1979, Geology, geochemistry and origin of mineral deposits in the Hill Gulch area, Jamestown, Colorado: Golden, Colo., Colorado School of Mines Ph. D. thesis, 220 p.

Joesting, H.R., and Byerly, P.E., 1958, Regional geophysical investigations of the Uravan area, Colorado: U.S. Geological Survey Professional Paper 316–A, p. 1–17.

Joesting, H.R., and Case, J.E., 1960, Salt anticlines and deep-seated structures in the Paradox basin, Colorado and Utah, *in* Short papers in the geological sciences: U.S. Geological Survey Professional Paper 400–B, p. B252–B256.

Joesting, H.R., and Plouff, Donald, 1958, Geophysical studies of the Upheaval Dome area, San Juan County, Utah, p. 86–92 *in* Sanborn, A.F., ed., International Association of Petroleum Geologists, Guidebook, Ninth Annual Field Conference: 308 p.

John, D.A., Nash, J.T., Clark, C.W., and Wulftange, N.H., 1991, Geology, hydrothermal alteration, and mineralization at the Paradise Peak gold-silver-mercury deposit, Nye County, Nevada, *in* Raines, G.L., Lisle, R.E., Schafer, R.W., and Wilkinson, W.H., eds., Geology and ore deposits of the Great Basin: Geological Society of Nevada Symposium Proceedings, v. 1, p. 1020–1050.

John, D.A., Thomason, R.E., and McKee, E.H., 1989, Geology and K-Ar geochronology of the Paradise Peak mine and the relationship of pre-basin and range extension to early Miocene precious metal mineralization in west-central Nevada: Economic Geology, v. 84, p. 631–649.

Johnson, A.M., and Pollard, D.D., 1973, Mechanics of growth of some laccolithic intrusions in the Henry Mountains, Utah—I, Field observations, Gilbert's model, physical properties, and flow of magma: Tectonophysics, v. 18, p. 261–309.

Johnson, C.M., 1991, Large-scale crust formation and lithosphere modification beneath middle to late Cenozoic calderas and volcanic fields, western North America: Journal of Geophysical Research, v. 96, no. B8, p. 13485–13507.

Johnson, C.M., and Fridrich, C.J., 1990, Non-monotonic chemical and O, Sr, Nd, and Pb isotope zonations and heterogeneities in the mafic- to silicic-composition magma chamber of the Grizzly Peak Tuff, Colorado: Contributions to Mineralogy and Petrology, v. 105, p. 677–690.

Johnson, C.M., Lipman, P.W., and Czamanske, G.K., 1990, H, O, Sr, Nd, and Pb isotope geochemistry of the Latir volcanic field and cogenetic intrusions, New Mexico, and relations between evolution of a continental magmatic center and modifications of the lithosphere: Contributions to Mineralogy and Petrology, v. 104, p. 99–124.

Johnson, R.C., 1985, Early Cenozoic history of the Uinta and Piceance Creek basins, Colorado and Utah, with special reference to the development of Lake Uinta, *in* Flores, R.M., and Kaplan, S.S., eds., Cenozoic paleogeography of the west-central United States: Rocky Mountain Paleogeography Symposium 3; Rocky Mountain Section, Society of Economic Paleontologists and Mineralogists, p. 247–276.

Johnson, R.C., and Finn, T.M., 1986, Cretaceous through Holocene history of the Douglas Creek Arch, Colorado and Utah, *in* Stone, D.S., ed., New interpretations of northwest Colorado geology: Denver, Rocky Mountain Association of Geologists, p. 77–95.

Johnson, S.Y., Schenk, C.J., and Karachewski, J.A., 1988, Pennsylvanian and Permian depositional systems and cycles in the Eagle basin, northwest Colorado, *in* Holden, G.S., ed., Geological Society of America field trip guidebook, 1988 centennial meeting, Denver, Colorado: Colorado School of Mines Professional Contributions 12, p. 156–174.

Johnson, V.C., 1983, Preliminary aeromagnetic interpretation of the Uncompahgre uplift and Paradox basin, west-central Colorado and east-central Utah, *in* Averett, W.R., ed., Northern Paradox basin–Uncompahgre uplift field trip: Grand Junction Geological Society, p. 67–70.

Kane, M.F., and Godson, R.H., 1989, A crust/mantle structural framework of the conterminous United States based on gravity and magnetic trends, *in* Pakiser, L.C., and Mooney, W.D., eds., Geophysical framework of the continental United States: Geological Society of America Memoir 172, p. 383–404.

Kane, W.T., Atkinson, W.W., Jr., and Snee, L.W., 1988, Trace element geochemistry and Ar/Ar geochronology of epithermal mineralization, Four Mile Canyon, Boulder County, Colorado: Geological Society of America Abstracts with Programs, v. 20, p. A353.

Karlstrom, K.E., Bowring, S.A., and Conway, C.M., 1987, Tectonic significance of an Early Proterozoic two-province boundary in central Arizona: Geological Society of America Bulletin, v. 99, p. 529–538.

Kay, B.D., 1986, Vein and breccia gold mineralization and associated igneous rocks at the Ortiz mine, New Mexico, U.S.A.: Golden, Colo., Colorado School of Mines M.S. thesis, 179 p.

Kay, F.W., and Kay, S.M., 1981, The nature of the lower continental crust—Inferences from geophysics, surface geology, and crustal xenoliths: Reviews of Geophysics and Space Physics, v. 19, p. 271–297.

Kelemen, P.B., Johnson, K.T.M., Kinzler, R.J., and Irving, A.J., 1991, High-field-strength element depletions in arc basalts due to mantle-magma interactions: Nature, v. 345, p. 521–524.

Keller, G.R., and Cordell, Lindrith, 1984, Bouguer gravity anomaly map of New Mexico, Overlay 1 *in* Geothermal Resources of New Mexico: Scientific Map Series, scale 1:500,000 (National Geophysical Data Center, Boulder, Colo., no. 84–TGB–05).

Kelley, V.C., 1955, Regional tectonics of the Colorado Plateau and relationship to origin and distribution of uranium: University of New Mexico Publications in Geology 5, 120 p.

Kelley, V.C., and Clinton, N.J., 1960, Fracture systems and tectonic elements of the Colorado Plateau: Albuquerque, N. Mex., University of New Mexico Press, 104 p.

Kelly, W.C., and Goddard, E.M., 1969, Telluride ores of Boulder County, Colorado: Geological Society of America Memoir 109, 237 p.

Kempton, P.D., Fitton, J.G., Hawkesworth, C.J., and Ormerod, D.S., 1991, Isotopic and trace element constraints on the composition and evolution of the lithosphere beneath the southwestern United States: Journal of Geophysical Research, v. 96, p. 13713–13735.

Kilinc, I.A., 1979, Melting relations in the quartz diorite–H_2O and quartz diorite–H_2O–CO_2 systems: Neues Jahrbuch für Mineralogie, 1979, no. 2, p. 67–72.

Kirchener, J.G., 1971, Petrology and petrography of the phonolite porphyry intrusions of the northern Black Hills, South Dakota: Iowa City, Iowa, University of Iowa Ph. D. thesis, 199 p.

Kirchner, G.L., 1982, Field relations, petrology, and mineralization of the Linster Peak dome, Fergus County, Montana: Missoula, Mont., University of Montana M.S. thesis, 115 p.

Kistler, R.W., 1990, Two different lithosphere types in the Sierra Nevada, California, *in* Anderson, J.L., ed., The nature of Cordilleran magmatism: Geological Society of America Memoir 174, p. 271–281.

Kleinkopf, M.D., Peterson, D.L., and Mattick, R.E., 1979, Gravity, magnetic, and seismic studies of the Silver Cliff and Rosita Hills volcanic area, Colorado: U.S. Geological Survey Professional Paper 726–F, p. F1–F17.

Klemme, H.D., 1980, Petroleum basins—Classification and characteristics: Journal of Petroleum Geology, v. 3, no. 2, p. 187–207.

Kluth, C.F., 1986, Plate tectonics of the ancestral Rocky Mountains, *in* Peterson, J.A., ed., Paleotectonics and sedimentation in the Rocky Mountain region, U.S.: American Association of Petroleum Geologists Memoir 41, p. 353–369.

Knoper, M.W., and Condie, K.C., 1988, Geochemistry and petrogenesis of early Proterozoic amphibolites, west-central Colorado: Chemical Geology, v. 67, p. 209–225.

Kohrt, P.B., 1991, Alkalic rocks of the Judith Mountains, central Montana, *in* Baker, D.W., and Berg, R.B., eds., Guidebook of the central Montana alkalic province: Montana Bureau of Mines and Geology Special Publication 100, p. 77–85.

Koschmann, A.H., 1949, Structural control of the gold deposits of the Cripple Creek district, Teller County, Colorado: U.S. Geological Survey Bulletin 955–B, p. 19–60.

Kowallis, B.J., and Best, M.G., 1990, Fission track ages from volcanic rocks in southwestern Utah and southeastern Nevada: Isochron/West, no. 55, p. 24–27.

Krantz, R.W., 1989, Laramide structures of Arizona, *in* Jenney, J.P., and Reynolds, S.J., eds., Geological evolution of Arizona: Tucson, Arizona Geological Society Digest 17, p. 463–483.

Kulander, B.R., Barton, C.C., and Dean, S.L., 1979, The application of fractography to core and outcrop fracture investigations: United States Department of Energy, Report METC/SP–79–3, 174 p.

Kuntz, M.A., 1992, A model-based perspective of basaltic volcanism, eastern Snake River Plain, Idaho, *in* Link, P.K., Kuntz, M.A., and Platt, L.B., eds., Regional geology of eastern Idaho and western Wyoming: Geological Society of America Memoir 179, p. 289–304.

Kuntz, M.A., Covington, H.R., and Schorr, L.J., 1992, An overview of basaltic volcanism of the eastern Snake River Plain, Idaho, *in* Link, P.K., Kuntz, M.A., and Platt, L.B., eds., Regional geology of eastern Idaho and western Wyoming: Geological Society of America Memoir 179, p. 227–267.

Kurisoo, Peter, 1991, Gold deposits of the Kendall mining district, Fergus County, Montana, *in* Baker, D.W., and Berg, R.B., eds., Guidebook of the central Montana alkalic province: Montana Bureau of Mines and Geology Special Publication 100, p. 39–44.

Lachenbruch, A.H., and Sass, J.H., 1978, Models of an extending lithosphere and heat flow in the Basin and Range province, *in* Smith, R.B., and Eaton, G.P., eds., Cenozoic tectonics and regional geophysics of the western Cordillera: Geological Society of America Memoir 152, p. 209–250.

Lachenbruch, A.H., Sass, J.H., and Morgan, Paul, 1994, Thermal regime of the southern Basin and Range province—2, Implications of heat flow for regional extension and metamorphic core complexes: Journal of Geophysical Research, v. 99, no B11, p. 22121–22133.

Lanphere, M.A., 1988, High resolution $^{40}Ar/^{39}Ar$ chronology of Oligocene volcanic rocks, San Juan Mountains, Colorado: Geochimica et Cosmochimica Acta, v. 52, p. 1425–1434.

Larsen, E.S., 1940, Petrographic province of central Montana: Geological Society of America Bulletin, v. 51, p. 887–948.

Larsen, R.K., 1977, Geology, alteration and mineralization of the northern Cutting stock, Lawrence County, South Dakota: Rapid City, S. Dak., South Dakota School of Mines and Technology M.S. thesis, 147 p.

Larson, E.E., Patterson, P.E., Curtis, G., Drake, R., and Mutschler, F.E., 1985, Petrologic, paleomagnetic, and structural evidence of a Proterozoic rift system in Oklahoma, New Mexico, Colorado and Utah: Geological Society of America Bulletin, v. 96, no. 11, p. 1364–1372.

Lawrence, R.D., 1976, Strike-slip faulting terminates the Basin and Range province in Oregon: Geological Society of America Bulletin, v. 87, p. 846–850.

Lee, Keenan, 1988, Remote sensing study in support of mineral resource appraisal of wilderness study areas near Moab, Utah—Dolores River Canyon wilderness study area, Montrose and San Miguel Counties, Colorado; Lost Spring Canyon wilderness study area, Grand County,

Utah; Behind the Rocks wilderness study area, Grand and San Juan Counties, Utah; Indian Creek wilderness study area, San Juan County, Utah; and Butler Wash wilderness study area, San Juan County, Utah: U.S. Geological Survey Open-File Report 88–219, 12 p.

Le Bas, M.J., Le Maitre, R.W., Streckeisen, A., and Zanettin, B., 1986, A chemical classification of volcanic rocks based on the total alkali-silica diagram: Journal of Petrology, v. 27, p. 745–750.

Leeman, W.P., and Harry, D.L., 1993, A binary source model for extension-related magmatism in the Great Basin, western North America: Science, v. 262, p. 1550–1554.

Li, Yianping, Geissman, J.W., Nur, Amos, Ron, Hagai, and Huang, Quing, 1990, Paleomagnetic evidence for counterclockwise block rotation in the north Nevada rift region: Geology, v. 18, p. 79–82.

Liberty, L.M., Heller, P.L., and Smithson, S.B., 1994, Seismic reflection evidence for two-phase development of Tertiary basins from east-central Nevada: Geological Society of America Bulletin, v. 106, p. 1621–1633.

Lidiak, E.G., and Zietz, Isidore, 1976, Interpretation of aeromagnetic anomalies between latitude 37° N and 38° N in eastern and central United States: Geological Society of America Special Paper 167, 37 p.

Liggett, M.A., and Childs, J.F., 1977, An application of satellite imagery to mineral exploration, *in* Woll, P.W., and Fischer, W.A., eds., Proceedings of the first annual William T. Pecora Memorial Symposium: U.S. Geological Survey Professional Paper 1015, p. 253–270.

Lindgren, Waldemar, Graton, L.C., and Gordon, C.H., 1910, The ore deposits of New Mexico: U.S. Geological Survey Professional Paper 68, 361 p.

Lindgren, Waldemar, and Ransome, F.L., 1906, Geology and gold deposits of the Cripple Creek district, Colorado: U.S. Geological Survey Professional Paper 54, 516 p.

Lindsey, D.A., 1982, Geologic map and discussion of selected mineral resources of the North and South Moccasin Mountains, Fergus County, Montana: U.S. Geological Survey Miscellaneous Investigations Series Map I–1362, scale 1:24,000.

———1985, A gold-mineralized breccia zone at Kendall, North Moccasin Mountains, Fergus County, Montana: U.S. Geological Survey Professional Paper 1301–C, p. 43–56.

Lindsey, D.A., and Fisher, F.S., 1985, Mineralized breccias in intrusive complexes of Late Cretaceous and Paleocene age, north-central Montana: U.S. Geological Survey Professional Paper 1301–A, p. 1–34.

Lindsey, D.A., and Naeser, C.W., 1985, Relation between igneous intrusion and gold mineralization in the North Moccasin Mountains, Fergus County, Montana: U.S. Geological Survey Professional Paper 1301–B, p. 35–41.

Lipman, P.W., 1969, Alkalic and tholeiitic basaltic volcanism related to the Rio Grande depression, southern Colorado and northern New Mexico: Geological Society of America Bulletin, v. 80, p. 1343–1354.

———1980, Cenozoic volcanism in the western United States—Implications for continental tectonics, *in* Studies in geophysics—Continental tectonics: Washington, D.C., National Academy of Sciences, p. 161–174.

———1981, Volcano-tectonic setting of Tertiary ore deposits, southern Rocky Mountains, *in* Dickinson, W.R., and Payne, W.D., eds., Relations of tectonics ore deposits in the southern Cordillera: Arizona Geological Society Digest, v. 14, p. 199–213. —N. New Mexico subjected to subduction-related magmatism and back-arc extension to WSW (see also Eaton, 1980)

———1984, The roots of ash flow calderas in western North America—Windows into the tops of granite batholiths: Journal of Geophysical Research, v. 89, p. 8801–8841.

———1988, Evolution of silicic magma in the upper crust—The mid-Tertiary Latir volcanic field and its cogenetic granitic batholith, northern New Mexico, U.S.A.: Transactions of the Royal Society of Edinburgh, v. 79, p. 265–288.

———1989, Excursion 16B, Oligocene-Miocene San Juan volcanic field, Colorado—Introduction, *in* Chapin, C.E., and Zidek, J., eds., Field excursions to volcanic terranes in the western United States—Volume I, Southern Rocky Mountain Region: New Mexico Bureau of Mines and Mineral Resources Memoir 46, p. 303–305.

———1992, Magmatism in the Cordilleran United States—Progress and problems, *in* Burchfiel, B.C., Lipman, P.W., and Zoback, M.L., eds., The Cordilleran orogen—Conterminous U.S.: Geological Society of America, The Geology of North America, v. G–3, p. 481–514.

Lipman, P.W., Doe, B.R., Hedge, C.E., and Steven, T.A., 1978, Petrologic evolution of the San Juan volcanic field, southwestern Colorado: Pb and Sr isotope evidence: Geological Society of America Bulletin, v. 89, p. 59–82.

Lipman, P.W., Fisher, F.S., Mehnert, H.H., Naeser, C.W., Luedke, R.G., and Steven, T.A., 1976, Multiple ages of mid-Tertiary mineralization and alteration in the western San Juan Mountains, Colorado: Economic Geology, v. 71, p. 571–588.

Lipman, P.W., and Glazner, A.F., 1991, Introduction to middle Tertiary Cordilleran volcanism: magma sources and relation to regional tectonics: Journal of Geophysical Research, v. 96 p. 13193–13199.

Lipman, P.W., Prostka, H.J., and Christiansen, R.L., 1972, Cenozoic volcanism and plate-tectonic evolution of the western United States—I, Early and middle Cenozoic: Philosophical Transactions of Royal Society of London, v. A.271, p. 217–248.

Lipman, P.W., Rowley, P.D., Mehnert, H.H., Evans, S.H., Jr., Nash, W.P., and Brown, F.H., 1978, Pleistocene rhyolite of the Mineral Mountains, Utah—Geothermal and archeological significance, *with sections on* Fission-track dating, by G.A. Izett and C.W. Naeser, *and on* Obsidian-hydration dating, by Irving Friedman: U.S. Geological Survey Journal of Research, v. 6, no. 1, p. 133–147.

Lister, G.S., and Baldwin, S.L., 1993, Plutonism and the origin of metamorphic core complexes: Geology, v. 21, p. 607–610.

Livacarri, R.F., Burke, Kevin, and Sengor, A.M.C., 1981, Was the Laramide orogeny related to subduction of an oceanic plateau?: Nature, v. 289, p. 276–278.

Livaccari, R.F., and Perry, F.V., 1993, Isotopic evidence for preservation of Cordilleran lithospheric mantle during the Sevier-Laramide orogeny, western United States: Geology, v. 21, p. 719–722.

Lohman, S.W., 1981, The geologic story of Colorado National Monument: U.S. Geological Survey Bulletin 1508, 142 p.

Loomis, T.A., and Alexander, D.L., 1990, The geology of the Foley Ridge gold mine: Proceedings of the Fourth Western Regional Conference on Precious Metals and the Environment, Lead, South Dakota; Society for Mining, Metallurgy, and Exploration (SME–AIME), p. 77–87.

Lorenz, J.C., 1985, Tectonic and stress histories of the Piceance Creek basin and the MWX site, from 75 million years ago to the present: Sandia National Laboratories Report SAND84–2503, 50 p.

Lorenz, J.C., and Finley, S.J., 1987, Differences in fracture characteristics and related production of natural gas in different zones of the Mesaverde Formation, northwestern Colorado: Society of Petroleum Engineers Annual Technical Conference and Exhibition, 62nd, Dallas, SPE Paper 16809, p. 589–593.

———1991, Regional fractures II—Fracturing of Mesaverde reservoirs in the Piceance Basin, Colorado: American Association of Petroleum Geologists Bulletin, v. 75, no. 11, p. 1738–1757.

Lorenz, J.C., Warpinski, N.R., Branagan, Paul, and Sattler, A.R., 1989, Fracture characteristics and reservoir behavior of stress-sensitive fracture systems in flat-lying lenticular formations: Journal of Petroleum Technology, v. 4, no. 1, p. 11–16.

Loughlin, G.F., 1927, Ore at deep levels in the Cripple Creek district, Colorado: American Institute of Mining and Metallurgical Engineers Technical Publication 13, 32 p.

Loughlin, G.F., and Koschmann, A.H., 1935, Geology and ore deposits of the Cripple Creek district, Colorado: Colorado Scientific Society Proceedings, v. 13, no. 6, p. 217–435.

Lowell, J.D., 1974, Plate tectonics of foreland basement deformation: Geology, v. 2, p. 275–278.

Lovering, T.S., 1930, The New World or Cooke City mining district, Park County, Montana: U.S. Geological Survey Bulletin 811–A, p. 1–87.

Lovering, T.S., and Goddard, E.N., 1950, Geology and ore deposits of the Front Range, Colorado: U.S. Geological Survey Professional Paper 223, 319 p.

Lucchitta, Ivo, 1974, Structural evolution of northwest Arizona and its relation to adjacent Basin and Range province structures, *in* Karlstrom, T.N.V., Swann, G.A., and Eastwood, R.L., eds., Geology of northern Arizona, with notes on archaeology and paleoclimate—Part I, Regional studies: Geological Society of America, Rocky Mountain Section Meeting [Guidebook for Field Trips, No. 27], p. 336–354.

———1979, Late Cenozoic uplift of the southwestern Colorado Plateau and adjacent lower Colorado River region: Tectonophysics, v. 61, p. 63–95.

———1990, Role of heat and detachment in continental extension as viewed from the eastern Basin and Range province in Arizona: Tectonophysics, v. 174, p. 77–114.

Lucchitta, Ivo, and Suneson, N.H., 1993, Dips and extension: Geological Society of America Bulletin, v. 105, p. 1346–1356.

Ludwig, K.R., Rasmussen, J.D., and Simmons, K.R., 1986, Age of uranium ores in collapse-breccia pipes in the Grand Canyon area, northern Arizona: Geological Society of America Abstracts with Programs, v. 18, no. 5, p. 392.

Ludwig, K.R., and Simmons, K.R., 1992, U-Pb dating of uranium deposits in collapse breccia pipes of the Grand Canyon region: Economic Geology, v. 87, p. 1747–1765.

Luedke, R.G., and Smith, R.L., 1981, Map showing distribution, composition, and age of late Cenozoic volcanic centers in California and Nevada: U.S. Geological Survey Miscellaneous Investigations Series Map I–1091–C, scale 1:1,000,000.

Lundin, E.R., 1989, Thrusting of the Claron Formation, Bryce Canyon region, Utah: Geological Society of America Bulletin, v. 101, p. 1038–1050.

Lundin, E.R., and Davis, G.H., 1987, Southeast vergent thrust faulting and folding of the Eocene(?) Claron Formation, Bryce Canyon National Park, Utah: Geological Society of America Abstracts with Programs, v. 19, no. 5, p. 317.

Lux, D.R., 1977, A major element geochemical study of Laramide igneous rocks of the Colorado mineral belt: Houston, Texas, Rice University M.S. thesis, 77 p.

Lynch, D.J., 1989, Neogene volcanism in Arizona—The recognizable volcanos, *in* Jenney, J.P., and Reynolds, S.J., eds., Geologic evolution of Arizona: Tucson, Arizona Geological Society Digest 17, p. 681–700.

Mabey, D.R., and Budding, K.E., 1987, High-temperature geothermal resources of Utah: Utah Geological and Mineral Survey Bulletin 123.

Mabey, D.R., and Webring, M.W., 1985, Regional geophysical studies in the Challis quadrangle: U.S. Geological Survey Bulletin 1658, Chapter E, p. 69–79.

Mabey, D.R., Zietz, Isidore, Eaton, G.P., and Kleinkopf, M.D., 1978, Regional magnetic patterns in part of the Cordillera in the western United States, *in* Smith, R.B., and Eaton, G.P., Cenozoic tectonics and regional geophysics of the western Cordillera: Geological Society of America Memoir 152, p. 93–106.

Macdonald, G.A., and Katsura, T., 1964, Chemical composition of Hawaiian lavas: Journal of Petrology, v. 5, p. 82–133.

Machette, M.N., 1985, Late Cenozoic geology of the Beaver basin, southwestern Utah: Brigham Young University Studies in Geology, v. 32, pt. 1, p. 19–37.

Machette, M.N., Steven, T.A., Cunningham, C.G., and Anderson, J.J., 1984, Geologic map of the Beaver quadrangle, Beaver and Piute Counties, Utah: U.S. Geological Survey Miscellaneous Investigations Series Map I–1520, scale 1:50,000.

Mackin, J.H., 1947, Some structural features of the intrusions in the Iron Springs district: Utah Geological Society Guidebook 2, 62 p.

———1960, Structural significance of Tertiary volcanic rocks in southwestern Utah: American Journal of Science, v. 258, p. 81–131.

MacLachlan, M.E., 1987, General geology of the Piceance basin, *in* Taylor, O.J., compiler, Oil shale, water resources, and valuable minerals of the Piceance basin, Colorado—The challenge and choices of development: U.S. Geological Survey Professional Paper 1310, p. 7–15.

MacLachlan, M.E., and Welder, F.E., 1987, Paleozoic and Mesozoic formations and their potential as ground-water reservoirs, *in* Taylor, O.J., compiler, Oil shale, water resources, and valuable minerals of the Piceance basin, Colorado—The challenge and choices of development: U.S. Geological Survey Professional Paper 1310, p. 95–106.

MacLeod, N.S., Walker, G.W., and McKee, E.H., 1976, Geothermal significance of eastward increase in age of upper Cenozoic rhyolitic domes in southeastern Oregon, *in* Second United Nations Symposium on the Development and Use of Geothermal Resources: Washington, D.C., U.S. Government Printing Office, v. 1, p. 465–474.

MacQuown, W.C., Jr., 1945, Structure of the White River Plateau near Glenwood Springs, Colorado: Geological Society of America Bulletin, v. 56, p. 877–892.

Maldonado, Florian, 1995, Decoupling of mid-Tertiary rocks, Red Hills–western Markagunt Plateau, southwestern Utah, chapter I *in* Scott, R.B., and Swadley, WC, eds., Geologic studies in the Basin and Range–Colorado Plateau transition in southeastern Nevada, southwestern Utah, and northwestern Arizona, 1992: U.S. Geological Survey Bulletin 2056, p. 232-254.

Maldonado, Florian, Sable, E.G., and Anderson, J.J., 1989, Evidence for shallow detachment faulting of mid-Tertiary rocks, Red Hills (Basin and Range), with implications for a more extensive detachment zone in the adjacent Markagunt Plateau (Colorado Plateau), southwest Utah [abs.]: Eos, v. 70, no. 43, p. 1336.

———1990, Shallow detachment of mid-Tertiary rocks, Red Hills (Basin and Range), with implications for a regional detachment zone in the adjacent Markagunt Plateau (Colorado Plateau), southwest Utah: Geological Society of America Abstracts with Programs, v. 22, no. 3, p. 63.

———1992, Evidence for a Tertiary low-angle shear zone, Red Hills, Utah, with implications for a regional shear zone in the adjacent Colorado Plateau, *in* Harty, K.M., ed., Engineering and environmental geology of southwestern Utah: Utah Geological Association Publication 21, p. 315–324.

Mandl, G., 1987, Tectonic deformation by rotating parallel faults—The "bookshelf" mechanism: Tectonophysics, v. 141, p. 177–316.

Mann, Paul, Hempton, M.R., Bradley, D.C., and Burke, Kevin, 1983, Development of pull-apart basins: Journal of Geology, v. 91, no. 5, p. 529–554.

Marvin, R.F., Hearn, B.C., Jr., Mehnert, H.H., Naeser, C.W., Zartman, R.E., and Lindsey, D.A., 1980, Late Cretaceous–Paleocene–Eocene igneous activity in north central Montana: Isochron/West, no. 29, p. 5–25.

Marvin, R.F., Witkind, I.J., Keefer, W.R., and Mehnert, H.H., 1973, Radiometric ages of intrusive rocks in the Little Belt Mountains, Montana: Geological Society of America Bulletin, v. 84, p. 1977–1986.

Marvin, R.F., Young, E.J., Mehnert, H.H., and Naeser, C.W., 1974, Summary of radiometric age determinations on Mesozoic and Cenozoic igneous rocks and uranium and base metal deposits in Colorado: Isochron/West, no. 11, 41 p.

Mattox, R.B., 1975, Upheaval Dome, a possible salt dome in the Paradox Basin, Utah, *in* Fassett, J.E., ed., Canyonlands Country: Four Corners Geological Society Guidebook, Eighth Field Conference, p. 225–234.

Mattox, S.R., 1991a, Origin of potassium-rich mafic lava flows, Marysvale volcanic field, Utah: Eos (American Geophysical Union) Program and Abstracts, v. 72, no. 44, p. 561.

———1991b, Petrology, age, geochemistry and correlation of the Tertiary volcanic rocks of the Awapa Plateau, Garfield, Piute, and Wayne Counties, Utah: Utah Geological Survey Miscellaneous Publication 91-5, 46 p.

———1992, Geochemistry, origin, and tectonics of mid-Tertiary and late Tertiary and Quaternary volcanic rocks, southern Marysvale volcanic field, Utah: Dekalb, Ill., Northern Illinois University Ph. D. thesis, 239 p.

Mattox, S.R., and Walker, J.A., 1989, Geochemistry and tectonism of lavas erupted during the transition from compression to extension, southern Marysvale complex, southwestern Utah: International Association of Volcanology and Chemistry of the Earth's Interior, Abstracts, New Mexico Bureau of Mines and Mineral Resources Bulletin 131, p. 178.

_____1990, Late Cenozoic lavas of the Utah transition zone: Geological Society of America Abstracts with Programs, v. 22, no. 3, p. 65.

May, S.J., Kelley, S.A., and Russell, L.R., 1994, Footwall unloading and rift shoulder uplifts in the Albuquerque Basin—Their relation to synrift fanglomerates and apatite fission-track ages, *in* Keller, G.R., and Cather, S.M., eds., Basins of the Rio Grande Rift—Structure, stratigraphy, and tectonic setting: Geological Society of America Special Paper 291, p. 125–134.

Maynard, S.R., Nelsen, C.J., Martin, K.W., and Schultz, J.L., 1990, Geology and gold mineralization of the Ortiz Mountains, Santa Fe County, New Mexico: Mining Engineering, v. 42, no. 8 (August), p. 1007–1011.

Maynard, S.R., Woodward, L.A., and Giles, D.L., 1991, Tectonics, intrusive rocks, and mineralization of the San Petro–Ortiz porphyry belt, north-central New Mexico: New Mexico Bureau of Mines and Mineral Resources Bulletin 137, p. 57–69.

McCarthy, Jill, and Parsons, Tom, 1994, Insights into the kinematic Cenozoic evolution of the Basin and Range–Colorado Plateau transition from coincident seismic refraction and reflection data: Geological Society of America Bulletin, v. 106, p. 747–759.

McDonald, R.E., 1976, Tertiary tectonics and sedimentary rocks along the transion—Basin and Range province to plateau and thrust belt province, Utah, *in* Hill, J.G., ed., Symposium on geology of the Cordilleran hingeline: Denver, Colo., Rocky Mountain Association of Geologists, p. 281–317.

McDowell, F.W., 1971, K-Ar ages of igneous rocks from the western United States: Isochron/West, no. 2, p. 1–16.

McEwan, C.J.A., 1986, Geological, stable isotope and fluid inclusion evidence for concealed epithermal precious metal mineralisation, Rosita Hills, Colorado: Geological Society of America Abstracts with Programs, v. 16, p. 688.

McGetchin, T.R., and Merrill, R.B., eds., 1979, Plateau uplift—Mode and mechanism: Tectonophysics (special issue), v. 61, 336 p.

McGetchin, T.R., and Silver, L.T., 1972, A crustal-upper-mantle model for the Colorado Plateau based on observations of crystalline rock fragments in the Moses Rock dike: Journal of Geophysical Research, v. 77, p. 7022–7037.

McGill, G.E., and Stromquist, A.W., 1979, The grabens of Canyonlands National Park, Utah—Geometry, mechanics, and kinematics: Journal of Geophysical Research, v. 84, no. B9, p. 4547–4563.

McIntosh, W.C., Chapin, C.E.,and Geissman, J.W., 1991, Stratigraphic framework for Eocene-Oligocene ignimbrites in the southwestern corner of New Mexico: Geological Society of America Abstracts with Programs, v. 23, p. 47.

McIntosh, W.C., Chapin, C.E., Ratté, J.C., and Sutter, J.F., 1992, Time-stratigraphic framework for the Eocene-Oligocene Mogollon-Datil volcanic field, southwest New Mexico: Geological Society of America Bulletin, v. 104, p. 851–871.

McIntosh, W.C., Geissman, J.W., Chapin, C.E., Kunk, M.J., and Henry, C.D., 1992a, Calibration of the latest Eocene-Oligocene geomagnetic polarity time scale using $^{40}Ar/^{39}Ar$ dated ignimbrites: Geology, v. 20, p. 459–463.

McKay, G.A., 1989, Partitioning of rare earth elements between major silicate minerals and basaltic melts, *in* Lipin, B.R., and McKay, G.A. eds., Geochemistry and mineralogy of the rare earth elements: Mineralogical Society of America Reviews in Mineralogy, v. 21, p. 45–78.

McKee, E.D., 1979, Arizona, *in* Craig, L.C., and Connor, C.W., coordinators, Paleotectonic investigations of the Mississippian System in the United States—Part 1, Introduction and regional analyses of the Mississippian System: U.S. Geological Survey Professional Paper 1010–L, p. 197–207.

McKee, E.D., and Gutschick, R.C., 1969, History of the Redwall Limestone of northern Arizona: Geological Society of America Memoir 114, 692 p.

McKee, E.H., 1971, Tertiary igneous chronology of the Great Basin of Western United States—Implications for tectonic models: Geological Society of America Bulletin, v. 892, p. 3497–3502.

McKee, E.H., and Blakely, R.J., 1990, Tectonic significance of linear, north-trending anomalies in north-central Nevada, *in* Geology and ore deposits of the Great Basin, program with abstracts: Geological Society of Nevada, p. 49.

McKee, E.H., and Noble, D.C., 1986, Tectonic and magmatic development of the Great Basin of western United States during late Cenozoic time: Modern Geology, v. 10, p. 39–49.

McKee, E.H., Noble, D.C., and Silverman, M.L., 1970, Middle Miocene hiatus in volcanic activity in the Great Basin area of the western United States: Earth and Planetary Science Letters, v. 8, no. 2, p. 93–96.

McKnight, E.T., 1940, Geology of area between Green and Colorado Rivers, Grand and San Juan Counties, Utah: U.S. Geological Survey Bulletin 908, 147 p.

McLain, J.P., 1965, Photographic analysis of lineaments in the San Francisco volcanic field, Coconino and Yavapai Counties, Arizona: Tucson, Ariz., University of Arizona M.S. thesis, 31 p.

McMillan, W.J., 1991, Porphyry deposits in the Canadian Cordillera: British Columbia Ministry of Energy, Mines and Petroleum Resources Paper 1991–1994, p. 253–276.

Meen, J.K., and Curtis, P.C., 1989, What are shoshonites? [abs.]: Eos, v. 70, p. 1421.

Menard, H.W., and Chase, T.E., 1970, Fracture zones, *in* The sea—Ideas and observations on progress in the study of the seas: New York, Wiley-Interscience, v. 4, pt. I, p. 421–443.

Menges, C.M., and Pearthree, P.A., 1989, Late Cenozoic tectonism in Arizona and its impact on regional landscape evolution, *in* Jenney, J.P., and Reynolds, S.J., eds., Geological evolution of Arizona: Tucson, Arizona Geological Society Digest 17, p. 649–680.

Merle, O.R., Davis, G.H., Nickelsen, R.P., and Gourlay, P.A., 1993, Relation of thin-skinned thrusting of Colorado Plateau strata in southwestern Utah to Cenozoic magmatism: Geological Society of America Bulletin, v. 105, p. 387–398.

Meyer, H.O.A., and Mitchell, R.H., 1988, Sapphire bearing ultramafic lamprophyre from Yogo, Montana—An ouachitite: Canadian Mineralogist, v. 26, p. 81–88.

Meyer, Jeff, and Foland, K.A.,1991, Magmatic-tectonic interaction during early Rio Grande rift extension at Questa, New Mexico: Geological Society of America Bulletin, v. 103, p. 993–1006.

Michel-Noel, G., Anderson, R.E., and Angelier, Jacques, 1990, Fault kinematics and estimates of strain partitioning of a Neogene extensional fault system in southeastern Nevada, *in* Wernicke, B.P., ed., Basin and Range extensional tectonics near the latitude of Las Vegas, Nevada: Geological Society of America Memoir 176, p. 155–180.

Milbert, D.G., 1991, GEOID90—A high-resolution geoid for the United States: Eos, v. 72, no. 49, p. 545, 554.

Miller, D.M., 1991, Mesozoic and Cenozoic tectonic evolution of the northeastern Great Basin, *in* Buffa, R.H., and Coyner, A.R., eds., Geology and ore deposits of the Great Basin: Reno, Nev., Geological Society of Nevada, p. 202–228.

Miller, D.M., and Allmendinger, R.W., 1991, Jurassic normal and strike-slip faults at Crater Island, northwestern Utah: Geological Society of America Bulletin, v. 103, no. 9, p. 1239–1251.

Miller, E.L., Gans, P.B., and Garing, John, 1983, The Snake Range decollement—An exhumed mid-Tertiary ductile-brittle transition: Tectonics, v. 2, no. 3, p. 239–263.

Minor, S.A., 1995, Superposed local and regional paleostresses—Fault-slip analysis of Neogene extensional faulting near coeval caldera complexes, Yucca Flat, Nevada: Journal of Geophysical Research, v. 100, no. B6, p. 10507–10528.

Molenaar, C.M., and Baars, D.L., 1985, eds., Field and river trip guide to Canyonlands country, Utah, SEPM Guidebook for field trip 7: Rocky Mountain section, Society of Economic Paleontologists and Mineralogists, 74 p.

Mollard, J.D., 1957, Aerial mosaics reveal fracture patterns on surface materials in southern Saskatchewan and Manitoba: Oil in Canada, v. IX; no. 40, p. 26–50.

Moos, Daniel, and Zoback, M.D., 1993, State of stress in the Long Valley caldera, California: Geology, v. 21, p. 837–840.

Morgan, C.D., Allison, M.L., and Yonkee, W.A., 1992, Optimum directions for horizontal drilling in fractured Cane Creek Shale, Paradox basin, Utah: American Association of Petroleum Geologists Bulletin, v. 76, no. 8, p. 1265.

Morgan, P., and Swanberg, C.A., 1985, On the Cenozoic uplift and tectonic stability of the Colorado Plateau: Journal of Geodynamics, v. 3, p. 39–63.

Morgan, W.J., 1972, Deep mantle convection plume and plate motions: American Association of Petroleum Geologists Bulletin, v. 56, p. 203–312.

Moye, F.J., 1988, Extensional control of Eocene volcanism and plutonism, Idaho and Montana: Geological Society of America Abstracts with Programs, v. 20, p. 434.

Moyer, T.C., and Nealey, L.D., 1989, Regional compositional variations of late Tertiary bimodal rhyolite lavas across the Basin Range/Colorado Plateau boundary in western Arizona: Journal of Geophysical Research, v. 94, p. 7799–7816.

Muehlberger, W.R., compiler, 1992, Tectonic map of North America: Tulsa, Okla., American Association of Petroleum Geologists, scale 1:5,000,000 (2 sheets).

Mutschler, F.E., Griffin, M.E., Stevens, D.S., and Shannon, S.S., Jr., 1985, Precious metal deposits related to alkaline rocks in the North American Cordillera—An interpretive review: Geological Society of South Africa Transactions, v. 88, p. 355–377.

Mutschler, F.E., Johnson, D.C., and Mooney, T.C., 1991, A speculative plate kinematic model for the Montana alkaline province and related gold deposits, *in* Baker, D.W., and Berg, R.B., eds., Guidebook of the central Montana alkalic province: Montana Bureau of Mines and Geology Special Publication 100, p. 121–124.

Mutschler, F.E., Johnson, D.C., and Mooney, T.C., 1993, ALK.BIB—A selected bibliography of alkaline igneous rocks and related mineral deposits, with an emphasis on western North America: U.S. Geological Survey Open-File Report.

Mutschler, F.E., Larson, E.E., and Bruce, R.M., 1987, Laramide and younger magmatism in Colorado—New petrologic and tectonic variations on old themes: Colorado School of Mines Quarterly, v. 82, no. 4, p. 1–47.

Mutschler, F.E., and Mooney, T.C., 1993, Precious metal deposits related to alkaline igneous rocks—Provisional classification, grade-tonnage data, and exploration frontiers, *in* Kirkham, R.V., Sinclair, W.D., Thorpe, R.I., and Duke, J.M., eds., Mineral deposit modeling: Geological Association of Canada Special Paper 40.

Mutschler, F.E., Mooney, T.C., and Johnson, D.C., 1990, Precious metal deposits related to alkaline igneous rocks—A space-time trip through the Cordillera: Proceedings of the Fourth Western Regional Conference on Precious Metals and the Environment, Lead, South Dakota; Society for Mining, Metallurgy, and Exploration (SME–AIME), p. 43–59.

Mutschler, F.E., Rougon, D.J., Lavin, O.P., and Hughes, R.D., 1981, PETROS—A data bank of major-element chemical analyses of igneous rocks (version 6–1); Boulder, Colo., NOAA National Geophysical Data Center (magnetic tape).

Naeser, C.W., 1978, Fission track dating: U.S. Geological Survey Open-File Report 76–190, 87 p.

Naeser, C.W., Zimmerman, R.A., and Cebula, G.T., 1981, Fission track dating of apatite and zircon—An interlaboratory comparison: Nuclear Tracks, v. 5, p. 65–72.

Nash, J.T., and Cunningham, C.G., Jr., 1973, Fluid-inclusion studies of the fluorspar and gold deposits, Jamestown district, Colorado: Economic Geology, v. 68, p. 1247–1262.

Nealey, L.D., Maldonado, Florian, Unruh, D.M., and Budahn, J.R., 1994, Tectonic implications of Quaternary volcanism in the western Markagunt Plateau and Red Hills area, southwestern Utah—Geochemical and geochronological evidence, *in* Blackett, R.E., and Moore, J.N., eds., Cenozoic geology and geothermal systems of southwestern Utah, 1994 Field Symposium: Utah Geological Association Publication 23, p. 117–124.

Nealey, L.D., and Sheridan, M.F., 1989, Post-Laramide volcanic rocks of Arizona and northern Sonora, Mexico, and their inclusions, *in* Jenney, J.P., and Reynolds, S.J., eds., Geologic evolution of Arizona: Arizona Geological Society Digest, v. 17, p. 609–647.

Nelson, B.K., and DePaolo, D.J., 1984, 1700 Ma greenstone volcanic successions in southwestern North America and isotopic evolution of Proterozoic mantle: Nature, v. 312, p, 143–146.

———1985, Rapid production of continental crust 1.7 to 1.9 b.y. ago: Nd isotopic evidence from the basement of the North American mid continent: Geological Society of America Bulletin, v. 96, p. 746–754.

Nelson, S.E., 1989, Geology, alteration, and mineral deposits of the Cresson diatreme, Cripple Creek district, Colorado: Fort Collins, Colo., Colorado State University M.S. thesis, 147 p.

———1990, Geology, alteration and Au-Ag mineralization of the Cresson diatreme, Cripple Creek district, Colorado: Geological Society of America Abstracts with Programs, v. 22, no. 6, p. 40–41.

Nelson, S.T., 1991, Mid-Tertiary magmatism of the Colorado Plateau—The Henry and La Sal Mountains, Utah: Los Angeles, Calif., University of California Ph. D. thesis, 318 p.

Nelson, S.T., and Davidson, J.P., 1993, Interactions between mantle-derived magmas and mafic crust, Henry Mountains, Utah: Journal of Geophysical Research, v. 98, p. 1837–1852.

Nelson, S.T., Davidson, J.P., and Sullivan, K.R., 1992, New age determinations of central Colorado Plateau laccoliths, Utah—Recognizing disturbed K-Ar systematics and re-evaluating tectonomagmatic relationships: Geological Society of America Bulletin, v. 104, p. 1547–1560.

Nelson, S.T., Heizler, M.T., and Davidson, J.P., 1992a, New $^{40}Ar/^{39}Ar$ ages of intrusive rocks from the Henry and La Sal Mountains, Utah: Utah Geological Survey Miscellaneous Publication 92–2, 24 p.

Nelson, S.T., Sullivan, K.R., and Davidson, J.P., 1992b, New age determinations of central Colorado Plateau laccoliths, Utah—Recognizing disturbed K-Ar systematics and re-evaluating tectonomagmatic relationships: Geological Society of America Bulletin, v. 104, p. 1547–1560.

New Mexico Geological Society, 1982, New Mexico highway geologic map: Socorro, N. Mex., New Mexico Bureau of Mines and Mineral Resources, scale 1:1,000,000.

Nickelsen, R.P., 1975, "New basement tectonics" evaluated in Salt Lake City: Geotimes, v. 20, no. 10, p. 16–17.

Nickelsen, R.P., and Merle, O., 1991, Structural evolution at the tip line of a mid-Tertiary compressional event in southwestern Utah: Geological Society of America Abstracts with Programs, v. 23, no. 1, p. 109.

Nielson, D.L., Evans, S.H., Jr., and Sibbett, B.S., 1986, Magmatic, structural, and hydrothermal evolution of the Mineral Mountains intrusive complex, Utah: Geological Society of America Bulletin, v. 97, p. 765–777.

Noble, J.A., 1948, High-potash dikes in the Homestake mine, Lead, South Dakota: Geological Society of America Bulletin, v. 59, p. 927–940.

Norman, M.D., and Mertzman, S.A., 1991, Petrogenesis of Challis volcanics from central and southwestern Idaho—Trace element and Pb isotopic evidence: Journal of Geophysical Research, v. 96, p. 13279–13293.

North, R.M., and McLemore, V.T., 1988, A classification of the precious metal deposits of New Mexico, *in* Schafer, R.W., Cooper, J.J., and Vikre, P.G., eds., Bulk mineable precious metal deposits of the western United States: Reno, Nev., Nevada Geological Society, p. 625–659.

Norton, J.J., 1989, Bald Mountain gold mining region, northern Black Hills, South Dakota: U.S. Geological Survey Bulletin 1857–C, p. C1–C13.

Nutt, C.J., and Thorman, C.H., 1992, Pre-late Eocene structures and their control of gold ore at the Drum Mine, west-central Utah, *in* Thorman, C.H., ed., Application of structural geology to mineral and energy resources of the Central and Western United States: U.S. Geological Survey Bulletin 2012, p. F1–F7.

———1993, Eocene sedimentary and volcanic rocks and their use in dating Mesozoic and Tertiary structures in the southern Deep Creek Range, Nevada and Utah: Geological Society of America Abstracts with Programs, v. 25, no. 5, p. 128.

———1994, Eocene or older low-angle attenuation faults in the eastern Great Basin—Features of regional extension or contraction?: Geological Society of America Abstracts with Programs, v. 26, no. 2, p. 78.

Nutt, C.J., Thorman, C.H., and Brooks, W.E., 1994, Geologic setting of the southern part of the Goshute Indian Reservation, Nevada and Utah, *in* Thorman, C.H., Nutt, C.J., and Potter, C.J., eds., Dating of pre-Tertiary attenuation structures in upper Paleozoic and Mesozoic rocks and the Eocene history in northeast Nevada and northwest Utah: Reno, Nev., Nevada Petroleum Society, Sixth Annual Field Trip, Guidebook, p. 47–57.

Nutt, C.J., Thorman, C.H., Snee, L.W., and Zimmermann, R.A., 1992, Mesozoic to early Tertiary low-angle younger-over-older faults in the Drum Mountains, west-central Utah: Geological Society of America Abstracts with Programs, v. 24, no. 6, p. 56.

Ogilvie, I.H., 1908, Some igneous rocks from the Ortiz Mountains, New Mexico: Journal of Geology, v. 16, p. 230–238.

O'Hara, M.H., 1968, The bearing of phase equilibria studies in synthetic and natural systems on the origin and evolution of basic and ultrabasic rocks: Earth Science Reviews, v. 4, p. 69–133.

Oldow, J.S., Bally, A.W., AvéLallemant, H.G., and Leeman, W.P., 1989, Phanerozoic evolution of the North American Cordillera; United States and Canada, *in* Bally, A.W., and Palmer, A.R., eds., The geology of North America—An overview: Boulder, Colo., Geological Society of America, The Geology of North America, v. A, p. 139–232.

Olmore, S.D., 1991, Geologic summary of the Big Ben molybdenum deposit near Neihart, Montana, *in* Baker, D.W., and Berg, R.B., eds., Guidebook of the central Montana alkalic province: Montana Bureau of Mines and Geology Special Publication 100, p. 124–127.

Olsen, K.H., Baldridge, W.S., and Callender, J.F., 1987, Rio Grande rift—An overview: Tectonophysics, v. 143, p. 119–139.

O'Neill, J.M., and Lopez, D.A., 1985, Character and significance of Great Falls tectonic zone, east-central Idaho and west-central Montana: American Association of Petroleum Geologists Bulletin, v. 69, p. 437–447.

Ormerod, D.S., Hawkesworth, C.J., Rogers, N.W., Leeman, W.P., and Menzies, M.A., 1988, Tectonic and magmatic transitions in the western Great Basin, U.S.A.: Nature, v. 333, p. 349–353.

O'Sullivan, R.B., 1991, Correlation of Middle Jurassic San Rafael Group and related rocks from Kane Springs, Utah to Uravan, Colorado: U.S. Geological Survey Oil and Gas Investigations Chart OC–134.

O'Sullivan, R.B., and Beikman, H.M., 1963, Geology, structure, and uranium deposits of the Shiprock quadrangle, New Mexico and Arizona: U.S. Geological Survey Miscellaneous Geologic Investigations Map I–345, scale 1:250,000.

Otton, J.K., 1995, Western frontal fault of the Canyon Range—Is it the breakaway zone of the Sevier Desert detachment?: Geology, v. 23, p. 547–550.

Overtoom, G.J., and Bartley, J.M., 1994, Evidence from the Golden Gate Range for Oligocene synvolcanic stress reorientation along the Blue Ribbon lineament in east-central Nevada: Geological Society of America Abstracts with Programs, v. 26, no. 2, p. 79.

Overtoom, G.J., Bartley, J.M., and Geissman, J.W., 1993, Structural and paleomagnetic constraints on the origin and nature of W–NW striking structures in the Blue Ribbon lineament—Evidence from the northern Golden Gate Range, Nevada [abs.]: Eos, v. 74, no. 43, p. 207.

Oviatt, C.G., 1988, Evidence for Quaternary deformation in the Salt Valley anticline, southeastern Utah: Utah Geological Survey Bulletin 122, p. 61–76.

Page, W.R., 1995, Low-angle normal faults in Devonian rocks of the southern Delamar Mountains, Lincoln County, Nevada, *in* Scott, R.B., and Swadley, W C, eds., Geologic studies in the Basin and Range–Colorado Plateau transition in southeastern Nevada, southwestern Utah, and northwestern Arizona, 1992: U.S. Geological Survey Bulletin 2056–G, p. 203–218.

Page, W.R., and Scott, R.B., 1991, Ramping extensional faults in the Devonian Guilmette Formation, southern Nevada: Geological Society of America Abstracts with Programs, v. 23, no. 4, p. 55.

Pakiser, L.C., and Mooney, W.D., eds., 1989 Geophysical framework of the continental United States: Geological Society of America Memoir 172, 826 p.

Pang, Ming, and Nummedal, Dag, 1995, Flexural subsidence and basement tectonics of the Cretaceous Western Interior basin, United States: Geology, v. 23, no. 1, p. 173–176.

Parker, J.M., 1981, Lisbon Field area, San Juan County, Utah, *in* Wiegand, D.L., ed., Geology of the Paradox basin: Rocky Mountain Association of Petroleum Geologists, p. 89–101.

Parsons, Tom, Howie, J.M., and Thompson, G.A., 1992, Seismic constraints on the nature of lower crustal reflectors beneath the extending southern transition zone of the Colorado Plateau, Arizona: Journal of Geophysical Research, v. 97, p. 12391–12407.

Parsons, Tom, and McCarthy, Jill, 1995, The active southwest margin of the Colorado Plateau—Uplift of mantle origin: Geological Society of America Bulletin, v. 107, no. 2, p. 139–147.

Parsons, Tom, Sleep, N.H., and Thompson, G.A., 1992, Host rock rheology controls on the emplacement of tabular intrusions—Implications for underplating of extending crust: Tectonics, v. 11, p. 1348–1356.

Parsons, Tom, and Thompson, G.A., 1991, The role of magma overpressure in suppressing earthquakes and topography—Worldwide examples: Science, v. 253, p. 1399–1402.

Parsons, Tom, Thompson, G.A., and Sleep, N.H., 1994, Mantle plume influence on the Neogene uplift and extension of the U.S. western Cordillera?: Geology, v. 22, p. 83–85. —Much of western United States of America is more than 1 km above MSL—lithospheric mantle and (or) underlying asthenosphere must vary in thickness or density across the region. A major hot spot broke out during complex series of Cenozoic events that included lithospheric thickening, backarc extension, and transition for subduction to transform plate boundary. All this is attributable to Yellowstone plume.

Paterson, C.J., 1990, Magmatic-hydrothermal model for epithermal-mesothermal Au-Ag deposits in the northern Black Hills: Proceedings of the Fourth Western Regional Conference on Precious Metals and the Environment, Lead, South Dakota; Society for Mining, Metallurgy, and Exploration (SME–AIME), p. 89–102.

Paterson, C.J., Lisenbee, A.L., and Redden, J.A., 1988, Gold deposits in the Black Hills, South Dakota, *in* Wyoming Geological Association 1988 Field Conference Guidebook—Eastern Powder River basin–Black Hills, p. 295–304.

Paterson, C.J., Uzunlar, Nuri, Groff, J., and Longstaffe, F.J., 1989, A view through an epithermal-mesothermal precious metal system in the northern Black Hills, South Dakota—A magmatic origin for the ore-forming fluids, *in* Keays, R.R., Ramsay, W.R.H., and Groves, D.I., eds., The geology of gold deposits—The perspective in 1988: Economic Geology Monograph 6, p. 564–570.

Peale, A.C., 1877, Geological report on the Grand River district: U.S. Geological and Geographical Survey Territorial 9th Annual Report.

———1878, Geological report on the Grand River district: U.S. Geological and Geographical Survey Territorial 10th Annual Report, p. 163–182.

Perry, F.V., Baldridge, W.S., and DePaolo, D.J., 1987, Role of asthenosphere and lithosphere in the genesis of late Cenozoic basaltic rocks from the Rio Grande rift and adjacent regions of the southwestern United States: Journal of Geophysical Research, v. 92, p. 9193–9213.

———1988, Chemical and isotopic evidence for lithospheric thinning beneath the Rio Grande rift: Nature, v. 332, p. 432–434.

———1993, Neodymium isotopic evidence for decreasing crustal contributions to Cenozoic ignimbrites of the western United States—Implications for the thermal evolution of the Cordilleran crust: Geological Society of America Bulletin, v. 105, p. 872–882.

Perry, W.J., Jr., Grout, M.A., Tang, R.L., and Hainsworth, T.J., 1988, Wedge model for late Laramide basement-involved thrusting, Grand Hogback monocline and White River uplift, western Colorado: Geological Society of America Abstracts with Programs, v. 20, no. 7, p. A384–A385.

Peterson, J.A., 1989, Geology and petroleum resources, Paradox basin province: U.S. Geological Survey Open-File Report 88–450–U, 69 p.

Petit, J.P., 1987, Criteria for the sense of movement on fault surfaces in brittle rocks: Journal of Structural Geology, v. 9, p. 597–608.

Phair, George, 1952, Radioactive tertiary porphyries in the Central City district, Colorado, and their bearing upon pitchblende deposition: U.S. Geological Survey TEI–247, 53 p.

Phair, George, and Jenkins, L.B., 1975, Tabulation of uranium and thorium data on the Mesozoic-Cenozoic intrusive rocks of known chemical composition in Colorado: U.S. Geological Survey Open-File Report 75–501, 57 p.

Pierce, K.L., Adams, K.D., and Sturchio, N.C., 1991, Geologic setting of the Corwin Springs Known Geothermal Resources Area–Mammoth Hot Springs area in and adjacent to Yellowstone National Park, *in* Sorey, M.L., ed., Effects of potential geothermal development in the Corwin Springs Known Geothermal Resources Area, Montana, on the thermal features of Yellowstone National Park: U.S. Geological Survey Water-Resources Investigations Report 91–4052, p. C1–C37.

Pierce, K.L., Milbert, D.G., and Saltus, R.W., 1992: Geoid dome culminates on Yellowstone—Yellowstone hotspot fed by a thermal mantle plume? [abs.]: Eos, v. 73, no. 14, p. 284.

Pierce, K.L., and Morgan, L.A., 1992, The track of the Yellowstone hot spot—Volcanism, faulting, and uplift, *in* Link, P.K., Kuntz, M.A., and Platt, L.B., eds., Regional geology of eastern Idaho and western Wyoming: Geological Society of America Memoir 179, p. 1–53.

Pirsson, L.V., 1894, On some phonolite rocks from the Black Hills: American Journal of Science, 3rd series, v. 47, p. 341–346.

———1900, Petrography of igneous rocks of the Little Belt Mountains, Montana: U.S Geological Survey 20th Annual Report, part 3, p. 463–581.

Plouff, Donald, and Pakiser, L.C., 1972, Gravity study of the San Juan Mountains, Colorado: U.S. Geological Survey Professional Paper 800–B, p. B183–B190.

Pollard, D.D., and Johnson, A.M., 1973, Mechanics of growth of some laccolithic intrusions in the Henry Mountains, Utah—II, Bending and failure of the overburden and sill formation: Tectonophysics, v. 18, p. 311–354.

Pontius, J.A., 1991, Geology and gold mineralization of the Cresson deposit, Cripple Creek, Colorado: Northwest Mining Association 97th Annual Convention, Dec. 1–6, 1991, Spokane, Washington, Abstract Booklet, p. 26.

Porter, E.W., and Ripley, Edward, 1985, Petrologic and stable isotope study of the gold-bearing breccia pipe at the Golden Sunlight deposit, Montana: Economic Geology, v. 80, p. 1689–1706.

Potter, C.J., Tang, Rex, and Hainsworth, T.J., 1991, Late Paleozoic structure of the southern part of the Uinta basin, Utah, from seismic reflection data: U.S. Geological Survey Bulletin 1787–V, 10 p.

Price, D.E., and Bartley, J.M., 1992, Three-dimensional extensional structure of the southern Mineral Mountains, southwestern Utah: Geological Society of America Abstracts with Programs, v. 24, no. 6, p. 58.

Price, J.G., and Henry, C.D., 1984, Stress orientations during Oligocene volcanism in Trans-Pecos, Texas—Timing the transition from Laramide compression to Basin and Range tension: Geology, v. 12, p. 238–241.

Price, N.J., 1966, Fault and joint development in brittle and semi-brittle rock: London, Pergamon Press, 176 p.

———1974, The development of stress systems and fracture patterns in undeformed sediments, *in* Advances in Rock Mechanics—Proceedings of the Third Congress of the International Society of Rock Mechanics, Denver, Colo., Sept. 1–7, 1974: Washington, National Academy of Sciences, v. 1, pt. A, p. 487–496.

Priesmeyer, S.T., 1986, Geology and gold mineralization on the New Year mine and North Guiltedge properties, Fergus County, Montana: Moscow, Idaho, University of Idaho M.S. thesis, 84 p.

Prodehl, Claus, and Lipman, P.W., 1989, Crustal structure of the Rocky Mountain region, *in* Pakiser, L.C., and Mooney, W.D., eds., Geophysical framework of the continental United States: Geological Society of America Memoir 172, p. 249–284.

Proffett, J.M., Jr., 1977, Cenozoic geology of the Yerington district, Nevada, and implications for the nature and origin of Basin and Range faulting: Geological Society of America Bulletin, v. 88, p. 247–266.

Reches, Ze'ev, 1978, Development of monoclines— Part 1, Structure of the Palisades Creek branch of the East Kaibab monocline, Grand Canyon, Arizona, *in* Matthews, Vincent, III, ed., Laramide folding associated with basement block faulting in the western United States: Geological Society of America Memoir 151, p. 235–271.

Reed, J.C., Jr., 1993, compiler, Precambrian rocks of the conterminous United States, scale 1:5,000,000, plate 1 *in* Reed, J.C., Jr., Bickford, M.E., Houston, R.S., Link, P.K., Rankin, D.W., Sims, P.K., and Van Schmus, W.R., 1993, eds., The Geology of North America, vol. C–2, Precambrian; Conterminous U.S.: Boulder, Colo., Geological Society of America, 657 p.

Ren, X., Kowallis, B.J., and Best, M.G., 1987, Stress rotation in the eastern Basin and Range, Utah and Nevada from healed microcrack orientation data: Geological Society of America Abstracts with Programs, v. 19, no. 7, p. 816.

Reynolds, S.J., 1988, Geologic map of Arizona: Arizona Geological Survey Map 26, scale 1:1,000,000.

Rice, C.M., Harmon, R.S., and Shepherd, T.J., 1985, Central City, Colorado—The upper part of an alkaline porphyry molybdenum system: Economic Geology, v. 80, p. 1769–1796.

Rice, C.M., Lux, D.R., and Macintyre, R.M., 1982, Timing of mineralization and related intrusive activity near Central City, Colorado: Economic Geology, v. 77, p. 1655–1666.

Richards, M.A., Duncan, R.A., and Courtillot, V.E., 1989, Flood basalts and hot spot tracks; Plume heads and tails: Science, v. 246, p. 103–107.

Rickard, M.J., Harrington, H.J., and Williams, P.R., eds., 1992, Basement tectonics 9—Australia and other regions: Ninth International Conference on Basement Tectonics, Proceedings, Canberra, Australia, 272 p.

Roberts, R.J., 1964, Economic geology, *in* Mineral and water resources of Nevada: U.S. 88th Congress, 2d sess., Senate Doc. 87, p. 39–48.

———1966, Metallogenic provinces and mineral belts in Nevada, *in* AIME Pacific southwest mineral industry conf., Sparks, Nev., 1965, Papers, part A: Nevada Bureau of Mines Report 13, pt. A, p. 47–72.

Robertson, J.M., and Condie, K.C., 1989, Geology and geochemistry of early Proterozoic volcanic and subvolcanic rocks of the Pecos greenstone belt, Sangre de Cristo Mountains, New Mexico, *in* Grambling, J.A., and Tewksbury, B.J., eds., Proterozoic geology of the Southern Rocky Mountains: Geological Society of America Special Paper 235, p. 119–146.

Robinson, P.T., McKee, E.H., and Moiola, R.J., 1968, Cenozoic volcanism and sedimentation, Silver Peak region, western Nevada and adjacent California, *in* Coats, R.R., Hay, R.L., and Anderson, C.A., Studies in volcanology—A memoir in honor of Howel Williams: Geological Society of America Memoir 116, p. 577–611.

Rock, N.M.S., Groves, D.I., Perring, C.S., and Golding, S.D., 1989, Gold, lamprophyres, and porphyries—What does their association mean?, *in* Keays, R.R., Ramsay, W.R.H., and Groves, D.I., eds., The geology of gold deposits—The perspective in 1988: Economic Geology Monograph 6, p. 609–625.

Roemmel, J.S., 1982, A petrographic and economic evaluation of the White Cow intrusion, Little Rocky Mountains, Montana: Missoula, Mont., University of Montana M.S. thesis, 51 p.

Rogers, R.D., and Bird, D.K., 1987, Fracture propagation associated with dike emplacement at the Skaergaard intrusion, East Greenland: Journal of Structural Geology, v. 9, no. 1, p. 71–86.

Roller, J.A., 1987, Fracture history of the Redwall Limestone and lower Supai Group, western Hualapai Indian Reservation, northwestern Arizona: U.S. Geological Survey Open-File Report 87–359, 33 p.

———1989, Fracture history of the Redwall Limestone, lower Supai Group, and Tertiary units on the Hualapai Indian Reservation, northwestern Arizona—Additional data to Open-File Report 87–359: U.S. Geological Survey Open-File Report 89–463, 42 p.

Ronkos, C.J., 1991, Geology, alteration and gold mineralization in the White Oaks mining district, Lincoln County, New Mexico: Geological Society of America Abstracts with Programs, v. 23, no. 4, p. 88.

Rosendahl, B.R., 1987, Architecture of continental rifts with special reference to East Africa: Annual Reviews of Earth and Planetary Science, v. 15, p. 445–503.

Ross, M.L., 1992, Geology and petrology of the Oligocene North Mountain intrusive center, La Sal Mountains, Utah: Geological Society of America Abstracts with Programs, v. 24, no. 6, p. 59.

Ross, H.P., and Moore, J.N., 1994, Geophysical investigations of the Cove Fort–Sulphurdale geothermal system, Utah, *in* Blackett, R.E., and Moore, J.N., eds., Cenozoic geology and geothermal systems of southwestern Utah, 1994 field symposium: Utah Geological Association Publication 23, p. 45–59.

Rowan, L.C., and Wetlaufer, P.H., 1981, Relation between regional lineament systems and structural zones in Nevada: American Association of Petroleum Geologists Bulletin, v. 65, p. 1414–1432.

Rowley, P.D., 1968, Geology of the southern Sevier Plateau, Utah: Austin, Texas, University of Texas Ph. D. dissertation, 340 p.

Rowley, P.D., and Barker, D.S., 1978, Geology of the Iron Springs mining district, Utah, *in* Shawe, D.R., and Rowley, P.D., eds., Field excursion C–2, guidebook to mineral deposits of southwestern Utah: Utah Geological Association Publication 7, p. 49–58.

Rowley, P.D., Cunningham, C.G., Steven, T.A., Mehnert, H.H., and Naeser, C.W., 1988a, Geologic map of the Marysvale quadrangle, Piute County, Utah: Utah Geological and Mineral Survey Map 105, scale 1:24,000.

Rowley, P.D., Cunningham, C.G., Steven, T.A., Mehnert, H.H., and Naeser, C.W., 1988b, Geologic map of the Antelope Range quadrangle, Sevier and Piute Counties, Utah: Utah Geological and Mineral Survey Map 106, scale 1:24,000.

Rowley, P.D., Kellogg, K.S., Vennum, W.R., Laudon, T.S., Thomson, J.W., O'Neill, J.M., and Lidke, D.J., 1991, Tectonic setting of the English Coast, eastern Ellsworth Land, Antarctica, *in* Thomson, M.R.A., Crame, J.A., and Thomson, J.W., eds., Geological evolution of Antarctica: Cambridge, Cambridge University Press, p. 467–473.

Rowley, P.D., Kellogg, K.S., Williams, P.L., Willan, C.F.H., and Thomson, J.W., 1992, Geological map, Sheet 6, Southern Palmer Land and eastern Ellsworth Land: Cambridge, British Antarctic Survey BAS 500G series, scale 1:500,000.

Rowley, P.D., Lipman, P.W., Mehnert, H.H., Lindsey, D.A., and Anderson, J.J., 1978, Blue Ribbon lineament, an east-trending structural zone within the Pioche mineral belt of southwestern Utah and eastern Nevada: U.S. Geological Survey Journal of Research, v. 6, no. 2, p. 175–192.

Rowley, P.D., McKee, E.H., and Blank, H.R., Jr., 1989, Miocene gravity slides resulting from emplacement of the Iron Mountain pluton, southern Iron Springs mining district, Iron County, Utah [abs.]: Eos, v. 70, no. 43, p. 1309.

Rowley, P.D., Mehnert, H.H., Naeser, C.W., Snee, L.W., Cunningham, C.G., Steven, T.A., Anderson, J.J., Sable, E.G., and Anderson, R.E., 1994, Isotopic ages and stratigraphy of Cenozoic rocks of the Marysvale volcanic field and adjacent areas, west-central Utah: U.S. Geological Survey Bulletin 2071, 35 p.

Rowley, P.D., Nealey, L.D., Unruh, D.M., Snee, L.W., Mehnert, H.H., Anderson, R.E., and Gromme, C.S., 1995, Stratigraphy of Miocene ash-flow tuffs in and near the Caliente caldera complex, southeastern Nevada and southwestern Utah, *in* Scott, R.B., and Swadley, W C, eds., Geologic studies in the Basin and Range–Colorado Plateau transition in southeastern Nevada, southwestern Utah, and northwestern Arizona, 1992: U.S. Geological Survey Bulletin 2056–B, p. 42–88.

Rowley, P.D., Shroba, R.R., Simonds, F.W., Burke, K.J., Axen, G.J., and Olmore, S.D., 1994, Geologic map of the Chief Mountain quadrangle, Lincoln County, Nevada: U.S. Geological Survey Geologic Quadrangle Map GQ–1731, scale 1:24,000.

Rowley, P.D., Snee, L.W., Mehnert, H.H., Anderson, R.E., Axen, G.J., Burke, K.J., Simonds, F.W., Shroba, R.R., and Olmore, S.D., 1992, Structural setting of the Chief mining district, eastern Chief Range, Lincoln County, Nevada, *in* Thorman, C.H., ed., Application of structural geology to mineral and energy resources of the central and western United States: U.S. Geological Survey Bulletin 2012, p. H1–H17.

Rowley, P.D., Steven, T.A., Anderson, J.J., and Cunningham, C.G., 1979, Cenozoic stratigraphic and structural framework of southwestern Utah: U.S. Geological Survey Professional Paper 1149, 22 p.

Rowley, P.D., Steven, T.A., and Mehnert, H.H., 1981, Origin and structural implications of upper Miocene rhyolites in Kingston Canyon, Piute County, Utah: Geological Society of America Bulletin, pt. I, v. 92, p. 590–602.

Rowley, P.D., Williams, P.L., and Kaplan, A.M., 1986a, Geologic map of the Greenwich quadrangle, Piute County, Utah: U.S. Geological Survey Geologic Quadrangle Map GQ–1589, scale 1:24,000.

_____1986b, Geologic map of the Koosharem quadrangle, Sevier and Piute Counties, Utah: U.S. Geological Survey Geologic Quadrangle Map GQ–1590, scale 1:24,000.

Rupp, J.A., 1980, Tertiary rhyolite dikes and plutons of the northern Little Belt Mountains, Montana: Cheney, Wash., Eastern Washington University M.S. thesis, 136 p.

Rumsey, I.A.P., 1971, Relationship of fractures in unconsolidated superficial deposits to those in the underlying bedrock: Modern Geology, v. 3, p. 25–41.

Rushmer, Tracy, 1991, Partial melting of two amphibolites—Contrasting experimental results under fluid absent conditions: Contributions to Mineralogy and Petrology, v. 107, p. 41–59.

Russell, C.W., 1991, Gold mineralization in the Little Rocky Mountains, Montana, *in* Baker, D.W., and Berg, R.B., eds., Guidebook of the central Montana alkalic province: Montana Bureau of Mines and Geology Special Publication 100, p. 1–18.

Russell, C.W., and Gabelman, Joan, 1991, Field guide, Little Rocky Mountains, *in* Baker, D.W., and Berg, R.B., eds., Guidebook of the central Montana alkalic province: Montana Bureau of Mines and Geology Special Publication 100, p. 185–194.

Ryzak, David, 1990, Gold deposits of active mining areas, Little Rocky Mountains, Montana: Proceedings of the Fourth Western Regional Conference on Precious Metals and the Environment, Lead, South Dakota; Society for Mining, Metallurgy, and Exploration (SME–AIME), p. 61–75.

Sable, E.G., and Anderson, J.J., 1985, Tertiary tectonic slide megabreccia, Markagunt Plateau, southwestern Utah: Geological Society of America Abstracts with Programs, v. 17, no. 4, p. 263.

Saltus, R.W., 1988, Regional, residual, and derivative gravity maps of Nevada: Nevada Bureau of Mines and Geology Map 94B, scales 1:1,000,000 and 1:2,500,000 (4 sheets).

Saltus, R.W., and Jachens, R.C., 1995, Gravity and basin-depth maps of the Basin and Range province, Western United States: U.S. Geological Survey Geophysical Investigations Map GP–1012, scale 1:2,500,000.

Sauck, W.A., and Sumner, J.S., 1970a, Residual aeromagnetic map of Arizona: Tucson, Ariz., University of Arizona, scale 1:1,000,000.

———1970b, Total intensity aeromagnetic map of New Mexico: Tucson, Ariz., University of Arizona, scale 1:1,000,000.

Saunders, J.A., 1988, Textural and geochemical characteristics of gold mineralization from the Cresson mine, Cripple Creek district, Colorado, U.S.A.: Institution of Mining and Metallurgy Transactions, section B, v. 97, p. B36–B39.

———1991, Gold deposits of the Boulder County gold district, Colorado: U.S. Geological Survey Bulletin 1857–I, p. I37–I48.

Saunders, J.A., and May, E.R., 1986, Bessie G—A high-grade epithermal gold telluride deposit, La Plata County, Colorado, U.S.A., *in* Macdonald, A.J., ed., Proceedings of Gold '86, an International Symposium on the Geology of Gold: Toronto, Ont., p. 431–444.

Sawyer, D.A., Fleck, R.J., Lanphere, M.A., Warren, R.G., Broxton, D.E., and Hudson, M.R., 1994, Episodic caldera volcanism in the Miocene southwestern Nevada volcanic field—Revised stratigraphic framework, 40Ar/39Ar geochronology, and implications for magmatism and extension: Geological Society of America Bulletin, v. 106, p. 1304–1318.

Sbar, M.L., and Sykes, L.R., 1973, Contemporary compressive stress and seismicity in eastern North America—An example of intraplate tectonics: Geological Society of America Bulletin 84, p. 1861–1882.

Schmidt, D.L., and Rowley, P.D., 1986, Continental rifting and transform faulting along the Jurassic Transantarctic Rift, Antarctica: Tectonics, v. 5, p. 279–291.

Scholz, C.H., Barazangi, Muawia, and Sbar, M.L., 1971, Late Cenozoic evolution of the Great Basin, western United States, an ensialic interarc basin: Geological Society of America Bulletin, v. 82, p. 2979–2990.

Schroeter, T.G., Lund, C., and Carter, G., 1989, Gold production and reserves in British Columbia: British Columbia Ministry of Energy, Mines and Petroleum Resources Open File 1989–22, 86 p.

Schultz-Ela, D.D., Jackson, M.P.A., Hudec, M.R., Fletcher, R.C., Porter, M.L., and Watson, I.A., 1994, Constraints on the origin of Upheaval Dome, Paradox Basin, Utah: American Association of Petroleum Geologists 1994 Annual Convention, Denver, Colorado, June 12–15, 1994, Abstracts with Programs, p. 253.

Schutz, J.L., Fulp, M.S., Suchomel, B.J., and Woodward, L.A., 1988, Geology and precious-metal mineralization at the Lenny mine, Little Belt Mountains, central Montana, *in* Schafer, R.W., Cooper, J.J., and Vikre, P.G., eds., Bulk mineable precious metal deposits of the western United States: Reno, Nev., Geological Society of Nevada, p. 747.

Scott, R.B., Gromme, C.S., Best, M.G., Rosenbaum, J.G., and Hudson, M.R., 1995, Stratigraphic relationships of Tertiary volcanic rocks in central Lincoln County, southeastern Nevada, chapter A *in* Scott, R.B., and Swadley, W C, eds., Geologic studies in the Basin and Range–Colorado Plateau transition in southeastern Nevada, southwestern Utah, and northwestern Arizona, 1992: U.S. Geological Survey Bulletin 2056, p. 5–41.

Scott, R.B., Unruh, D.M., Snee, L.W., Harding, A.E., Nealey, L.D., Blank, H.R., Jr., Budahn, J.R., and Mehnert, H.H., 1995, Relation of peralkaline magmatism to heterogeneous extension during the middle Miocene, southeastern Nevada: Journal of Geophysical Research, v. 100, no. B6, p. 10381–10401.

Sears, J.W., 1973, Structural geology of the Precambrian Grand Canyon Series, Arizona: Laramie, Wyo., University of Wyoming M.S. thesis, 100 p.

Seccombe, J.C., and Decker, A.D., 1986, Geologic and reservoir characteristics of the Red Mountain coalbed methane test site in the Piceance basin: Chicago, Ill., Gas Research Institute Topical Report GRI–86/0109, 89 p.

Seedorff, Eric, 1991, Magmatism, extension, and ore deposits of Eocene to Holocene age in the Great Basin—Mutual effects and preliminary proposed genetic relationships, *in* Raines, G.L., Lisle, R.E., Schafer, R.W., and Wilkinson, W.H., eds., Geology and ore deposits of the Great Basin: Reno, Nev., Geological Society of Nevada, p. 133–177.

Segerstrom, K., and Ryberg, G.E., 1974, Geology and placer gold deposits of the Jicarilla Mountains, Lincoln County, New Mexico: U.S. Geological Survey Bulletin 1308, 25 p.

Segerstrom, K., Stotelmeyer, R.B., Williams, F.E., and Cordell, Lindrith, 1979, Mineral resources of the White Mountain Wilderness and adjacent areas, Lincoln County, New Mexico: U.S. Geological Survey Bulletin 1453, 135 p.

Severinghaus, Jeff, and Atwater, Tanya, 1990, Cenozoic geometry and thermal state of the subducting slabs beneath western North America, *in* Wernicke, B.P., ed., Basin and Range extensional tectonics near the latitude of Las Vegas, Nevada: Geological Society of America Memoir 176, p. 1–22.

Shand, H.S., 1951, Eruptive rocks, 4th edition: New York, John Wiley, 488 p.

Shannon, J.R., Epis, R.C., Naeser, C.W., and Obradovich, J.D., 1987, Correlation of intracauldron and outflow tuffs and an intrusive tuff dike related to the Oligocene Mount Aetna cauldron, central Colorado, *in* Drexler, J.W., and Larson, E.E., eds., Cenozoic volcanism in the Southern Rocky Mountains revisited: Colorado School of Mines Quarterly, v. 82, no. 4, p. 65–80.

Sharp, W.N., 1978, Geologic map of the Silver Cliff and Rosita volcanic centers, Custer County, Colorado: U.S. Geological Survey Miscellaneous Investigations Series Map I–1081, scale 1:24,000.

Sharp, W.N., and Naeser, C.W., 1986, Apatite fission-track age for the Bull Domingo boulder pipe, Custer County, Colorado: U.S. Geological Survey Bulletin 1622, p. 77–80.

Sharwood, W.J., 1911, Analyses of some rocks and minerals from the Homestake mine, Lead, South Dakota: Economic Geology, v. 6, p. 729–789.

Shawe, D.R., Simmons, G.C., and Archbold, N.L., 1968, Stratigraphy of Slick Rock district and vicinity, San Miguel and Dolores Counties, Colorado: U.S. Geological Survey Professional Paper 576–A, p. A1–A108.

Shawe, D.R., and Stewart, J.H., 1976, Ore deposits as related to tectonics and magmatism, Nevada and Utah: Transactions of American Institute of Mining, Metallic, and Petroleum Engineers, v. 260, p. 225–232.

Shearer, C.K., 1990, Geochemsitry of Laramide igneous intrusions, northern Black Hills, with implications for petrogenetic relationships and magma sources, *in* Peterson, C.J., and Lisenbee, A.L., eds., Metallogeny of gold in the Black Hills, South Dakota: Society of Economic Geologists Guidebook Series, v. 7, p. 35–42.

Shoemaker, E.M., 1956, Structural features of the central Colorado Plateau and their relation to uranium deposits: U.S. Geological Survey Professional Paper 300, p. 155–170.

Shoemaker, E.M., Case, J.E., and Elston, D.P., 1958, Salt anticlines of the Paradox basin, *in* Sanborn, A.F., ed., Guidebook to the geology of the Paradox basin: Intermountain Association of Petroleum Geologists, p. 39–59.

Shoemaker, E.M., and Henehoff, K.E., 1984, Upheaval Dome impact structure, Utah [abs.]: Lunar and Planetary Science Conference, v. 15, pt. 2, p. 778–779.

Shoemaker, E.M., Squires, R.L., and Abrams, M.J., 1978, Bright Angel and Mesa Butte fault systems of northern Arizona, *in* Smith, R.B., and Eaton, G.P., eds., 1978, Cenozoic tectonics and regional geophysics of the western Cordillera: Geological Society of America Memoir 152, p. 341–367.

Shuey, R.T., Schellinger, D.K., Johnson, E.H., and Alley, L.B., 1973, Aeromagnetics and the transition between the Colorado Plateau and Basin Range provinces: Geology, v. 1, no. 11, p. 107–110.

Sibson, R.H., 1989, Earthquake faulting as a structural process: Journal of Structural Geology, v. 11, p. 1–14.

Siders, M.A, and Shubat, M.A., 1986, Stratigraphy and structure of the northern Bull Valley Mountains and Antelope Range, Iron County, Utah, *in* Griffen, D.T., and Phillips, W.R., eds., Thrusting and extensional structures and mineralization in the Beaver Dam Mountains, southwestern Utah: Utah Geological Association Publication 15, p. 87–102.

Silberman, M.L., 1975, Ages and tectonic implications of the transition of calc-alkaline andesitic to basaltic volcanism in the western Great Basin and the Sierra Nevada: Geological Society of America Abstracts with Programs, v. 7, no. 3, p. 375.

Simons, F.S., Armbrustmacher, T.J., van Noy, R.M., Zilka, N.T., Federspiel, F.E., and Ridenour, James, 1979, Mineral resources of the Beartooth Primitive area and vicinity, Carbon, Park, Stillwater, and Sweet Grass Counties, Montana, and Park County, Wyoming: U.S. Geological Survey Bulletin 1391–F, p. F1–F125.

Simpson, R.W., Hildenbrand, T.G., Godson, R.H., and Kane, M.F., 1987, Digital colored Bouguer gravity, free-air gravity, station location, and terrain maps for the conterminous United States: U.S. Geological Survey Geophysical Investigations Map GP–953–B, scale 1:7,500,000 (2 sheets).

Simpson, R.W., Jachens, R.C., Blakely, R.J., and Saltus, R.W., 1986, A new isostatic residual gravity map of the conterminous United States, with a discussion on the significance of isostatic residual anomalies: Journal of Geophysical Research, v. 91, p. 8348–8372.

Size, W.B., 1967, The petrology of the Goose Lake Tertiary intrusive complex, Park County, Montana: DeKalb, Ill., Northern Illinois University M.S. thesis, 96 p.

Sleep, N.H., 1990, Hotspots and mantle plumes—Some phenomenology: Journal of Geophysical Research 95, v. B5, p. 6715–6736. —Relates geoidal and topo anomalies to plume heads worldwide.

Smith, R.B., 1978, Seismicity, crustal structure, and intraplate tectonics of the interior of the western Cordillera, *in* Smith, R.B., and Eaton, G.P., eds., Cenozoic tectonics and regional geophysics of the western Cordillera: Geological Society of America Memoir 152, p. 111–144.

Smith, R.B., and Braile, L.W., 1993, Topographic signature, space-time evolution, and physical properties of the Yellowstone–Snake River Plain volcanic system; The Yellowstone hotspot *in* Snoke, A.W., Steidtmann, J.R., and Roberts, S.M., eds., Geology of Wyoming: Geological Survey of Wyoming Memoir 5, p. 694–754.

Smith, R.B., Nagy, W.C., Julander, K.A., Viveros, J.J., Barker, C.A., and Gants, D.G., 1989, Geophysical and tectonic framework of the eastern Basin and Range–Colorado Plateau–Rocky Mountain transition, *in* Pakiser, L.C., and Mooney, W.D., eds., Geophysical framework of the continental United States: Geological Society of America Memoir 172, p. 205–233.

Smith, R.L., 1979, Ash-flow magmatism, *in* Chapin, C.E., and Elston, W.E., eds., Ash-flow tuffs: Geological Society of America Special Paper 180, p. 5–27.

Smith, R.L., and Luedke, R.G., 1984, Potentially active volcanic lineaments and loci in western conterminous United States, *in* Explosive volcanism—Inception, evolution, and hazards: Washington, D.C., National Academy Press, Studies in Geophysics, p. 47–66.

Sparlin, M.A., Braile, L.W., and Smith, R.B., 1982, Crustal structure of the eastern Snake River Plain determined from ray trace modeling of seismic refraction data: Journal of Geophysical Research, v. 87, no. B4, p. 2619–2633.

Spencer, J.E., and Reynolds, S.J., 1989, Middle Tertiary tectonics of Arizona and adjacent areas, *in* Jenney, J.P., and Reynolds, S.J., eds., Geologic evolution of Arizona: Tucson, Arizona Geological Society Digest 17, p. 539–574.

Spurney, J.C., 1984, Geology of the Iron Peak intrusion, Iron County, Utah: Kent, Ohio, Kent State University M.S. thesis, 84 p.

Spurr, J.E., and Garrey, G.H., 1908, Economic geology of the Georgetown quadrangle (together with the Empire district), Colorado: U.S. Geological Survey Professional Paper 63, 422 p.

Steenland, N.C., 1962, Gravity and aeromagnetic exploration in the Paradox basin: Geophysics, v. 27, p. 73–89.

Stein, H.J., Bankey, Viki, Cunningham, C.G., Zimbelman, D.R., Brickey, D.W., Shubat, Michael, Campbell, D.L., and Podwysocki, M.H., 1989, Tooele 1°×2° quadrangle, northwest Utah, a CUSMAP preassessment study: U.S. Geological Survey Open-File Report 89–467, 134 p.

Steiner, M.B., 1986, Rotation of the Colorado plateau: Tectonics, v. 5, p. 649–660. (Concludes central-western rotation of 11°±4° NET; multiple rotations likely. Pole of rotation is in southwestern part of Colorado Plateau. Also, southward translation of 3°–5° of lat.)

Stern, T.A., 1985, A back-arc basin formed within continental lithosphere—The central volcanic region of New Zealand: Tectonophysics, v. 112, p. 385–409.

Stern, T.W., Newell, M.F., Kistler, R.W., and Shawe, D.R., 1965, Zircon uranium-lead and thorium-lead ages and mineral potassium-argon ages of La Sal Mountains rocks, Utah: Journal of Geophysical Research, v. 70, p. 1503–1507.

Steven, T.A., 1975, Middle Tertiary volcanic field in the Southern Rocky Mountains, *in* Curtis, B.F., ed., Cenozoic history of the Southern Rocky Mountains: Geological Society of America Memoir 144, p. 75–94.

———1981, Three Creeks caldera, southern Pavant Range, Utah: Brigham Young University Geology Studies, v. 28, part 3, p. 1–7.

———1988, Mosaic faulting as a guide to mineral exploration in the Richfield 1°×2° quadrangle, western Utah, *in* Schindler, K.S., ed., USGS research on mineral resources—1989; program and abstracts: U.S. Geological Survey Circular 1035, p. 71.

Steven, T.A., Campbell, D.L., Cunningham, C.G., Mackie, R.L., and Rowley, P.D., in press, Comparison of major gravity and aeromagnetic features with geology in the Tushar Mountains and adjoining areas, Marysvale volcanic field, Utah: U.S. Geological Survey Bulletin.

Steven, T.A., Cunningham, C.G., and Anderson, J.J., 1984, Geologic history and uranium potential of the Big John caldera, southern Tushar Mountains, Utah, *in* Steven, T.A., ed., Igneous activity and related ore deposits in the western and southern Tushar Mountains, Marysvale volcanic field, west-central Utah: U.S. Geological Survey Professional Paper 1299–B, p. 23–33.

Steven, T.A., and Lipman, P.W., 1976, Calderas of the San Juan volcanic field, southwestern Colorado: U.S. Geological Survey Professional Paper 958, 35 p.

Steven, T.A., Lipman, P.W., Hail, W.J., Jr., Barker, Fred, and Luedke, R.G., 1974, Geologic map of the Durango quadrangle, southwestern

Colorado: U.S. Geological Survey Miscellaneous Investigations Series Map I–764, scale 1:250,000.

Steven, T.A., and Morris, H.T., 1983, Geologic map of the Cove Fort quadrangle, west-central Utah: U.S. Geological Survey Miscellaneous Investigations Map I–1481, scale 1:50,000.

———1987, Summary mineral resource appraisal of the Richfield 1°×2° quadrangle, west-central Utah: U.S. Geological Survey Circular 916, 24 p.

Steven, T.A., Morris, H.T., and Rowley, P.D., 1990, Geologic map of the Richfield 1°×2° quadrangle, west-central Utah: U.S. Geological Survey Miscellaneous Investigations Series Map I–1901, scale 1:250,000.

Steven, T.A., Rowley, P.D., and Cunningham, C.G., 1984, Calderas of the Marysvale volcanic field, west-central Utah: Journal of Geophysical Research, v. 89, no. B10, p. 8751–8764.

Stevenson, G.M., and Baars, D.L., 1986, The Paradox—A pull-apart basin of Pennsylvanian age, *in* Peterson, ed., Paleotectonics and sedimentation in the Rocky Mountain region, United States: American Association of Petroleum Geologists Memoir 41, p. 513–539.

Stevenson, G.M., and Baars, D.L, 1987, The Paradox—A pull-apart basin of Pennsylvanian age, *in* Campbell, J.A., editor, Geology of Cataract Canyon and vicinity: Four Corners Geological Society Guidebook, p. 31–50.

Stewart, J.H., 1971, Basin and range structure—A system of horsts and grabens produced by deep-seated extension: Geological Society of America Bulletin, v. 82, p. 1019–1044.

———1978, Basin-range structure in western North America—A review, *in* Smith, R.B., and Eaton, G.P., eds., Cenozoic tectonics and regional geophysics of the western Cordillera: Geological Society of America Memoir 152, p. 1–32.

———1980, Regional tilt patterns of late Cenozoic basin-range fault blocks, Western United States: Geological Society of America Bulletin, pt. I, v. 91, p. 460–464.

———1983, Cenozoic structure and tectonics of the northern Basin and Range province, California, Nevada, and Utah, *in* The role of heat in the development of energy and mineral resources in the northern Basin and Range province: Special Report 13, Davis, California, Geothermal Resources Council, p. 25–40.

Stewart, J.H., and Carlson, J.E., 1976, Cenozoic rocks of Nevada: Nevada Bureau of Mines and Geology Map 52, 4 sheets, scale 1:1,000,000.

———1977, Geologic map of Nevada, one million scale set: Nevada Bureau of Mines and Geology Map 57, scale 1:1,000,000.

Stewart, J.H., and Carlson, J.E., compilers, 1978, Geologic map of Nevada: Reston, Va., U.S. Geological Survey, scale 1:500,000 (2 sheets).

Stewart, J.H., Moore, W.J., and Zietz, Isidore, 1977, East-west patterns of Cenozoic igneous complexes, aeromagnetic anomalies, and mineral deposits, Nevada and Utah: Geological Society of America Bulletin, v. 88, p. 67–77.

Stewart, J.H., and Roldan-Quintana, Jaime, 1994, Tilt domains and large-scale segmentation of the Basin and Range province with emphasis on new data from Sonora, Mexico: Geological Society of America Abstracts with Programs, v. 26, no. 2, p. 96.

Stewart, J.H., and Suczek, C.A., 1977, Cambrian and latest Precambrian paleogeography and tectonics in western United States, *in* Stewart, J.H., Stevens, C.H., and Fritsche, A.E., eds., Paleozoic paleogeography of the western United States: Pacific Coast Palaeogeography Symposium I; Society of Economic Palaeontologists and Mineralogists, p. 1–18.

Stewart, J.H., Walker, G.W., and Kleinhampl, F.J., 1975, Oregon-Nevada lineament: Geology, v. 3, p. 265–268.

Stock, J., and Molnar, P., 1988, Uncertainties and implications of the late K and Tertiary position of North America relative to the Forallon, Kula and Pacific plates: Tectonics I, p. 1339–1384.

Stoeser, D.B., 1993, Tertiary calderas and regional extension of the east-central part of the Tintic–Deep Creek mineral belt, eastern Great Basin, Utah, chapter A *in* Scott, R.W., Jr., Detra, P.S., and Berger, B.R., eds., Advances related to United States and international mineral resources—Developing frameworks and explorations technologist: U.S. Geological Survey Bulletin 2039, p. 5–24.

Stokes, W.L., 1948, Geology of the Utah-Colorado salt dome region, with emphasis on Gypsum Valley, *in* Guidebook to the geology of Utah, 2: Utah Geological Society, 50 p.

———1982, Geologic comparisons and contrasts, Paradox and Arapien Basins, *in* Nielson, D.L., ed., Overthrust belt of Utah: 1982 Symposium and Field Conference, Salt Lake City, Utah Geological Association Publication 10, p. 1–12.

Stone, D.S., 1969, Wrench faulting and Rocky Mountain tectonics: The Mountain Geologist, v. 6, no. 2, p. 67–79.

———1977, Tectonic history of the Uncompahgre Uplift, *in* Exploration frontiers of the central and southern Rockies: Rocky Mountain Association of Geologists Guidebook, p. 23–30.

———1986, Wrench faulting and Rocky Mountains tectonics: The Mountain Geologist, v. 6, p. 67–79.

Strobell, J.D., Jr., 1956, Geology of the Carrizo Mountains area in northeastern Arizona and northwestern New Mexico: U.S. Geological Survey Oil and Gas Investigations Map OM–106, scale 1:48,000.

Sturgul, J.R., and Irwin, T.D., 1971, Earthquake history of Arizona and New Mexico, 1850–1966: Arizona Geological Society Digest, v. 9, p. 1–22.

Sullivan, K.R., 1987, Igneous intrusions in southeastern Utah—Relation to space-time-composition patterns of Cenozoic igneous activity in Nevada, Utah, and Colorado: Geological Society of America Abstracts with Programs, v. 19, p. 337.

Sullivan, K.R., Kowallis, B.J., and Mehnert, H.H., 1991, Isotopic ages of igneous intrusions in southeastern Utah—Evidence for a Mid-Cenozoic Reno–San Juan magmatic zone: Brigham Young University Geology Studies, v. 37, p. 139–144.

Sumner, J.S., 1985, Crustal geology of Arizona as interpreted from magnetic, gravity, and geologic data, *in* Hinze, W.J., ed., The utility of regional gravity and magnetic anomaly maps: Tulsa, Okla., Society of Exploration Geophysicists, p. 164–180.

Sun, M.S., and Baldwin, Brewster, 1958, Volcanic rocks of the Cienega area, Santa Fe County, New Mexico: New Mexico Bureau of Mines and Mineral Resources Bulletin 54, 80 p.

Sun, S.S., 1980, Lead isotopic study of young volcanic rocks from mid-ocean ridges, ocean islands, and island arcs: Philosophical Transactions of the Royal Society of London, v. A27, p. 409–445.

Suppe, J., Powell, J.C., and Berry, R., 1975, Regional topography, seismicity, Quaternary volcanism and the present-day tectonics of the western United States: American Journal of Science, v. 275–A, p. 397–436.

Sutphin, H.B., 1986, Occurrence and structural control of collapse features on the southern Marble Plateau, Coconino County, Arizona: Flagstaff, Ariz., Northern Arizona University M.S. thesis, 139 p.

Sutphin, H.B., and Wenrich, K.J., 1983, Structural control of breccia pipes on the southern Marble Plateau, Arizona: U.S. Geological Survey Open-File Report 83–908, 7 p., 2 plates, scale 1:50,000.

———1988, Map showing structural control of breccia pipes on the southern Marble Plateau, north-central Arizona: U.S. Geological Survey Miscellaneous Investigations Series Map I–1778, 2 sheets, scale 1:50,000.

Sutphin, H.B., Wenrich, K.J., and Verbeek, E.R., 1983, Structural control of breccia pipes on the southern Marble Plateau, Arizona: Geological Society of America Abstracts with Programs, v. 115, no. 5, p. 376.

Swadley, W C, Page, W.R., Scott, R.B., and Pampeyan, E.H., 1994, Geologic map of the Delamar 3 SE quadrangle, Lincoln County, Nevada: U.S. Geological Survey Geologic Quadrangle Map GQ–1754, scale 1:24,000.

Szabo, E., and Wengerd, S.A., 1975, Stratigraphy and tectogenesis of the Paradox basin, *in* Fassett, J.E., ed., Canyonlands Country Guidebook: Four Corners Geological Society Field Conference, 8th, p. 193–210.

Tanaka, K.L., Shoemaker, E.M., Ulrich, G.E., and Wolfe, E.W., 1986, Migration of volcanism in the San Francisco volcanic field, Arizona: Geological Society of America Bulletin, v. 97, p. 129–141.

Tatsumi, Y., Hamilton, D.L., and Nesbitt, R.W., 1986, Chemical characteristics of fluid phase released from a subducted lithosphere and origin of arc magmas—Evidence from high-pressure experiments and natural rocks: Journal of Volcanology and Geothermal Research, v. 29, 293–309.

Taylor, D.J., and Huffman, A.C., Jr., 1988, Overthrusting in the northwestern San Juan Basin, New Mexico—A new interpretation of the Hogback Monocline, *in* Carter, L.M.H.,ed., USGS Research on Energy Resources, 1988: U.S. Geological Survey Circular 1025, p. 60–61.

Taylor, W.J., and Bartley, J.M., 1992, Prevolcanic extensional Seaman breakaway fault and its geologic implications for eastern Nevada and western Utah: Geological Society of America Bulletin, v. 104, p. 255–266.

Taylor, W.J., Bartley, J.M., Lux, D.R., and Axen, G.J., 1989, Timing of Tertiary extension in the Railroad Valley–Pioche transect, Nevada—Constraints from $^{40}Ar/^{39}Ar$ ages of volcanic rocks: Journal of Geophysical Research, v. 94, no. B6, p. 7757–7774.

Taylor, S.R., and McLennan, S.M., 1985, The continental crust—Its composition and evolution—An examination of the geochemical record preserved in sedimentary rocks: Oxford, Blackwell Scientific, 312 p.

Thompson, G.A., and Burke, D.B., 1974, Regional geophysics of the Basin and Range province: Earth and Planetary Science Annual Reviews, v. 2, p. 213–238.

Thompson, G.A., Parsons, Tom, and Smith, Richard, 1990, Examples of magma overpressure suppressing normal faulting and inhibiting seismicity—Snake River Plain, Idaho, Yucca Mountain, Nevada, and Mono Craters, California [abs.]: Eos, v. 71, no. 43, p. 1622.

Thompson, G.A., and Zoback, M.L., 1979, Regional geophysics of the Colorado Plateau: Tectonophysics, v. 61, p. 149–181.

Thompson, R.A., Milling, M.E., Jr., Fleck, R.J., Wright, L.A., Rogers, N.W., and Johnson, C.M., in press, Temporal, spatial, and compositional constraints on extension-related volcanism in central Death Valley, California: Journal of Geophysical Research.

Thompson, T.B., 1968, Hydrothermal alteration and mineralization of the Rialto stock, Lincoln County, New Mexico: Economic Geology, v. 63, p. 943–949.

———1972, Sierra Blanca igneous complex, New Mexico: Geological Society of America Bulletin, v. 83, p. 2341–2356.

———1991, The Bonito-Nogal district, Lincoln County, New Mexico: Geological Society of America Abstracts with Programs, v. 23, no. 4, p. 99.

Thompson, T.B., Trippel, A.D., and Dwelley, P.C., 1985, Mineralized veins and breccias of the Cripple Creek district, Colorado: Economic Geology, v. 80, p. 1669–1688.

Throckmorton, C.K., and Verbeek, E.R., 1995, Joint networks in the Tiva Canyon and Topopah Spring Tuffs of the Paintbrush Group, southwestern Nevada: U.S. Geological Survey Open-File Report 95–2, 182 p.

Tingey, D.G., Christiansen, E.H., and Best, M.G., 1990, Late Oligocene and Miocene minette and olivine nephelinite dikes, Wasatch Plateau, Utah: Geological Society of America Abstracts with Programs, v. 22, no. 3, p. 88.

Tingey, D.G., Christiansen, E.H., Best, M.G., Ruiz, Juaquin, and Lux, D.R., 1991, Tertiary minette and melanephelinite dikes, Wasatch Plateau, Utah—Records of mantle heterogeneities and changing tectonics: Journal of Geophysical Research, v. 96, no. B8, p. 13529–13544.

Tisot, P.R., and Murphy, W.I.R., 1962, Physical structure of Green River oil shale from Colorado: U.S. Bureau of Mines Report of Investigations 6184, 24 p.

Tisot, P.R., and Sohns, H.W., 1971, Structural deformation of Green River oil shale as it relates to in-situ retorting: U.S. Bureau of Mines Report of Investigations 7576, 27 p.

Tobisch, O.T., Saleeby, J.B., and Fiske, R.S., 1986, Structural history of continental volcanic arc rocks, eastern Sierra Nevada, California—A case for extensional tectonics: Tectonics, v. 5, p. 65–94.

Tooker, E.W., 1990, Gold in the Bingham district, Utah: U.S. Geological Survey Bulletin 1857–E, p. E1–E16.

Tweto, O.L., 1968, Geologic setting and interrelationships of mineral deposits in the mountain province of Colorado and south-central Wyoming, *in* Ridge, J.D., ed., Ore deposits of the United States, 1933–1967 (Graton-Sales Volume): New York, American Institute of Mining, Metall., and Petroleum Engineers, v. 1, p. 551–558.

———1979, The Rio Grande rift system in Colorado, *in* Riecker, R.E., ed., Rio Grande rift—Tectonics and magmatism: Washington, D.C., American Geophysical Union, p. 33–56.

———1980, Precambrian geology of Colorado, *in* Kent, H.C., and Porter, K.W., eds., Colorado Geology: Denver, Rocky Mountain Association of Geologists, p. 37–46. —Trends of Precambrian in Colorado = NNW; in NM = NNE. Noted repeated utilization of Precambrian trends, in late Precambrian, Laramide and late Cenozoic.

———1987, Rock units of the Precambrian basement in Colorado: U.S. Geological Survey Professional Paper 1321–A, 54 p.; with colored geologic map, scale 1:1,000,000.

Tweto, O.L., and Case, J.E., 1972, Gravity and magnetic features as related to geology in the Leadville 30-minute quadrangle, Colorado: U.S. Geological Survey Professional Paper 726–C, 31 p.

Tweto, O.L., and Sims, P.K., 1963, Precambrian ancestry of the Colorado Mineral Belt: Geological Society of America Bulletin, v. 74, p. 991–1014.

Tweto, O.L., Steven, T.A., Hail, W.J., Jr., and Moench, R.H., 1976, Preliminary geologic map of the Montrose 1°×2° quadrangle, southwestern Colorado: U.S. Geological Survey Miscellaneous Field Studies Map MF–761, scale 1:250,000.

Tyler, Noel, and Ethridge, F.G., 1983, Depositional setting of the Salt Wash Member of the Morrison Formation, southwest Colorado: Journal of Sedimentary Petrology, v. 53, p. 67–82.

Ulrich, G.E., Billingsley, G.H., Hereford, Richard, Wolfe, E.W., Nealey, L.D., and Sutton, R.L., 1984, Map showing geology, structure, and uranium deposits of the Flagstaff 1°×2° quadrangle, Arizona: U.S. Geological Survey Miscellaneous Investigations Series Map I–1446, 2 sheets, scale 1:250,000.

Van der Voo, Rob, 1989, Paleomagnetism of North America; the craton, its margins, and the Appalachian Belt, *in* Pakiser, L.C., and Mooney, W.D., eds., Geophysical framework of the Continental United States: Geological Society of America Memoir 172, p. 449–470.

Van Gosen, B.S., and Wenrich, K.J., 1988, Ground magnetometer surveys over known and suspected breccia pipes on the Coconino Plateau, northwestern Arizona: U.S. Geological Survey Bulletin 1683–C, 31 p.

Van Kooten, G.K., 1988, Structure and hydrocarbon potential beneath the Iron Springs laccolith, southwestern Utah: Geological Society of America Bulletin, v. 100, no. 10, p. 1533–1540.

Varga, R.J., and Smith, B.M., 1984, Evolution of the early Oligocene Bonanza caldera, northeast San Juan volcanic field, Colorado: Journal of Geophysical Research, v. 89. p. 8679–8694.

Verbeek, E.R., and Grout, M.A., 1983, Fracture history of the northern Piceance Creek basin, northwestern Colorado, *in* Gary, J.H., ed., 16th Oil Shale Symposium, Proceedings: Golden, Colo., Colorado School of Mines Press, p. 26–44.

———1984a, Fracture studies in Cretaceous and Paleocene strata in and around the Piceance basin, Colorado—Preliminary results and their bearing on a fracture-controlled natural-gas reservoir at the MWX site: U.S. Geological Survey Open-File Report 84–156, 30 p.

———1984b, Prediction of subsurface fracture patterns from surface studies of joints—An example from the Piceance Creek basin,

Colorado, *in* Spencer, C.W., and Keighin, C.W., eds., Geologic studies in support of the U.S. Department of Energy Multiwell Experiment, Garfield County, Colorado: U.S. Geological Survey Open-File Report 84–757, p. 75–86.

———1986, Cenozoic stress rotation, northeastern Colorado Plateau, *in* Stone, D.S., and Johnson, K.S., eds., New interpretations of northwest Colorado geology: Rocky Mountain Association of Geologists, p. 97.

———1991, Some general properties of joints and joint networks in horizontally layered sedimentary and volcanic rocks—An overview, *in* Gomberg, Joan, ed., U.S. Geological Survey Committee for the Advancement of Science in the Yucca Mountain Project Symposium on "Fractures, Hydrology, and Yucca Mountain"—Abstracts and Summary: U.S. Geological Survey Open-File Report 91–125, p. 10–12.

———1992, Structural evolution of gilsonite dikes, eastern Uinta basin, Utah, *in* Fouch, T.D., Nuccio, V.F., and Chidsey, T.C., Jr., eds., Hydrocarbon and mineral resources of the Uinta basin, Utah and Colorado: Utah Geological Association Guidebook 20, p. 237–255.

———1993, Geometry and structural evolution of gilsonite dikes in the eastern Uinta basin, Utah: U.S. Geological Survey Bulletin 1787–HH, 42 p.

Verbeek, E.R., Grout, M.A., and Van Gosen, B.S., 1988, Structural evolution of a Grand Canyon breccia pipe—The Ridenour Copper-Vanadium-Uranium Mine, Hualapai Indian Reservation, Coconino County, Arizona: U.S. Geological Survey Open-File Report 88–006, 75 p.

Vogel, J.D., 1960, Geology and ore deposits of the Klondike Ridge area, Colorado: U.S. Geological Survey Open-File Report, 206 p.

Voight, Barry, and St. Pierre, H.P.B., 1974, Stress history and rock stress, *in* Advances in Rock Mechanics—Proceedings of the Third Congress of the International Society of Rock Mechanics, Denver, Colo., Sept. 1–7, 1974: Washington, National Academy of Sciences, v. 2, pt. A, p. 580–582.

Von Tish, D.B., Allmendinger, R.W., and Sharp, J.W., 1985, History of Cenozoic extension in central Sevier Desert, west-central Utah, from COCORP seismic reflection data: American Association of Petroleum Geologists Bulletin, v. 69, p. 1077–1087.

Waechter, N.B., and Johnson, W.E., 1986, Pennsylvanian-Permian paleostructure and stratigraphy as interpreted from seismic data in the Piceance basin, Colorado, *in* Stone, D.S., ed., New interpretations of northwest Colorado geology: Denver, Colo., Rocky Mountain Association of Geologists, p. 51–64.

Walker, G.E., 1991, Geology of the Barker mining district, Judith Basin and Cascade Counties, Montana, *in* Baker, D.W., and Berg, R.B., eds., Guidebook of the central Montana alkalic province: Montana Bureau of Mines and Geology Special Publication 100, p. 28–37.

Walker, J.A., and Mattox, S.R., 1989, The influences of subduction on mid-late Cenozoic volcanism in southwestern Utah: Geological Society of America Abstracts with Programs, v. 21, no. 6, p. A57.

Walker, J.A., Mattox, S.R., and Feigenson, M.D., 1991, Temporal changes in the source region of Cenozoic lavas of the Utah transition zone: Geological Society of America Abstracts with Programs, v. 23, no. 4, p. 102.

Wallace, A.R., 1989, Gold in the Central City mining district, Colorado: U.S. Geological Survey Bulletin 1857–C, p. C38–C47.

Wallace, R.E., 1984, Patterns and timing of late Quaternary faulting in the Great Basin province and relation to some regional tectonic features: Journal of Geophysical Research, v. 89, no. B7, p. 5763–5769.

Wallace, S.R., 1953, The petrology of the Judith Mountains, Fergus County, Montana: Ann Arbor, Mich., University of Michigan Ph. D. thesis, 245 p.

Walter, R.C., and Geller, B.A., 1992, Age of telluride-stage mineralization from single-crystal $^{40}Ar/^{39}Ar$ dating of vein-deposited adularia: Eos, v. 73, no. 14, Spring Meeting Supplement, p. 362.

Ward, P.L., 1991, On plate tectonics and the geological evolution of southwestern North America: Journal of Geophysical Research, v. 96, p. 12479–12496.

Wargo, J.G., 1964, Geology of a disseminated copper deposit near Cerrillos, New Mexico: Economic Geology, v. 59, p. 1525–1538.

Warner, L.A., 1978, The Colorado lineament—A middle Precambrian wrench fault system: Geological Society of America Bulletin, v. 89, p. 161–171.

———1980, The Colorado lineament, *in* Kent, H.C., and Porter, K.W., eds., Colorado geology: Denver, Colo., Rocky Mountain Association of Geologists, p. 11–21.

Weaver, B.L., and Tarney, John, 1984, Empirical approach to estimating the composition of the continental crust: Nature, v. 310, p. 375–377.

Webster, I.D., Taylor, R.P., and Bean, C., 1992, Water-rich pantellerite magma—Analysis of glass inclusions from Fantale volcano, Ethiopia: Eos, v. 73, no. 14, Spring Meeting Supplement, p. 368.

Weed, W.H., and Pirsson, L.V., 1896, Geology of the Little Rocky Mountains: Journal of Geology, v. 4, p. 399–428.

———1898, Geology and mineral resources of the Judith Mountains of Montana: U.S. Geological Survey 18th Annual Report, part 3, p. 437–616.

Weir, G.W., and Puffett, W.P., 1960, Preliminary geologic map and sections of the Mount Peale 2NE quadrangle, San Juan County, Utah: U.S. Geological Survey Mineral Investigations Field Studies Map MF–141, scale 1:24,000.

———1981, Incomplete manuscript on stratigraphy and structural geology and uranium-vanadium and copper deposits of the Lisbon Valley area, Utah-Colorado: U.S. Geological Survey Open-File Report 81-39, 292 p.

Weir, G.W., Dodson, C.L., and Puffet, W.P., 1960, Preliminary geologic map and section of the Mount Peale 2 SE quadrangle, San Juan County, Utah: U.S. Geological Survey Mineral Investigations Field Studies Map MF–143, scale 1:24,000.

Weir, G.W., Puffett, W.P., and Dodson, C.L., 1961, Preliminary geologic map and section of the Mount Peale 4 NW quadrangle, San Juan County, Utah: U.S. Geological Survey Mineral Investigations Field Studies Map MF–151, 1 sheet, scale 1:24,000.

Weiss, S.I., Noble, D.C., Worthington, J.E., IV, and McKee, E.H., 1993, Neogene tectonism from the southwestern Nevada volcanic field to the White Mountains, California—Part I, Miocene volcanic stratigraphy, paleotopography, extensional faulting and uplift between northern Death Valley and Pahute Mesa, *in* Lahren, M.M., Trexler, J.H., Jr., and Spinosa, Claude, Crustal evolution of the Great Basin and the Sierra Nevada: Geological Society of America Cordilleran/Rocky Mountain Sections Joint Meeting, Reno, Nev., 1993, Field Trip Guidebook: Reno, Nev., University of Nevada, Mackay School of Mines, p. 353–369.

Wells, M.L., Dallmeyer, R.D., and Allmendinger, R.W., 1990, Late Cretaceous extension in the hinterland of the Sevier thrust belt, northwestern Utah and southern Idaho: Geology, v. 18, p. 929–933.

Wenrich, K.J., 1985, Mineralization of breccia pipes in northern Arizona: Economic Geology, v. 80, no. 6, p. 1722–1735.

Wenrich, K.J., Billingsley, G.H., and Huntoon, P.W., 1986, Breccia pipe and geologic map of the northeastern Hualapai Indian Reservation and vicinity, Arizona: U.S. Geological Survey Open-File Report 86–458A, 1:48,000, 2 plates, with 29 p. text.

———1987, Breccia pipe and geologic map of the northwestern Hualapai Indian Reservation and vicinity, Arizona: U.S. Geological Survey Open-File Report 86–458C, 1:48,000, 2 plates, with 32 p. text.

———1995a, Breccia pipe and geologic map of the northwestern Hualapai Indian Reservation and vicinity, Arizona: U.S. Geological Survey Miscellaneous Investigations Series Map I–2522, 32 p., 2 plates, scale 1:48,000.

———1995b, Breccia pipe and geologic map of the northeastern Hualapai Indian Reservation and vicinity, Arizona: U.S. Geological Survey

Miscellaneous Investigations Series Map I–2440, 60 p., 2 plates, scale 1:48,000.

Wenrich, K.J., Chenoweth, W.L., Finch, W.I., and Scarborough, R.B., 1989, Uranium in Arizona, *in* Jenney, J.P., and Reynolds, S.J., eds., Geologic evolution of Arizona: Arizona Geological Society Digest, v. 17, p. 759–794.

Wenrich, K.J., and Sutphin, H.B., 1989a, Lithotectonic setting necessary for formation of a uranium-rich, solution-collapse breccia-pipe province, Grand Canyon region, Arizona: U.S. Geological Survey Open-File Report 89–173, 33 p.

———1989b, Lithotectonic setting necessary for formation of a uranium-rich, solution-collapse breccia-pipe province, Grand Canyon region, Arizona, *in* Metallogenesis of uranium deposits (Proceedings of a Technical Meeting, March 9–12, 1987, Vienna): International Atomic Energy Agency, p. 307–344.

Werle, J.L., Ikramuddin, M., and Mutschler, F.E., 1984, Allard stock, La Plata Mountains, Colorado—An alkaline rock-hosted porphyry copper-precious metal deposit: Canadian Journal of Earth Sciences, v. 21, p. 630–641.

Wernicke, B.P., 1981, Low-angle normal faults in the Basin and Range province—Nappe tectonics in an extending orogen: Nature, v. 291, p. 645–648.

———1985, Uniform-sense normal simple shear of the continental lithosphere: Canadian Journal of Earth Sciences, v. 22, no. 1, p. 108–125.

———1992, Cenozoic extensional tectonics of the U.S. Cordillera, *in* Burchfiel, B.C., Lipman, P.W., and Zoback, M.L., eds., The Cordilleran orogen—Conterminous U.S.: Geological Society of America, The Geology of North America, v. G–3, p. 553–581.

Wernicke, B.P., and Axen, G.J., 1988, On the role of isostasy in the evolution of normal fault systems: Geology, v. 16, p. 848–851.

Wernicke, B.P., Axen, G.J., and Snow, J.K., 1988, Basin and Range extensional tectonics at the latitude of Las Vegas, Nevada: Geological Society of America Bulletin, v. 100, p. 1738–1757.

Wernicke, B.P., Christiansen, R.L., England, P.C., and Sonder, L.J., 1987, Tectonomagmatic evolution of Cenozoic extension in the North American Cordillera, *in* Coward, M.P., Dewey, J.F., and Hancock, P.L., eds., Continental extensional tectonics: Geological Society Special Publication 28, p. 203–221.

Wernicke, B.P., Walker, J.D., and Beaufait, M.S., 1985, Structural discordance between Neogene detachments and frontal Sevier thrusts, central Mormon Mountains, southern Nevada: Tectonics, v. 4, no. 2, p. 213–246.

Western Geographics/Colorado Geological Survey, 1991, Colorado geological highway map: Canon City, Colo., Western Geographics, scale 1:1,000,000.

Wheeler, R.L., and Dixon, J.M., 1980, Intensity of systematic joints—Methods and applications: Geology, v. 8, p. 230–233.

Wheeler, R.L., and Holland, S.M., 1981, Style elements of systematic joints—An analytical procedure with a field example, *in* O'Leary, D.W., and Earle, J.L., eds., Third International Conference on Basement Tectonics, Proceedings: Basement Tectonics Committee, Denver, Colorado, p. 93–104.

Wheeler, R.L., and Stubbs, J.L., Jr., 1976, Style elements of systematic joints—Statistical analysis of size, spacing, and other characteristics, *in* Podwysocki, M.H., and Earle, J.L., eds., Second International Conference on Basement Tectonics, Proceedings,: Basement Tectonics Committee, Denver, Colorado, p. 491–499.

White, M.A., and Jacobson, M.I., 1983, Structures associated with the southwest margin of the ancestral Uncompahgre uplift, *in* Averett, W.R., ed., Northern Paradox basin—Uncompahgre uplift: Grand Junction Geological Society Guidebook, p.33–39.

White, P.J., and Lawless, J.V., 1989, Geochemistry of porphyry-hosted Au-Ag deposits in the Little Rocky Mountains, Montana—A discussion: Economic Geology, v. 84, p. 970–971.

White, R.S., and McKenzie, Dan, 1989, Magmatism at rift zones—The generation of volcanic continental margins and flood basalts: Journal of Geophysical Research, v. 94, p. 7685–7729.

White, R.S., Spence, G.D., Fowler, S.R., McKenzie, D.P., Westbrook, G.K., and Bowen, A.N., 1987, Magmatism at rifted continental margins: Nature, v. 330, p. 439–444.

Whitney, J.W., and Andrews, E.D., 1983, Past and present geomorphic activity in the Piceance Creek drainage basin, northwestern Colorado, *in* Gary, J.H., ed., 16th Oil Shale Symposium, Proceedings: Golden, Colo., Colorado School of Mines Press, p. 566–577.

Wiebe, R.A., 1993, Basaltic injections into floored silicic magma chambers: Eos, v. 74, no. 1, p. 1–3.

Williams, P.L., 1964, Geology, structure, and uranium deposits of the Moab quadrangle, Colorado and Utah: U.S. Geological Survey Miscellaneous Investigations Series Map I–360, sheet 1, scale 1:250,000.

Williams, P.L., and Hackman, R.J., 1971, Geology, structure, and uranium deposits of the Salina quadrangle, Utah: U.S. Geological Survey Miscellaneous Geologic Investigations Map I–591, scale 1:250,000.

Willis, G.C., 1985, Revisions to the geochronology and source areas of early Tertiary formations in the Salina area of Sevier Valley, central Utah: Geological Society of America Abstracts with Programs, v. 17, no. 4, p. 272.

———1991, Geologic map of the Harley Dome quadrangle, Grand County, Utah: Utah Geological Survey Open-File Report 216, 70 p., scale 1:24,000.

———1993, Geologic map of the Dry Canyon quadrangle, Grand County, Utah: Utah Geological Survey Open-File Report, scale 1:24,000.

Willis, G.F., and Tosdal, R.M., 1992, Formation of gold veins and breccias during dextral strike-slip faulting in the Mesquite mining district, southeastern California: Economic Geology, v. 87, p. 2002–2022.

Wilson, E.D., Moore, R.T., and Cooper, J.R., 1969, Geologic map of Arizona, scale 1:500,000: U.S. Geological Survey, Washington, D.C.

Wilson, J.T., 1965, A new class of faults and their bearing on continental drift: Nature, v. 207, p. 343–347.

———1973, Mantle plumes and plate motions: Tectonophysics, v. 19, p. 149–164.

Wilson, M.R., and Kyser, T.K., 1988, Geochemistry of porphyry-hosted Au-Ag deposits in the Little Rocky Mountains, Montana: Economic Geology, v. 83, p. 1329–1346.

———1989, Geochemistry of porphyry-hosted Au-Ag deposits in the Little Rocky Mountains, Montana—A reply: Economic Geology, v. 84, p. 971–973.

Wise, D.U., 1963, An outrageous hypothesis for the tectonic pattern of the North American Cordillera: Geological Society of America Bulletin, v. 74, p. 357–362.

Witkind, I.J., 1964, Geology of the Abajo Mountains area, San Juan County, Utah: U.S. Geological Survey Professional Paper 453, 110 p.

———1970, Composite dikes in the Little Belt Mountains, central Montana: U.S. Geological Survey Professional Paper 700–C, p. C82–C88.

———1973, Igneous rocks and related mineral deposits of the Barker quadrangle, Little Belt Mountains, Montana: U.S. Geological Survey Professional Paper 752, 58 p.

———1975, The Abajo Mountains—An example of the laccolithic groups on the Colorado Plateau, *in* Fassett, J.E., editor, Canyonlands Country: Four Corners Geological Society Guidebook, p. 245–252.

———1991, Implications of distinctive fault sets in the San Rafael Swell and adjacent areas, east-central Utah Utah Geological Association Publication 19, p. 141–148.

Witkind, I.J., and Weiss, M.P., 1991, Geologic map of the Nephi 30'×60' quadrangle, Carbon, Emery, Juab, Sanpete, Utah, and Wasatch Counties, Utah: U.S. Geological Survey Miscellaneous Investigations Series Map I–1937, scale 1:100,000.

Wolf, L.W., and Cipar, J.J., 1993, Through thick and thin—A new model for the Colorado Plateau from seismic refraction data from Pacific to Arizona crustal experiment: Journal of Geophysical Research, ser. B, v. 98, p. 19881–19894.

Wong, I.G., and Humphrey, J.R., 1989, Contemporary seismicity, faulting, and the state of stress in the Colorado Plateau: Geological Society of America Bulletin, v. 101, no. 9, p. 1127–1146.

Woodward-Clyde Consultants, 1983, Overview of the regional geology of the Paradox basin study area, ONWI–92: Prepared for Office of Nuclear Waste Isolation, Battelle Memorial Institute, Columbus, Ohio, 433 p.

Woodward, L.A., 1991, Metallic mineralization in the Yogo and Running Wolf mining districts, Little Belt Mountains, Montana, *in* Baker, D.W., and Berg, R.B., eds., Guidebook of the central Montana alkalic province: Montana Bureau of Mines and Geology Special Publication 100, p. 19–27.

Wright, A., 1983, The Ortiz gold deposit (Cunningham Hill)—Geology and exploration: Nevada Bureau of Mines and Geology Report 36, p. 42–51.

Wright, Lauren, 1976, Late Cenozoic fault patterns and stress fields in the Great Basin and westward displacement of the Sierra Nevada block: Geology, v. 4, p. 489–494.

Zhang, Xiaomao, and Spry, P.G., 1991, Mineralogical and fluid inclusion characteristics of the epithermal Gies gold-silver telluride deposit, Judith Mountains, Montana—A preliminary study, *in* Baker, D.W., and Berg, R.B., eds., Guidebook of the central Montana alkalic province: Montana Bureau of Mines and Geology Special Publication 100, p. 63–76.

Zhao, Guozhu, and Johnson, A.M., 1992, Sequence of deformations recorded in joints and faults, Arches National Park, Utah: Journal of Structural Geology, v. 14, no. 2, p.225–236.

Zhong, Shijie, and Gurnis, Michael, 1995, Mantle convection with plates and mobile, faulted plate margins: Science, v. 267, 10 February 1995, p. 838–843. —Models show subducted plate spreading out on core-mantle interface.

Zietz, Isidore, Gilbert, F.P., and Kirby, J.R., 1978, Aeromagnetic map of Nevada—Color coded intensities: U.S. Geological Survey Geophysical Investigations Map GP–922, scale 1:1,000,000.

Zietz, Isidore, and Kirby, J.R., Jr., 1972a, Aeromagnetic map of Colorado: U.S. Geological Survey Geophysical Investigations Map GP–836, scale 1:500,000.

———1972b, Aeromagnetic map of Colorado: U.S. Geological Survey Geophysical Investigations Map GP–880, scale 1:1,000,000.

Zietz, Isidore, Shuey, Ralph, and Kirby, J.R., Jr., 1976, Aeromagnetic map of Utah: U.S. Geological Survey Geophysical Investigations Map GP–907, scale 1:1,000,000.

Zietz, Isidore, Stewart, J.H., Gilbert, F.P., and Kirby, J.R., Jr., 1977, Aeromagnetic map of Nevada: U.S. Geological Survey Miscellaneous Field Studies Map MF–902, scale 1:500,000 (2 sheets, with accompanying text).

Zoback, M.L., 1989, State of stress and modern deformation of the northern Basin and Range province: Journal of Geophysical Research, v. 94, no. B6, p. 7105–7128.

Zoback, M.L., Anderson, R.E., and Thompson, G.A., 1981, Cenozoic evolution of the state of stress and style of tectonism in the Basin and Range province of the western United States: Royal Society of London Philosophical Transactions, v. A300, p. 407–434.

Zoback, M.L., McKee, E.H., and Blakely, R.J., in press, Significance of middle Miocene tectonic and igneous features in the western United States—Large-scale rotations in northern Nevada?: Geological Society of America Bulletin.

Zoback, M.L., McKee, E.H., Blakely, R.J., and Thompson, G.A., 1994, The northern Nevada rift—Regional tectono-magmatic relations and middle Miocene stress direction: Geological Society of America Bulletin, v. 106, no. 3, p. 371–382.

Zoback, M.L., and Thompson, G.A., 1978, Basin and Range rifting in northern Nevada—Clues from a mid-Miocene rift and its subsequent offsets: Geology, v. 6, p. 111–116.

Zoback, M.L., and Zoback, M.D., 1980, State of stress in the conterminous United States: Journal of Geophysical Research, v. 85, no. B11, p. 6113–6156.

———1989, Tectonic stress field of the continental United States, *in* Pakiser, L.C., and Mooney, W.D., eds., Geophysical framework of the continental United States: Geological Society of America Memoir 172, p. 523–540.

Manuscript approved for publication June 4, 1996
Published in the Central Region, Denver, Colorado
Graphics and photocomposition by Gayle M. Dumonceaux
Edited by P.L. Martin; completed for printing by L.M. Carter

☆ U.S. GOVERNMENT PRINTING OFFICE: 1997 - 673-048 / 20090 REGION NO. 8